VOLUME NINETY SEVEN

VITAMINS AND HORMONES

Nociceptin Opioid

VOLUME NINETY SEVEN

VITAMINS AND HORMONES

Nociceptin Opioid

Series Editor

GERALD LITWACK, PhD

Toluca Lake, North Hollywood, California

ELSEVIER

AMSTERDAM • BOSTON • HEIDELBERG • LONDON
NEW YORK • OXFORD • PARIS • SAN DIEGO
SAN FRANCISCO • SINGAPORE • SYDNEY • TOKYO
Academic Press is an imprint of Elsevier

Cover photo credit:
Lohman, R.-J., Harrison, R.S., Ruiz-Gómez, G.G., Hoang, H.N., Shepherd, N.E., Chow, S., Hill, T.A., Madala, P.K., Fairlie, D.P.
Helix-Constrained Nociceptin Peptides Are Potent Agonists and Antagonists of ORL-1 and Nociception
Vitamins and Hormones (2015) **97**, pp. 1–56

Academic Press is an imprint of Elsevier
225 Wyman Street, Waltham, MA 02451, USA
525 B Street, Suite 1800, San Diego, CA 92101-4495, USA
125 London Wall, London, EC2Y 5AS, UK
The Boulevard, Langford Lane, Kidlington, Oxford OX5 1GB, UK

First edition 2015

ISBN: 978-0-12-802443-0
ISSN: 0083-6729

CONTENTS

CONTRIBUTORS

Ouagazzal Abdel-Mouttalib
IGBMC, (UMR7104), CNRS, Illkirch, France

Christina Bergqvist
Department of Neuroscience, Unit of Pharmacology, Science for Life Laboratory, Uppsala University, Uppsala, Sweden

Girolamo Calo
Department of Medical Sciences, Section of Pharmacology and National Institute of Neuroscience, University of Ferrara, Ferrara, Italy

Shiao Chow
Institute for Molecular Bioscience, University of Queensland, Brisbane, Queensland, Australia

Lauren Dalhousay
Department of Biological Sciences, California State University, Long Beach, California, USA

Iris Ucella de Medeiros
Department of Biophysic and Pharmacology, Federal University of Rio Grande do Norte, Natal, Brazil

Beáta H. Deák
Department of Pharmacodynamics and Biopharmacy, University of Szeged, Szeged, Hungary

Eszter Ducza
Department of Pharmacodynamics and Biopharmacy, University of Szeged, Szeged, Hungary

Ko Eto
Department of Biological Sciences, Graduate School of Science and Technology, Kumamoto University, Kumamoto City, Kumamoto, Japan

David P. Fairlie
Institute for Molecular Bioscience, University of Queensland, Brisbane, Queensland, Australia

Allison Jane Fulford
Centre for Comparative and Clinical Anatomy, University of Bristol, Bristol, BS2 8EJ, United Kingdom

Elaine C. Gavioli
Department of Biophysic and Pharmacology, Federal University of Rio Grande do Norte, Natal, Brazil

Róbert Gáspár
Department of Pharmacodynamics and Biopharmacy, University of Szeged, Szeged, Hungary

Rosemary S. Harrison
Institute for Molecular Bioscience, University of Queensland, Brisbane, Queensland, Australia

Timothy A. Hill
Institute for Molecular Bioscience, University of Queensland, Brisbane, Queensland, Australia

Huy N. Hoang
Institute for Molecular Bioscience, University of Queensland, Brisbane, Queensland, Australia

Seiji Ito
Department of Medical Chemistry, Kansai Medical University, Hirakata, Japan

Anna Klukovits
Department of Pharmacodynamics and Biopharmacy, University of Szeged, Szeged, Hungary

Dan Larhammar
Department of Neuroscience, Unit of Pharmacology, Science for Life Laboratory, Uppsala University, Uppsala, Sweden

Rink-Jan Lohman
Institute for Molecular Bioscience, University of Queensland, Brisbane, Queensland, Australia

Praveen K. Madala
Institute for Molecular Bioscience, University of Queensland, Brisbane, Queensland, Australia

Marta C. Monteiro
Laboratory of Clinical Microbiology and Immunology, Faculty of Pharmacy, Federal University of Pará, Belém, Brazil

Emilia Naydenova
Department of Organic Chemistry, University of Chemical Technology and Metallurgy, Sofia, Bulgaria

Emiko Okuda-Ashitaka
Department of Biomedical Engineering, Osaka Institute of Technology, Osaka, Japan

Pedro R.T. Romão
Laboratory of Immunology, Department of Basic Health Sciences, Federal University of Health Sciences of Porto Alegre, Rua Sarmento Leite, Porto Alegre, Brazil

Gloria G. Ruiz-Gómez
Institute for Molecular Bioscience, University of Queensland, Brisbane, Queensland, Australia

Nayna Sanathara
Department of Pharmacological Sciences, University of California, Irvine, California, USA

Nicholas E. Shepherd
Institute for Molecular Bioscience, University of Queensland, Brisbane, Queensland, Australia

Kevin Sinchak
Department of Biological Sciences, California State University, Long Beach, California, USA

Craig W. Stevens
Department of Pharmacology and Physiology, Oklahoma State University Center for Health Sciences, Tulsa, Oklahoma, USA

Görel Sundström
Department of Neuroscience, Unit of Pharmacology, Science for Life Laboratory, Uppsala University, Uppsala, Sweden

Kornélia Tekes
Department of Pharmacodynamics, Semmelweis University, Budapest, Hungary

Petar Todorov
Department of Organic Chemistry, University of Chemical Technology and Metallurgy, Sofia, Bulgaria

Xinmin (Simon) Xie
AfaSci Research Laboratories, Redwood City, and Department of Anesthesia, Stanford University School of Medicine, Stanford, California, USA

Rositza Zamfirova
Institute of Neurobiology, Bulgarian Academy of Sciences, Sofia, Bulgaria

PREFACE

Nociceptin/orphanin FQ is a 17-amino acid-containing peptide and is the agonist of the NOP/ORL-1 receptor, the latest member of the opioid receptor family, consisting of the mu-, delta-, and kappa receptors. However, this receptor has actions opposite to some of the actions of the classical opioid receptors and induces a variety of biological activities that would be predicted from its wide distribution in the human body. As nociceptin is the agonist of NOP, another related peptide, nocistatin, is an antagonist of the NOP receptor. Recent research suggests that there will be a wide range of clinical therapies that could be developed from this system. Much of the basic chemistry, biology, physiology, and therapeutic information is described in this volume. The chapters below deal first with the more basic aspects followed by biological information and finally clinically related material.

The first chapter is by R.-J. Lohman, R.S. Harrison, G.G. Ruiz-Gómez, H.N. Hoang, N.E. Shepherd, S. Chow, T.A. Hill, and D. Fairlie on "Potent ORL-1 Peptide Agonists and Antagonists of Nociceptin Using Helix Constraints." This is followed by "Bioinformatics and Evolution of Vertebrate Nociceptin and Opioid Receptors" by C.W. Stevens. D. Larhammar, C. Bergqvist, and G. Sundström review "Ancestral Vertebrate Complexity of the Opioid System." This section is completed by "Synthesis and Biological Activity of Small Peptides as NOP and Opioid Receptors' Ligands—View on Current Developments" by E. Naydenova, P. Todorov, and R. Zamfirova.

Initiating the biological information is "Pain Regulation Induced by Nocistatin-Targeting Molecules: G Protein-Coupled-Receptor and Nocistatin-Interacting Protein" by E. Okuda-Ashitaka and S. Ito. Next, K. Eto describes "Nociceptin and Meiosis During Spermatogenesis in Postnatal Testes." "Orphanin FQ-ORL-1 Regulation of Reproduction and Reproductive Behavior in the Female" is the contribution of K. Sinchak, L. Paaske, and N. Sanathara. R. Gáspár, B.H. Deák, A. Klukovits, E. Ducza, and K. Tekes report on the "Effects of Nociceptin and Nocistatin on Uterine Contraction."

With regard to the more clinically relevant information, E.C. Gavioli, I. Ucella de Medeiros, M.C. Monteiro, G. Calo, and P.R.T. Romão describe "Nociceptin/Orphanin FQ-NOP Receptor System in Inflammatory and Immune-Mediated Diseases." A.J. Fulford reports on "Endogenous Nociceptin System Involvement in Stress Responses and

Anxiety Behavior." Related to this topic, X. (S) Xie offers "The Neuronal Circuit Between Nociceptin/Orphanin FQ and Hypocretins/Orexins Coordinately Modulates Stress-Induced Analgesia and Anxiety-Related Behavior." Finally, O. Abdel-Mouttalib reviews "Nociceptin/Orphanin-FQ Modulation of Learning and Memory."

Helene Kabes is the mediator between my work and the production process in the development of these volumes. My appreciation goes to her, Mary Ann Zimmerman, and Vignesh Tamilselvvan who contributed to various aspects of the publication of this Series.

The illustration on the cover of this book is taken from Figure 1 of the chapter entitled "Helix-Constrained Nociceptin Peptides Are Potent Agonists and Antagonists of ORL-1 and Nociception" by R.-J. Lohman, R.S. Harrison, G.G. Ruiz-Gómez, H.N. Hoang, N.E. Shepherd, S. Chow, T.A. Hill, and D. Fairlie.

Gerald Litwack
North Hollywood, California
September 17, 2014

CHAPTER ONE

Helix-Constrained Nociceptin Peptides Are Potent Agonists and Antagonists of ORL-1 and Nociception

Rink-Jan Lohman[1], Rosemary S. Harrison[1], Gloria G. Ruiz-Gómez, Huy N. Hoang, Nicholas E. Shepherd, Shiao Chow, Timothy A. Hill, Praveen K. Madala, David P. Fairlie[2]

Institute for Molecular Bioscience, University of Queensland, Brisbane, Queensland, Australia
[2]Corresponding author: e-mail address: d.fairlie@imb.uq.edu.au

Contents

[1] Joint first authors

Vitamins and Hormones, Volume 97
ISSN 0083-6729
http://dx.doi.org/10.1016/bs.vh.2014.10.001

Abstract

Nociceptin (orphanin FQ) is a 17-residue neuropeptide hormone with roles in both nociception and analgesia. It is an opioid-like peptide that binds to and activates the G-protein-coupled receptor opioid receptor-like-1 (ORL-1, NOP, orphanin FQ receptor, kappa-type 3 opioid receptor) on central and peripheral nervous tissue, without activating classic delta-, kappa-, or mu-opioid receptors or being inhibited by the classic opioid antagonist naloxone. The three-dimensional structure of ORL-1 was recently published, and the activation mechanism is believed to involve capture by ORL-1 of the high-affinity binding, prohelical C-terminus. This likely anchors the receptor-activating N-terminus of nociception nearby for insertion in the membrane-spanning helices of ORL-1. In search of higher agonist potency, two lysine and two aspartate residues were strategically incorporated into the receptor-binding C-terminus of the nociceptin sequence and two Lys(*i*) → Asp(*i*+4) side chain–side chain condensations were used to generate lactam cross-links that constrained nociceptin into a highly stable α-helix in water. A cell-based assay was developed using natively expressed ORL-1 receptors on mouse neuroblastoma cells to measure phosphorylated ERK as a reporter of agonist-induced receptor activation and intracellular signaling. Agonist activity was increased up to 20-fold over native nociceptin using a combination of this helix-inducing strategy and other amino acid modifications. An NMR-derived three-dimensional solution structure is described for a potent ORL-1 agonist derived from nociceptin, along with structure–activity relationships leading to the most potent known α-helical ORL-1 agonist (EC_{50} 40 pM, pERK, Neuro-2a cells) and antagonist (IC_{50} 7 nM, pERK, Neuro-2a cells). These α-helix-constrained mimetics of nociceptin(1–17) had enhanced serum stability relative to unconstrained peptide analogues and nociceptin itself, were not cytotoxic, and displayed potent thermal analgesic and antianalgesic properties in rats (ED_{50} 70 pmol, IC_{50} 10 nmol, s.c.), suggesting promising uses *in vivo* for the treatment of pain and other ORL-1-mediated responses.

1. NOCICEPTION IN BRIEF

Nociception is a term used to describe the ability of organisms to detect noxious stimuli (Wall & Melzack, 2000). It involves neural processing of external stimuli, signaling through receptors on neurons, that may damage the organism, enabling it to sense pain and take action to evade damage. In higher organisms, nociception is a series of exquisitely complex neural events involving neurons of the peripheral and central nervous system (CNS) that allow an organism to sense pain or algesia (Wall & Melzack, 2000). Noxious stimuli can be mechanical (pressure or sharp objects), thermal (temperatures above 45 °C or extreme cold), and chemical (acids, environmental irritants such as capsaicin), which are detected by an array of specialized receptors (termed nociceptors) on the terminals of spinal nerve afferents that have their cell bodies in ganglia positioned outside of the spinal cord. These pain-sensing neurons (canonically unmyelinated, slow conduction velocity C-fibers and myelinated moderate conduction velocity Aδ-fibers) are generally considered part of the peripheral nervous system and send signals after detection of noxious stimuli via their extraspinal ganglia to the dorsal horn of the spinal cord *en route* to the brain for processing of conscious pain perception (Wall & Melzack, 2000). This ultimately allows the organism to act to avoid further damage by removing itself from the noxious stimuli or cause tissue injury, and allow healing. To add to the complexity, the initial response to pain avoidance is usually considered a reflex action, with the withdrawal response not initially involving the brain (Wall & Melzack, 2000).

Aside from the classical descriptions of pain in uninjured tissue via specialized nociceptors globally referred to as mechanoceptors, thermoceptors, and chemoceptors (with obvious nomenclature), pain can be promoted by endogenous inflammatory mediators released from various inflammatory cells (Wall & Melzack, 2000). These mediators are detected by diverse classes of chemoceptors that respond to many exogenous and endogenous chemicals, including histamine (Harasawa, 2000; Rosa & Fantozzi, 2013) (H1 receptors: Akdis & Simons, 2006; possibly others, H2: Hasanein, 2011; Mobarakeh et al., 2005; and H3: Cannon & Hough, 2005; Smith, Haskelberg, Tracey, & Moalem-Taylor, 2007), neuropeptides (Abrams & Recht, 1982) such as substance P (Munoz & Covenas, 2011), enkephalins (Bodnar, 2013), and bradykinins (Jaggi & Singh, 2011; Maurer et al., 2011) via various receptors including the NK1 and transient receptor potential channel families (Brederson, Kym, & Szallasi, 2013; Salat, Moniczewski, & Librowski, 2013). Even various proteases (such as tryptase) acting at protease-activated receptors (Bao, Hou, & Hua, 2014; Bunnett, 2006; Vergnolle et al., 2001) can signal pain. These substances via their receptors can contribute to a heightened pain sensation, referred to as hyperalgesia, which describes when a normally painful stimulus becomes excessively painful. However, if persistent it can lead to allodynia, when a normally nonpainful stimulus becomes painful to the individual (Wall & Melzack, 2000). These can both be symptoms of normal inflammatory pain and can be of benefit to an organism by warning the individual of tissue damage. However, when pain becomes chronic, it can seriously interfere with the quality of life of the individual, leading to significant morbidity. Such pain is considered neuropathic if it becomes either ongoing or episodic in nature, the cause of which may be in absence of a known or precipitating inflammatory condition or lesion. Such chronic pain is commonly treated with opiates, a name given to a family of alkaloids, such as morphine or codeine, derived from the opium poppy (*Papaver somniferum*), or their synthetic counterparts, the opioids, all of which act through G-protein-coupled receptors of the opioid receptor family (delta (δ1–2), kappa (κ1–3), and mu (μ1–3); Wall & Melzack, 2000). However, the actions of the opiate alkaloids at their receptors can produce significant and unwanted effects such as respiratory depression, physical dependence, sedation, hallucinations, and other dissociative effects that may significantly impact on an individual's well-being and contribution to society if taken for extended periods, as generally required for chronic pain sufferers. Likewise, once they are no longer needed due to resolution of the condition, withdrawal symptoms precipitated by their dependence effects may result, and these are not only unpleasant, but can be devastating to patients and their families if dependence becomes abuse. This limits their effectiveness as drugs for the greater population, and thus there is a requirement for potent antinociceptive compounds that target the opioid receptors without the side effects of the classical alkaloid opiates.

1.1 Opioid receptor-like receptor—ORL-1

A relatively recent addition to the GPCR opioid receptor family is the opioid receptor-like-1 (ORL-1 or NOP) receptor (Fig. 1). It was named because of high homology with the classical opioid receptors, but it was not affected by classical opioid receptor antagonists such as naloxone. The "orphan" receptor ORL-1 was initially identified from mRNA transcripts taken from mouse and rat CNSs, and deorphanized with the discovery of nociceptin as an endogenous ligand (Bunzow et al., 1994; Chen et al., 1994; Meunier et al., 1995; Mollereau et al., 1994; Salvadori, Guerrini, Calo, & Regoli, 1999; Wang et al., 1994; Wick, Minnerath, Roy, Ramakrishnan, & Loh, 1995). The location of the ORL-1 receptor has since been confirmed, and receptor-binding assays and *in situ* hybridization techniques have been used to pinpoint ORL-1 to the cortex, anterior olfactory nucleus, lateral septum, hypothalamus, hippocampus, amygdala, and other regions of the brain. Interestingly, ORL-1 transcripts have also been identified in nonneuronal peripheral organs such as intestine, vas deferens, kidney, and the spleen (Osinski, Pampusch, Murtaugh, & Brown, 1999; Wang et al., 1994) and in unexpected cell types, such as mouse sphenic lymphocytes (Halford, Gebhardt, & Carr, 1995) as well as various human immune cells (Peluso et al., 1998).

the main cellular functions of ORL-1 are the inhibition of adenylate cyclase resulting in suppression of cAMP formation (Butour, Moisand, Mazarguil, Mollereau, & Meunier, 1997; Meunier et al., 1995; Vaudry, Stork, Lazarovici, & Eiden, 2002), inhibition of voltage-gated calcium channel opening, and activation (opening) of K^+ rectifying channels (Beedle et al., 2004; Knoflach, Reinscheid, Civelli, & Kemp, 1996), the net effect being suppression of neuronal (or other cell type) excitability. Other intracellular signaling pathways affected by ORL-1 feature MAPK, ERK, and JUN activation (Armstead, 2006; Chan & Wong, 2000; Zhang et al., 1999). In neuronal systems, these signaling effects result in modulated release of neurotransmitters like glutamate, catecholamines, and tachykinins, much like other opioid receptors.

Several recent *in vivo* studies on ORL-1 have shown that it has modulatory roles in a multitude of complex central neurobiological processes involved with neuroplasticity commonly associated with the limbic system. Thus, functional roles of ORL-1 are not only restricted to nociception, but also extend to behavioral manifestations involved with feeding and satiety (Glass, Billington, & Levine, 1999), reward, addiction (Bodnar, 2013; Munoz & Covenas, 2011; Shoblock, 2007; Ubaldi, Bifone, & Ciccocioppo, 2013; Zaveri, 2011), fear, stress, anxiety, mood and depression (Chiou et al., 2007; Gavioli et al., 2003; Knoflach et al., 1996), seizure and epilepsy (Armagan et al., 2012; Bregola et al., 2002), and learning and memory (Bodnar, 2013; Meunier, 1997; Redrobe, Calo, Guerrini, Regoli, & Quirion, 2000). This list is expected to expand substantially in the near future given the roles of other opioid receptors.

Roles for ORL-1 are not as clear in the peripheral tissues, such as in the cardiovascular system and the immune system (Chiou et al., 2007). For example, ORL-1-deficient mice do not show any immunological abnormalities (Nishi et al., 1997), despite the fact that ORL-1 on wild-type immune cells appears to be functional, with both immunosuppressant (Halford et al., 1995; Nemeth et al., 1998; Peluso, Gaveriaux-Ruff, Matthes, Filliol, & Kieffer, 2001) and proinflammatory actions (Kimura et al., 2000; Serhan, Fierro, Chiang, & Pouliot, 2001). The functions of ORL-1, both centrally and peripherally, are interesting and need further investigation.

The development of potent and selective agonists and antagonists may lead to drugs with marketable effects against relatively common, debilitating, and usually under-managed conditions, such as neuropathic pain, epilepsy, drug and alcohol addiction, eating disorders, and possibly cardiovascular disease. Targeting ORL-1 may lack the addictive and dependence properties of the μ-, κ-, and δ-opioid receptors (Lin & Ko, 2013; Yu et al., 2011), thus agonists of ORL-1 may not have the potential for abuse like most clinically used opiates and opioids. Furthermore, the potential for ORL-1 ligands has been highlighted for treatment of addiction to opioids and other agents (Bodnar, 2013; Robinson, 2002; Shoblock, 2007; Ubaldi et al., 2013; Zaveri, 2011). These properties make ORL-1 an attractive drug discovery target for various clinical conditions additional to those involving pain. However, ORL-1 activation can have contrasting effects when agonists/antagonists are administered either centrally or peripherally, with supraspinal delivery of agonists producing unexpected hyperalgesic effects in experimental models contradictory to when the same agonists are administered peripherally (Calo, Rizzi, et al., 1998). Likewise, the central administration of ORL-1 antagonists has been shown to enhance opiate-induced analgesia (Rizzi et al., 2000), which highlights the complexity of ORL-1 pharmacology. Thus, ORL-1 agonists that do not enter the brain may be best for clinically treating ORL-1 mediated chronic pain.

1.2 Nociceptin

ORL-1 was deorphanized in 1995 upon isolation and characterization of nociceptin, also called orphanin FQ. In the mature form, it is a 17-residue peptide that binds to ORL-1 (Meunier et al., 1995). The cDNA region that encodes nociceptin shows dibasic amino acids and an endopeptidase recognition site, suggesting that nociceptin is proteolytically processed from the pre/pro form (176 amino acids) to a mature 17-amino acid peptide with a free carboxyl terminus [nociceptin(1–17)-OH)] (Wang et al., 1994). Despite high sequence similarity between ORL-1 and other opioid receptors, nociceptin(1–17)-OH has no significant cross-reactivity with endogenous opioid peptides or selective μ-, δ-, or κ-agonists (Mollereau et al., 1994; Wang et al., 1994) at their receptors. It has been found that mature nociceptin is also highly conserved across mammalian species (Fig. 2A) and has sequence and possibly structural similarities to human dynorphins A and B and alpha-neoendorphin (Fig. 2B).

Small-molecule agonist ligands for ORL-1 have been discovered, including buprenorphine (nonselective for opioid receptors; Lutfy et al., 2003; Robinson, 2002), norbuprenorphine (also nonselective; Robinson, 2002), SCH-221,510 (Lin & Ko, 2013; Varty et al., 2008), NNC 63-0532 (Guerrini et al., 2004; Thomsen & Hohlweg, 2000), Ro64-6198

```
A
H.sapiens            ----GMEEAGEMEQKQLQKRFGGFTGARKSARKLANQKRF - 149
B.taurus             ----GLEEVGEIEQKQLQKRFGGFTGARKSARKLANQKRF - 149
P.scrofa             --------------------FGGFTGARKSARKLANQ--- - 17
M.musculus           DAEPGADDAEEVEQKQLQKRFGGFTGARKSARKLANQKRF - 160
R.rattus             ----VADEADEVEQKQLQKRFGGFTGARKSARKLANQKRF - 154
                                         *****************

B
Nociceptin           EPEPGMEEAGEMEQKQLQKRFGGFTGARKSARKLANQKRF - 149
Dynorphin A          SEVAGEGDGDSMGHEDLYKRYGGFLRRIRPKLKWDNQKRY - 226
Dynorphin B          RYGGFLRRIRPKLKWDNQKR[illegible]RSQED-- - 243
Alpha-neoendorphin   AMETGTLYLAEEDPKEQVKR[illegible]GEGD - 194
```

Figure 2 Sequence alignment of mature nociceptin(1–17)-OH and related peptides. (A) Nociceptin from human and other vertebrate species and (B) nociceptin with other human opioid peptides highlighting the relatively conserved N-terminus. Sequences obtained from SwissProt using accession numbers O62647, P55791, Q64387, Q62923, Q13519, and P01213.

(Lin & Ko, 2013; Shoblock, 2007; Smith & Moran, 2001), and TH-030418 (Yu et al., 2011) to name a few. Antagonists have also been developed, including J-113,397 (Smith et al., 2008), SB-612111, and JTC-801 (Shinkai et al., 2000; Yamada, Nakamoto, Suzuki, Ito, & Aisaka, 2002). Most small molecules are still in preclinical or early clinical development (Fig. 3D–G, see review: Lambert, 2008) or are used as pharmacological research tools. These selective ORL-1 ligands can be classified into four main groups: 4-aminoquinolines, benzimidazopiperidines, aryl-piperidines, and spiropiperidines.

Peptide-based drugs tend to be unstable *in vivo*, being rapidly degraded by proteases in the gut, blood, and cells, and rapidly cleared from the circulation. Many research groups have developed nociceptin peptides with structural stability in attempts to make them more suitable as drugs (Arduin et al., 2007; Bigoni et al., 2002; Bobrova et al., 2003; Calo, Guerrini, et al., 1998, 2000; Calo et al., 2005, 2002; Carra et al., 2005; Chen et al., 2004; Chen, Wang, et al., 2002; Chiou, Fan, Guerrini, & Calo, 2002; Chiou, Liao, Guerrini, & Calo, 2005; Guerrini et al., 2004, 2003; Harrison et al., 2010; Kapusta et al., 2005; Kitayama et al., 2003; Kuo, Liao, Guerrini, Calo, & Chiou, 2008; McDonald et al., 2002; Okawa et al., 1999; Redrobe et al., 2000; Rizzi, Rizzi, Bigoni, et al., 2002; Wright et al., 2003), some having been shown to significantly increase stability and potency relative to native nociceptin (Carra et al., 2005; Harrison et al., 2010; Kuo et al., 2008). These are discussed in more detail here.

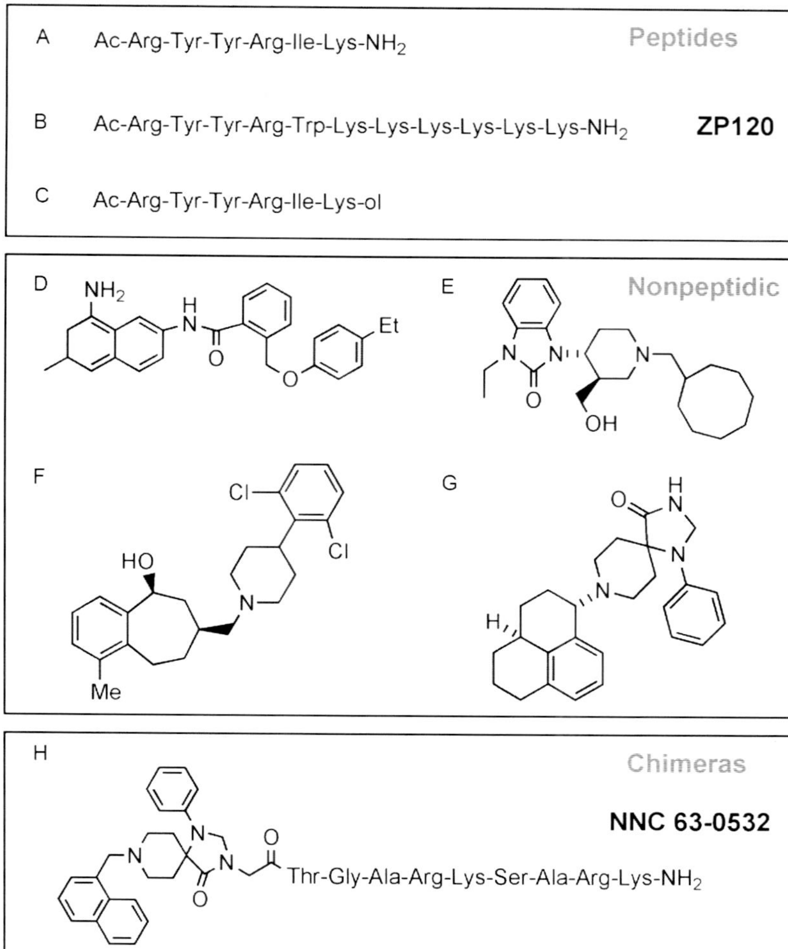

Figure 3 Miscellaneous peptide, nonpeptide, and chimeric modulators of ORL-1. (A, B) Hexapeptide ligand Ac-RYYRIK-NH_2 and its poly-lysine derivative; (C) when the C-terminus is an alcohol, the hexapeptide is an antagonist. Nonpeptidic compounds in development: (D) JTC-801, Japan Tobacco (antagonist; 4-aminoquinolines); (E) J-113393, Banyu (antagonist; benzimidazopiperidines); (F) SB-612111, GlaxoSmithKline (antagonist, aryl-piperidines); and (G) Ro64-6198, Roche (agonist; spiropiperidines). (H) Chimeric molecule NNC 63-0532-nociceptin(5–13)-NH_2. *Adapted in part from Lambert (2008).*

1.3 Interrogating the activation and address domains of nociceptin(1–17)

Sequence similarities between nociceptin and dynorphin A have been previously described (Chavkin & Goldstein, 1981; Meunier et al., 1995). For nociceptin(1–17), the native N-terminal tetrapeptide (FGGF) was identified as the *message* domain, essential for activating biological responses following receptor binding, while the remainder of its sequence, termed the *address* domain, likely confers high-affinity binding (Fig. 4). The address domain of nociceptin (i.e., residues 7–17) contains basic amino acid residues that likely bind to acidic residues present in the second extracellular loop of the ORL-1 receptor (Thompson et al., 2012).

Nociceptin (nociceptin(1–17)-OH) is equipotent with its amidated form (nociceptin(1–17)-NH_2; Guerrini et al., 1997), yet truncation of the nociceptin sequence possessing either a free acid or an amidated C-terminus resulted in substantial changes in binding affinity for ORL-1 (Butour et al., 1997; Calo et al., 1997; Dooley & Houghten, 1996; Guerrini et al., 1997). C-terminal truncation of nociceptin-(1–17)-OH, for instance, induced lower binding affinity and biological potency at the ORL-1 receptor (Butour et al., 1997; Reinscheid et al., 1996). In contrast, C-terminal truncation of nociceptin(1–17)-NH_2 to nociceptin(1–13)-NH_2 did not reduce potency. Only when truncated beyond the 12th residue did potency progressively decrease, with all activity being lost upon truncation beyond Ser10 (Calo et al., 1997; Dooley & Houghten, 1996; Guerrini et al., 1997). Thus, while dynorphin A(1–17) could be truncated to residues 1–7 with retention of high affinity for its receptor (Mansour, Hoversten, Taylor, Watson, & Akil, 1995), nociceptin(1–7)-NH_2 was completely inactive at the ORL-1 receptor. Such differences in potency of truncated peptides may relate to introduction of a negatively charged carboxylate at the C-terminus (Chavkin & Goldstein, 1981), which likely altered peptide–receptor interactions that are critical for biological signaling.

The biological activity for nociceptin at ORL-1 was further characterized by alanine mutagenesis of the nociceptin(1–17) peptide (Dooley & Houghten, 1996; Orsini et al., 2005; Reinscheid et al., 1996; Fig. 4). This method investigates the contribution of each amino acid side chain to

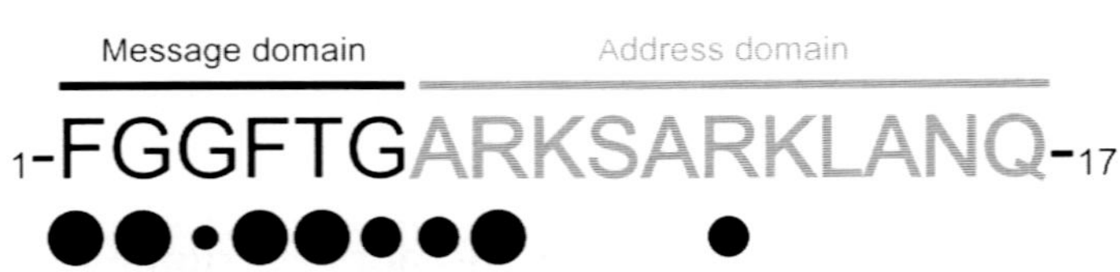

Figure 4 Sequence of nociceptin(1–17) and summary of mutagenesis data. The "message" and "address" domains are shown. Dots represent amino acids important in functional activity of nociceptin; their size reflects relative functional importance. *Data taken from Reinscheid, Ardati, Monsma, and Civelli (1996).*

potency by individually replacing them with the uncharged, chemically inert, and small alanine side chain via peptide synthesis. N-terminal residues Phe1, Gly2, Phe4, Thr5 and Arg8, were of greater importance for biological activity than C-terminal residues 9–17, based on inhibition of cAMP (Reinscheid et al., 1996). Although truncation of nociceptin(1–17) to less than 13 residues led to significantly reduced potency, the alanine-scan indicated that residues 9–17 were unimportant (except for Arg12) for function at the ORL-1 receptor (Fig. 4). It was therefore concluded that these residues were associated with binding affinity for ORL-1 rather than for agonist activity *per se*.

2. PROSPECTING THE IMPORTANCE OF THE N-TERMINAL TETRAPEPTIDE OF NOCICEPTIN(1–17)

Changes to the N-terminal tetrapeptide of nociceptin(1–17) (i.e., FGGF), including amino acid substitutions, deletions, and peptide bond modifications (Calo, Guerrini, et al., 1998; summarized in Fig. 5A–D), have revealed the importance of the physical distance between Phe1 and Phe4 for nociceptin activity. This was highlighted by a Gly2- and Gly3-deleted analogue [desGly2,3]nociceptin(1–13)-NH_2 that effectively lost ORL-1 activity, and replacement of these glycine residues with either L-Phe or D-Phe prevented activation of ORL-1. Further modifications of Phe1 and Gly2 in nociceptin were also made by replacing the carboxyl-group of Phe1 with a CH_2 moiety (Calo, Guerrini, et al., 1998; Fig. 5). This produced the first reported ORL-1 antagonist, [Phe1Ψ(CH_2-NH)Gly2]nociceptin(1–13)-NH_2, yet similar modifications to dynorphin resulted only in a loss of agonist potency at the κ-opioid receptor (Meyer et al., 1995). Replacing CO by CH_2 removed a hydrogen bond acceptor, increased flexibility of the Phe1-Gly2 bond, and introduced a basic amine. Subsequent research in other tissue preparations and CHO-transfected cells, however, suggested that [Phe1Ψ(CH_2-NH)Gly2]nociceptin(1–13)-NH_2 may act as a full or partial agonist (Rizzi et al., 1999), possibly indicating that differences in experimental approaches or in responses of cell types can produce different biological effects.

To better understand the antagonist properties of [Phe1Ψ(CH_2-NH)-Gly2]nociceptin(1–13)-NH_2, a series of Phe1 modifications were made and assessed for both agonist and antagonist activity (Guerrini et al., 2000). It was shown that the analogue [BzlGly1]nociceptin(1–13)-NH_2

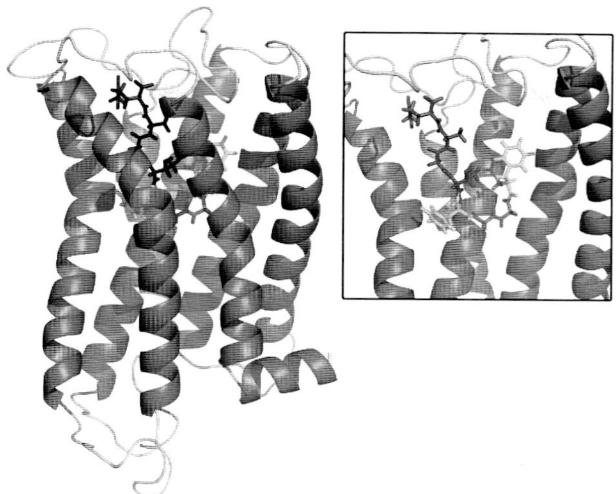

Figure 1 Modeled structure of ORL-1 and putative binding of nociceptin. The membrane-spanning domain of ORL-1 is typical of other rhodopsin-like GPCRs. The nociceptin-binding site for ORL-1 consists of two adjacent hydrophobic pockets in a crevice formed by transmembrane helices 3, 5, 6, and 7, corresponding to the conserved opioid-binding site in opioid receptors. Further profiling has identified the absence of lipid-facing charged residues in TM helices 2, 3, and 4 in ORL-1, which is atypical for GPCRs (Topham, Mouledous, Poda, Maigret, & Meunier, 1998). The transmembrane helical domains are represented by red ribbon while extracellular and intracellular loops are depicted as green tube. Nociceptin is depicted in lines (black: address domain, blue: message domain) where Phe1 and Phe4 from the message domain is depicted in green. Inset; Phe1 (green) docks deep into one of the pockets and interacts with Asp130 (highlighted in black), a highly conserved residue in TM2. By contrast, Thr5 and Gly6 bind to a nonconserved region in EL2, where the basic side chain of Thr5 makes favorable contact with Gln286 in the acidic EL2 loop. Residues 5 and 6 thus might serve as one of the determinants of selective binding of ORL-1 as other more classical opioids will encounter unfavorable binding due to the presence of cationic residues at the same positions (Mollereau et al., 1999; Topham et al., 1998). The crystal structure of ORL-1 in complex with a peptidomimetic was recently reported (Thompson et al., 2012). (See the color plate.)

Considering the distribution of ORL-1 in the CNS and its relationship to other opioid receptors, ORL-1 was hypothesized to play a role in a variety of CNS functions including nociception, motor control, reward reinforcement, stress responses, sexual behavior, and aggression and possibly contributing to autonomic control of physiological systems (Chiou et al., 2007; Neal et al., 1999). Of particular interest is the role of nociceptin in pain regulation, which is not surprising given the structural relationship of ORL-1 to other opioid receptors. Multiple studies have shown that

A Phe-Gly-Gly-Phe-Thr-Gly-Ala-Arg-Lys-Ser-Ala-Arg-Lys-Leu-Ala-Asn-Glu-NH_2

B Phe-Gly-Gly-Phe-Thr-Gly-Ala-Arg-Lys-Ser-Ala-Arg-Lys-NH_2

C Phe-**φ(CH_2-NH)-Gly**-Gly-Phe-Thr-Gly-Ala-Arg-Lys-Ser-Ala-Arg-Lys-NH_2

D **BzlGly**-Gly-Gly-Phe-Thr-Gly-Ala-Arg-Lys-Ser-Ala-Arg-Lys-NH_2

E Phe-Gly-Gly-**(pF)Phe**-Thr-Gly-Ala-Arg-Lys-Ser-Ala-Arg-Lys-**Arg-Lys**-Asn-Glu-NH_2

F **BzlGly**-Gly-Gly-**(pF)Phe**-Thr-Gly-Ala-Arg-Lys-Ser-Ala-Arg-Lys-**Arg-Lys**-Asn-Glu-NH_2

G **BzlGly**-Gly-Gly-Phe-Thr-Gly-Ala-Arg-Lys-Ser-Ala-Arg-Lys-**Arg-Lys**-Asn-Glu-NH_2

Figure 5 Summary of truncated nociceptin peptides and their substituted analogues. (A) Nociceptin(1–17)-NH_2; (B) nociceptin(1–13)-NH_2; and (C) [Phe1Ψ(CH_2-NH)Gly2] nociceptin(1–13)-NH_2 showed partial agonist activity at ORL-1. (D) [Phe1ΔBzlGly1] nociceptin(1–13)-NH_2 was an antagonist at ORL-1. Nociceptin(1–17)-NH_2 analogues with enhanced potency (E) [(pF)Phe4,Arg14,Lys15]nociceptin(1–17) was a "super-agonist" of ORL-1. (F) [BzlGly1,(pF)Phe4,Arg14,Lys15]nociceptin(1–17)-NH_2 was an antagonist with potential agonist activity. (G) [BzlGly1,Arg14,Lys15]nociceptin(1–17)-NH_2 was an antagonist at ORL-1. (Berger, Calo, Albrecht, Guerrini, & Bienert, 2000; Calo, Guerrini, et al., 1998; Calo, Rizzi, et al., 1998; Chen et al., 2004; Chen, Chang, et al., 2002; Chen, Wang, et al., 2002; Guerrini et al., 2000, Guerrini, Calo, Bigoni, Rizzi, Regoli, et al., 2001; Guerrini, Calo, Bigoni, Rizzi, Rizzi, et al., 2001; Guerrini et al., 2003; Meyer et al., 1995; Redrobe et al., 2000; Rizzi et al., 1999; Sasaki, Kawano, Kohara, Watanabe, & Ambo, 2006).

with a C→N shift in the Phe1 side chain was a moderate antagonist (Fig. 5D). Although this new antagonist [BzlGly1]nociceptin1–13-NH_2 had 100-fold lower binding affinity for ORL-1 (K_i 125 nM) relative to native [Phe1]nociceptin(1–13)-NH_2 peptide (K_i 0.80 nM), it was an antagonist of nociceptin both *in vitro* (Berger et al., 2000; Guerrini, Calo, Bigoni, Rizzi, Regoli, et al., 2001; Rizzi et al., 1999) and *in vivo* (Chen, Chang, et al., 2002; Redrobe et al., 2000). Attempts to increase antagonist potency of [BzlGly1]nociceptin(1–13)-NH_2 with alternative modifications to Phe1 were unsuccessful (Chen, Chang, et al., 2002; Guerrini, Calo, Bigoni, Rizzi, Regoli, et al., 2001; Redrobe et al., 2000; Sasaki et al., 2006).

Further scrutiny suggested that the Phe1-Gly2 *peptide bond* was not crucial for nociceptin activity (Chen et al., 2004; Guerrini et al., 2003), but that orientation and conformation of the Phe side chain were important in receptor activation (Guerrini et al., 2003). Furthermore, analogues with Gly2 and Gly3 residues replaced by sarcosine (*N*-methylglycine) decreased flexibility of the N-terminus and increased hydrophobicity (Chen et al.,

2004; Guerrini et al., 2003). These substitutions at position 2 reduced potency at ORL-1 (and decreased selectivity at other opioid receptors), whereas at position 3 they completely eliminated all activity at ORL-1, confirming the importance of flexibility and backbone orientation of the N-terminal FGGF tetrapeptide component.

3. OTHER MODIFICATIONS TO NOCICEPTIN(1–17)

Some modifications to nociceptin(1–17)-OH (Okada et al., 2000) were based on its two basic residue pairs Arg-Lys at positions 8, 9 and 12, 13 possibly interacting with acidic residues on extracellular loop 2 of ORL-1 (Guerrini et al., 1997). Relative to nociceptin, the peptide [Arg14,Lys15]nociceptin(1–17)-NH_2 had 3-fold higher affinity for ORL-1 and 17-fold increased agonist activity, which was attributed to either cation–π interactions with aromatic groups in the receptor, or additional electrostatic interactions with the acidic cluster on ORL-1. This information was used to create the ORL-1 antagonist [BzlGly1] nociceptin(1–17)-NH_2, with Leu14Arg and Ala15Lys mutations to produce [BzlGly1,Arg14,Lys15]nociceptin(1–17)-NH_2 (Fig. 5G), this being one of the most potent ORL-1 antagonists reported (Calo et al., 2005; McDonald, Calo, Guerrini, & Lambert, 2003; Nazzaro et al., 2007).

The functional role of Phe4 in nociceptin(1–13)-NH_2 was investigated, by modifying either the aromaticity or side chain length or conformational constraints of Phe4, or substitution of the phenyl ring (Guerrini, Calo, Bigoni, Rizzi, Rizzi, et al., 2001; Fig. 6). Only the latter approach improved agonist potency. Two equipotent analogues [(pF)Phe4] nociceptin(1–13)-NH_2 and [(pNO_2)Phe4]nociceptin(1–13)-NH_2 were agonists, being 1.5–6-fold more potent than nociceptin(1–17)-NH_2 against mouse vas deferens, mouse forebrain membranes, and forskolin-stimulated cAMP in CHO_{hOP4} cells (Guerrini, Calo, Bigoni, Rizzi, Rizzi, et al., 2001). There was a strong correlation between agonist affinity/potency and the electron-withdrawing properties of the group in the *para*-position of Phe4 that was inversely proportional to its size (r^2 values 0.47–0.72).

In vitro, [(pF)Phe4]nociceptin(1–13)-NH_2 and [(pNO_2)Phe4] nociceptin(1–13)-NH_2 were equipotent and by far the most active agonists in the series tested. Both compounds were antagonized by [BzlGly1]nociceptin(1–13)-NH_2 (Bigoni et al., 2002). Furthermore, both

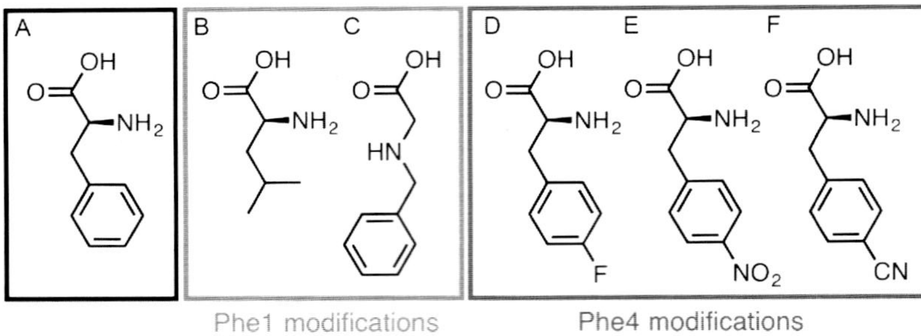

Figure 6 N-terminal modifications of nociceptin(1–17)-NH_2 and nociceptin(1–13)-NH_2. (A) Phenylalanine. (B) Leucine. (C) *N*-benzyl-glycine. (D) *para*-Fluoro-phenylalanine. (E) *para*-Nitro-phenylalanine. (F) *para*-Cyano-phenylalanine.

[(pF)Phe4]nociceptin(1–13)-NH_2 and [(pNO_2)Phe4]nociceptin(1–13)-NH_2 were less selective at the μ-opioid receptor, and [(pF)Phe4] nociceptin(1–13)-NH_2 showed reduced selectivity at κ- and δ-opioid receptors, although selectivity was still 1000-fold greater at other opioid receptors (Bigoni et al., 2002). [(pF)Phe4]nociceptin(1–13)-NH_2 was more potent than nociceptin(1–17)-NH_2 and longer lasting *in vivo* (spontaneous locomotor activity, tail-withdrawal test, hemodynamic measurements, food intake) and was antagonized by [BzlGly1]nociceptin(1–13)-NH_2 in the locomotor activity test (Rizzi, Salis, et al., 2002). By combining the (pF) Phe4 substitution with Arg14 and Lys15 substitutions, the first ORL-1 "super-agonist" was developed (Fig. 5E; Carra et al., 2005). For this [BzlGly1]nociceptin(1–17)-NH_2 series, the improvement in binding affinity was similar to that observed for nociceptin(1–17)-NH_2 analogues: [BzlGly1,(pF)Phe4,Arg14,Lys15] > [BzlGly1,Phe4,Arg14,Lys15] > [BzlGly1,(pF)Phe4] > [BzlGly1]nociceptin(1–17)-NH_2 (Guerrini et al., 2005; full sequences in Fig. 5F and 5G). However, in the functional assays, compounds containing the Phe4Δ(pF)Phe4 substitution had residual agonist activity at higher concentrations, whereas those without the (pF)Phe4 substitution did not show any residual agonist activity (Guerrini et al., 2005). Chang et al. investigated a similar series of compounds as well as 1-aminoisobutyric acid (Aib)-substituted peptides (Chang et al., 2005; Table 1). Aib is known to stabilize helical structure, albeit 3_{10}- rather than α-helicity. In a biological assay of electrically stimulated mouse vas deferens, the most potent agonists were Aib-substituted peptides ([Aib7]-, [Aib11]- and [Aib7,11]-nociceptin(1-17)-NH2.

Table 1 Summary of effect of Aib/Leu mutation on agonist potency of nociceptin(1–17)-NH_2 in a biological assay[a] (Chang et al., 2005; Tancredi et al., 2005)

Compound	−log EC_{50} (CI 95%)	EC_{50} (nM)
Nociceptin(1–17)-NH_2	7.97 (7.93–8.01)	11
[Aib7]nociceptin(1–17)-NH_2	8.35 (8.24–8.46)	4
[Aib11]nociceptin(1–17)-NH_2	8.38 (8.07–8.29)	4
[Aib7,11]nociceptin(1–17)-NH_2	8.18 (8.07–8.29)	7
[Leu7,11]nociceptin(1–17)-NH_2	7.60 (7.55–7.65)	25
[Leu11,15]nociceptin(1–17)-NH_2	8.13 (8.08–8.18)	7
[Leu11,15,Glu16]nociceptin(1–17)-NH_2	7.64 (7.59–7.69)	23
[Glu16]nociceptin(1–17)-NH_2	7.44 (7.20–7.68)	36

[a]EC_{50}: concentration of peptide which reduces the maximal possible electrically induced twitch responses to 50% in an organ bath-isolated mouse vas deferens.

4. THE IMPORTANCE OF STRUCTURE IN NOCICEPTIN ANALOGUES

4.1 Importance of helicity

Various attempts to resolve the three-dimensional structure of nociceptin(1–17)-NH_2 and related peptides nociceptin(1–13)-NH_2 (amide terminus, biologically active), nociceptin(1–13)-OH (carboxy terminus, biologically inactive), and nociceptin(1–11)-OH (COOH terminus, biologically inactive; Klaudel, Legowska, Brzozowski, Silberring, & Wojcik, 2004) using circular dichroism (CD) spectroscopy and nuclear magnetic resonance (NMR) spectroscopy were made (Amodeo et al., 2002, 2000). Spectra recorded under different conditions suggested that nociceptin had little structure in water, but may be in an α-helical conformation under membranous conditions. Specifically, it was predicted that the *address* domain of nociceptin(7–17) may adopt an amphipathic α-helical (Zhang, Miller, Valenzano, & Kyle, 2002) conformation upon binding to ORL-1, due to the regularly spaced alanine resides and basic Arg-Lys pairs within the full-length nociceptin(1–17) [FGGFTG**A**RKS**A**RKL**A**NQ].

A small library of lactam bridge-constrained nociceptin(1–13)-NH_2 peptides was published as summarized in Table 2 (Charoenchai, Wang, Wang, & Aldrich, 2008). Specific lactam bridges (Asp6→Lys10) or

Table 2 Binding affinity (K_i) and agonist potency (EC_{50}) of nociceptin(1–13)-NH_2 and lactam-constrained analogues (Charoenchai et al., 2008)

Compound	$K_i \pm$ SEM (nM)	$EC_{50} \pm$ SEM (nM)[a]
Nociceptin(1–13)-NH_2	0.62 ± 0.23	7.82 ± 1.88
Cyclo-[Asp6,Lys10]nociceptin(1–13)-NH_2	0.34 ± 0.10	4.12 ± 1.21
[Asp6,Lys10]nociceptin(1–13)-NH_2	0.54 ± 0.05	22.3 ± 5.9
cyclo-[D-Asp7,Lys10]nociceptin(1–13)-NH_2	0.27 ± 0.03	1.60 ± 0.45
[D-Asp7,Lys10]nociceptin(1–13)-NH_2	1.79 ± 0.10	106 ± 23

[a]EC_{50} measured in [^{35}S]GTPγS activity assay.

(D-Asp7 → Lys10) were incorporated into the peptide and these were investigated for [^{35}S]GTPγS activity and binding affinity, relative to their linear unconstrained analogues. The mutations Gly6Asp and Ser10Lys did not significantly affect binding affinity but deleteriously affected receptor activation by 2–3-fold. The negative effect of the Gly6Asp mutation supports the suggestion that large bulky groups are not tolerated in the hinge region of nociceptin (FGGF**TGA**RKSARKLANQ; Tancredi et al., 2005). Nevertheless, the constraint in cyclo-[Asp6,Lys10]nociceptin(1–13)-NH_2 improved binding affinity twofold and activation of [^{35}S]GTPγS fivefold over the linear analogue. By comparison, the Gly6(D-Asp) mutation decreased binding affinity and agonist activity up to 14-fold, probably due to an altered conformation induced by the D-amino acid. Constraining the peptide with an $i \rightarrow i+3$ lactam bridge also improved binding and activity over the linear analogue and was the most active compound in the series (Table 2). Even though the structure for these peptides was not investigated, it is not expected that the $D(i) \rightarrow K(i+4)$ linkages promote any significant α-helicity (Shepherd, Hoang, Abbenante, Fairlie, 2005) and a $i \rightarrow i+3$ linkage would be expected to stabilize a β-turn rather than an α-helix.

4.2 Other nociceptin derivatives

Shortly after the discovery of nociceptin, combinatorial libraries and positional scanning were used to identify five new hexapeptide ligands for the ORL-1 receptor (Dooley et al., 1997), all with commonality in structure (i.e., Ac-RYY(R/K)(W/I)(R/K)-NH_2). In all assays tested (stimulation of [^{3}S]GTPγS binding, inhibition of forskolin-stimulated cAMP, and inhibition of electrically induced contractions in the mouse vas deferens), these compounds were partial agonists. They were further developed to Ac-RYYRWKKKKKKK-NH_2 (ZP120) with high affinity for peptide is the paradigm that most proteolytic enzymes that degrade peptides need to recognize an extended nonhelical conformation in their active sites in order catalyze proteolysis, while the α-helical conformation is simply too big to fit into most human protease active site grooves (Fairlie et al., 2000; Madala, Tyndall, Nall, & Fairlie, 2010; Tyndall, Nall, & Fairlie, 2005).

6. THE DEVELOPMENT OF NEW HELIX-CONSTRAINED NOCICEPTIN ANALOGUES

6.1 Design of helix-constrained nociceptin analogues

Structure–activity relationship studies involving peptide truncations, as well as alanine and D-residue scanning have revealed that the message sequence is very sensitive to substitution with changes to Phe1, Gly2, and Phe4 resulting in complete loss of activity. The address domain (residues 7–17) is less sensitive to substitution, but replacing Arg8 abolished activity. Modifying the N-terminal Phe1 residue to *N*-benzyl-glycine (BzlGly) conferred a functional shift from agonist to pure antagonist activity (Calo, Guerrini, et al., 2000; Guerrini, Calo, Bigoni, Rizzi, Rizzi, et al., 2001). The available NMR structure in sodium dodecyl sulfate solution suggests that residues 4–17 may have some helical propensity, whereas in water the N-terminal pentapeptide appeared to be significantly more flexible and is not likely to be helical. Three nociceptin analogues, each containing two lactam bridges with different spacing, have recently been studied. These contain (1) back-to-back, (2) separated, or (3) overlapping lactam bridges (Fig. 7). In all strategies, critical residues Phe1, Phe4, and Arg8 are conserved. Due to the lack of three-dimensional structural information for the ORL-1 receptor in complex with nociceptin, it is uncertain whether the introduction of lactam bridges would interfere with receptor binding, so all three approaches for helix constraints were utilized in synthesized peptides. Based on the literature, conversion of the cyclic agonists to antagonists was expected to be possible by removing the N-terminal message domain or replacing the N-terminal phenylalanine (Phe1) with *N*-benzyl-glycine (BzlGly1). It was also expected that these changes would improve the potency of cyclic agonist and antagonist peptides by making Phe4(pF)Phe, Leu14Arg, and Ala15Lys substitutions previously reported in the literature (Fig. 5).

Compounds **1–23** (Table 3) were synthesized using solid-phase techniques. Peptide concentration was assessed using absorbance at $\lambda = 258$ nm

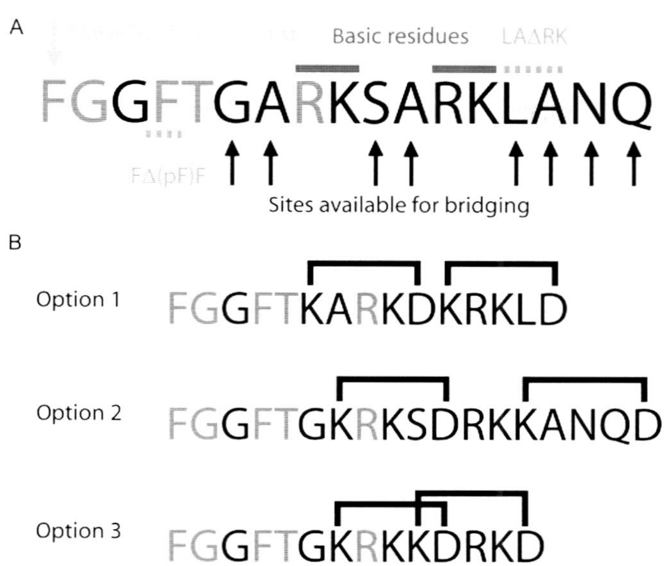

Figure 7 Design strategies for nociceptin(1–17)-NH_2 analogues. (A) Amino acids important for activity are highlighted in red and the modification sites for Phe1BzlGly (green), Phe4(pF)Phe, Leu12Arg, and Ala13Lys (orange). (B) Three prospective cyclization strategies show the relative bridge positions and conservation of Phe1, Gly2, Phe4, Thr5, and Arg8. (See the color plate.)

or an NMR method called PULCON (Wider & Dreier, 2006; Table 3). The presence of chromophores pF(Phe) or BzlGly altered the emission spectrum at 258 nm, requiring assessment of peptide concentration by NMR methods.

6.2 Helical structure of nociceptin(1–17)-NH_2 analogues in water

The structures of nociceptin peptides summarized in Table 3 were investigated by CD spectroscopy in aqueous buffer (i.e., 10 mM phosphate buffer, pH 7.4, 22 °C). All linear or unconstrained peptides (**1**–**10**) showed minimal helicity (<6%) in this buffer (e.g., Fig. 8A and B), and addition of the helix-favoring solvent 2,2,2-trifluoroethanol to the buffer only marginally increased helicity, in agreement with previous reports (Amodeo et al., 2002, Klaudel et al., 2000, 2004). On the other hand, incorporating two lactam bridges induced appreciable helicity (32–77%), as assessed by CD spectra by monitoring molar ellipticity at $\lambda = 222$ nm (Fig. 8C and D). The use of the lactam bridges did not, however, induce 100% helicity, since in most compounds only a third to half of the peptide sequence became helical (Fig. 8). This is not surprising since residues 1–6 (or 1–5 in **11** and **19**) are

Table 3 Summary of a subset of nociceptin compounds synthesized and investigated

Compound	Sequence	% Peptide
1	H-FGGFTGARKSARKLANQ-OH	0.86[a]
2	H-FGGFTGARKSARKLANQ-NH_2	0.80[a]
3	H-FGGFTGARKSARK-NH_2	0.68
4	BzlGlyGGFTGARKSARK-NH_2	0.81[a]
5	Ac-TGARKSARK-NH_2	0.65[a]
6	H-FGG(pF)FTGARKSARKLKNQ-NH_2	0.76[a]
7	BzlGlyGG(pF)FTGARKSARKLKNQ-NH_2	0.75[a]
8	H-FGGFTGKRKSNRKKANQN-NH_2	0.86[a]
9	H-FGGFTGARKSARKAANQA-NH_2	0.82[a]
10	BzlGlyGGFTGARKSARKLKNQ-NH_2	0.80[a]
11	H-FGGFT[**K**ARK**D**][**K**RKL**D**]-NH_2	0.55
12	H-FGGFTG[**K**RKS**D**]RK[**K**ANQ**D**]-NH_2	0.78
13	H-FGGFTG[**K**RK[**KD**]RK**D**]-NH_2	0.40
14	H-FGGFTG[**K**RKS**D**]RK-NH_2	0.49
15	H-FGGFTG[**d**RKS**K**]RK-NH_2	0.56
16	H-FGGFTG[**d**RK**K**]ARK-NH_2	0.52
17	H-FGG(pF)FTG[**K**RKS**D**]RK[**K**KNQ**D**]-NH_2	0.68[a]
18	Ac-T[**K**ARK**D**][**K**RKL**D**]-NH_2	0.82[a]
19	BzlGlyGGFT[**K**ARK**D**][**K**RKL**D**]-NH_2	0.66[a]
20	BzlGlyGGFTG[**K**RKS**D**]RK[**K**ANQ**D**]-NH_2	0.82[a]
21	BzlGlyGG(pF)FTG[**K**RKS**D**]RK[**K**KNQ**D**]-NH_2	0.70[a]
22	BzlGlyGGFTG[**K**RKS**D**]RK[**K**KNQ**D**]-NH_2	0.74[a]
23	Ac-[**K**RKS**D**]RK[**K**KNQ**D**]-NH_2	0.76[a]

[a]% Peptide refers to the amount of peptide in the material excluding water and salts adsorbed during the purification process. This value is taken into consideration when calculating molar ellipticity and bioactivity, determined by Abs_{258}.

not constrained and would not be expected to be in a helical conformation as they feature amino acids that have little or no known propensity to adopt helical structure in proteins. In peptides **11** and **13**, for example, only the respective 66% and 57% of the residues are constrained in the lactam bridges, which are proportionally reflected (Fig. 8) in fractional helicities of 54% and 52%.

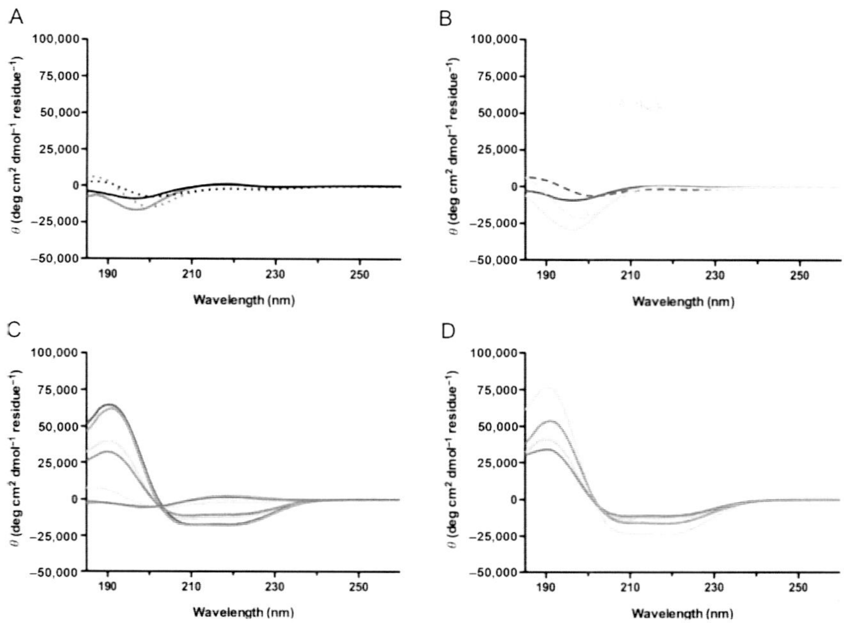

Figure 8 CD spectra of nociceptin peptides. (A) Unconstrained agonists of **2** in buffer (—) and 30% TFE (···), and compound **3** in buffer (—) and 30% TFE (···). (B) Unconstrained antagonists of compound **4** in buffer (—) and 30% TFE ( ). Compound **5** in buffer (—) and 30% TFE (– –). (C) Constrained compounds **11** (—), **12** (—), **13** (—), **14** ( ), **15** ( ), **17** (—), and **16** (—). (D) Constrained compounds **18** (—), **19** ( ), **20** (—), and **21** (—). (See the color plate.)

To determine whether the exocyclic regions were the sole cause of low helicity, compound **23** without the N-terminal message domain was investigated further. This 12-residue peptide features eight residues (K, R, Q) known to favor α-helicity in proteins, and yet possessed only 60% helicity in phosphate buffer, as determined by CD spectral analysis (see Table 4 ahead). The serine residue is conserved within the sequence of many peptides in the series and is usually considered to disrupt alpha helicity.

For comparison with the K(i)→D($i+4$)-constrained peptides, compounds **15** and **16** (both previously reported as ORL-1 agonists) with either a (D-Asp6)→K(10) or (D-Asp7)→K(10) lactam bridge position were characterized. Data suggested that **15** and **16** have similar structure, but neither were α-helical, as both have a maximum at 220 nm and a minimum at 200 nm in their CD spectra (Fig. 8). Furthermore, the signal was very weak therefore difficult to interpret. The single D-amino acid complicates interpretation of CD spectra. Compound **16** was further investigated by NMR spectroscopy and was found to possess coupling constants greater than, rather than less than, 6 Hz (data not shown) which is incompatible with appreciable α-helical structure.

6.3 Nuclear magnetic resonance spectra-derived structures

The CD spectra suggesting that compounds **11**, **12**, and **17** were significantly α-helical (Fig. 8) was confirmed by 1D and 2D ^{1}H NMR spectra of the compounds in 90% H_2O:10% D_2O at 288–308 K. Data for all three structures are shown for comparison of differences that might correlate with their activities. Two-dimensional total correlation spectroscopy (2D TOCSY) of **11** was used to identify resonances for each amino acid and to assign individual *spin* systems (not shown). Nuclear Overhauser effect spectroscopy (NOESY) spectra were used to identify sequential connectivity and intraresidue NH–NH and NH–CHα crosspeak correlations (Fig. 9).

The coupling constants of all backbone amide protons were examined for evidence in support of α-helicity ($^3J_{NHCH\alpha} \leq 6$ Hz). Of 17 amide NH protons, only 12 were visible or sufficiently dispersed to characterize a coupling constant. All coupling constants were below 6 Hz, with the exception of Phe4 ($^3J_{NHCH\alpha}$ 6.3 Hz). This supports α-helical character throughout the peptide. Residues Lys6 to Asp15 have Hα chemical shift differences greater than −0.1 ppm relative to their random coil value (Fig. 10A), supporting α-helicity through this region of the peptide. Phe4 and Thr5 have smaller chemical shift differences from random coil (−0.083 and −0.076 ppm, respectively), suggesting that helicity does not extend into the N-terminal region of the peptide. In the temperature range 288–308 K, only five protons had low-temperature-dependent amide chemical shifts ($\Delta\delta/T \leq 4$ ppb/K) and these were all in the C-terminal region (Lys11, Arg12, Lys13, Leu14, and NT; Fig. 10B). This suggests that these residues are involved in hydrogen bonding and, given the evidence of α-helicity from the coupling constants and chemical shift data, we assume that these are hydrogen bonding in an $i \rightarrow i+4$ position.

2D NOESY spectra were measured for compound **11**. They showed numerous sequential NOEs ($d_{NN(i,i+1)}$; $d_{\alpha N(i,i+1)}$) and numerous dominant medium-long range NOEs, indicative of α-helicity (Fig. 11). In particular, the $d_{\alpha N(i,i+3)}$ and $d_{\alpha N(i,i+4)}$ correlations between Gly2 and Asp18 are consistent with α-helicity. A solution structure was calculated for compound **11** in 90% H_2O:10% D_2O, using a dynamic simulated annealing and energy minimization protocol in X-PLOR, from 72 NOE distance restraints (29 sequential, 43 medium-long range) and 11 backbone ϕ-dihedral angle

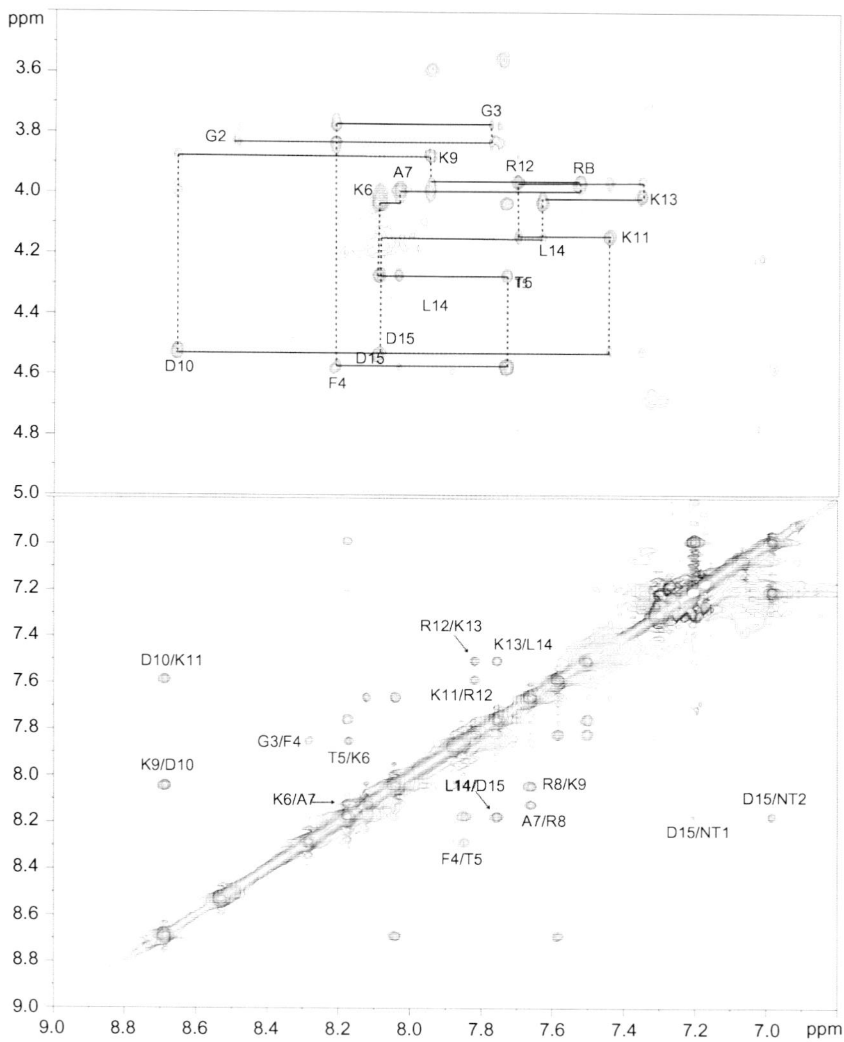

Figure 9 Section from the 600 MHz NOESY spectra for compound **11** (90% H_2O:10% D_2O, 298 K, mixing time 250 ms) showing NH–Cα (top panel) and NH–NH (bottom panel) connectivity.

restraints. The structure was calculated with and without five hydrogen bond restraints (for amide protons with $\Delta\delta/T \leq 4$ ppb/K), and the 20 lowest energy structures of **11** show a tight α-helix (RMSD 0.305 Å^2) between residues Lys6-Asp15. These structures were refined and had no ϕ-dihedral angle (>5°) or distance (>0.2 Å) violations. Three well-defined α-helical

ORL-1 (Kocsis et al., 2004; Rizzi, Rizzi, Marzola, et al., 2002), and Ac-RYYRIK-ol (with a reduced primary alcohol lysinol terminus) was found to be an antagonist (Kocsis et al., 2004; Fig. 3A–C). It is not clear how these peptides interact with ORL-1, but it was assumed that positively charged side chains interact with the acid cluster on extracellular loop 2 of ORL-1 (Dooley et al., 1997).

While the agonist effects of nociceptin could be inhibited by the ORL-1 selective peptidic antagonist [BzlGly1]GGFTGARKSARKRKNQ-NH_2, but not the small-molecule antagonist naloxone, a nonpeptide agonist, corresponding to the left hand end only of NNC63-0532 (Fig. 3H), was inhibited by naloxone but not by [BzlGly1]GGFTGARKSARKRKNQ-NH_2. When conjugated to residues 5–13 or 5–17 of nociceptin (i.e., the *address* domain), the inhibitory effects of resulting chimeras NNC 63-0532-nociceptin(5–13)-NH_2 and NNC 63-0532-nociceptin(5–17)-NH_2 could be modulated (albeit only by a small amount) by 1 μM [BzlGly1]GGFTGARKSARKRKNQ-NH_2 and by 1 μM naloxone producing an unusual dose–response curve. These results suggest a possible use for the peptide address domain of nociceptin(1–17) as a template for directing nonpeptidic compounds to a specific site on ORL-1.

5. RECENT ADVANCES IN ORL-1 ACTIVE NOCICEPTIN PEPTIDES

α-Helical constrained nociceptin(1–17)-NH_2 and its peptide analogues may show enhanced functional activity (agonist or antagonist) as well as higher stability in serum if constrained to adopt a water-stable α-helical structure. The basis for the assertion of enhanced activity is that nociceptin(1–17)-NH_2 itself has negligible α-helicity in water, but has some helical propensity in nonaqueous solvents. Since the binding site on ORL-1 may be a hydrophobic membrane-spanning region, it seemed likely that a helical conformation for nociceptin(1–17) may be favored and important. Structural characterization of nociceptin(1–17)-NH_2 by proton NMR spectroscopy (Orsini et al., 2005) had suggested that the address domain (residues 4–17) does adopt an α-helical conformation in a SDS/water solution. Although this was only circumstantial evidence for helical propensity in a hydrophobic environment, the results of others have suggested the importance of this region for binding of the address domain, which supports the presence of a high-affinity helical binding motif attached to the message domain. The basis for the assertion of higher stability in plasma for a helical

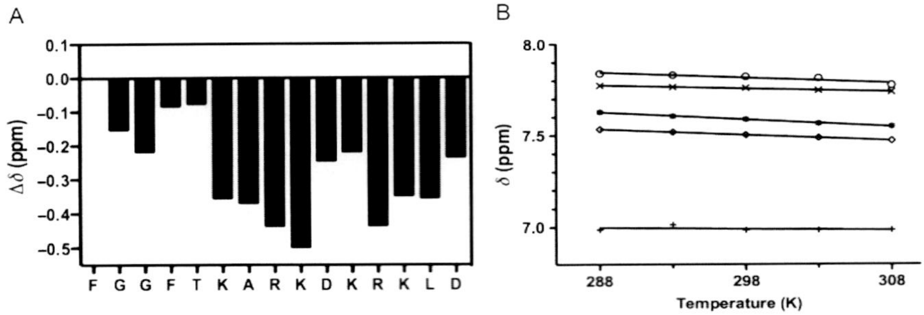

Figure 10 Chemical shift index ($\Delta\delta = \delta - \delta_{ref}$) of Hα and variable temperature data for compound **11**. (A) Chemical shift differences for residues 1–15 in **11**. Consecutive negative Δδ(Hα) values are characteristic of α-helical structure. (B) Temperature dependence of amide NH chemical shifts for compound **11** in 90% H_2O:10% D_2O. Gly2 (−8 ppb/K), Gly3 (−6 ppb/K), Phe4 (−8 ppb/K), Thr5 (−5 ppb/K), Lys6 (−7 ppb/K), Ala7 (−6 ppb/K), Arg8 (−7 ppb/K), Lys9 (−6 ppb/K), Asp10 (−9 ppb/K), Lys11 (−4 ppb/K) (∗), Arg12 (−3 ppb/K) (○), Lys13 (−3 ppb/K) (◇), Leu14 (−2 ppb/K) (×), Asp15 (−6 ppb/K), and NT (−1 ppb/K) (+).

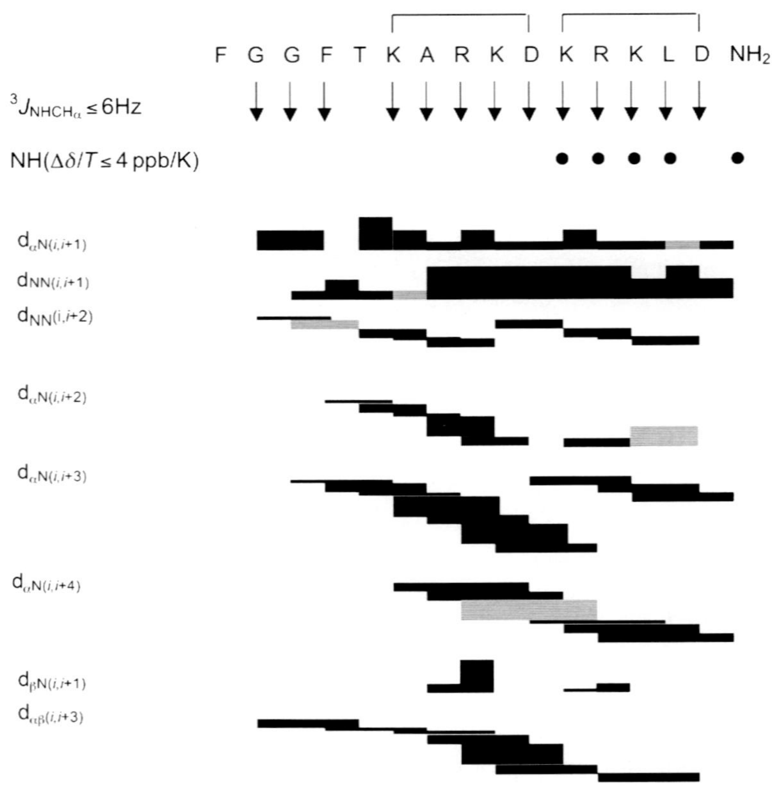

Figure 11 NOE summary diagram for compound **11** in 90% H_2O:10% D_2O at 298 K. Sequential, short- and medium-range NOE intensities were classified as strong (upper distance constraint 2.7 Å), medium (3.5 Å), weak (5.0 Å), and very weak (6.0 Å) and are proportional to bar thickness; gray bars indicate overlapping signals. $^3J_{NHCH\alpha}$ coupling constants <6 Hz are indicated by ↓. Amide NHs for which chemical shifts changed by <4 ppb/K are indicated by •.

turns were present within the bis-cycle (Lys6 to Asp15). Residues outside the bis-cycle had little structure, similar to the native nociceptin structure previously described in SDS micelles. The present structure was, however, determined in water alone and is a much more stable conformation than nociceptin(1–17)-NH_2.

The NMR-derived structures of compounds **12** and **17** were very similar (Fig. 12), with minor sequence changes at position 4 (Phe → (pF)Phe) and position 14 (Ala → Lys). Both compounds have significant $i \rightarrow i+3$ and $i \rightarrow i+4$ crosspeaks between Lys7 to Asp18, suggesting that the cyclized region is α-helical. By comparison, the N-terminal hexapeptide segment has numerous $i \rightarrow i+2$ and $i \rightarrow i+3$ crosspeaks, suggesting that this region is in a turn-like conformation. This finding is not surprising given that the Gly2-Gly3 segment is known to form turns (Datta, Shamala, Banerjee, & Balaram, 1997; Yang, Hitz, & Honig, 1996).

Ramachandran plots of all three compounds (not shown) indicated that the constrained regions of the peptides were α-helical, whereas residues 1–5 were less helical. Considering the presence of three glycine residues in the first six residues, it is not unexpected that this region has any structure as glycine is well known to fall outside the designated phi/psi regions defined by the Ramachandran plot.

Together, all of the structural data suggest that the $K(i) \rightarrow D(i+4)$ lactam bridges can encourage an α-helical structure in this small library of nociceptin-like agonists and antagonists. The three different linker strategies induced similar levels of α-helicity in agonists and antagonists, but surprisingly, the constrained bicyclic peptides had relatively low fractional α-helicity (32–52%) by CD spectral measurements (Table 4). Most sequences have an N-terminal message domain that was not expected to be helical.

The sequence of the model compound **17** (i.e., H-FGG(pF)FTG [**K**RKS**D**]RK[**K**KNQ**D**]-NH_2), with a low fractional helicity of 45%, has Arg8, Lys9, Arg12, Lys13, and Lys15 located on the same face on the helix, destabilizing the helix due to cationic repulsions. This region, Arg8-Glu17 is constrained into an α-helix in all of our bicyclic compounds, forcing Arg8, Lys9, Arg12, Lys13, as well as Lys15 (as in compounds **17** and **21**) onto the same face. It is thus not overly surprising that the α-helix is less stable than initially expected. The three model compounds (**11**, **12**, and **17**) with the most appreciable helicity are shown in the NMR-derived structures in Fig. 13.

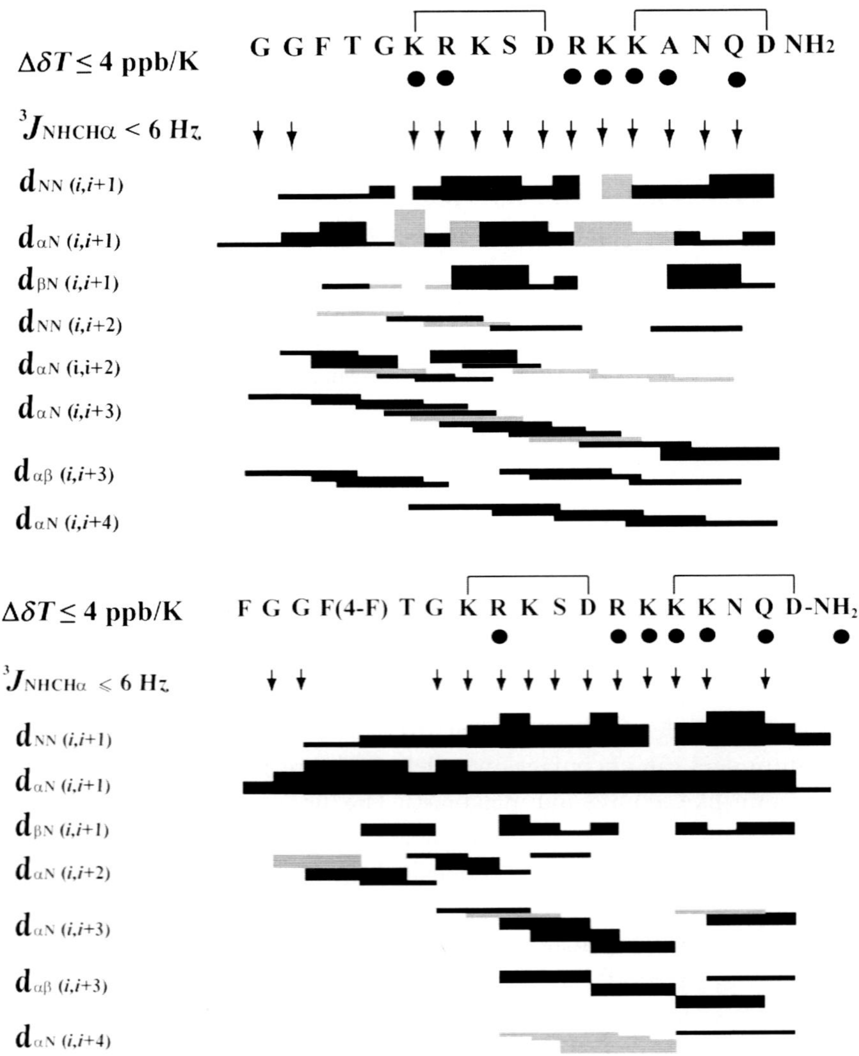

Figure 12 NOE summary diagram for (upper panel) **12** and (lower panel) **17** in 90% H_2O:10% D_2O at 288 K. Sequential, short- and medium-range NOE intensities were classified as strong (upper distance constraint 2.7 Å), medium (3.5 Å), weak (5.0 Å), and very weak (6.0 Å) and are proportional to bar thickness; gray bars indicate overlapping signals. $^3J_{NHCH\alpha}$ coupling constants <6 Hz are indicated by ↓. Amide NHs for which chemical shifts changed by <4 ppb/K are indicated by •.

Table 4 CD spectral data of nociceptin analogues in 10 mM phosphate buffer (pH 7.4) at 22 °C

Compound	Sequence	f_H (buffer)	$\theta_{222/208}$ (buffer)	f_H (TFE[a])
1	H-FGGFTGARKSARKLANQ-OH	0.02	−0.18	0.15
2	H-FGGFTGARKSARKLANQ-NH_2	0.02	−0.16	0.10
3	H-FGGFTGARKSARK-NH_2	0.01	−0.32	0.10
4	BzlGlyGGFTGARKSARK-NH_2	0.03	−0.10	0.10
5	Ac-TGARKSARK-NH_2	0.06	0.10	0.16
6	H-FGG(pF)FTGARKSARKLKNQ-NH_2	0.02	−0.10	0.20
7	BzlGlyGG(pF)FTGARKSARKLKNQ-NH_2	0.03	0.02	0.25
8	H-FGGFTGKRKSNRKKANQN-NH_2	0.03	0.03	0.13
9	H-FGGFTGARKSARKAANQA-NH_2	0.02	−0.11	0.21
10	BzlGlyGGFTGARKSARKLKNQ-NH_2	0.04	0.05	0.24
11	H-FGGFT[**K**ARK**D**][**K**RKL**D**]-NH_2	0.54	1.00	nd[b]
12	H-FGGFTG[**K**RKS**D**]RK[**K**ANQ**D**]-NH_2	0.32	0.91	nd
13	H-FGGFTG[**K**RK[**KD**]RK**D**]-NH_2	0.52	1.00	nd
14	H-FGGFTG[**K**RKS**D**]RK-NH_2	0.09	0.45	nd
15	H-FGGFTG[**d**RKS**K**]RK-NH_2	−0.05	−1.85	nd
16	H-FGGFTG[**d**RK**K**]ARK-NH_2	−0.01	−1.03	nd
17	H-FGG(pF)FTG[**K**RKS**D**]RK[**K**KNQ**D**]-NH_2	0.45	0.90	nd
18	Ac-T[**K**ARK**D**][**K**RKL**D**]-NH_2	0.50	1.01	nd
19	BzlGlyGGFT[**K**ARK**D**][**K**RKL**D**]-NH_2	0.77	1.00	nd
20	BzlGlyGGFTG[**K**RKS**D**]RK[**K**ANQ**D**]-NH_2	0.34	0.96	nd

Continued

Table 4 CD spectral data of nociceptin analogues in 10 mM phosphate buffer (pH 7.4) at 22 °C—cont'd

Compound	Sequence	f_H (buffer)	$\theta_{222/208}$ (buffer)	f_H (TFE)
21	BzlGlyGG(pF)FTG[**K**RKS**D**]RK [**K**KNQ**D**]-NH_2	0.39	0.94	nd
22	BzlGlyGGFTG[**K**RKS**D**]RK [**K**KNQ**D**]-NH_2	0.41	0.93	nd
23	Ac-[**K**RKS**D**]RK[**K**KNQ**D**]-NH_2	0.60	0.93	nd

[a]30% TFE used in these experiments. Bold residues reflect lactam-forming residues. *d* = D-aspartic acid.
[b]nd, not determined; f_H calculated based on θ at 222 nm using the equation described in Luo and Baldwin (1997).

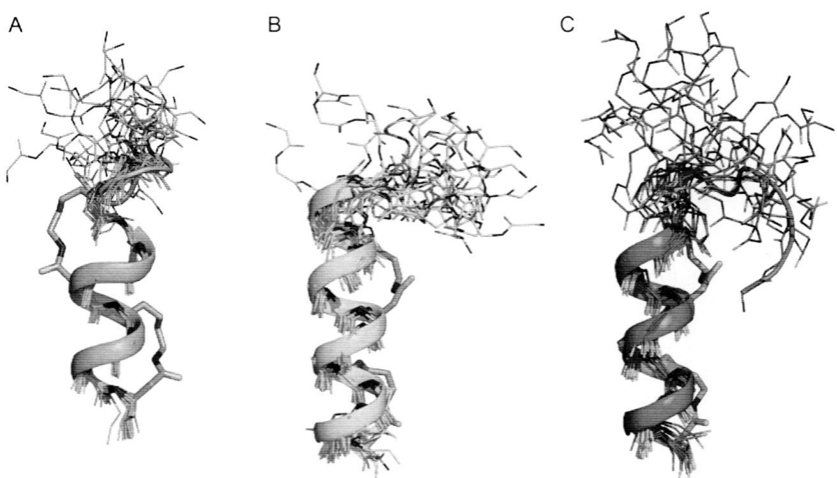

Figure 13 NMR-derived solution structures for compounds **11**, **12**, and **17**. Shown are the 20 lowest energy conformations of (A) **11** (RMSD [K6-D15] 0.31 Å^2), (B) **12** (RMSD [K7-D18] 0.75 Å^2), and (C) **17** (RMSD [K7-D18] 0.68 Å^2). Lactam bridges are shown as gray sticks. (See the color plate.)

7. BIOLOGICAL PROPERTIES OF HELICAL NOCICEPTIN MIMETICS

7.1 Cellular expression of ORL-1 and ERK phosphorylation

Transfected cell lines or primary tissue isolates are conventionally used to measure ligand–receptor binding and functions. However, a common difficulty in the culture of primary tissue is that the origin of the cells is frequently not uniform and leads to irreproducible data. Moreover, over-expression of receptors on null cells can give a false impression of the affinity and functions of ligands on native or primary cells. The most attractive method to test compounds is therefore to develop an assay using a non-transfected native, yet immortalized, cell line. ORL-1 has been identified to be present on a number of immune and neuronal immortalized cell lines, including U937 and THP-1 (monocytes; Peluso et al., 2001, 1998), EBV 1–4 (B cells; Peluso et al., 2001, 1998), MOLT-4 (T cells; Peluso et al., 2001, 1998), SH-SY5Y (neuroblastoma; Cheng, Standifer, Tublin, Su, & Pasternak, 1995; Peluso et al., 2001), SK-N-SH (neuroblastoma; Cheng et al., 1997), SK-N-BE (neuroblastoma; Spampinato, Di Toro, & Qasem, 2001), and NG108-15 (neuroblastoma × glioma; Morikawa et al., 1998). The ORL-1 receptor has also been found on primary immune cell lines (peripheral blood lymphocytes (Arjomand, Cole, & Evans, 2002; Pampusch et al., 2000; Peluso et al., 1998), PMNs (Fiset, Gilbert, Poubelle, & Pouliot, 2003; Peluso et al., 1998), and macrophages (Arjomand et al., 2002; Pampusch et al., 2000)) as well as on nervous system cell types (rat neurons (Buzas, Rosenberger, & Cox, 1998) and astrocytes (Buzas, Rosenberger, Kim, & Cox, 2002; Buzas, Symes, & Cox, 1999; Takayama & Ueda, 2005)). The use of assays to interrogate the intracellular secondary messenger signaling pathways involved in receptor signaling is well developed for the nociceptin–ORL-1 system, for which most studies use primary tissue isolates (Calo, Guerrini, et al., 1998; Guerrini, Calo, Bigoni, Rizzi, Rizzi, et al., 2001; Guerrini et al., 1998; Rizzi et al., 1999) or transfected cell lines (Guerrini et al., 2000; Ibba et al., 2008; Kitayama et al., 2007). Researchers have previously identified elements of the nociceptin signaling system and used them to measure ORL-1 activation or inhibition (summarized in Table 5).

Using three of these model cell lines (human monocytic U937, human neuroblastoma SHSY5Y, mouse neuroblastoma Neuro-2a), different signaling mechanisms were briefly examined and partially characterized. First, exposure of monocytic U937 cells to nociception did not trigger intracellular calcium release or inhibit forskolin-induced cAMP, but did lead to phosphorylation of ERK under conditions tested. The activation of ORL-1 on SH-SY5Y did not trigger intracellular calcium release, contrary to a previous report (Connor et al., 1996). These differences in activity may be due to specific assay conditions such as intra- and extracellular levels of calcium, as suggested previously by work investigating muscarinic (M_3) regulation of

Table 5 Summary of published ORL-1 functional and binding assays using native cell lines

Assay type	Cell line	Treatments	G-protein	Refs
Inhibition of cAMP	SH-SY5Y	None	$G_{i/o}$	Chan et al. (1998)
	SH-SY5Y	Differentiated	$G_{i/o}$ and $G_{i/o}$-independent component	Chan et al. (1998)
	SK-N-SH	None	$G_{i/o}$	Cheng et al. (1995)
Activation of [^{35}S]GTPγS	SK-N-SH	None	$G_{i/o}$	Cheng et al. (1995)
	SH-SY5Y	None	Not investigated	Peluso et al. (2001)
	U937	None	Not investigated	Peluso et al. (2001)
	CEM	None	Not investigated	Peluso et al. (2001)
	MOLT-4	None	Not investigated	Peluso et al. (2001)
Activation of MAPK pathway	NG108-15	None	(p38) G_i/G_o	Zhang et al. (1999)
	NG108-15	None	(JNK) $G_{i/o}$ and $G_{i/o}$-independent component	Chan and Wong (2000)
Inhibition of Ca^{2+} current channels	NG108-15	None	$G_{i/o}$	Morikawa et al. (1998)
	SH-SY5Y	Differentiated	$G_{i/o}$	Connor, Yeo, and Henderson (1996)
Intracellular Ca release	SH-SY5Y	1 μM carbachol	$G_{i/o}$	Connor et al. (1996)
Binding experiments				
H^3-nociceptin(1–17)-NH_2	SK-N-BE	None		Spampinato et al. (2001)
	SH-SY5Y	None		Peluso et al. (2001)

Table 5 Summary of published ORL-1 functional and binding assays using native cell lines—cont'd

Assay type	Cell line	Treatments	G-protein	Refs
	U937	None		Peluso et al. (2001)
	CEM	None		Peluso et al. (2001)
	MOLT-4	None		Peluso et al. (2001)
	SK-N-SH	None		Peluso et al. (2001)

adenylate cyclase (Hirst & Lambert, 1995). Furthermore, activation of SH-SY5Y ORL-1 did not activate pERK above background levels but was able to inhibit forskolin-induced cAMP. Finally, the activation of Neuro-2a ORL-1 did not trigger intracellular calcium release, but effectively inhibited forskolin-induced cAMP and as well-activated phosphorylation of ERK (summarized in Table 5 and Fig. 14).

Although ORL-1 signaling was identified in three cell lines, the Neuro-2a cell line was selected to further test our constrained nociceptin compounds due to its more robust and repeatable signaling patterns. This cell line was found responsive to nociceptin(1–17)-OH through both the pERK and cAMP pathways, allowing further activator-mediated response assays in the future. Thus, we continued by assessing the ability of nociceptin(1–17)-OH (compound **1**) for pERK activation in Neuro-2a cells using a second primary antibody targeting both pERK1 and pERK2 (rabbit IgG phospho-ERK1[T202/Y204]/ERK2[T185/Y187]). Nociceptin(1–17)-OH predominately induced phosphorylation of ERK2, although had some effect on ERK1 (Fig. 15C). However, as Western blots are notoriously noisy, an antibody-based assay (AlphaScreen Surefire) was used, allowing a more accurate determination of pERK activity. Phosphorylation of ERK was mediated through pertussis toxin-sensitive $G_{i/o}$-proteins and downstream MEK pathways. Nociceptin(1–17)-OH-induced pERK activation with an EC_{50} of 11 nM ($n \geq 3$). The specific ORL-1 antagonist compound **4** was challenged with 100 nM native nociceptin(1–17)-NH_2 and had an IC_{50} of 570 nM ($n \geq 3$; Fig. 15D).

G-protein coupling to ORL-1 in Neuro-2a cells is likely to be important for the downstream signaling events and physiological responses of

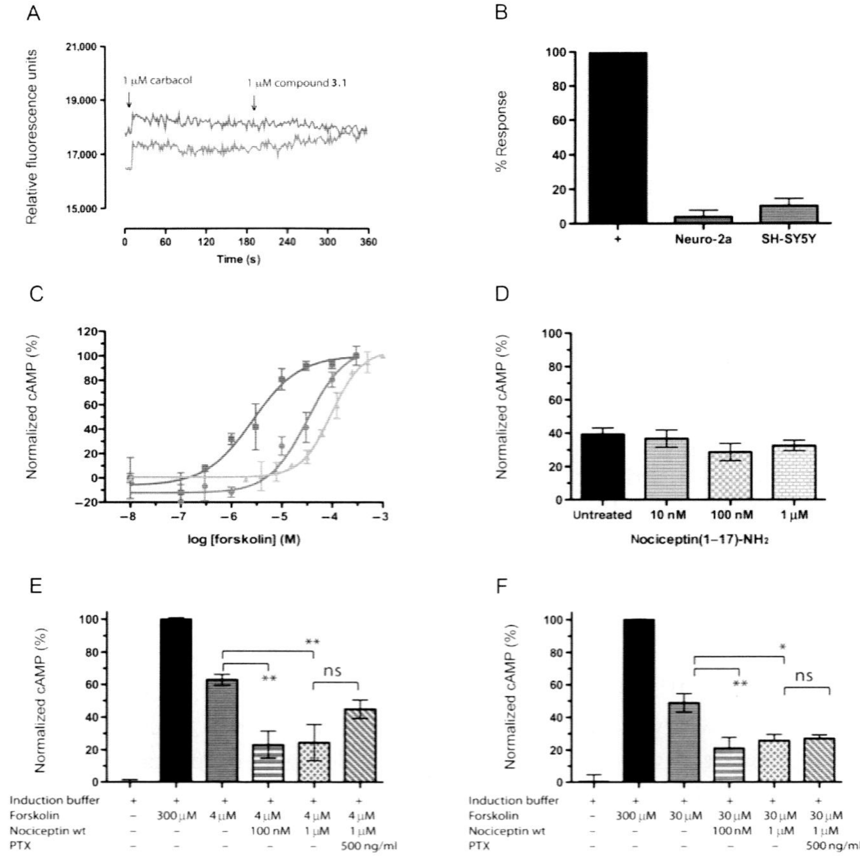

Figure 14 Signaling of nociceptin(1–17)-OH activation of ORL-1. (A, B) No intracellular calcium release was induced in Neuro-2a (blue) or SH-SY5Y cells (red) by carbachol or nociceptin(1–17)-OH (compound **1**). (C) Dose–response curves to forskolin in U937 cells (green), Neuro-2a (blue), and SH-SY5Y (red) with EC_{50} 94, 32, and 2.6 μM, respectively, $n=2$. (D) U937 preincubated with increasing concentrations of nociceptin(1–17)-OH showed no inhibit forskolin-induced cAMP (20 μM). (E) Neuro-2a cells preincubated with increasing concentration of nociceptin(1–17)-OH inhibited 4 μM forskolin-induced cAMP. (F) SH-SY5Y preincubated with nociceptin(1–17)-OH inhibited 30 μM forskolin-induced cAMP. Treatment with PTX was not able to modulate any cAMP effects. $^{*}p \leq 0.05$; $^{**}p \leq 0.01$; one-way ANOVA with Tukey's multiple comparison. ns, not significant. (See the color plate.)

nociceptin. Based on these findings and data from the literature, a signaling mechanism for nociceptin-activated ORL-1 in Neuro-2a cells has been proposed (Harrison, 2009). Many opioid receptors, including ORL-1, are generally not thought to couple to G_s (Connor & Christie, 1999; Law, Wong, & Loh, 2000) and, to our knowledge, ORL-1 has not been reported

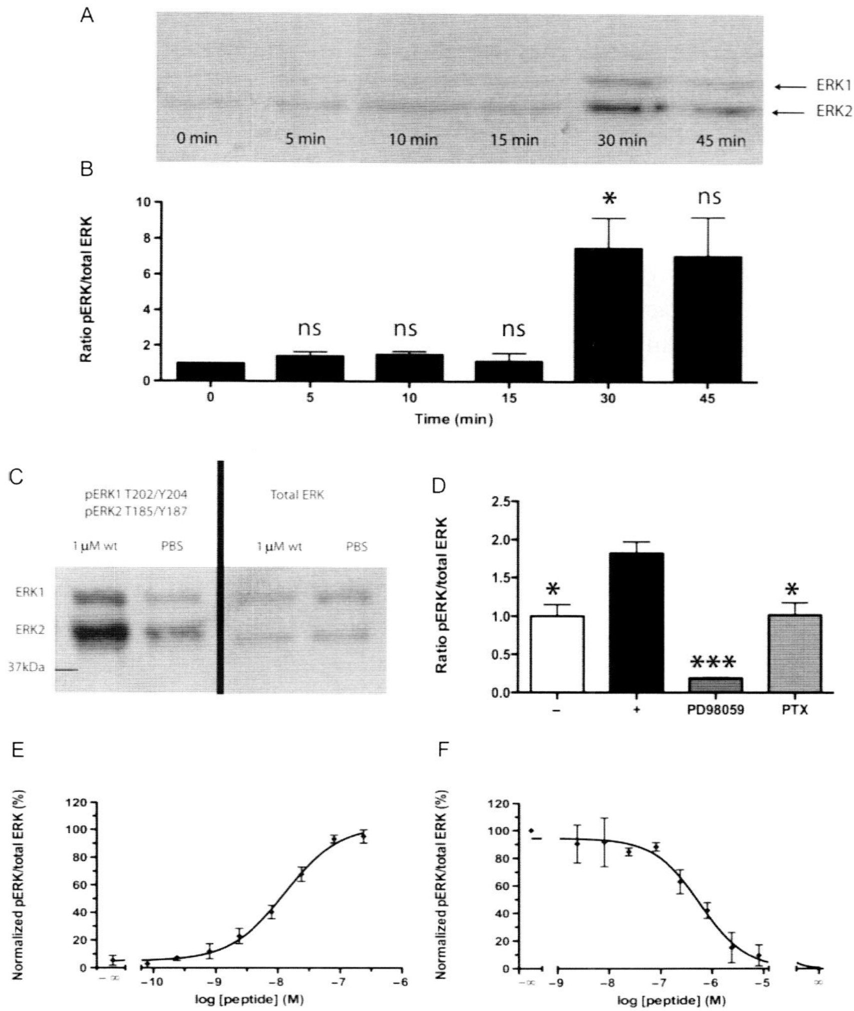

Figure 15 Serum-starved Neuro-2a pERK response to nociceptin(1–17)-NH_2 (compound **2**). (A) Western blot of time-dependent pERK response to peptide **2** (rabbit polyclonal ERK1 + 2 [pTpY185/187]) and (B) quantification of the pERK response showing maximal response at 30–45 min ($n = 2$). (C) Western blot of pERK response to 1 μM peptide **2** (rabbit IgG phospho-ERK1[T202/Y204]/ERK2[T185/Y187]). (D) Peptide **2** pERK response (300 nM) mediated through MEK and pertussis toxin-sensitive G-proteins. (E) Peptide **2** dose response with an EC_{50} of 11 nM ($n \geq 3$). (F) Antagonist response of [BzlGly1]nociceptin(1–13)-NH_2 with an IC_{50} of 570 nM ($n \geq 3$), challenged with 100 nM nociceptin(1–17)-NH_2. pERK assessed through AlphaScreen Surefire Kit. (See the color plate.)

to effectively couple to G_s in either native or recombinant cell lines. Involvement of $G_{12/13}$ in ORL-1 signaling has not been reported previously in any immortalized cell line natively expressing ORL-1 but has been characterized in transfected cells, where the ORL-1 receptor could couple with G_{12}, G_{14}, G_{16}, and G_z (Calo, Bigoni, et al., 2000; Chan & Wong, 2000; Chan et al., 1998; Yung et al., 1999). Others have shown in cells cotransfected with both ORL-1 and G_z, ORL-1 can utilize the G_z subunit to inhibit cAMP production. Overall, the data available suggest that nociceptin can activate ERK phosphorylation through the $G_{i/o}$ family of G-proteins and inhibit cAMP through PTX-insensitive G-proteins. While the specific G-protein involved in cAMP inhibition has not been identified, it is possible that G_z may be involved in the process. It is known that G_z is present in native cells expressing ORL-1 (namely retinoic differentiated SH-SY5Y (Ammer & Schulz, 1994; Chan et al., 1998)) and, although part of the $G_{i/o}$ family, G_z is insensitive to PTX.

7.2 Agonist and antagonist activity of nociceptin(1–17)-NH_2 and analogues

The pERK assay was further used to investigate the activity of constrained nociceptin analogues, expecting that α-helicity of the analogues would correlate with agonist/antagonist potency (data summarized in Tables 6 and 7 for agonists and antagonists, respectively). Compounds were tested as agonists and as antagonists challenged with nociceptin(1–17)-NH_2. The ratio of [pERK:total ERK] was determined by Western blot, quantification of densitometry proving to be quite noisy so data were interpreted in a qualitative rather than quantitative manner. In general, the results agreed with predicted activities. For potential agonists, an increase in basal pERK/total ERK ratio was sought (from control). Those residues with phenylalanine at position 1 appeared to have agonist activity, whereas those with *N*-tetrapeptide truncation or Phe1BzlGly substitution exhibited no agonist activity. For antagonists, a decrease in the pERK/total ERK ratio from nociceptin was sought. Residues with Phe1BzlGly substitution appeared to have antagonist activity, whereas those with native Phe1 residue were not antagonists (independent of agonist response). Interestingly, compound **15** had been expected to show agonist activity and no antagonist activity based on sequence alone; however, the reverse was seen. Additionally, compound **18** with an N-terminal truncation was expected to show antagonist activity but had none.

Fifteen compounds were tested further for agonist activity over a range of concentrations (Table 6). Wild-type nociceptin(1–17)-OH (compound **1**)

Table 6 Nociceptin compounds and their agonist activity (pERK, Neuro2 cells)

Compound	Sequence	Activity[a]	−log EC_{50} ± SEM	EC_{50} (nM)	E_{max} (%)
1	H-FGGFTGARKSARKLANQ-OH	Agonist	7.94 ± 0.16	11	90 ± 7
2	H-FGGFTGARKSARKLANQ-NH_2	Agonist	7.89 ± 0.09	13	103 ± 6
3	H-FGGFTGARKSARK-NH_2	Agonist	7.94 ± 0.14	11	90 ± 8
6	H-FGG(pF)FTGARKSARKLKNQ-NH_2	Agonist	9.44 ± 0.12	0.36	108 ± 9
7	BzlGlyGG(pF)FTGARKSARKLKNQ-NH_2	Antagonist		IA[b]	
8	H-FGGFTGK(Ac)RKSNRKK(Ac)ANQN-NH_2	Agonist	6.45 ± 0.08	360	99 ± 8
9	H-FGGFTGARKSARKAANQA-NH_2	Agonist	8.42 ± 0.15	4	93 ± 7
11	H-FGGFT[**K**ARK**D**][**K**RKL**D**]-NH_2	Agonist	8.51 ± 0.14	3	96 ± 8
12	H-FGGFTG[**K**RKS**D**]RK[**K**ANQ**D**]-NH_2	Agonist	9.22 ± 0.08	0.61	94 ± 3
13	H-FGGFTG[**K**RK[**KD**]RK**D**]-NH_2	Agonist	7.86 ± 0.08	14	92 ± 5
14	H-FGGFTG[**K**RKS**D**]RK-NH_2	Agonist	7.52 ± 0.18	30	80 ± 15
15	H-FGGFTG[**d**RKS**K**]RK-NH_2	Inactive	IA	IA[c]	–
16	H-FGGFTG[**d**RK**K**]ARK-NH_2	Agonist	7.73 ± 0.16	18	98 ± 13
17	H-FGG(pF)FTG[**K**RKS**D**]RK[**K**KNQ**D**]-NH_2	Agonist	10.40 ± 0.11	0.04	101 ± 6
21	BzlGlyGG(pF)FTG[**K**RKS**D**]RK[**K**KNQ**D**]-NH_2	Antagonist		IA[b]	

[a]Based on a single-point 1 μM concentration assay, detected by Western blot (Fig. 18A).
[b]Maximum concentration tested 1 μM; *K*(Ac) is a lysine amino acid where the free amide terminus of the sidechain is capped with an acetyl group.
[c]Maximum concentration tested 100 nM.

was equipotent with the amidated analogues (**2** and **3**), and modifications PheΔ(pF)Phe and AlaΔ15Lys (**6**) improved potency 30-fold. The three different cyclization strategies used to induce helicity in the address domain significantly affected the extent of ORL-1 activation as measured by ERK phosphorylation. Compound **11** (back-to-back lactam bridge) showed improved activity over the unconstrained **2**, whereas spacing the $K(i) \rightarrow D(i+4)$ bridges apart further improved activity. By comparison, compound **13** (crossover lactam bridges) had no improvement in agonist activity relative to **2**. The cyclization strategy of compound **12** was identified as most favorable for agonist activity, thus **12** was further modified to include PheΔ(pF)Phe and AlaΔ15Lys substitutions (producing **17**), changes which has been shown by ourselves (i.e., **6** vs. **2**) and others to improve activity (Carra et al., 2005). These substitutions significantly increased agonist activity relative to **12** and its unconstrained analogue compound **6** (Fig. 16).

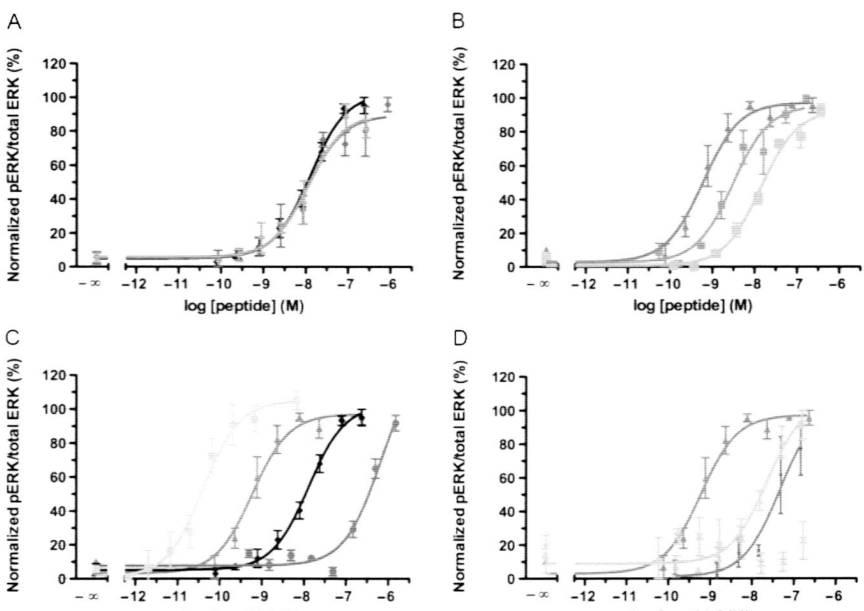

Figure 16 pERK concentration–response profiles for agonist compounds. (A) Compounds **1** (—), **2** (—), and **3** (—). (B) Compounds **11** ( ), **12** (—), and **13** ( ). (C) Compounds **17** ( ), **12** (—), **2** (—), and **8** (—). (D) Compounds **12** (—), **14** (—), **15** ( ), and **16** (—). EC_{50} for all compounds reported in Table 8. All data normalized to PBS control (0%) and 1 μM compound **1** (100%), all data points represent $n \geq 3$. For illustration purposes only, hillslope was set to 1.0. (See the color plate.)

To investigate the potential for monocyclic K→D lactam-constrained peptides, compound **14** was designed by truncating **12** to include only residues 1–13. The rationale is based on the *unconstrained* truncated analogue that is widely regarded as equipotent to the nociceptin(1–17)-NH_2. In addition, compound **16** with a modified d(*i*)→K(*i*+3) lactam bridge was synthesized for comparison with the K(*i*)→D(*i*+4) analogues (Charoenchai et al., 2008). Compound **15** with a d(*i*)→K(*i*+4) lactam bridge was synthesized to investigate the role of a D-amino acid in position *i*. Monocyclic compounds **14** and **16** had a similar potency to compound **2** (whereas compound **15** showed no recordable agonist activity). Compounds **21** and **7** were also assessed for agonist activity due to evidence these compounds have residual agonist activity despite the BzlGly moiety at position 1 (Guerrini et al., 2005). Constrained **21** and the linear analogue **7** were tested at a single-point concentration (1 μM) and neither showed significant agonist activity (Fig. 17).

The specificity of the most potent agonist, compound **17**, was tested against compound **4**, previously shown to be a specific ORL-1 antagonist (Guerrini et al., 2004). The potency of **17** was significantly reduced from EC_{50} 40 pM to 1 nM ($p \leq 0.001$), clearly suggesting that agonist activity of **17** is likely mediated through the ORL-1 receptor (Table 6).

To establish that the increased agonist activity for compounds **12** and **17** was not due to replacement of amino acids, compounds **8** and **9** were designed where those amino acids involved in K(*i*)→D(*i*+4) lactam bridges (amino acids 7, 11, 14, and 18) were replaced with either alanines (compound **9**) or Lys/Asn (compound **8**). The positively charged amine of lysine side chain was capped with an acetyl group removing the charge to more effectively mimic the lactam bridging residues. Asparagine rather than aspartic acid was used to maintain the length of the side chain, while removing the negative charge of the carboxylic acid moiety. By replacing these residues with alanines in **9**, agonist activity was altered relative to compound **2**, but this difference was not statistically significant. Replacement with Ac-Lys/Asn on the other hand significantly decreased the potency of the compound ($p \leq 0.001$, Table 6), suggesting that replacement of these charged residues to form the lactam bridge did not adversely affect activity, but that the Lys and Asn (isosteric to Asp) residues may actually have hindered the activity due to the large side chains or potential destabilizing influence on the α-helix.

Potential antagonists were tested using full dose–response curves (Fig. 17 and Table 7). Compound **4** was significantly less potent than **10**. Compound

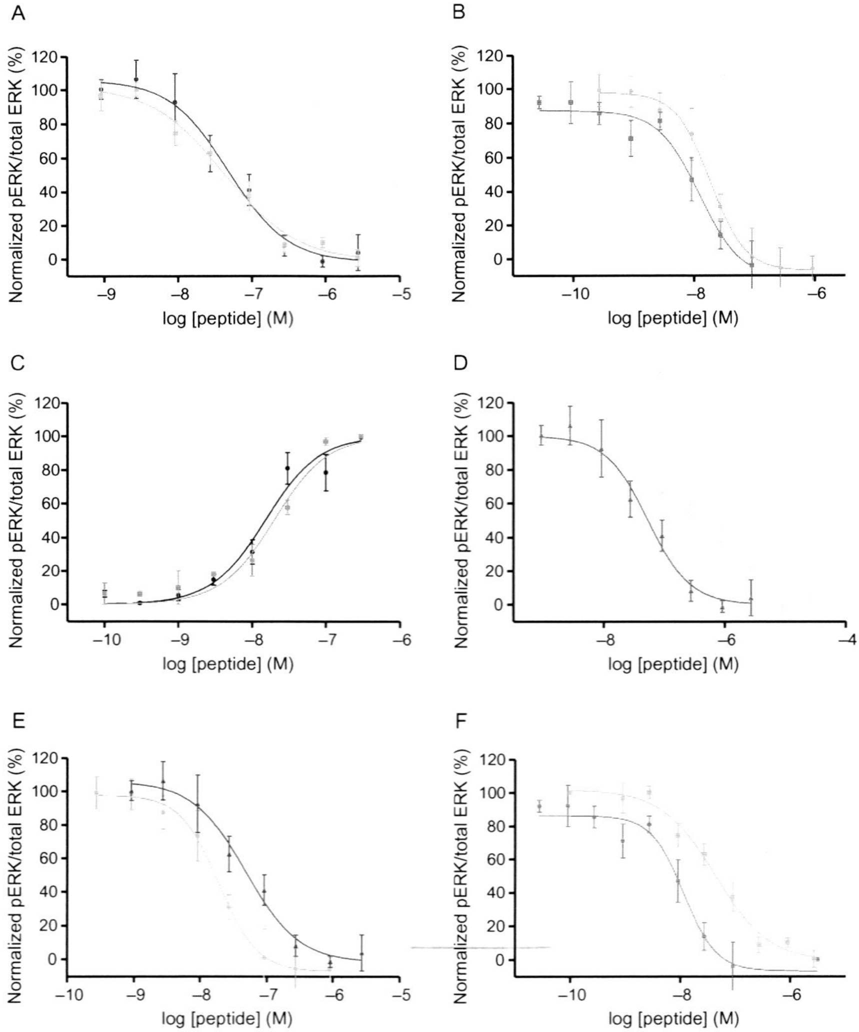

Figure 17 Comparison of nociceptin(1–17)-NH_2 and its truncated analogues for agonism, or antagonism. (A) Compound **10** (---), compound **7** (---); (B) compound **21** (---), compound **22** (---); (C) nociceptin(1–17)-NH_2 (compound **2** (---) versus nociceptin(1–13)-NH_2 compound **3** (---)); (D) [BzlG[1]]nociceptin(1–17)-NH_2 (---) versus [BzlG[1]]nociceptin(1–13)-NH_2 ( ); (E) compound **22** (---), compound **10** (---); (F) compound **21** (---), compound **7** (---). All data normalized to PBS control (0%) and 100 nM compound **2** (100%), and all data points represent $n \geq 3$. Statistical analysis ($p \leq 0.001$) using Student's *t*-test (Prism 5, Graphpad Software). (See the color plate.)

Table 7 Nociceptin compounds and their antagonist pERK activity

Compound	Sequence	Activity[a]	$-\log IC_{50} \pm SEM$	IC_{50} (nM)
4	BzlGlyGGFTGARKSARK-NH_2	Antagonist	-6.24 ± 0.15	570
5	Ac-TGARKSARK-NH_2	Antagonist	NT	
7	BzlGlyGG(pF)FTGARKSARKLKNQ-NH_2	Antagonist	-7.49 ± 0.09	32
10	BzlGlyGGFTGARKSARKLKNQ-NH_2	Antagonist	-7.40 ± 0.11	40
15	H-FGGFTG[**d**RKS**K**]RK-NH_2	?	NT	
18	Ac-T[**K**ARK**D**][**K**RKL**D**]-NH_2	Inactive	IA	IA
19	BzlGlyGGFT[**K**ARK**D**][**K**RKL**D**]-NH_2	Inactive	IA	IA
20	BzlGlyGGFTG[**K**RKS**D**]RK[**K**ANQ**D**]-NH_2	Antagonist	NT	
21	BzlGlyGG(pF)FTG[**K**RKS**D**]RK[**K**KNQ**D**]-NH_2	Antagonist	-8.13 ± 0.11	7
22	BzlGlyGGFTG[**K**RKS**D**]RK[**K**KNQ**D**]-NH_2	Antagonist	-7.91 ± 0.10	12
23	Ac-[**K**RKS**D**]RK[**K**KNQ**D**]-NH_2	Inactive	IA	IA

[a]Antagonist/agonist activity as tested in cell assays based on a single timepoint 10M concetration assay, detected by Western blot.
Based on a single-point 10 μM concentration assay, detected by Western blot (Fig. 18B); NT, not tested; IA, inactive (maximum concentration tested: 3 μM); ?, unknown agonist or antagonist charactierisation, as it was untested (i.e. NT).

10 has an additional four residues ([BzlGly1]GGFTGARKSARK**LKNQ**-NH_2) that, according to these results, are significant contributors to antagonist activity. Second, the *para*-fluoro-phenylalanine substitution, shown to improve agonist activity of nociceptin analogues, does not enhance the *antagonist* activity of compound **10** over **7** (Fig. 17).

The cyclization strategies employed showed a large influence on antagonist activity. Specifically, consecutive K6 → D10, K11 → D15 lactam bridges (as in **19**) abolished reasonable antagonist activity, whereas the alternative arrangement, K14 → D18 in compound **22** enhanced antagonist activity relative to the unconstrained **10** (Table 7). While differences in activity were expected between the cyclization strategies, we did not expect the complete loss of antagonist activity of compound **19** based on the agonist data set (Table 7). Nevertheless, comparing unconstrained **10** and constrained **22**, improvements in antagonist activity were threefold. Similarly, constrained **21** was over fourfold more potent than the unconstrained **7**. Based on the alanine mutants in the agonist data set, the enhancement in activity is likely due to the induction of helical structure in **21** and **22**, rather than substitution of residues at positions 7, 11, 14, and 18.

7.3 Effects of helical constraint on biological activity in Neuro-2a cells

Even though compounds **11–13** had similar fractional helicities, their ability to activate ERK varied significantly (Table 8), the lactam-constrained agonist demonstrated enhanced activity up to 20-fold, suggesting a free energy saving of 7.7 kJ/mol K (Table 8, compound **12**). Based on the similar fractional helicity of compounds **11–13**, it is clear that the differences in agonist activity that we observe cannot be explained entirely by induction of α-helicity.

Replacement of certain amino acids or the relative positions of the lactam bridges may be responsible for differences in ERK phosphorylation. Mutagenesis studies suggested that substitution of either Gly6Ala or Ala7Ser would reduce activity of nociceptin peptides by 140-fold each (Reinscheid et al., 1996). The current series of compounds showed substitutions at both of these sites may deleteriously affect activity. Difference in activity between compounds **11–13** may also be explained by relative positioning of lactam bridges and potential steric clashes with the ORL-1 receptor. However, at least for residues 7, 11, 14, and 18, we have shown that amino acid replacement does not significantly enhance activity of nociceptin(1–17)-NH_2 (**2**, **8**, and **9**), but may in fact decrease agonist ORL-1 activity. Without additional detailed mutagenesis and structural information about **11–13**, it is difficult to

Table 8 Summary of fractional helicity versus agonist activity for nociceptin(1–17)-NH_2 analogues

	Sequence	f_H	EC_{50}/IC_{50}	$\Delta\Delta G$
2	H-FGGFTGARKSARKLANQ-NH_2	0.02	13	–
8	H-FGGFTGK(Ac)RKSNRKK(Ac)ANQN-NH_2	0.03	360	−8.56[a]
9	H-FGGFTGARKSARKAANQA-NH_2	0.02	4	ns[a]
11	H-FGGFT[**K**ARK**D**][**K**RKL**D**]-NH_2	0.54	3	3.78[a]
12	H-FGGFTG[**K**RKS**D**]RK[**K**ANQ**D**]-NH_2	0.32	0.61	7.72[a]
13	H-FGGFTG[**K**AK[**KD**]RK**D**]-NH_2	0.52	14	ns[a]
Conversion to antagonist				
7	BzlGlyGG(pF)FTGARKSARKLKNQ-NH_2	0.03	32	–
21	BzlGlyGG(pF)FTG[**K**RKS**D**]RK[**K**KNQ**D**]-NH_2	0.45	7	3.92[b]
Phe4 and Leu14 substitution				
6	H-FGG(pF)FTGARKSARKLKNQ-NH_2	0.02	0.36	9.25[a]
17	H-FGG(pF)FTG[**K**RKS**D**]RK[**K**KNQ**D**]-NH_2	0.45	0.04	14.91[a]
14	H-FGGFTG[**K**RKS**D**]RK-NH_2	0.09	30	ns[a]
Effect of truncation				
15	H-FGGFTG[**d**RKS**K**]RK-NH_2	−0.05	IA	Less than −5.26[a]
16	H-FGGFTG[**d**RK**K**]ARK-NH_2	−0.01	18	ns[a]

[a]$\Delta\Delta G$ values (kJ/mol K; free energy calculation based on EC_{50} values) are reported relative to compound **3.2**.
[b]$\Delta\Delta G$ values (kJ/mol K; free energy calculation based on IC_{50} values) are reported relative to compound **3.7**. $\Delta\Delta G$ values calculated using $K=310$, $R=8.315$ J/mol. ns used where the differences in values were deemed insignificant and thus $\Delta\Delta G$ values meaningless; K(Ac) is a lysine amino acids where the free amide terminus of the sidechain is capped with an acetyl group.

determine specific reasons for differences in pERK activity. However, the improved potency of **12**, compared to **2**, is likely due to the induction of α-helicity through residues 7–17, and not due to favorable amino acid substitutions *per se*.

Using the most favorable lactam bridge arrangement identified in compound **12**, the potency through amino acid substitutions alone was improved. Using two substitutions identified in the literature (Phe4(pF)Phe and Ala15Lys), agonist activity improved by 30-fold (compound **1** vs. **6**). These substitutions [F4(pF)F and A15K] applied to a K(7)→D(11); K (14)→D(18) scaffold, agonist potency further increased to 40 pM (**17**). As suggested above, the improvement in activity is likely due to the induction of α-helicity, rather than solely amino acid substitution. Similar enhancement of agonist activity between compounds **6** and **17** (compared to **2** and **12**, respectively) has been reported previously (Arduin et al., 2007; Bigoni et al., 2002; Guerrini, Calo, Bigoni, Rizzi, Regoli, et al., 2001; Guerrini et al., 2005; Okada et al., 2000).

Compound **12** (18 amino acids) could not be truncated to a monocyclic analogue (**14**, 13 amino acids) and retain agonist activity. The reason for is most likely due to the fact that compound **14**, in comparison to the α-helical bis-lactam-containing peptides, has a significantly lower fractional helicity (10%). The $\theta_{222}/\theta_{208}$ ratio of 0.45 suggests that the peptide is not in an α-helical conformation, more likely a 3_{10} helix. Compound **16** recently reported by Charoenchai et al. to have improved activity at the ORL-1 receptor (Charoenchai et al., 2008) had a similar activity to linear compound **2** and our structural characterization suggests compound **16** had little, if any, helicity. It is important to note that the differences in relative activity between our data and that reported by Charoenchai et al. may be due to the different assay systems used or the promiscuous activation of other GPCRs (Charoenchai et al., 2008).

Modified agonist peptides created a small library of ORL-1 antagonists. Key substitutions in compound **7** (Phe4Δ(pF)Phe) did not increase the potency of the compound relative to **10** (Table 8), contrary to previous reports suggesting this single substitution can improve the antagonist activity 11-fold (*in vivo* assay using electrically stimulated guinea pig ileum; Guerrini et al., 2005). pERK signaling has not been used previously to investigate ORL-1 antagonists and thus direct comparison cannot be made with results described in literature. The K(*i*)→D(*i*+4) technology was also applied to the nociceptin antagonist sequences using two cyclization strategies (**19** and **21**). Compound **19** (with consecutive "back-to-back" lactam bridges) was inactive, whereas **22** (with spaced lactam bridges) had nanomolar antagonist activity. There are two potential reasons for these differences: (1) additional residues in **22** may provide better binding affinity to the ORL-1 receptor, thus improving potency, or (2) the "back-to-back" cyclization strategy of **19** negatively affected the compound potency due to either steric interference with the receptor or replacement of key amino acids.

By comparing the potency of unconstrained **7** to constrained **21** and the potency of unconstrained **10** to constrained **22**, it is clear that the induction of α-helicity can improve the antagonist activity up to fivefold (relating to 3.9 kJ/mol K; Table 8). Although this increase is substantial, it is relatively small compared to the 20-fold improvement in the agonist series (**12** vs. **2**; Table 8). The reason for this difference is not obvious but may relate to *N*-benzyl-glycine substitution at position 1. Relative binding affinities of nociceptin(1–17)-NH_2 and [BzlGly1]nociceptin(1–17)-NH_2 on recombinant human ORL (CHO cells; Guerrini et al., 2000, 2005) showed that [BzlGly1]nociceptin(1–17)-NH_2 had 100-fold lower binding affinity for ORL-1 than nociceptin(1–17)-NH_2. Based on these data, [BzlGly1] nociceptin(1–17)-NH_2 may bind to ORL-1 in a slightly different orientation relative to nociceptin(1–17)-NH_2, possibly explaining different enhancement in activity between the current agonist and antagonist data set.

7.4 Stability and cell toxicity of helix-constrained versus unconstrained peptides

The biological stability of cyclic peptide **12** was much greater in normal human serum, as shown by the significant increase in serum half-life of **12** compared to native nociceptin, compound **1** ($t_{1/2} > 6$ h vs. $t_{1/2} \approx 1$ h, respectively; Fig. 18A). Neither the native unconstrained nociceptin(1–17)-OH (**1**) nor either of the most potent cyclic peptides (compounds **12** and **17**) had any measurable hemolytic activity compared to positive lysis controls (Fig. 18B; Table 8).

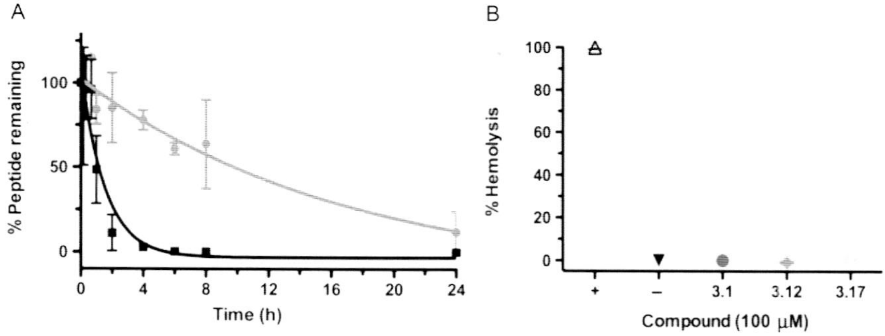

Figure 18 Stability and hemolytic studies of active nociceptin compounds. (A) Time-dependent degradation of **1** (—) and **12** (—) in normal human serum at 37 °C. Data represent $n = 2$, $p \leq 0.01$ at the 6-h time point. (B) Negligible hemolysis of human erythrocytes by compounds **1**, **12**, and **17** at 37 °C over 1 h, data normalized using 1% Triton X-100 (100% hemolysis) and PBS (0% hemolysis). Data represent $n \geq 3$. (See the color plate.)

7.5 *In vivo* activity of helix-constrained versus unconstrained nociceptin analogues

The most potent helical peptide (**17**) was compared with its unconstrained analogue (**6**) and native nociception in an *in vivo* mouse model of thermal nociception, the hot plate paw withdrawal assay, performed at 53 °C (Fig. 19). Studies in C57/B6 mice administered with native nociceptin-NH_2 in its amide form (compound **2**) showed dose-dependent increases in paw withdrawal latency response (maximal at 400 pmol [**2**], ED_{50} 185 pmol), clearly demonstrating antithermal nociceptive actions of the native peptide **2**. Similar results have also previously been reported in other *in vivo* nociception studies using the native nociceptin peptide. However, this increased thermal tolerance dissipated quickly within 5 minutes postadministration, suggesting rapid clearance of this peptide from the receptor and tissue, which is consistent with its low serum stability (Fig. 18). The linear unconstrained peptide (compound **6**) showed a comparable potency (maximal withdrawal latency increase at 200 pmol, ED_{50} 160 pmol) to the native nociceptin, with evidence of some nonsignificant increased duration of effect, and antinociceptive effects were detectable at 5 min only for the higher doses. However, effects were no longer recordable after 15 min postadministration of this peptide, suggesting that even though *in vivo* stability was improved over the native peptide, compound **6** still had rapid clearance (Fig. 19). The helix-constrained peptide (**17**) also showed significant increases in paw withdrawal latency at all doses tested (maximal 200 pmol); however, it showed significant dose-dependent improvement in both potency and stability, with effects still detectable at 1 h postadministration at the highest dose tested (200 pmol). Compound **17** was thus found to have significant improvements in both potency (ED_{50} 70 pmol) and stability over both native nociceptin and linear unconstrained peptide (compounds **2** and **6**, respectively; Fig. 19). Of interest, peptide **17** appeared to have a slower on-rate to reach maximal effect at all doses compared to the other peptides tested, since the maximal response did not occur until 5 min postadministration (compared to 2 min for the other peptides; Fig. 19). From these data, it can be concluded that the helix-constrained nociceptin peptide was more potent, more stable, and likely had a slower receptor on- and off-rate at ORL-1 than the native and linear constrained peptides *in vivo*; however, this should be confirmed in *in vitro* receptor-binding assays. The presented *in vivo* data correlate well with the *in vitro* data and highlight the potential for such constrained peptides targeting the ORL-1 receptor to be used as analgesics when given

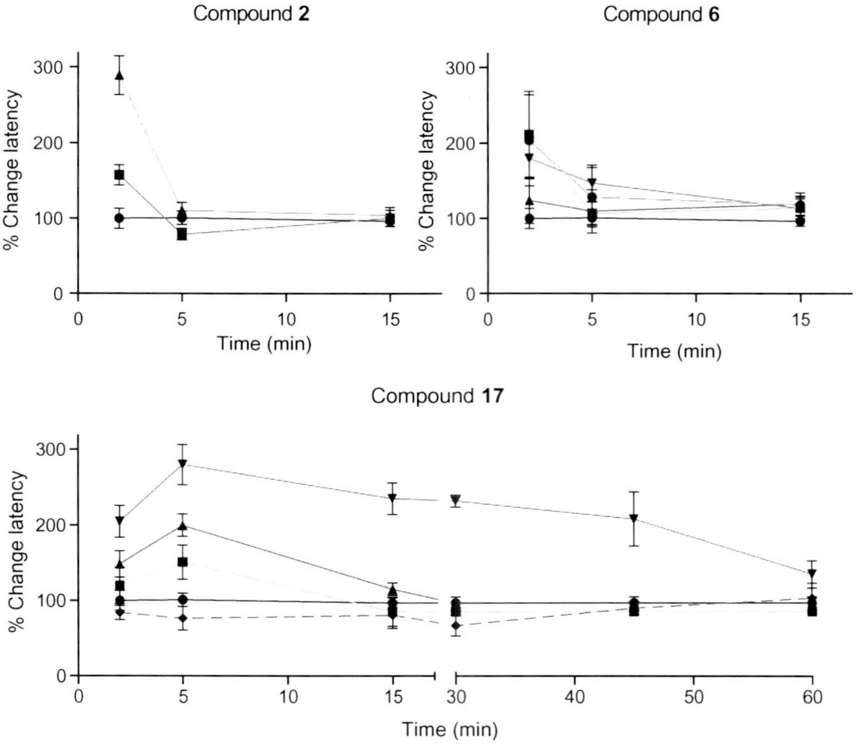

Figure 19 Antinociception activity of constrained peptides *in vivo*. Dose-dependent paw withdrawal latency effects in rats exposed to 53 °C hot plate at various time points postcompound administration. Compounds **2** and **6** have similar potencies and duration of activity at the doses given (EC_{50} 185 and 160 pmol, respectively), whereas compound **17** showed significant improvements in both potency (EC_{50} 70 pmol, $p < 0.05$) and duration of activity, with activity still recordable after 60 min at the highest dose. The antagonist compound **21** (200 nmol) when coadministered with agonist compound **17** (200 pmol) blocked all antinociceptive activity induced by the agonist **17**. These data clearly show an increased potency and stability of the agonist **17** at the ORL-1 receptor *in vivo* compared to the native nociceptin (compound **2**) and the linear constrained peptide **6**. It also provides evidence that the antagonist, compound **21**, has efficacy *in vivo*. All compounds were administered intraplantar in sterile saline (20 μL). Agonist compounds **2**, **6**, and **17**; saline control (0 pmol peptide, —), 50 pmol (—), 100 pmol (—), 200 pmol (—), and 400 pmol (—). Antagonist compound **21** (200 nmol, - -) administered together with agonist compound **21** (200 pmol). (See the color plate.)

peripherally. Thus, such helix-constrained peptides need not be administered intrathecally (i.e., directly into spinal canal or subarachnoid space), as previously suggested may have to be the case for unconstrained nociceptin analogues, and therefore helix-constrained peptides may be better suited for clinical use.

8. CONCLUDING REMARKS

Induction of α-helicity in nociceptin can enhance the agonist and antagonist potency at the ORL-1 receptor in cells, including Neuro-2a, providing that lactam bridges are optimally located and appropriate function-determining amino acid substitutions are also made. The data suggested that neither compound **12** nor **17** were hemolytic and that **12** had a significantly improved stability profile in human serum, making this series of compounds useful either for *in vivo* or *in vitro* studies. It may also be possible to use the α-helix stabilized address domain of compound **12** H-(1,5–8,13-cyclo)-[**K**RKS**D**]RK[**K**ANQ**D**]-NH_2 as a template for the addition of nonpeptidic molecules that may mimic the N-terminal tetrapeptide. This concept suggests that the address domain may provide binding affinity, while the nonpeptidic N-terminal unit may enhance potency as either an agonist or antagonist with improved pharmacokinetic profiles. The use of multiple $K(i) \rightarrow D(i+4)$ lactam bridges to induce α-helicity in nociceptin and improve potency at the ORL-1 receptor has proven very successful and we are further developing these leads for potential therapeutic use. The helix-constrained compound **17** in particular proved to be an exquisitely potent and plasma stable agonist with superior properties to native nociceptin both *in vitro* in human cells and *in vivo* in a thermal nociception assay in mice. Further investigation is required to prove the value of this compound in assays other than basic thermal (or even mechanical) nociception, such as in animal models more relevant to human neuropathic pain. Such research will be of significant benefit not only for the understanding of ORL-1 signaling in neurobiology, but also for the development of peptides or peptidomimetics active at ORL-1 intended for clinical use for treatment of neuropathic pain, if not addiction, eating disorders, and other manifestations in which ORL-1 has been implicated.

REFERENCES

Abrams, G. M., & Recht, L. (1982). Neuropeptides and their role in pain and analgesia. *Acupuncture & Electro-Therapeutics Research*, *7*(2–3), 105–121.

Akdis, C. A., & Simons, F. E. (2006). Histamine receptors are hot in immunopharmacology. *European Journal of Pharmacology*, *533*(1–3), 69–76. http://dx.doi.org/10.1016/j.ejphar.2005.12.044.

Ammer, H., & Schulz, R. (1994). Retinoic acid-induced differentiation of human neuroblastoma SH-SY5Y cells is associated with changes in the abundance of G proteins. *Journal of Neurochemistry*, *62*(4), 1310–1318.

Amodeo, P., Guerrini, R., Picone, D., Salvadori, S., Spadaccini, R., Tancredi, T., et al. (2002). Solution structure of nociceptin peptides. *Journal of Peptide Science*, *8*(9), 497–509.

Amodeo, P., Lopez Mendez, B., Guerrini, R., Salvadori, S., Temussi, P. A., & Tancredi, T. (2000). Pain peptides. Solution structure of orphanin FQ2. *FEBS Letters*, *473*(2), 157–160.

Arduin, M., Spagnolo, B., Calo, G., Guerrini, R., Carra, G., Fischetti, C., et al. (2007). Synthesis and biological activity of nociceptin/orphanin FQ analogues substituted in position 7 or 11 with Calpha, alpha-dialkylated amino acids. *Bioorganic & Medicinal Chemistry*, *15*(13), 4434–4443.

Arjomand, J., Cole, S., & Evans, C. J. (2002). Novel orphanin FQ/nociceptin transcripts are expressed in human immune cells. *Journal of Neuroimmunology*, *130*(1–2), 100–108, S0165572802002175 [pii].

Armagan, G., Bojnik, E., Turunc, E., Kanit, L., Gunduz Cinar, O., Benyhe, S., et al. (2012). Kainic acid-induced changes in the opioid/nociceptin system and the stress/toxicity pathways in the rat hippocampus. *Neurochemistry International*, *60*(6), 555–564. http://dx.doi.org/10.1016/j.neuint.2012.02.015.

Armstead, W. M. (2006). Differential activation of ERK, p38, and JNK MAPK by nociceptin/orphanin FQ in the potentiation of prostaglandin cerebrovasoconstriction after brain injury. *European Journal of Pharmacology*, *529*(1–3), 129–135. http://dx.doi.org/10.1016/j.ejphar.2005.08.059, S0014-2999(05)00919-2 [pii].

Bao, Y., Hou, W., & Hua, B. (2014). Protease-activated receptor 2 signalling pathways: A role in pain processing. *Expert Opinion on Therapeutic Targets*, *18*(1), 15–27. http://dx.doi.org/10.1517/14728222.2014.844792.

Beedle, A. M., McRory, J. E., Poirot, O., Doering, C. J., Altier, C., Barrere, C., et al. (2004). Agonist-independent modulation of N-type calcium channels by ORL1 receptors. *Nature Neuroscience*, 7(2), 118–125. http://dx.doi.org/10.1038/nn1180, nn1180 [pii].

Berger, H., Calo, G., Albrecht, E., Guerrini, R., & Bienert, M. (2000). [Nphe(1)]NC(1-13)NH(2) selectively antagonizes nociceptin/orphanin FQ-stimulated G-protein activation in rat brain. *The Journal of Pharmacology and Experimental Therapeutics*, *294*(2), 428–433.

Bigoni, R., Rizzi, D., Rizzi, A., Camarda, V., Guerrini, R., Lambert, D. G., et al. (2002). Pharmacological characterisation of [(pX)Phe4]nociceptin(1-13)amide analogues. 1. In vitro studies. *Naunyn-Schmiedeberg's Archives of Pharmacology*, *365*(6), 442–449.

Bobrova, I., Vlaskovska, M., Kasakov, L., Surovoy, A., Egorova, N., Johansson, L., et al. (2003). A structure-activity study of nociceptin-(1-13)-peptide amide. Synthesis of analogues substituted in positions 0, 1, 3, 4 and 10. *European Journal of Medicinal Chemistry*, *38*(7–8), 687–694.

Bodnar, R. J. (2013). Endogenous opiates and behavior: 2012. *Peptides*, *50C*, 55–95. http://dx.doi.org/10.1016/j.peptides.2013.10.001.

Brederson, J. D., Kym, P. R., & Szallasi, A. (2013). Targeting TRP channels for pain relief. *European Journal of Pharmacology*, *716*(1–3), 61–76. http://dx.doi.org/10.1016/j.ejphar.2013.03.003.

Bregola, G., Zucchini, S., Frigati, L., Candeletti, S., Romualdi, P., Reinscheid, R., et al. (2002). Involvement of the neuropeptide orphanin FQ/nociceptin in kainate and kindling seizures and epileptogenesis. *Epilepsia*, *43*(Suppl. 5), 18–19.

Bunnett, N. W. (2006). Protease-activated receptors: How proteases signal to cells to cause inflammation and pain. *Seminars in Thrombosis and Hemostasis*, *32*(Suppl. 1), 39–48. http://dx.doi.org/10.1055/s-2006-939553.

Bunzow, J. R., Saez, C., Mortrud, M., Bouvier, C., Williams, J. T., Low, M., et al. (1994). Molecular cloning and tissue distribution of a putative member of the rat opioid receptor gene family that is not a mu, delta or kappa opioid receptor type. *FEBS Letters*, *347*(2–3), 284–288. http://dx.doi.org/10.1016/0014-5793(94)00561-3.

Butour, J. L., Moisand, C., Mazarguil, H., Mollereau, C., & Meunier, J. C. (1997). Recognition and activation of the opioid receptor-like ORL 1 receptor by nociceptin, nociceptin analogs and opioids. *European Journal of Pharmacology*, *321*(1), 97–103.

Buzas, B., Rosenberger, J., & Cox, B. M. (1998). Activity and cyclic AMP-dependent regulation of nociceptin/orphanin FQ gene expression in primary neuronal and astrocyte cultures. *Journal of Neurochemistry*, *71*(2), 556–563.

Buzas, B., Rosenberger, J., Kim, K. W., & Cox, B. M. (2002). Inflammatory mediators increase the expression of nociceptin/orphanin FQ in rat astrocytes in culture. *Glia*, *39*(3), 237–246. http://dx.doi.org/10.1002/glia.10106.

Buzas, B., Symes, A. J., & Cox, B. M. (1999). Regulation of nociceptin/orphanin FQ gene expression by neuropoietic cytokines and neurotrophic factors in neurons and astrocytes. *Journal of Neurochemistry*, *72*(5), 1882–1889.

Calo, G., Bigoni, R., Rizzi, A., Guerrini, R., Salvadori, S., & Regoli, D. (2000). Nociceptin/orphanin FQ receptor ligands. *Peptides*, *21*(7), 935–947.

Calo, G., Guerrini, R., Bigoni, R., Rizzi, A., Bianchi, C., Regoli, D., et al. (1998). Structure-activity study of the nociceptin(1-13)-NH_2 N-terminal tetrapeptide and discovery of a nociceptin receptor antagonist. *Journal of Medicinal Chemistry*, *41*(18), 3360–3366.

Calo, G., Guerrini, R., Bigoni, R., Rizzi, A., Marzola, G., Okawa, H., et al. (2000). Characterization of [Nphe(1)]nociceptin(1-13)NH(2), a new selective nociceptin receptor antagonist. *British Journal of Pharmacology*, *129*(6), 1183–1193.

Calo, G., Guerrini, R., Rizzi, A., Salvadori, S., Burmeister, M., Kapusta, D. R., et al. (2005). UFP-101, a peptide antagonist selective for the nociceptin/orphanin FQ receptor. *CNS Drug Reviews*, *11*(2), 97–112.

Calo, G., Rizzi, A., Bodin, M., Neugebauer, W., Salvadori, S., Guerrini, R., et al. (1997). Pharmacological characterization of nociceptin receptor: An in vitro study. *Canadian Journal of Physiology and Pharmacology*, *75*(6), 713–718.

Calo, G., Rizzi, A., Marzola, G., Guerrini, R., Salvadori, S., Beani, L., et al. (1998). Pharmacological characterization of the nociceptin receptor mediating hyperalgesia in the mouse tail withdrawal assay. *British Journal of Pharmacology*, *125*(2), 373–378.

Calo, G., Rizzi, A., Rizzi, D., Bigoni, R., Guerrini, R., Marzola, G., et al. (2002). [Nphe1, Arg14, Lys15]nociceptin-NH_2, a novel potent and selective antagonist of the nociceptin/orphanin FQ receptor. *British Journal of Pharmacology*, *136*(2), 303–311.

Cannon, K. E., & Hough, L. B. (2005). Inhibition of chemical and low-intensity mechanical nociception by activation of histamine H3 receptors. *The Journal of Pain*, *6*(3), 193–200. http://dx.doi.org/10.1016/j.jpain.2004.12.003.

Carra, G., Rizzi, A., Guerrini, R., Barnes, T. A., McDonald, J., Hebbes, C. P., et al. (2005). [(pF)Phe4, Arg14, Lys15]N/OFQ-NH_2 (UFP-102), a highly potent and selective agonist of the nociceptin/orphanin FQ receptor. *The Journal of Pharmacology and Experimental Therapeutics*, *312*(3), 1114–1123.

Chan, A. S., & Wong, Y. H. (2000). Regulation of c-Jun N-terminal kinase by the ORL(1) receptor through multiple G proteins. *The Journal of Pharmacology and Experimental Therapeutics*, *295*(3), 1094–1100.

Chan, J. S., Yung, L. Y., Lee, J. W., Wu, Y. L., Pei, G., & Wong, Y. H. (1998). Pertussis toxin-insensitive signaling of the ORL1 receptor: Coupling to Gz and G16 proteins. *Journal of Neurochemistry*, *71*(5), 2203–2210.

Chang, M., Peng, Y. L., Dong, S. L., Han, R. W., Li, W., Yang, D. J., et al. (2005). Structure-activity studies on different modifications of nociceptin/orphanin FQ: Identification of highly potent agonists and antagonists of its receptor. *Regulatory Peptides*, *130*(3), 116–122.

Charoenchai, L., Wang, H., Wang, J. B., & Aldrich, J. V. (2008). High affinity conformationally constrained nociceptin/orphanin FQ(1-13) amide analogues. *Journal of Medicinal Chemistry*, *51*(15), 4385–4387.

Chavkin, C., & Goldstein, A. (1981). Specific receptor for the opioid peptide dynorphin: Structure-activity relationships. *Proceedings of the National Academy of Sciences of the United States of America*, *78*(10), 6543–6547.

Chen, Y., Chang, M., Wang, Z. Z., Chen, L. X., Yang, Q., Qi, Y. M., et al. (2002). [Nphe1] nociceptin(1-13)-NH_2 antagonizes nociceptin-induced hypotension, bradycardia, and hindquarters vasodilation in the anesthetized rat. *Canadian Journal of Physiology and Pharmacology*, *80*(1), 31–35.

Chen, Y., Fan, Y., Liu, J., Mestek, A., Tian, M., Kozak, C. A., et al. (1994). Molecular cloning, tissue distribution and chromosomal localization of a novel member of the opioid receptor gene family. *FEBS Letters*, *347*(2–3), 279–283. http://dx.doi.org/10.1016/0014-5793(94)00560-5.

Chen, L. X., Fang, Q., Chen, Q., Guo, J., Wang, Z. Z., Chen, Y., et al. (2004). Study in vitro and in vivo of nociceptin/orphanin FQ(1-13)NH_2 analogues substituting N-Me-Gly for Gly2 or Gly3. *Peptides*, *25*(8), 1349–1354.

Chen, L. X., Wang, Z. Z., Wu, H., Fang, Q., Chen, Y., & Wang, R. (2002). Effects of nociceptin(13-17) in pain modulation at supraspinal level in mice. *Neuroscience Letters*, *331*(2), 95–98.

Cheng, Z. J., Fan, G. H., Zhao, J., Zhang, Z., Wu, Y. L., Jiang, L. Z., et al. (1997). Endogenous opioid receptor-like receptor in human neuroblastoma SK-N-SH cells: Activation of inhibitory G protein and homologous desensitization. *Neuroreport*, *8*(8), 1913–1918.

Cheng, J., Standifer, K. M., Tublin, P. R., Su, W., & Pasternak, G. W. (1995). Demonstration of kappa 3-opioid receptors in the SH-SY5Y human neuroblastoma cell line. *Journal of Neurochemistry*, *65*(1), 170–175.

Chiou, L. C., Fan, S. H., Guerrini, R., & Calo, G. (2002). [Nphe(1)]N/OFQ-(1-13)-NH(2) is a competitive and selective antagonist at nociceptin/orphanin FQ receptors mediating K(+) channel activation in rat periaqueductal gray slices. *Neuropharmacology*, *42*(2), 246–252.

Chiou, L. C., Liao, Y. Y., Fan, P. C., Kuo, P. H., Wang, C. H., Riemer, C., et al. (2007). Nociceptin/orphanin FQ peptide receptors: Pharmacology and clinical implications. *Current Drug Targets*, *8*(1), 117–135.

Chiou, L. C., Liao, Y. Y., Guerrini, R., & Calo, G. (2005). [Nphe1, Arg14, Lys15]N/OFQ-NH_2 is a competitive antagonist of NOP receptors in the periaqueductal gray. *European Journal of Pharmacology*, *515*(1–3), 47–53.

Connor, M., & Christie, M. D. (1999). Opioid receptor signalling mechanisms. *Clinical and Experimental Pharmacology & Physiology*, *26*(7), 493–499.

Connor, M., Yeo, A., & Henderson, G. (1996). The effect of nociceptin on Ca^{2+} channel current and intracellular Ca^{2+} in the SH-SY5Y human neuroblastoma cell line. *British Journal of Pharmacology*, *118*(2), 205–207.

Datta, S., Shamala, N., Banerjee, A., & Balaram, P. (1997). Conformational variability of Gly-Gly segments in peptides: A comparison of the crystal structures of an acyclic pentapeptide and an octapeptide. *Biopolymers*, *41*(3), 331–336.

Dooley, C. T., & Houghten, R. A. (1996). Orphanin FQ: Receptor binding and analog structure activity relationships in rat brain. *Life Sciences*, *59*(1), L23–L29.

Dooley, C. T., Spaeth, C. G., Berzetei-Gurske, I. P., Craymer, K., Adapa, I. D., Brandt, S. R., et al. (1997). Binding and in vitro activities of peptides with high affinity for the nociceptin/orphanin FQ receptor, ORL1. *The Journal of Pharmacology and Experimental Therapeutics*, *283*(2), 735–741.

Fairlie, D. P., Tyndall, J. D., Reid, R. C., Wong, A. K., Abbenante, G., Scanlon, M. J., et al. (2000). Conformational selection of inhibitors and substrates by proteolytic enzymes: Implications for drug design and polypeptide processing. *Journal of Medicinal Chemistry*, *43*(7), 1271–1281.

Fiset, M. E., Gilbert, C., Poubelle, P. E., & Pouliot, M. (2003). Human neutrophils as a source of nociceptin: A novel link between pain and inflammation. *Biochemistry*, *42*(35), 10498–10505. http://dx.doi.org/10.1021/bi0300635.

Gavioli, E. C., Marzola, G., Guerrini, R., Bertorelli, R., Zucchini, S., De Lima, T. C., et al. (2003). Blockade of nociceptin/orphanin FQ-NOP receptor signalling produces antidepressant-like effects: Pharmacological and genetic evidences from the mouse forced swimming test. *The European Journal of Neuroscience*, *17*(9), 1987–1990.

Glass, M. J., Billington, C. J., & Levine, A. S. (1999). Opioids and food intake: Distributed functional neural pathways? *Neuropeptides*, *33*(5), 360–368. http://dx.doi.org/10.1054/npep.1999.0050.

Guerrini, R., Calo, G., Bigoni, R., Rizzi, D., Regoli, D., & Salvadori, S. (2001). Structure-activity relationship of [Nphe1]-NC-(1-13)-NH_2, a pure and selective nociceptin/orphanin FQ receptor antagonist. *The Journal of Peptide Research*, *57*(3), 215–222.

Guerrini, R., Calo, G., Bigoni, R., Rizzi, D., Rizzi, A., Zucchini, M., et al. (2001). Structure-activity studies of the Phe(4) residue of nociceptin(1-13)-NH(2): Identification of highly potent agonists of the nociceptin/orphanin FQ receptor. *Journal of Medicinal Chemistry*, *44*(23), 3956–3964.

Guerrini, R., Calo, G., Bigoni, R., Rizzi, A., Varani, K., Toth, G., et al. (2000). Further studies on nociceptin-related peptides: Discovery of a new chemical template with antagonist activity on the nociceptin receptor. *Journal of Medicinal Chemistry*, *43*(15), 2805–2813.

Guerrini, R., Calo, G., Lambert, D. G., Carra, G., Arduin, M., Barnes, T. A., et al. (2005). N- and C-terminal modifications of nociceptin/orphanin FQ generate highly potent NOP receptor ligands. *Journal of Medicinal Chemistry*, *48*(5), 1421–1427.

Guerrini, R., Calo, G., Rizzi, A., Bianchi, C., Lazarus, L. H., Salvadori, S., et al. (1997). Address and message sequences for the nociceptin receptor: A structure-activity study of nociceptin-(1-13)-peptide amide. *Journal of Medicinal Chemistry*, *40*(12), 1789–1793.

Guerrini, R., Calo, G., Rizzi, A., Bigoni, R., Bianchi, C., Salvadori, S., et al. (1998). A new selective antagonist of the nociceptin receptor. *British Journal of Pharmacology*, *123*(2), 163–165.

Guerrini, R., Carra, G., Calo, G., Trapella, C., Marzola, E., Rizzi, D., et al. (2004). Non-peptide/peptide chimeric ligands for the nociceptin/orphanin FQ receptor: Design, synthesis and in vitro pharmacological activity. *The Journal of Peptide Research*, *63*(6), 477–484.

Guerrini, R., Rizzi, D., Zucchini, M., Tomatis, R., Regoli, D., Calo, G., et al. (2003). Nociceptin/Orphanin FQ(1-13)NH_2 analogues modified in the Phe1-Gly2 peptide bond. *Bioorganic & Medicinal Chemistry Letters*, *13*(3), 365–368.

Halford, W. P., Gebhardt, B. M., & Carr, D. J. (1995). Functional role and sequence analysis of a lymphocyte orphan opioid receptor. *Journal of Neuroimmunology*, *59*(1–2), 91–101.

Harasawa, K. (2000). Does the histaminergic system play a role in spinal nociception? *Hokkaido Igaku Zasshi*, *75*(4), 275–281.

Harrison, R. S. (2009). *Alpha helix mimetics*. PhD thesis, The University of Queensland.

Harrison, R. S., Ruiz-Gomez, G., Hill, T. A., Chow, S. Y., Shepherd, N. E., Lohman, R. J., et al. (2010). Novel helix-constrained nociceptin derivatives are potent agonists and antagonists of ERK phosphorylation and thermal analgesia in mice. *Journal of Medicinal Chemistry*, *53*(23), 8400–8408. http://dx.doi.org/10.1021/jm101139f.

Hasanein, P. (2011). Two histamine H2 receptor antagonists, zolantidine and cimetidine, modulate nociception in cholestatic rats. *Journal of Psychopharmacology*, *25*(2), 281–288. http://dx.doi.org/10.1177/0269881109106912.

Hirst, R. A., & Lambert, D. G. (1995). Adenylyl cyclase in SH-SY5Y human neuroblastoma cells is regulated by intra- and extracellular calcium. *Biochemical Pharmacology*, *49*(11), 1633–1640. http://dx.doi.org/10.1016/0006-2952(95)00075-B.

Ibba, M., Kitayama, M., McDonald, J., Calo, G., Guerrini, R., Farkas, J., et al. (2008). Binding of the novel radioligand [(3)H]UFP-101 to recombinant human and native rat nociceptin/orphanin FQ receptors. *Naunyn-Schmiedeberg's Archives of Pharmacology*, *378*(6), 553–561.

Jaggi, A. S., & Singh, N. (2011). Therapeutic targets for the management of peripheral nerve injury-induced neuropathic pain. *CNS & Neurological Disorders: Drug Targets*, *10*(5), 589–609.

Kapusta, D. R., Thorkildsen, C., Kenigs, V. A., Meier, E., Vinge, M. M., Quist, C., et al. (2005). Pharmacodynamic characterization of ZP120 (Ac-RYYRWKKKKKKK-NH_2), a novel, functionally selective nociceptin/orphanin FQ peptide receptor partial agonist with sodium-potassium-sparing aquaretic activity. *The Journal of Pharmacology and Experimental Therapeutics*, *314*(2), 652–660.

Kimura, T., Kitaichi, K., Hiramatsu, K., Yoshida, M., Ito, Y., Kume, H., et al. (2000). Intradermal application of nociceptin increases vascular permeability in rats: The possible involvement of histamine release from mast cells. *European Journal of Pharmacology*, *407*(3), 327–332, S0014299900007469 [pii].

Kitayama, M., Barnes, T. A., Carra, G., McDonald, J., Calo, G., Guerrini, R., et al. (2003). Pharmacological profile of the cyclic nociceptin/orphanin FQ analogues c[Cys10,14]N/OFQ(1-14)NH_2 and c[Nphe1, Cys10,14]N/OFQ(1-14)NH_2. *Naunyn-Schmiedeberg's Archives of Pharmacology*, *368*(6), 528–537.

Kitayama, M., McDonald, J., Barnes, T. A., Calo, G., Guerrini, R., Rowbotham, D. J., et al. (2007). In vitro pharmacological characterisation of a novel cyclic nociceptin/orphanin FQ analogue c[Cys(7,10)]N/OFQ(1-13)NH (2). *Naunyn-Schmiedeberg's Archives of Pharmacology*, *375*(6), 369–376.

Klaudel, L., Legowska, A., Brzozowski, K., Silberring, J., & Wojcik, J. (2004). Solution conformational study of nociceptin and its 1–13 and 1–11 fragments using circular dichroism and two-dimensional NMR in conjunction with theoretical conformational analysis. *Journal of Peptide Science*, *10*(11), 678–690.

Knoflach, F., Reinscheid, R. K., Civelli, O., & Kemp, J. A. (1996). Modulation of voltage-gated calcium channels by orphanin FQ in freshly dissociated hippocampal neurons. *The Journal of Neuroscience*, *16*(21), 6657–6664.

Kocsis, L., Orosz, G., Magyar, A., Al-Khrasani, M., Kato, E., Ronai, A. Z., et al. (2004). Nociceptin antagonism: Probing the receptor by N-acetyl oligopeptides. *Regulatory Peptides*, *122*(3), 199–207.

Kuo, C. J., Liao, Y. Y., Guerrini, R., Calo, G., & Chiou, L. C. (2008). Quantitative study of [(pF)Phe4, Arg14, Lys15]nociceptin/orphanin FQ-NH_2 (UFP-102) at NOP receptors in rat periaqueductal gray slices. *European Journal of Pharmacology*, *579*(1–3), 110–115.

Lambert, D. G. (2008). The nociceptin/orphanin FQ receptor: A target with broad therapeutic potential. *Nature Reviews. Drug Discovery*, 7(8), 694–710.

Law, P. Y., Wong, Y. H., & Loh, H. H. (2000). Molecular mechanisms and regulation of opioid receptor signaling. *Annual Review of Pharmacology and Toxicology*, *40*, 389–430. http://dx.doi.org/10.1146/annurev.pharmtox.40.1.389.

Lin, A. P., & Ko, M. C. (2013). The therapeutic potential of nociceptin/orphanin FQ receptor agonists as analgesics without abuse liability. *ACS Chemical Neuroscience*, *4*(2), 214–224. http://dx.doi.org/10.1021/cn300124f.

Luo, P., & Baldwin, R. L. (1997). Mechanism of helix induction by trifluoroethanol: A framework for extrapolating the helix-forming properties of peptides from trifluoroethanol/water mixtures back to water. *Biochemistry*, *36*, 8413–8421.

Lutfy, K., Eitan, S., Bryant, C. D., Yang, Y. C., Saliminejad, N., Walwyn, W., et al. (2003). Buprenorphine-induced antinociception is mediated by mu-opioid receptors and compromised by concomitant activation of opioid receptor-like receptors. *The Journal of Neuroscience*, *23*(32), 10331–10337.

Madala, P. K., Tyndall, J. D., Nall, T., & Fairlie, D. P. (2010). Update 1 of: Proteases universally recognize beta strands in their active sites. *Chemical Reviews*, *110*(6), PR1–PR31. http://dx.doi.org/10.1021/cr900368a.

Mansour, A., Hoversten, M. T., Taylor, L. P., Watson, S. J., & Akil, H. (1995). The cloned mu, delta and kappa receptors and their endogenous ligands: Evidence for two opioid peptide recognition cores. *Brain Research*, *700*(1–2), 89–98.

Maurer, M., Bader, M., Bas, M., Bossi, F., Cicardi, M., Cugno, M., et al. (2011). New topics in bradykinin research. *Allergy*, *66*(11), 1397–1406. http://dx.doi.org/10.1111/j.1398-9995.2011.02686.x.

McDonald, J., Barnes, T. A., Calo, G., Guerrini, R., Rowbotham, D. J., & Lambert, D. G. (2002). Effects of [(pF)Phe(4)]nociceptin/orphanin FQ-(1-13)NH(2) on GTPgamma(35) S binding and cAMP formation in Chinese hamster ovary cells expressing the human nociceptin/orphanin FQ receptor. *European Journal of Pharmacology*, *443*(1–3), 7–12.

McDonald, J., Calo, G., Guerrini, R., & Lambert, D. G. (2003). UFP-101, a high affinity antagonist for the nociceptin/orphanin FQ receptor: Radioligand and GTPgamma(35)S binding studies. *Naunyn-Schmiedeberg's Archives of Pharmacology*, *367*(2), 183–187.

Meunier, J. C. (1997). Nociceptin/orphanin FQ and the opioid receptor-like ORL1 receptor. *European Journal of Pharmacology*, *340*(1), 1–15, S0014-2999(97)01411-8 [pii].

Meunier, J. C., Mollereau, C., Toll, L., Suaudeau, C., Moisand, C., Alvinerie, P., et al. (1995). Isolation and structure of the endogenous agonist of opioid receptor-like ORL1 receptor. *Nature*, *377*(6549), 532–535.

Meyer, J. P., Davis, P., Lee, K. B., Porreca, F., Yamamura, H. I., & Hruby, V. J. (1995). Synthesis using a Fmoc-based strategy and biological activities of some reduced peptide bond pseudopeptide analogues of dynorphin A1. *Journal of Medicinal Chemistry*, *38*(18), 3462–3468.

Mobarakeh, J. I., Takahashi, K., Sakurada, S., Nishino, S., Watanabe, H., Kato, M., et al. (2005). Enhanced antinociception by intracerebroventricularly administered orexin A in histamine H1 or H2 receptor gene knockout mice. *Pain*, *118*(1–2), 254–262. http://dx.doi.org/10.1016/j.pain.2005.08.024.

Mollereau, C., Parmentier, M., Mailleux, P., Butour, J. L., Moisand, C., Chalon, P., et al. (1994). ORL1, a novel member of the opioid receptor family. Cloning, functional expression and localization. *FEBS Letters*, *341*(1), 33–38. http://dx.doi.org/10.1016/0014-5793(94)80235-1.

Mollereau, C., Mouledous, L., Lapalu, S., Cambois, G., Moisand, C., Butour, J. L., et al. (1999). Distinct mechanisms for activation of the opioid receptor-like 1 and kappa-opioid receptors by nociceptin and dynorphin A. *Molecular pharmacology*, *55*, 324–331.

Morikawa, H., Fukuda, K., Mima, H., Shoda, T., Kato, S., & Mori, K. (1998). Nociceptin receptor-mediated Ca^{2+} channel inhibition and its desensitization in NG108-15 cells. *European Journal of Pharmacology*, *351*(2), 247–252, S0014-2999(98)00306-9 [pii].

Munoz, M., & Covenas, R. (2011). NK-1 receptor antagonists: A new paradigm in pharmacological therapy. *Current Medicinal Chemistry*, *18*(12), 1820–1831.

Nazzaro, C., Rizzi, A., Salvadori, S., Guerrini, R., Regoli, D., Zeilhofer, H. U., et al. (2007). UFP-101 antagonizes the spinal antinociceptive effects of nociceptin/orphanin FQ: Behavioral and electrophysiological studies in mice. *Peptides*, *28*(3), 663–669.

Neal, C. R., Jr., Mansour, A., Reinscheid, R., Nothacker, H. P., Civelli, O., Akil, H., et al. (1999). Opioid receptor-like (ORL1) receptor distribution in the rat central nervous system: Comparison of ORL1 receptor mRNA expression with (125)I-[(14)Tyr]-orphanin FQ binding. *The Journal of Comparative Neurology*, *412*(4), 563–605. http://dx.doi.org/10.1002/(SICI)1096-9861(19991004)412:4<563::AID-CNE2>3.0.CO;2-Z.

Nemeth, J., Helyes, Z., Oroszi, G., Than, M., Pinter, E., & Szolcsanyi, J. (1998). Inhibition of nociceptin on sensory neuropeptide release and mast cell-mediated plasma extravasation in rats. *European Journal of Pharmacology*, *347*(1), 101–104, S0014-2999(98)00216-7 [pii].

Nishi, M., Houtani, T., Noda, Y., Mamiya, T., Sato, K., Doi, T., et al. (1997). Unrestrained nociceptive response and disregulation of hearing ability in mice lacking the nociceptin/orphanin FQ receptor. *The EMBO Journal*, *16*(8), 1858–1864. http://dx.doi.org/10.1093/emboj/16.8.1858.

Okada, K., Sujaku, T., Chuman, Y., Nakashima, R., Nose, T., Costa, T., et al. (2000). Highly potent nociceptin analog containing the Arg-Lys triple repeat. *Biochemical and Biophysical Research Communications*, *278*(2), 493–498.

Okawa, H., Nicol, B., Bigoni, R., Hirst, R. A., Calo, G., Guerrini, R., et al. (1999). Comparison of the effects of [Phe1psi(CH_2-NH)Gly2]nociceptin(1-13)NH_2 in rat brain, rat vas deferens and CHO cells expressing recombinant human nociceptin receptors. *British Journal of Pharmacology*, *127*(1), 123–130.

Orsini, M. J., Nesmelova, I., Young, H. C., Hargittai, B., Beavers, M. P., Liu, J., et al. (2005). The nociceptin pharmacophore site for opioid receptor binding derived from the NMR structure and bioactivity relationships. *The Journal of Biological Chemistry*, *280*(9), 8134–8142.

Osinski, M. A., Pampusch, M. S., Murtaugh, M. P., & Brown, D. R. (1999). Cloning, expression and functional role of a nociceptin/orphanin FQ receptor in the porcine gastrointestinal tract. *European Journal of Pharmacology*, *365*(2–3), 281–289.

Pampusch, M. S., Serie, J. R., Osinski, M. A., Seybold, V. S., Murtaugh, M. P., & Brown, D. R. (2000). Expression of nociceptin/OFQ receptor and prepro-nociceptin/OFQ in lymphoid tissues. *Peptides*, *21*(12), 1865–1870, S0196978100003326 [pii].

Peluso, J., Gaveriaux-Ruff, C., Matthes, H. W., Filliol, D., & Kieffer, B. L. (2001). Orphanin FQ/nociceptin binds to functionally coupled ORL1 receptors on human immune cell lines and alters peripheral blood mononuclear cell proliferation. *Brain Research Bulletin*, *54*(6), 655–660, S0361923001004828 [pii].

Peluso, J., LaForge, K. S., Matthes, H. W., Kreek, M. J., Kieffer, B. L., & Gaveriaux-Ruff, C. (1998). Distribution of nociceptin/orphanin FQ receptor transcript in human central nervous system and immune cells. *Journal of Neuroimmunology*, *81*(1–2), 184–192, S0165-5728(97)00178-1 [pii].

Redrobe, J. P., Calo, G., Guerrini, R., Regoli, D., & Quirion, R. (2000). [Nphe(1)]-nociceptin(1-13)-NH(2), a nociceptin receptor antagonist, reverses nociceptin-induced spatial memory impairments in the Morris water maze task in rats. *British Journal of Pharmacology*, *131*(7), 1379–1384.

Reinscheid, R. K., Ardati, A., Monsma, F. J., Jr., & Civelli, O. (1996). Structure-activity relationship studies on the novel neuropeptide orphanin FQ. *The Journal of Biological Chemistry*, *271*(24), 14163–14168.

Rizzi, A., Bigoni, R., Calo, G., Guerrini, R., Salvadori, S., & Regoli, D. (1999). [Nphe(1)] nociceptin-(1-13)-NH(2) antagonizes nociceptin effects in the mouse colon. *European Journal of Pharmacology*, *385*(2–3), R3–R5, S0014-2999(99)00730-X [pii].

Rizzi, A., Bigoni, R., Marzola, G., Guerrini, R., Salvadori, S., Regoli, D., et al. (2000). The nociceptin/orphanin FQ receptor antagonist, [Nphe1]NC(1-13)NH_2, potentiates morphine analgesia. *Neuroreport*, *11*(11), 2369–2372.

Rizzi, D., Rizzi, A., Bigoni, R., Camarda, V., Marzola, G., Guerrini, R., et al. (2002). [Arg(14), Lys(15)]nociceptin, a highly potent agonist of the nociceptin/orphanin FQ receptor: In vitro and in vivo studies. *The Journal of Pharmacology and Experimental Therapeutics*, *300*(1), 57–63.

Rizzi, A., Rizzi, D., Marzola, G., Regoli, D., Larsen, B. D., Petersen, J. S., et al. (2002). Pharmacological characterization of the novel nociceptin/orphanin FQ receptor ligand, ZP120: In vitro and in vivo studies in mice. *British Journal of Pharmacology*, *137*(3), 369–374.

Rizzi, A., Salis, M. B., Ciccocioppo, R., Marzola, G., Bigoni, R., Guerrini, R., et al. (2002). Pharmacological characterisation of [(pX)Phe4]nociceptin(1-13)NH_2 analogues. 2. In vivo studies. *Naunyn-Schmiedeberg's Archives of Pharmacology*, *365*(6), 450–456.

Robinson, S. E. (2002). Buprenorphine: An analgesic with an expanding role in the treatment of opioid addiction. *CNS Drug Reviews*, *8*(4), 377–390.

Rosa, A. C., & Fantozzi, R. (2013). The role of histamine in neurogenic inflammation. *British Journal of Pharmacology*, *170*(1), 38–45. http://dx.doi.org/10.1111/bph.12266.

Salat, K., Moniczewski, A., & Librowski, T. (2013). Transient receptor potential channels—Emerging novel drug targets for the treatment of pain. *Current Medicinal Chemistry*, *20*(11), 1409–1436.

Salvadori, S., Guerrini, R., Calo, G., & Regoli, D. (1999). Structure-activity studies on nociceptin/orphanin FQ: From full agonist, to partial agonist, to pure antagonist. *Farmaco*, *54*(11–12), 810–825.

Sasaki, Y., Kawano, S., Kohara, H., Watanabe, H., & Ambo, A. (2006). ORL1 and opioid receptor preferences of nociceptin and dynorphin A analogues with Dmp substituted for N-terminal aromatic residues. *Bioorganic & Medicinal Chemistry*, *14*(7), 2433–2437.

Serhan, C. N., Fierro, I. M., Chiang, N., & Pouliot, M. (2001). Cutting edge: Nociceptin stimulates neutrophil chemotaxis and recruitment: Inhibition by aspirin-triggered-15-epi-lipoxin A4. *The Journal of Immunology*, *166*(6), 3650–3654.

Shepherd, N. E., Hoang, H. N., Abbenante, G., & Fairlie, D. P. (2005). Single turn alpha helical peptides with exceptional stability in water. *Journal of the American Chemical Society*, 2974–2983.

Shinkai, H., Ito, T., Iida, T., Kitao, Y., Yamada, H., & Uchida, I. (2000). 4-Aminoquinolines: Novel nociceptin antagonists with analgesic activity. *Journal of Medicinal Chemistry*, *43*(24), 4667–4677.

Shoblock, J. R. (2007). The pharmacology of Ro 64-6198, a systemically active, nonpeptide NOP receptor (opiate receptor-like 1, ORL-1) agonist with diverse preclinical therapeutic activity. *CNS Drug Reviews*, *13*(1), 107–136. http://dx.doi.org/10.1111/j.1527-3458.2007.00007.x.

Smith, E. D., Ariane Vinson, N., Zhong, D., Berrang, B. D., Catanzaro, J. L., Thomas, J. B., et al. (2008). A new synthesis of the ORL-1 antagonist 1-[(3R,4R)-1-cyclooctylmethyl-3-hydroxymethyl-4-piperidinyl]-3-ethyl-1,3-dihydro-2H-benzimidazol-2-one (J-113397) and activity in a calcium mobilization assay. *Bioorganic & Medicinal Chemistry*, *16*(2), 822–829. http://dx.doi.org/10.1016/j.bmc.2007.10.023.

Smith, F. M., Haskelberg, H., Tracey, D. J., & Moalem-Taylor, G. (2007). Role of histamine H3 and H4 receptors in mechanical hyperalgesia following peripheral nerve injury. *Neuroimmunomodulation*, *14*(6), 317–325. http://dx.doi.org/10.1159/000125048.

Smith, P. A., & Moran, T. D. (2001). The nociceptin receptor as a potential target in drug design. *Drug News & Perspectives*, *14*(6), 335–345.

Spampinato, S., Di Toro, R., & Qasem, A. R. (2001). Nociceptin-induced internalization of the ORL1 receptor in human neuroblastoma cells. *Neuroreport*, *12*(14), 3159–3163.

Takayama, N., & Ueda, H. (2005). Morphine-induced overexpression of prepro-nociceptin/orphanin FQ in cultured astrocytes. *Peptides*, *26*(12), 2513–2517. http://dx.doi.org/10.1016/j.peptides.2005.05.005, S0196-9781(05)00245-7 [pii].

Tancredi, T., Carra, G., Guerrini, R., Arduin, M., Calo, G., Regoli, D., et al. (2005). The interaction of highly helical structural mutants with the NOP receptor discloses the role of the address domain of nociceptin/orphanin FQ. *Chemistry*, *11*(7), 2061–2070.

Thompson, A. A., Liu, W., Chun, E., Katritch, V., Wu, H., Vardy, E., et al. (2012). Structure of the nociceptin/orphanin FQ receptor in complex with a peptide mimetic. *Nature*, *485*(7398), 395–399. http://dx.doi.org/10.1038/nature11085.

Thomsen, C., & Hohlweg, R. (2000). (8-Naphthalen-1-ylmethyl-4-oxo-1-phenyl-1,3,8-triaza-spiro[4.5]dec-3-yl)-acetic acid methyl ester (NNC 63-0532) is a novel potent nociceptin receptor agonist. *British Journal of Pharmacology*, *131*(5), 903–908. http://dx.doi.org/10.1038/sj.bjp.0703661.

Topham, C. M., Mouledous, L., Poda, G., Maigret, B., & Meunier, J. C. (1998). Molecular modelling of the ORL1 receptor and its complex with nociceptin. *Protein engineering*, *11*, 1163–1179.

Tyndall, J. D., Nall, T., & Fairlie, D. P. (2005). Proteases universally recognize beta strands in their active sites. *Chemical Reviews*, *105*(3), 973–999. http://dx.doi.org/10.1021/cr040669e.

Ubaldi, M., Bifone, A., & Ciccocioppo, R. (2013). Translational approach to develop novel medications on alcohol addiction: Focus on neuropeptides. *Current Opinion in Neurobiology, 23*(4), 684–691. http://dx.doi.org/10.1016/j.conb.2013.04.009.

Varty, G. B., Lu, S. X., Morgan, C. A., Cohen-Williams, M. E., Hodgson, R. A., Smith-Torhan, A., et al. (2008). The anxiolytic-like effects of the novel, orally active nociceptin opioid receptor agonist 8-[bis(2-methylphenyl)methyl]-3-phenyl-8-azabicyclo[3.2.1] octan-3-ol (SCH 221510). *The Journal of Pharmacology and Experimental Therapeutics, 326*(2), 672–682. http://dx.doi.org/10.1124/jpet.108.136937.

Vaudry, D., Stork, P. J., Lazarovici, P., & Eiden, L. E. (2002). Signaling pathways for PC12 cell differentiation: Making the right connections. *Science, 296*(5573), 1648–1649. http://dx.doi.org/10.1126/science.1071552, 296/5573/1648 [pii].

Vergnolle, N., Bunnett, N. W., Sharkey, K. A., Brussee, V., Compton, S. J., Grady, E. F., et al. (2001). Proteinase-activated receptor-2 and hyperalgesia: A novel pain pathway. *Nature Medicine*, 7(7), 821–826. http://dx.doi.org/10.1038/89945.

Wall, P. D., & Melzack, R. M. (2000). *Textbook of pain*. London, UK: Churchill Livingstone.

Wang, J. B., Johnson, P. S., Imai, Y., Persico, A. M., Ozenberger, B. A., Eppler, C. M., et al. (1994). cDNA cloning of an orphan opiate receptor gene family member and its splice variant. *FEBS Letters, 348*(1), 75–79. http://dx.doi.org/10.1016/0014-5793(94) 00557-5.

Wick, M. J., Minnerath, S. R., Roy, S., Ramakrishnan, S., & Loh, H. H. (1995). Expression of alternate forms of brain opioid 'orphan' receptor mRNA in activated human peripheral blood lymphocytes and lymphocytic cell lines. *Brain Research Molecular Brain Research, 32*(2), 342–347, 0169328X9500096B [pii].

Wider, G., & Dreier, L. (2006). Measuring protein concentrations by NMR spectroscopy. *Journal of the American Chemical Society, 128*(8), 2571–2576. http://dx.doi.org/10.1021/ja055336t.

Wright, K. E., McDonald, J., Barnes, T. A., Rowbotham, D. J., Guerrini, R., Calo, G., et al. (2003). Assessment of the activity of a novel nociceptin/orphanin FQ analogue at recombinant human nociceptin/orphanin FQ receptors expressed in Chinese hamster ovary cells. *Neuroscience Letters, 346*(3), 145–148.

Yamada, H., Nakamoto, H., Suzuki, Y., Ito, T., & Aisaka, K. (2002). Pharmacological profiles of a novel opioid receptor-like1 (ORL(1)) receptor antagonist, JTC-801. *British Journal of Pharmacology, 135*(2), 323–332. http://dx.doi.org/10.1038/sj.bjp.0704478.

Yang, A. S., Hitz, B., & Honig, B. (1996). Free energy determinants of secondary structure formation: III. Beta-turns and their role in protein folding. *Journal of Molecular Biology, 259*(4), 873–882. http://dx.doi.org/10.1006/jmbi.1996.0364, S0022-2836(96) 90364-2 [pii].

Yu, G., Yan, L. D., Li, Y. L., Wen, Q., Dong, H. J., & Gong, Z. H. (2011). TH-030418: A potent long-acting opioid analgesic with low dependence liability. *Naunyn-Schmiedeberg's Archives of Pharmacology, 384*(2), 125–131. http://dx.doi.org/10.1007/s00210-011-0652-8.

Yung, L. Y., Joshi, S. A., Chan, R. Y., Chan, J. S., Pei, G., & Wong, Y. H. (1999). GalphaL1 (Galpha14) couples the opioid receptor-like1 receptor to stimulation of phospholipase C. *The Journal of Pharmacology and Experimental Therapeutics, 288*(1), 232–238.

Zaveri, N. T. (2011). The nociceptin/orphanin FQ receptor (NOP) as a target for drug abuse medications. *Current Topics in Medicinal Chemistry, 11*(9), 1151–1156.

Zhang, C., Miller, W., Valenzano, K. J., & Kyle, D. J. (2002). Novel, potent ORL-1 receptor agonist peptides containing alpha-Helix-promoting conformational constraints. *Journal of Medicinal Chemistry, 45*(24), 5280–5286.

Zhang, Z., Xin, S. M., Wu, G. X., Zhang, W. B., Ma, L., & Pei, G. (1999). Endogenous delta-opioid and ORL1 receptors couple to phosphorylation and activation of p38 MAPK in NG108-15 cells and this is regulated by protein kinase A and protein kinase C. *Journal of Neurochemistry, 73*(4), 1502–1509.

CHAPTER TWO

Bioinformatics and Evolution of Vertebrate Nociceptin and Opioid Receptors

Craig W. Stevens[1]
Department of Pharmacology and Physiology, Oklahoma State University Center for Health Sciences, Tulsa, Oklahoma, USA
[1]Corresponding author: e-mail address: cw.stevens@okstate.edu

Contents

Abstract

G protein-coupled receptors (GPCRs) are ancestrally related membrane proteins on cells that mediate the pharmacological effect of most drugs and neurotransmitters. GPCRs are the largest group of membrane receptor proteins encoded in the human genome. One of the most famous types of GPCRs is the opioid receptors. Opioid family receptors consist of four closely related proteins expressed in all vertebrate brains and spinal cords examined to date. The three classical types of opioid receptors shown unequivocally to mediate analgesia in animal models and in humans are the *mu-* (MOR), *delta-* (DOR), and *kappa-*(KOR) opioid receptor proteins. The fourth and most recent member of the opioid receptor family discovered is the nociceptin or orphanin FQ receptor (ORL). The role of ORL and its ligands in producing analgesia is not as clear, with both analgesic and hyperalgesic effects reported. All four opioid family receptor genes were cloned from

Vitamins and Hormones, Volume 97
ISSN 0083-6729
http://dx.doi.org/10.1016/bs.vh.2014.10.002

expressed mRNA in a number of vertebrate species, and there are enough sequences presently available to carry out bioinformatic analysis. This chapter presents the results of a comparative analysis of vertebrate opioid receptors using pharmacological studies, bioinformatics, and the latest data from human whole-genome studies. Results confirm our initial hypotheses that the four opioid receptor genes most likely arose by whole-genome duplication, that there is an evolutionary vector of opioid receptor type divergence in sequence and function, and that the *hMOR* gene shows evidence of positive selection or adaptive evolution in *Homo sapiens*.

1. INTRODUCTION

The concept of "receptor" was first elaborated by Paul Erlich in 1900 in connection with his "side-chain theory" in studies on immunological reactions (Prüll, 2003). From that purely conceptual beginning, receptor theory, mechanism, and structure have become the major area of study in modern pharmacology. Among all receptor types, the G protein-coupled receptors (GPCRs) have garnered the most attention. GPCRs represent the largest group of cell membrane proteins encoded in the human genome. There are at least 2000 GPCRs within the rhodopsin-like superfamily of proteins recognized in the genome, with more than a third of these receptors classified as "orphan GPCRs" in that their endogenous ligand is not known (Attwood, 2001; Flower & Attwood, 2004; Perez, 2003, 2005). Genes that encode GPCRs represent about 5% of the known coding sequences and are the most promising targets for current and new drug development in the pharmaceutical industry (Bockaert & Pin, 1999). It is estimated that about 36% of drugs on the global pharmaceutical market target about 80 or so identified GPCRs (Rask-Andersen, Almen, & Schioth, 2011). GPCRs are classified by their sequence homology which in most part parallels earlier receptor classification schemes based solely on receptor function and ligand selectivity (Fredriksson, Lagerstrom, Lundin, & Schiöth, 2003; Fredriksson & Schiöth, 2005).

Given the numerous members of the GPCR superfamily present in all vertebrate animals, it is of great interest to understand the molecular evolution of these receptor proteins. Molecular evolution is the process driving the changes that occur in the amino acid sequence of the receptor protein over evolutionary time and among the different classes of vertebrate species. Ultimately, we would like to be able to describe the "vector" of this molecular evolution such that trends might be detected in the selectivity of receptor proteins or in evolving receptor efficacy or other aspects of receptor function. One such vector of molecular evolution we recently discovered is the increase in type selectivity among opioid receptor proteins throughout vertebrate evolution, as discussed further below.

The pursuit to understand GPCR and specifically opioid receptor evolution is not only cerebral onanism. The correlation between receptor amino acid sequence and receptor function, while obvious from first principles: sequence determines structure determines function, is at the crux of fully understanding receptor selectivity and ensuing pharmacological action. Additionally, a detailed understanding of receptor sequence impacting type selectivity and the direction of vertebrate receptor evolution is needed for upcoming gene therapy, which will undoubtedly include expression of exogenous GPCR proteins.

1.1 The origin of G protein-coupled receptors

Before examining the evolution of vertebrate GPCRs, the origin of all GPCR proteins is of interest. All life on earth is classified into three domains or superkingdoms: Archea, Bacteria, and Eucaryotes, although there are other schemes with up to six kingdoms (Cavalier-Smith, 2004). This classification does not include viruses; some taxonomists seem to ignore them or unable to reach a conclusion whether they meet the definition of a life form or not. Viruses are simply self-replicating bits of DNA or RNA, and one wonders if self-replication is sufficient by itself to serve as the sole criterion for life. Prions self-replicate also, but do not use RNA or DNA to do so; maybe heritable information transfer is the key to life. Of course, this leads us to think that more complex behaving species somehow better fit the definition of life, but at the end of the day, self-replication is really the only thing that even *Homo sapiens* accomplishes.

The origin of viruses is not known. Two main theories exists: that viruses were precellular and arose before the time of the most primitive cells, or that viruses are the results of a massively reduced cellular parasite to its core replication machinery (Bamford, 2003; Iyer, Balaji, Koonin, & Aravind, 2006). These scenarios are not mutually incompatible as some viruses may have arisen by either route. In any event, there are viruses that encode GPCR-like proteins. The transcription of these genes produces receptors not destined for use by the virus but to be embedded in the host cell's membrane. Most notably, human herpesviruses encode for various chemokine receptors that may help the virus gain a foothold in the host organism

(Maussang, Vischer, Leurs, & Smit, 2009; Rosenkilde, Smit, & Waldhoer, 2008). Thus, if viruses are precellular ancestors of cells, it is possible that genes for GPCR proteins in all life forms may have arisen from initial viral transfection, akin to life itself on Earth originating due to extraterrestrial panspermia.

The famous seven-transmembrane structure (7TM) of all GPCRs could have initially arisen through the duplication and fusion of a gene that encoded a three helix transmembrane (3TM) receptor protein (Taylor & Agarwal, 1993). It is hypothesized that gene duplication occurred in the evolution of an ancestral GPCR gene, such that helices 5–7 originated as duplicates of helices 1–3. This gives rise to the amino acid similarities between helices 1–3 and 5–7 of many GPCRs (Taylor & Agarwal, 1993). This suggests that ancestral GPCRs may not have been as structurally complete as extant GPCRs found today. Regardless of the ultimate origin of ancestral GPCR, the extensive duplication, expansion, and utilization of GPCRs by all vertebrates suggest that this superfamily has and is undergoing positive selection to generate the GPCRs expressed in all animals today. The next sections introduce the opioid receptors in general and studies on opioid systems in nonmammalian species in particular.

1.2 A brief history of opioid receptors

The classical opioid receptors are membrane proteins of the Type A or rhodopsin-like GPCRs (the R group in Fredriksson's "GRAFS" classification) mediate the analgesic effects of opioid agents like morphine. Opioid receptors also mediate substance abuse with drugs like heroin as well as the growing epidemic of prescription opioid abuse. The natural ligands for opioid receptors are endogenous opioid peptides; most notably *beta*-endorphin, met- and leu-enkephalin, dynorphin, and endomorphin. The role of opioid receptors in physiological processes is vast; nearly any PubMed search with the Boolean parameters of "*any physiological process* and opioid" brings up pages of references. This is true for terms like reproduction, growth, development, respiration, blood pressure, renal function, thermoregulation, endocrinology, seizures, stress, immunology, pregnancy, and aging. It is also true when considering behaviors using terms like eating, drinking, sex, learning, memory, and locomotion. While there appears to be no limit to the effects of opioids, both endogenous and exogenous, on any system in the brain and body, underlying all opioid effects is the initial event of an opioid molecule binding to an opioid receptor protein.

Early in receptor pharmacology, only the structure–activity relationships of various ligands defined the types of GPCRs. In this way, Martin and colleagues initially proposed multiple opioid receptors based on the differential action of morphine and ketocyclazocine in producing analgesia in a spinal dog model (Martin, Eades, Thompson, Huppler, & Gilbert, 1976). Martin defined two receptors and called them *mu-* (Greek "m" for morphine, MOR) and *kappa-* (for ketocyclazocine, KOR) opioid receptors. (Martin actually proposed a third type of opioid receptor, called sigma, for the drug SKF10047, but it proved later not to be a type of opioid receptor.) Kosterlitz and colleagues, working with the newly discovered enkephalin peptides in the mouse vas deferens preparation, proposed a *delta-* (for deferens, DOR) opioid receptor type (Lord, Waterfield, Hughes, & Kosterlitz, 1977). Much work at toward the end of the last century solidified the structure–activity relationship of these three opioid receptors, aided in large part by the development of highly selective (type-specific) opioid antagonists by Takemori and Portoghese. These highly selective opioid antagonists used were *beta*-funaltrexamine (*beta*-FNA, for MOR) (Portoghese, Sultana, & Takemori, 1988), naltrindole (NTI, for DOR) (Takemori, Ho, Naeseth, & Portoghese, 1988), and nor-binaltorphimine (nor-BNI, for KOR) (Ward, Portoghese, & Takemori, 1982). Pharmacological evidence of three separate opioid receptors mediating analgesia in mammalian models was confirmed using these antagonists.

Evidence of stereospecific opioid binding sites from radioligand binding studies using rodent brain homogenates did not emerge until 1973 from the labs of Snyder, Simon, and Terenius (Pert & Snyder, 1973; Simon, Hiller, & Edelman, 1973; Terenius, 1973). Ongoing refinements of such "grind-and-bind" studies clearly established three main opioid receptor binding profiles (Loh & Smith, 1990; Simon, 1981, 1986) and a number of pharmacological subtypes for each of the main receptors (Wollemann, Benyhe, & Simon, 1993; Zaki et al., 1996).

The ultimate discovery and isolation of the cDNA for the mouse *delta*-opioid receptor (mDOR[1]) from the labs of Evans in the United States and Kieffer in France in 1992 (Evans, Keith, Morrison, Magendzo, & Edwards,

[1] The convention for naming opioid receptor proteins here follows a hybrid system. For example, the mu opioid receptor in humans, mice, and rats is hMOR, mMOR, and rMOR, but for all other species, a representation of the Linnaean binomial taxonomy is used such that in the zebrafish, *Danio rerio*, the receptor is drMOR; in the leopard frog, *Rana pipiens*, rpMOR; and in the rough-skinned newt, *Taricha granulosa*, tgMOR, likewise, for the DOR, KOR, and ORL proteins. For simplification, the genes for these proteins are referred to using italicized acronyms; i.e., the gene that codes for hMOR is *hMOR*.

1992; Kieffer, Befort, Gaveriaux-Ruff, & Hirth, 1992) for the first time linked an opioid receptor sequence to a pharmacologically defined opioid receptor. Once the mDOR cDNA sequence was known, homology cloning led to the identification of receptor sequences for numerous rodent, monkey, and the human opioid receptors, hMOR (Wang, Johnson, Persico, et al., 1994), hDOR (Knapp et al., 1994; Simonin et al., 1994), and hKOR (Mansson, Bare, & Yang, 1994; Simonin et al., 1995; Zhu et al., 1995). At the same time, cloning studies revealed another "opioid receptor-like" (ORL) sequence from numerous labs that was similar to an opioid receptor but did not look like one from pharmacological studies (Bunzow et al., 1994; Chen et al., 1994; Fukuda et al., 1994; Lachowicz, Shen, Monsma, & Sibley, 1995; Mollereau et al., 1994; Wang, Johnson, Imai, et al., 1994; Wick et al., 1994). This opioid receptor-like protein remained an "orphan" receptor until the endogenous ligand for ORL was identified as the neuropeptide named nociceptin by the lab of Meunier et al. (1995) or named orphanin FQ from the team assembled by Civelli (Reinscheid et al., 1995). This made the nociceptin (or orphanin FQ) receptor (ORL) the fourth member of the opioid receptor family expressed in the nervous system of mammals (Fredriksson et al., 2003; Meunier, 1997). However, the behavioral effects of nociceptin or orphanin FQ binding to the ORL protein were not clear as behavioral studies show hyperalgesia (the origin of the name *nociceptin*) or analgesia, or sometimes no effect at all (Mogil & Pasternak, 2001). Additionally, the vast majority of opioids do not bind well to ORL receptors, nor do the opioid antagonists, naloxone or naltrexone. For this reason, MOR, DOR, and KOR are considered the classical opioid receptor types of the family and ORL a first-cousin. *In vivo* studies using antisense oligodeoxynucleotides designed from MOR, DOR, and KOR cDNA sequences confirm that each one of the three opioid receptor proteins can independently produce analgesia in mammalian models (Chen et al., 1996; Pasternak & Standifer, 1995; Wang, Kampine, & Tseng, 1996).

There is evidence of opioid receptor subtypes in mammalian species (e.g., mu_1, mu_2, $delta_1$, $delta_2$, $kappa_1$, $kappa_2$, and $kappa_3$), which may result from alternative splicing of receptor mRNA (Pan et al., 1995, 1999, Pasternak, 2005; Uhl, Childers, & Pasternak, 1994). Interestingly, it appears that many more splice variants are noted for MOR than for the other types of vertebrate opioid receptors. Driven by the prodigious efforts of Pan and Pasternak, there are now at least 20 alternative splice variants of mMOR (Pan et al., 2005) and at least 6 variants of hMOR (Pan et al., 2003). In contrast, only a few alternative splice variants are reported for KOR (Gaveriaux-Ruff et al., 1997; Wei, Law, & Loh, 2004) and even less for DOR (Gaveriaux-Ruff et al., 1997) and ORL (Pan, Xu, Wan, Zuckerman, & Pasternak, 1998). This finding may reflect sampling bias as MOR is the most clinically relevant receptor type and may have been examined to a greater degree, but more likely supports the notion that MOR is very different than the other types of opioid receptors, as discussed further below.

1.3 Evidence for opioid receptors in nonmammalian vertebrates

Early evidence for the existence of opioid receptors in nonmammalian vertebrates came from behavioral and radioligand binding studies using morphine and other opioids. The best studied nonmammalian vertebrates with regard to opioid systems come from a behavioral model of analgesia in amphibians, so the focus is on these studies. These studies using amphibians provide at least one model comparison of opioid receptors in earlier-evolved species to those in later-evolved species.

Initial studies of the analgesic or antinociceptive effects of opioids in amphibians (*Rana pipiens*) were conducted using nonselective opioid agonists, endogenous opioid peptides, and antagonists (Pezalla & Stevens, 1984; Stevens & Pezalla, 1983, 1984; Stevens, Pezalla, & Yaksh, 1987). Tolerance to the analgesic effects of daily morphine administration was documented (Stevens & Kirkendall, 1993) and stress-induced release of endogenous opioids produced analgesia which was potentiated by enkephalinase inhibitors and blocked by naltrexone (Pezalla & Dicig, 1984; Stevens, Sangha, & Ogg, 1995). These studies showed that both exogenous opioid agonists and endogenous opioid peptides could raise the nociceptive (pain) threshold in amphibians by an action at opioid receptors.

The relative antinociceptive potency of *mu*-, *delta*-, or *kappa*-opioid agonists after systemic, intraspinal, or intracerebroventricular administration in amphibians was highly correlated to that observed in typical mammalian models and to the relative analgesic potency of opioid analgesics in human clinical studies (Stevens, 1996; Stevens, Klopp, & Facello, 1994; Stevens & Rothe, 1997). These data established the amphibian model as a robust and predictive alternative or adjunct nonmammalian model for the testing of opioid analgesics (Stevens, 1992, 1995, 1997, 2004, 2008). In the one study examining the behavioral effects of nociceptin in a nonmammalian model, nociceptin produced antinociceptive effects following spinal administration in *R. pipiens*, which was blocked by ORL antagonists but not by opioid antagonists (Stevens, Martin, & Stahlheber, 2009).

In a study examining the antagonism of highly selective MOR, DOR, and KOR antagonists in amphibians, a surprising finding occurred. *Beta*-FNA, NTI, and nor-BNI blocked the antinociceptive effects produced by all three types of selective opioid agonists in amphibians (Stevens & Newman, 1999). For example, *beta*-FNA, a highly selective MOR antagonist in mammalian models, blocked the effects of selective MOR, DOR, and KOR agonists in amphibians. This provided the initial data for our original hypothesis that opioid receptors are less type selective in earlier-evolved vertebrates compared to humans and other mammals, a hypothesis further supported by additional data as described below.

Radioligand binding studies also supported the existence of functional opioid receptors in nonmammalian vertebrates, including fish (Bird, Jackson, Baker, & Buckingham, 1988; Brooks, Standifer, Cheng, Ciszewska, & Pasternak, 1994; Buatti & Pasternak, 1981; Gonzalez-Nunez, Barrallo, Traynor, & Rodriguez, 2006), amphibians (Benyhe et al., 1990; Simon et al., 1982; Simon, Benyhe, Borsodi, Szucs, & Wollemann, 1985; Simon et al., 1987; Zawilska, Lajtha, & Borsodi, 1988), and reptiles (Bakalkin, Pivovarov, Kobylyansky, Nesterenko, & Yarygin, 1989; Pert, Aposhian, & Snyder, 1974; Xia & Haddad, 2001). Many reports examining opioid binding sites using the European water frog, *Rana esculenta*, were produced from the Hungarian research group headed by Wolleman, Simon, Borsodi, and Benyhe (Benyhe et al., 1997, 1999; Benyhe, Simon, Borsodi, Wollemann, & Barnard, 1994; Benyhe, Szucs, Borsodi, & Wollemann, 1992; Moitra, Öktem, & Borsodi, 1995; Simon et al., 1990; Wollemann, Farkas, Toth, & Benyhe, 1994, 1999). The main difference observed between amphibians and mammals was a greater affinity of *mu*- and *delta*-selective opioids for the amphibian *kappa*-like site and a lesser affinity for *kappa*-selective opioids, compared to mammalian *kappa*-opioid receptors (Benyhe et al., 1990; Mollereau et al., 1988). These results provided the first hint from radioligand studies that opioid receptors in nonmammalian species may be less selective than opioid receptors expressed in mammals.

In our laboratory, brain and spinal cord membranes from *R. pipiens* showed a single high-affinity site [^{3}H]-naloxone binding site which was displaced by *mu*-, *delta*-, and *kappa*-selective opioid agonists with apparent affinities ranging from 1.86 nM to 31 μM. Surprisingly, the highly selective opioid antagonists (*beta*-FNA, NTI, and nor-BNI) displaced [^{3}H]-naloxone binding with equal affinity to opioid receptors in brain and spinal cord tissue, each with an apparent affinity of about 3.0 nM (Newman, Wallace, & Stevens, 2000a, 2000b). This finding was consistent with behavioral studies showing nonselectivity of type-selective antagonists and also supports the hypothesis that opioid receptors from earlier-evolved vertebrates are less selective than mammalian receptors (Stevens & Newman, 1999).

However, using the selective opioid agonist radioligands, [^{3}H]-DAMGO (MOR), [^{3}H]-DPDPE (DOR), and [^{3}H]-U65953 (KOR), three distinct opioid binding sites were identified based on different binding densities and selective competitive displacement of agonist radioligand by *mu*-, *delta*-, and *kappa*-opioid ligands (Newman, Sands, Wallace, & Stevens, 2002). As opposed to their equipotent displacement of [^{3}H]-naloxone binding, the type-selective opioid antagonists (*beta*-FNA, nor-BNI, and NTI) were highly selective in displacing the binding of their respective *mu*-, *delta*-, and *kappa*-opioid agonist radioligands (Newman et al., 2002).

A summary of the above binding studies is that brain tissue of amphibians contains the three types of classical opioid binding sites, similar to those characterized in mammalian tissues, but with significant differences in the binding of selective opioid ligands. There is one study characterizing ORL sites in amphibians by the Hungarian group showing high-affinity, saturable binding of labeled nociceptin to brain homogenates (Benyhe et al., 1999).

2. THE VERTEBRATE OPIOID RECEPTOR SEQUENCE DATABASE

Prior to the identification of any nonmammalian, full-length opioid receptor sequences, a study from Evans group utilized PCR and degenerate primers isolated opioid receptor-like fragments from genomic DNA obtained from each of the major vertebrate classes but not in any of the invertebrates tested (Li, Keith, & Evans, 1996a, 1996b). The first full-length clone sequenced from a nonmammalian species was ccMOR from the brain of the white suckerfish, *Catostomus commersoni* (Darlison et al., 1997). Other nonmammalian species with full-length opioid receptors cloned are the zebrafish, *Danio rerio*, by Rodriguez and colleagues (Alvarez et al., 2006; Barrallo, González-Sarmiento, Alvar, & Rodriguez, 2000; Barrallo, González-Sarmiento, Porteros, Gracia-Isídoro, & Rodríguez, 1998; Porteros, Garcia-Isodoro, Barrallo, González-Sarmiento, & Rodriguez,

1999) and the rough-skinned newt, *Taricha granulosa*, in the laboratory of Moore (Bradford, Walthers, Searcy, & Moore, 2005; Bradford, Walthers, Stanley, Baugh, & Moore, 2006; Walthers, Bradford, & Moore, 2005). Using the same sets of primers that Evans group used in the first phylogenetic study of opioid receptors (Li et al., 1996b), our group was able to clone all four opioid receptor types expressed in *R. pipiens* brain tissue: rpMOR, rpDOR, rpKOR, and rpORL (Stevens, Brasel, & Mohan, 2007). These were the first opioid receptors cloned from Class Amphibia and, on the amino acid level, showed 70–84% identity with their orthologous mammalian counterparts.

At present, the vertebrate dataset of opioid receptor cDNA sequences yields 24 nucleotide and 24 protein sequences for phylogenetic analysis. As shown in Table 1, at present there is a dataset of six vertebrates from three different classes that have all four opioid receptor sequences from brain tissue cDNA deposited in GenBank. As the focus of this chapter is the relationship between amino acid sequence and pharmacological action throughout evolution, the bioinformatic analyses were done using the protein data.

2.1 Alignment of protein sequences

A shown in Fig. 1, a ClustalW (Thompson, Higgins, & Gibson, 1994) alignment of the present dataset of vertebrate opioid receptors was made using the Mega software (Kumar, Tamura, & Nei, 2004). The alignment shows a high degree of sequence similarity both within types of opioid receptors and among all 24 sequences. Key residues thought to be important for the function of GPCRs are present, such as the DRY motif at alignment positions 189–191 (Gether, 2002; Minami & Satoh, 1995) and the conserved Cys-Cys bridge between EL1 and EL2 provided by residues at 165 and 245 (Palczewski et al., 2000; Shahrestanifar & Howells, 1996). Na^+ is known to regulate the agonist binding of GPCRs and is dependent on an allosteric binding site at the conserved Asp (134) in the TM2 domain (Christopoulos & Kenakin, 2002; Lomize, Pogozheva, & Mosberg, 1999). There is high overall homology among existing vertebrate MOR, DOR, KOR, and ORL proteins, with nearly identical sequences among the seven-transmembrane helical regions. Intracellular loop domains were also highly conserved and the N-terminus regions were most divergent, followed by the C-terminus and extracellular loop domains.

Table 1 Existing vertebrate species with cloned cDNA sequences for *MOR*, *DOR*, *KOR*, and *ORL* receptors

			Access #s			
Vertebrate Class	**Common name**	**Genus species**	**MOR**	**DOR**	**KOR**	**ORL**
Pisces	Zebrafish	*Danio rerio*	AAK01143[a]	CAA04862	AAG60607	AAN46747
Amphibia	Leopard frog	*Rana pipiens*	AAQ09991	AAQ09992	ABY82593	AAR08905
Amphibia	Newt	*Taricha granulosa*	AAV28689	AAV28690	AAU15126	AAU26067
Mammalia	Rat	*Rattus norvegicus*	AAA41630	AAA19939	AAA18261	AAA50827
Mammalia	Mouse	*Mus musculus*	AAA86878	AAA37522	AAA39363	CAA62922
Mammalia	Human	*Homo sapiens*	AAA20580	AAA83426	AAC50158	AAA84913

[a]Access numbers are given for the NCBI protein database.

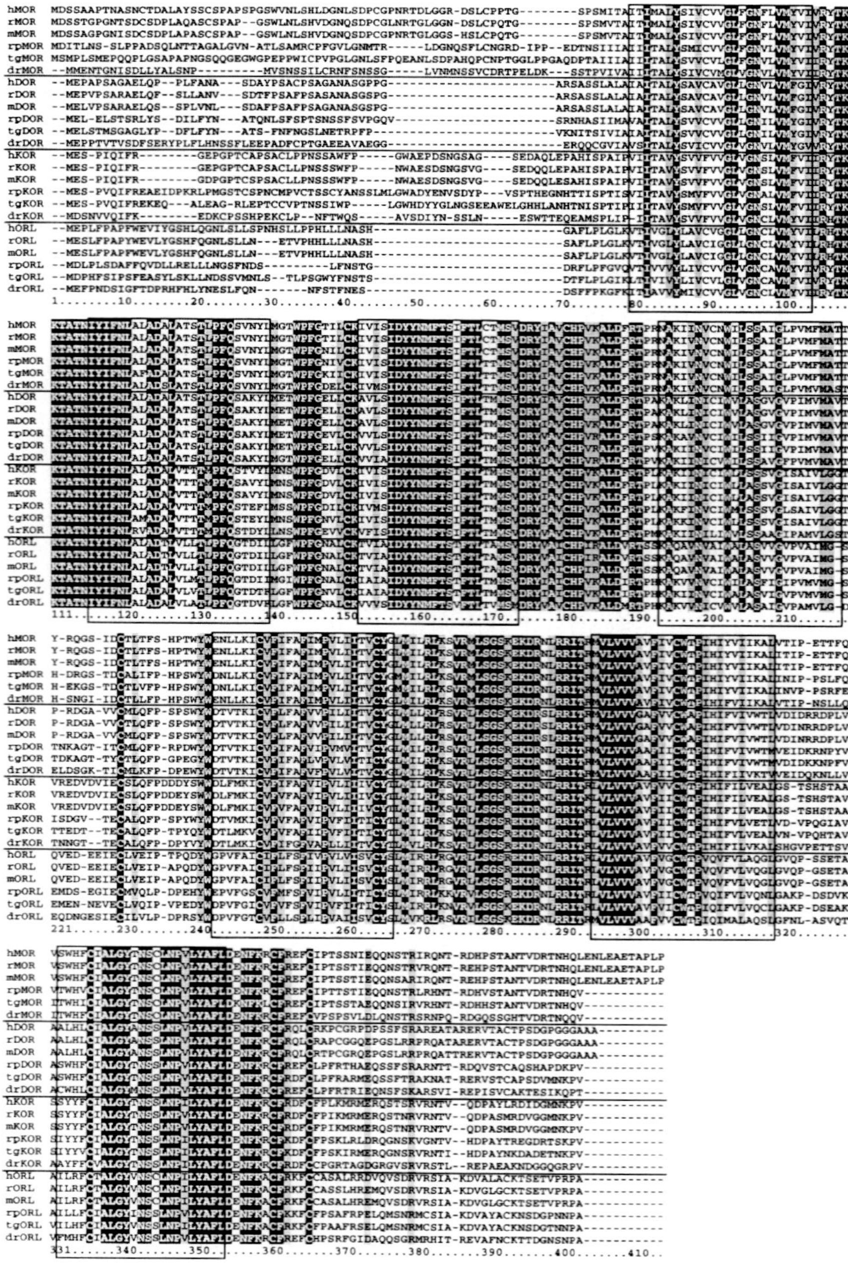

Figure 1 Alignment of mRNA-derived protein sequences of the four opioid receptor types in six vertebrates. Protein alignment was done using ClustalW with default values. Identical sites are indicated by white text on black background, conservative substitutions noted by gray background by using Boxshade. Boxed-in domains indicate the seven-transmembrane regions of the receptor proteins and numbers below the sequences are the alignment positions. See Table 1 for species and protein sequence information.

There are also obvious differences between mammalian and nonmammalian sequences. For example, mammalian DOR contains an extended C-terminus sequence compared to the nonmammalian species. More striking is the 11 amino acid C-terminus extension of mammalian MOR (C-terminus, alignment positions 419–429) compared to the shorter sequence in earlier-evolved vertebrates. Phosphorylation sites along this extended sequence include Thr (position 425) which is critical for internalization and desensitization of rat *mu*-opioid receptors (Wang et al., 2002). Interestingly, this "add-on" piece of protein represents translation of exon 4, which is lacking in some hMOR and mMOR splice variants (Pan et al., 2005, 2003). The differences observed in amphibians with regard to selective opioid ligands in behavioral and radioligand binding studies may result from other substitutions in the primary amino acid sequence. The *mu*-opioid antagonist, *beta*-FNA, is dependent on Lys (234) in EL2 for covalent attachment and selective *mu*-opioid blockade (Law, Wong, & Loh, 1999). The amphibian sequences show conservation of this residue in rpMOR, rpDOR, and rpKOR, which may account for the pan-antagonism of the type-selective antagonist cited above. The action of the *kappa*-selective opioid antagonist, nor-BNI, is dependent on the *kappa*-specific Glu (335) in EL3 (Ferguson, Kramer, Metzger, Law, & Portoghese, 2000) and the amphibian sequence contains a Val in this position which may account for diminished type selectivity in amphibians.

2.2 Phylogenetic analysis of vertebrate opioid receptors

The analyses below are based on the sequence of the canonical MOR, DOR, KOR, and ORL proteins for each species listed in Table 1. Each type of opioid receptor (each set of orthologs) provided a pattern of vertebrate evolution consistent with established fossil evidence and phenotypic characteristics (see Fig. 2). Using the MEGA software, the radial tree was rooted with the sequences of rhodopsin (RHO, lower left) from each species with a sequence available in GenBank. Overall, the four groups of opioid receptor sequences formed a dyad, with MOR and DOR sequences sharing a common ancestor (node) and KOR and ORL sharing a different common node. A similar, but unrooted tree was generated after all four types of opioid receptors were cloned and sequenced in another amphibian, the newt (Bradford et al., 2006).

The graphical view of MOR and DOR sharing one fork of the tree and KOR and ORL sharing the other is supported by an earlier study comparing

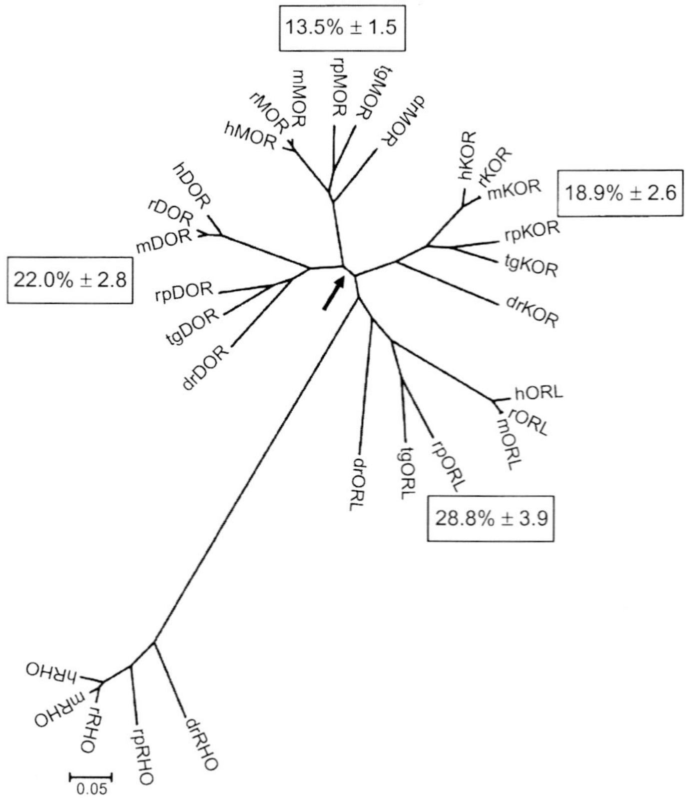

Figure 2 Phylogenetic analysis of MOR, DOR, KOR, and ORL protein sequences in six vertebrates. Bioinformatics software (MEGA4) was used to generate a radial phylogenetic tree using the neighbor-joining method, rooted with the available matching sequences of rhodopsin (RHO). Key: humans, *Homo sapiens* (h); mouse, *Mus musculus* (m); rat, *Rattus norvegicus* (r); leopard frog, *Rana pipiens* (rp); rough-skinned newt, *Taricha granulosa* (tg); and the zebrafish, *Danio rerio* (dr). The arrow shows the bifurcation of MOR and DOR sequences from KOR and ORL. Values in boxes by each opioid receptor type are the mean ± SEM of the pairwise distance (% divergence) among members of each type. Branch length is equal to the proportional difference among the sequences (scale bar = 0.05% or 5% difference in amino acid sequence).

mouse and rat opioid receptor protein sequences by Chaturvedi, Christoffers, Singh, and Howells (2000). They noted that the pairings of MOR and DOR, and KOR and ORL, were the most closely related among the rodent sequences. Using the present vertebrate dataset and simple bioinformatics, these findings were confirmed and extended to all available vertebrates (Stevens, 2005; Stevens et al., 2007).

2.3 Divergence and convergence of opioid receptor types

As pharmacological selectivity is correlated to similarity of amino acid sequences at the GPCR family level (e.g., opioid receptors vs. muscarinic receptors), type selectivity within family members (e.g., MOR, DOR, or KOR) is also correlated with percent identity or similarity. The results of pairwise Blast analysis (Altschul, Gish, Miller, Myers, & Lipman, 1990) for the three classical types of opioid receptor sequences within each species yielded a rank order of divergence such that MOR, DOR, and KOR proteins were more closely related to each other in earlier-evolved vertebrates than in humans and other mammals (see Table 2). Insofar as divergence of molecular sequence is related to the greater type selectivity of opioid receptors, this finding supported our hypothesis that opioid receptors are more type selective in mammals than in nonmammalian species.

To state our hypothesis in another way, the molecular evolution of opioid receptors has a "vector" of increased type selectivity as reflected in the greatest divergence of MOR, DOR, and KOR proteins in mammalian species, especially in *H. sapiens*. In other words, opioid receptors in earlier-evolved vertebrates are less type selective. Given the balancing constraints of maintaining the common sequence for opioid receptor family membership and that of adaptive evolution for greater type selectivity, it is likely that even minor differences in MOR, DOR, and KOR sequences between vertebrate species could be detected in the binding and efficacy of selective opioid ligands. Looking back, our hypothesis is supported by data from studies of the first nonmammalian opioid receptor, ccMOR, cloned from the white suckerfish, *C. commersoni* (Darlison et al., 1997). In this study, ccMOR expressed in HEK cells bound the nonselective opioid antagonist naloxone with high affinity to HEK cell membranes; however, the *mu*-selective opioid agonist, DAMGO, displaced naloxone with surprisingly low affinity. Additionally, nonselective opioid ligands bound well to zebrafish, *D. rerio delta*-opioid receptor (drDOR) expressed in HEK cells, but low affinity was observed for selective *mu*-, *delta*-, or *kappa*-opioid ligands (Rodríguez et al., 2000). Studies of opioid receptors cloned from a second amphibian species, the rough-skinned newt, *T. granulosa*, led to the conclusion that the opioid type selectivity of newt opioid receptors was less stringent than that in mammals (Bradford et al., 2006). Recent *in vitro* data from our lab comparing amphibian (rpMOR) and human (hMOR) opioid receptors expressed in CHO cells showed significant less affinity of selective opioid ligands with rpMOR compared to hMOR (Brasel, Sawyer, & Stevens, 2008).

Table 2 Comparison of vertebrate MOR, DOR, and KOR protein sequences within vertebrate species and by group

Group/species	MOR vs. DOR	MOR vs. KOR	DOR vs. KOR	Species mean (SEM)[b]	Group mean (SEM)
Percent amino acid identity[a]					
Nonmammals					
Danio rerio	70	62	65	65.7 (2.3)	
Rana pipiens	73	65	63	67.0 (3.1)	
Taricha granulosa	70	66	68	68.0 (1.2)	66.9 (1.2)†
Mammals					
Rattus norvegicus	66	61	61	62.7 (1.7)	
Mus musculus	61	60	61	60.7 (0.3)	
Homo sapiens	62	60	59	60.3 (0.9)	61.2 (0.7)
Percent amino acid similarity					
Nonmammals					
Danio rerio	82	79	79	80.0 (1.0)*	
Rana pipiens	85	81	77	81.0 (2.3)*	
Taricha granulosa	82	80	82	81.3 (0.7)*	80.8 (0.7)†
Mammals					
Rattus norvegicus	77	75	75	75.6 (0.7)	
Mus musculus	72	75	74	73.6 (0.9)	
Homo sapiens	73	74	72	73.0 (0.6)	74.1 (0.5)

*Denotes significantly different % similarity than rat, mouse, and human mean values ($p < 0.05$, one-way ANOVA followed by post hoc Newman–Kuels test).

†Denotes group means ($N = 9$) different for identity and similarity at $p < 0.01$, Student's t-test.

[a]ver. BLASTP 2.2.14, settings: matrix = Blossum62, gap open = 11, gap extension = 1, x-drop-off = 50, expect = 10.00, word size = 3, and filter off.

[b]Standard error of the mean.

Besides comparing the MOR, DOR, and KOR sequences to each other within species, analysis of receptor types as a group shows that vertebrate MOR proteins are more similar to each other than ORL proteins (convergence). As shown in Fig. 2 (boxed values), the mean distance of the vertebrate MOR group was 13.5% (which means that after aligning all six vertebrate MOR sequences, 13.5% of the sites were not identical), 18.9% for the KOR group, and 22.0% for the DOR group. For the group of vertebrate ORL sequences, 28.8% of sites were not identical. Figure 3 is a plot of the divergence values for each receptor type, showing that both MOR and KOR are significantly less divergent, thus more convergent, than the group of vertebrate ORL proteins.

The conclusion at this point from our simple bioinformatic analyses is that vertebrate opioid receptors are less divergent in earlier-evolved vertebrates; however, the groups of MOR and KOR proteins show signs of convergence throughout the vertebrate spectrum. Vertebrate MOR proteins are more similar to DOR proteins, with KOR more similar to ORL. Vertebrate ORL is most similar to RHO. Next, data from the human genome and other results from the world of evolutionary biology are applied to the story of the evolution of vertebrate opioid receptors.

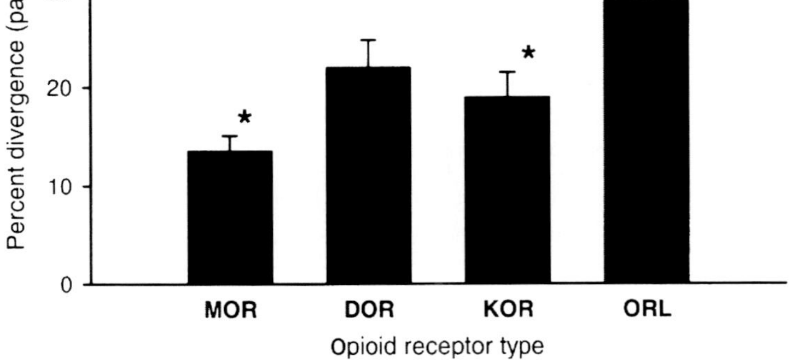

Figure 3 Percent divergence of vertebrate opioid receptor protein sequences plotted by opioid receptor type. Pairwise distance was calculated for each group of six amino acid sequences after alignment by ClustalW using Mega4 (Poisson correction, uniform rate, and gaps deleted). Mean plus SEM are plotted for each group. Asterisks indicate that the MOR and the KOR group significantly less divergent than the ORL group (one-way ANOVA followed by a post hoc Newman–Keuls test, $p = 0.004$; SigmaStat v. 3.1).

3. THE HUMAN GENOME AND THE EVOLUTION OF OPIOID RECEPTORS

This section reviews the recent data available since the completion of the human genome project, including background information on the identification and characteristics of duplicate genes, opioid receptor gene chromosomal location, and quantitative aspects of opioid receptor gene variation and single nucleotide polymorphisms (SNPs).

3.1 Duplicated opioid family receptor genes in the human genome

The three classical opioid receptor genes and the fourth member of the opioid receptor gene family, the nociceptin or *hORL* receptor gene, are mapped to four different chromosomes in the human genome (Fredriksson et al., 2003). The classical opioid and ORL receptor gene loci are part of a broader set of paralogs, or common suites of genes across chromosomes that may indicate gene duplication, which is presented next.

Although not widely known among most scientists, it is accepted by evolutionary biologists that two rounds of genome-wide duplication (paleoploidization) occurred early in vertebrate evolution, originally stated by Ohno and called the 2R hypothesis (Lundin, Larhammar, & Hallbook, 2003; Ohno, 1970). Besides the initial explosion of quadrupled genes at the origin of vertebrate evolution, it is thought that duplicate genes (paralogs) have a gene-birth rate of one new pair of duplicates every million years (Zhang, 2003). The human genome contains at least 5% of its protein-coding genes as paralogs, and paralogous regions on chromosomes are identified as they contain suites of similar genes aligned on two or more different chromosomes (Hokamp, McLysaght, & Wolfe, 2003; McLysaght, Hokamp, & Wolfe, 2002). Most paralogous genes do not survive, becoming pseudogenes or otherwise nonfunctional DNA sequence (Krylov, Wolf, Rogozin, & Koonin, 2003; Prince & Pickett, 2002). The fate of the duplicate genes that do survive is characterized by an asymmetrical sequence divergence such that one member of the pair continues as before and the other is "free" to undergo adaptive evolution (Jordan, Wolf, & Koonin, 2004; Kondrashhov, Rogozin, Wolf, & Koonin, 2002).

The phylogenetic pattern of (AB)(CD) in the "tree" of vertebrate opioid receptor proteins (shown in Fig. 2) is expected from the application of the 2R hypothesis to a single ancestral gene (Makalowski, 2001). Applied to the family of opioid receptor genes, the 2R hypothesis supports the idea that the genes encoding MOR, DOR, KOR, and ORL within a species are paralogs, i.e., genes related by duplication. The evidence for gene duplication for opioid receptors is also supported by the location of MOR, DOR, KOR, and ORL genes on paralogous regions of human chromosomes (Fredriksson et al., 2003). Human MOR and DOR genes are mapped to chromosomes 6 and 1, respectively, while KOR and ORL genes map to chromosomes 8 and 20. The specific pairing of MOR with DOR, and KOR with ORL on separate branches of the sequence tree is supported by a genomic study whereby the greatest number of gene duplicates was found between paralogous regions on chromosomes 1 and 6, and between paralogous regions on chromosomes 8 and 20 (McLysaght et al., 2002). Evidence for 2R duplication of the family of opioid receptor genes (and endogenous opioid peptides) is also supported by studies of chromosomal synteny (Dreborg, Sundstrom, Larsson, & Larhammar, 2008; Josefsson, 2000; Larhammar, Dreborg, Larsson, & Sundstrom, 2009; Lundin et al., 2003; Sundstrom, Dreborg, & Larhammar, 2010).

3.2 Variation in human opioid receptor genes

The vast information available from human genomic data is beginning to allow studies of *intraspecific* variation due to differences in the nucleotide code from one human to the next. Generically called polymorphisms, variation data are obtaining by sampling the genome of numerous individuals from different population groups and cataloging sites where single nucleotide bases differ, i.e., SNPs. Analysis of SNPs, or sets of associated SNPs called haplotypes, provides an insight into the evolution of particular genes and is especially useful in determining genetic defects leading to disease states or medical conditions (Hoehe, 2003). Although the majority of SNPs are found among intronic, repeatable sequences, or other lengths of noncoding DNA, SNPs found in coding or exonic DNA will either be classified as *synonymous* (mutating the nucleotide base does not change the amino acid, due to degeneracy in the codon table) or *nonsynonymous* (changing the amino acid at that place on the protein).

Studies of polymorphism in human *mu*-opioid receptor genes (*hMOR*) have focused on the association of particular SNPs with opioid-dependent and alcohol-dependent populations (Bond et al., 1998; Hoehe et al., 2000; LaForge, Yuferov, & Kreek, 2000; Rommelspacher, Smolka, Schmidt, Samochowiec, & Hoehe, 2001). Most studies have focused on the

A118G allele of *hMOR*, which changes the amino acid Asn to Asp (N40D) in the N-terminus of the hMOR protein (Campa, Gioia, Tomei, Poli, & Barale, 2008; Ding et al., 2013; Dlugos et al., 2011; Gong et al., 2013; Hastie et al., 2012; Hernandez-Avila et al., 2007; Hishimoto et al., 2008; Manini, Jacobs, Vlahov, & Hurd, 2013; Matsunaga et al., 2009; Menon et al., 2012; Olsen et al., 2012; Ray et al., 2011; Weerts et al., 2013). The location of numerous *hMOR* SNPs to the 5′-untranslated region (5′UTR) and upstream promoter regions as well as 3′UTR regions of hMOR do not affect the receptor structure but do alter the expression of hMOR and may explain individual differences in pain sensitivity, drug dependency, and the clinical efficacy of morphine (Klepstad, Dale, Skorpen, Borchgrevink, & Kaasa, 2005; Mayer & Hollt, 2006; Uhl, Sora, & Wang, 1999). Altering the expression of genes, by adaptive evolution in noncoding DNA, is a viable outlet for Darwinian or positive selection (Gu & Gu, 2003; Jordan, Marino-Ramirez, & Koonin, 2005). There are fewer studies of the *hKOR* polymorphism (Gerra et al., 2007; LaForge et al., 2000; Saito et al., 2003; Yuferov et al., 2004), *hDOR* (Beer et al., 2013; Crist, Ambrose-Lanci, et al., 2013; Crist, Clarke, et al., 2013; Kobayashi et al., 2006; Nelson et al., 2012; Zhang et al., 2010), and less than a handful of studies examining SNPs in *hORL* (Huang, Young, Pletcher, Heilig, & Wahlestedt, 2008; Ito, Xie, Maruyama, & Palmer, 2000; Xuei et al., 2008).

Variation of the human genome at the resolution of individual genes can be obtained by accessing the 1000 genome database. As shown in Fig. 4A, the *hMOR* gene has the greatest number of total variants in its gene sequence. These include slice acceptor/donor, frameshift, initiator codon, insertion/deletion, synonymous SNP (sSNP), nonsynonymous SNP (nsSNP), and 5′- and 3′-variants. Total variants for the four human opioid receptor genes were 39,766 for *hMOR*; 1102 for *hDOR*; 3538 for *hKOR*; and 1788 for *hORL* (Fig. 4A). Among these variants in coding regions, there are a much greater number of sSNPs and nsSNPs for *hMOR* compared to the other human opioid receptor genes (see Fig. 4B).

The high degree of polymorphism in the *hMOR* gene resulting in amino acid changes compared to the other types of opioid receptor genes suggests that *hMOR* is under greater selection pressure and undergoing adaptive evolution. Under the nearly neutral theory of evolution, all four opioid receptor genes should show rates of nonsynonymous mutations much less than synonymous substitution yielding a non-synSNP/synSNP ration of $\ll 1.0$ (Iwama & Gojobori, 2002; McPartland, Norris, & Kilpatrick, 2007; Rana

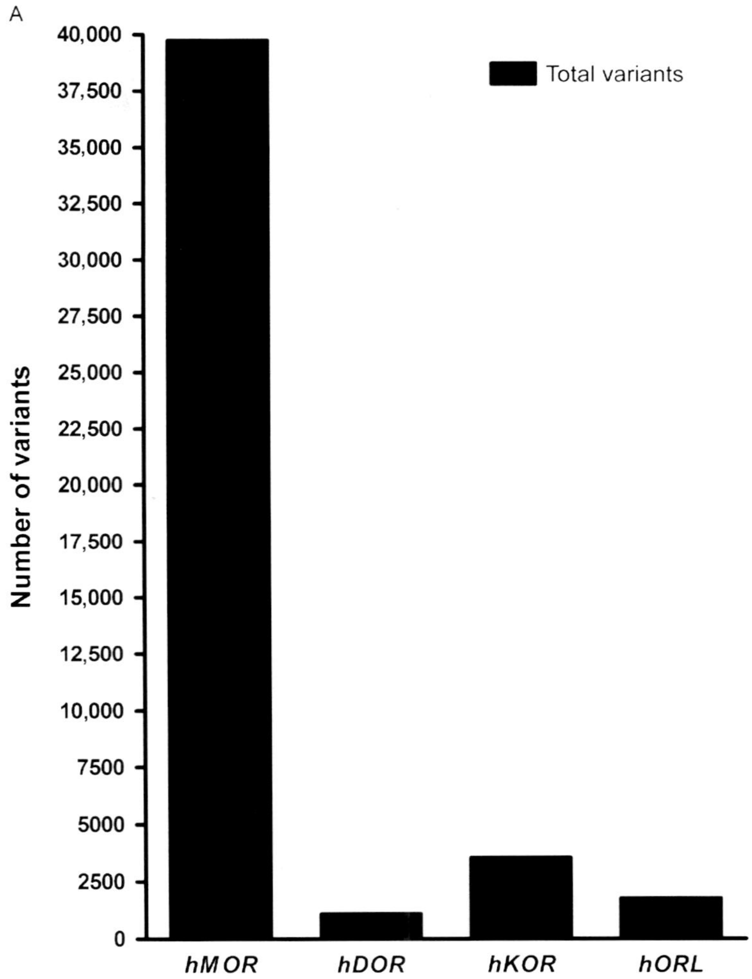

Figure 4 (A) Total variants for each type of opioid receptor gene in humans. Data obtained from 1092 individual whole genomes of various ethnicities available on the 1000 Genomes website (http://www.1000genomes.org).

et al., 1999). Relatively speaking, the alleles in *hMOR* have not been fixed by purifying selection to the same degree as they are in *hDOR*, *hKOR*, and *hORL*. Comparing the *interspecific* differences (divergence) in vertebrate MOR, DOR, KOR, and ORL with the *intraspecific* variation (polymorphism) of *H. sapiens hMOR*, *hDOR*, *hKOR*, and *hORL* is hampered by the lack of comprehensive SNP data from the five other vertebrate species. It is interesting that the most studied nsSNP in *hMOR* (A118G) is also found

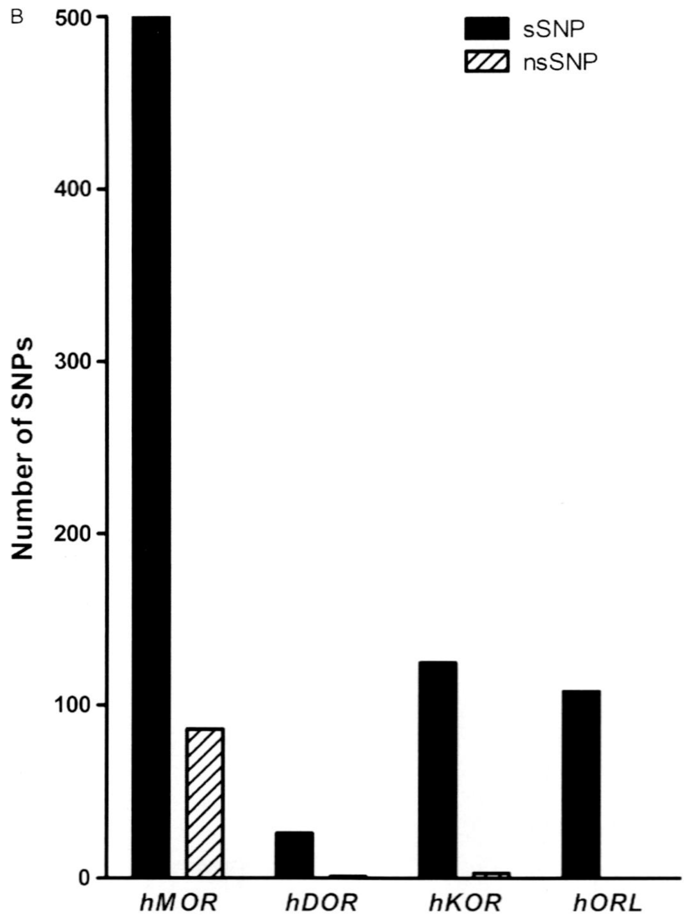

Figure 4—Cont'd (B) Number of synonymous and nonsynonymous SNPs in each of the opioid receptor genes in humans. Values of the bars for sSNP/nsSNP are equal to 500/86, 26/1, 125/3, and 108/0, for human *MOR*, *DOR*, *KOR*, and *ORL* genes, respectively.

in the same alignment position as an nsSNP in the MOR gene of the rhesus monkey (Ikeda et al., 2005). This suggests that SNP data from the population studies of nonhuman and nonmammalian genomes will be a fruitful approach for comparative analysis of interspecific variation. It can be stated that in the most recently evolved species examined (humans), *hMOR* is the most polymorphic opioid receptor gene, followed by *hKOR*, *hDOR*, and *hORL*. This finding suggests greater forces of adaptive evolution operating on MOR genes since the time of its birth by whole-genome duplication, as outlined in the next section.

4. THE MOLECULAR EVOLUTION OF VERTEBRATE OPIOID FAMILY RECEPTORS

A synthesis of the results from the above bioinformatic analyses comparing the sequences of vertebrate opioid receptors, by species and by type of opioid receptor, along with behavioral and binding data of opioids in nonmammalian species and recent data gleaned from human genome studies, suggests the following scenario for the evolution of vertebrate opioid receptors (see Fig. 5). Early in vertebrate evolution, there was a single opioid unireceptor gene. The unireceptor gene was likely a duplicate of a proto-unireceptor gene, perhaps from the time of the arthropod and chordate split, long before the posited 2R whole-genome duplication at the root of the vertebrate evolution. The first round of genome duplication early in chordate evolution produced the ancestral *DOR/MOR* and *ORL/KOR* genes. A second round of genome duplication led to the four opioid receptors present in all extant vertebrates today.

That the initial duplication led to an ancestral *DOR/MOR* gene is supported by the finding that the sequences of MOR and DOR are most identical to each other as shown by above by pairwise comparison and graphically by the bifurcating pattern of the phylogenetic analysis shown in Fig. 2. Likewise, the idea that KOR and ORL shared a common ancestral gene (*KOR/ORL* in Fig. 5) is supported by the above analysis as well as other data from ligand binding that suggests that KOR and ORL maintain

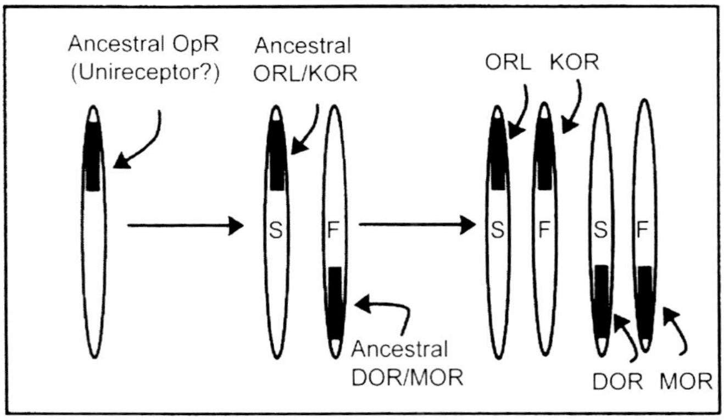

Figure 5 The molecular evolution of vertebrate opioid receptors. For simplicity, the genes are referred to by the same acronym as the opioid receptor proteins they encode. "S" denotes slow and "F" fast rate of adaptive evolution. See text for further details.

close functional ties. As mentioned above, it is known that duplicate genes undergo asymmetrical divergence such that one gene is under relaxed constraint, showing an increased rate of adaptive evolution (positive selection), while the other gene maintains ancestral structure and function (Kondrashhov et al., 2002; Merritt & Quattro, 2001; Prince & Pickett, 2002). This asymmetry for the opioid receptor gene duplicates is noted by an "F" for faster rate of adaptive evolution or positive selection (letter on the inside the chromosome) and conversely by an "S" for the gene duplicate with a slower rate of evolution.

The gene encoding hMOR (*OPRM1*), and not any of the other opioid receptor types, was one of the only nine genes controlling brain size or behavior that showed a significantly increased rate of molecular evolution in the *H. sapiens* genome compared to primate and rodent genomes (Dorus et al., 2004). Thus, *MOR* is assigned an "F" and conversely *DOR* is the "S" member of the pair (see Fig. 5). Likewise, because ORL protein is most closely related to the RHO sequences in the dendrogram, it is assigned an "S" (maintaining more ancestral characteristics) and *KOR* is the faster-evolving member of the pair. This is also supported by the finding that the vertebrate MORs, as a group, are more closely related to each other than vertebrate ORL (i.e., the least divergent, see Fig. 3). Likewise, vertebrate KOR proteins share more common sequence as a group than do the vertebrate ORL proteins. Stated another way, the least divergence among all vertebrates is seen in the MOR proteins, followed by the KOR group, suggesting that adaptive evolution is greatest in the *MOR* genes (confirmed with the human genome studies above), less so for the *KOR* gene, and the least for *DOR* and *ORL*. Going back to the first duplication event, the finding that vertebrate ORL maintained most ancestral characteristics (most closely related to RHO) supports the assignment of the *ORL/KOR* duplicate gene as the "slow" ancestor gene and *DOR/MOR* as the "fast" duplicate (Fig. 5).

Although many GPCR and other gene families in the human genome appear deficient in an even number of duplicate genes due to gene deletion or mutation into pseudogenes (Wolfe & Li, 2003), it is possible that the opioid receptor gene family avoided this fate due to a gene dosage effect (increased signaling by doubling opioid receptor expression) that provided a selective advantage, as noted for other gene families (Kondrashhov et al., 2002). It is of greater interest to note that the target of most clinically used opioids is the MOR and selective *mu*-opioid agonists are the most efficacious type of opioid analgesics in the clinic and in animal models. The rank order of opioid receptor efficacy in a number of animal models (MOR ≫ KOR > DOR ≫ ORL) approximates the rank order of the protein sequence identity of opioid receptor types among vertebrates, suggesting adaptive evolution of MOR (and to a lesser extent, KOR) to be a "better" opioid receptor. Thus, the molecular evolution of vertebrate opioid receptors may provide a striking example of Darwinian or positive selection at the receptor level with convergent evolution in all vertebrate species leading to the most efficacious opioid receptor in the shape of the MOR protein.

The hMOR protein is unique among the members of the family of opioid receptors for many reasons. As mentioned above, hMOR was the only opioid receptor (and one of the few GPCRs) to show accelerated protein evolution since the time of common ancestors with rodents and nonhuman primates (Dorus et al., 2004). The *hMOR* gene contains many more total variants, and both nsSNPs and sSNPs than the other opioid receptor genes. Additionally, recent reports associate the hMOR gene (OPRM1) with adaptations within hominid lineage. For example, variations in *hMOR* are linked to skin pigmentation differences in Native Americans and Europeans (Quillen et al., 2012) and in adaptations to humans living at high altitudes (Ji et al., 2012).

5. FUTURE DIRECTIONS

The results presented here are only a rough sketch of the evolution of opioid receptors as the opioid receptor sequence database is limited. Complete cDNA sequences and the conceptual translation of the protein amino acids for MOR, DOR, KOR, and ORL are only available at present from six species representing three vertebrate classes. Specifically, opioid receptor sequences cloned from brain tissue cDNA of key species representing Class Reptilia or Class Aves are not available, not to mention the earliest extant class of vertebrates, Class Agnatha (hagfish and lamprey). While new animal genomes are appearing online with increasing rapidity, these results are not always of high fidelity nor confirmed by expression of putative opioid receptor mRNA in the brain tissue of that species.

Understanding the evolution of vertebrate opioid receptor proteins contributes to the fundamental model of molecular attraction between opioid drugs and their receptor proteins. The primary structure of the protein sequence must be deterministic in each aspect of measured opioid function including ligand binding, conformational change, and signal transduction. The conceit of our evolutionary approach is not only to explain differences observed in the present opioid receptor proteins expressed in various vertebrate species but also to seek the pattern of opioid receptor evolution in the millions of years to come. This will allow the design of engineered opioid receptors, most likely a variant of the MOR protein (super-MOR) that could be usurped to provide unsurpassed analgesia, perhaps solely from endogenous opioid activation, once the inevitable arrival of gene therapy is secured.

Experiments can be performed to support the hypothesis of opioid receptor by gene duplication by using ancestral sequence analysis, as was done for steroid receptors (Ortlund, Bridgham, Redinbo, & Thornton, 2007). Using bioinformatics and available sequences, the most likely DOR/MOR protein sequence can be determined, a synthetic *DOR/MOR* gene made, transfected in CHO cells, and the ancestral receptor protein characterized. Predictions would include the binding of both MOR and DOR opioid agonists (but not KOR or ORL agonists) but perhaps less robust binding to the DOR/MOR receptor than to either MOR or DOR protein alone.

Many issues and analyses remain. The relationship between a gene's number of splice variants and SNP is not known. The *MOR* gene has more of both compared to the other opioid receptor genes. It is not known what specific amino acids determine the divergence in type selectivity. The correlation of *mu*-, *kappa*-, and *delta*-opioid selectivity to the primary amino acid structure of individual opioid receptor proteins is not known. Specifically, it is not known what exact residues or domains of the primary amino acid sequence differ in each of the three proteins (MOR, DOR, and KOR) to produce *mu*-, *delta*-, or *kappa*-opioid agonist type selectivity. Continued analysis of the vertebrate dataset coupled with experimental data should provide some answers soon.

Much of the selective pressures on opioid receptor genes, and especially MOR, are not known. It is clear that opioid receptor genes are pleiotropic, as they appear to mediate and modulate so many functions in the organism. What type of adaptive evolution is occurring? Certainly not purifying selection for all opioid receptor genes, but does the presence of a large number of sSNPs and nsSNPs in *hMOR* signal positive or Darwinian selection, and if so, by balancing selection or diversifying selection? What are the selection pressures driving the convergent evolution of opioid receptors in different species, especially MOR and to a lesser extent KOR, and how are they balanced against the evolutionary vector of divergence of opioid receptor types? Do the different types of opioid receptor genes evolve at different rates? These and other questions must wait for now.

Future bioinformatic analyses include evolutionary trace studies, whereby single amino acid sites are compared among different species and significance levels applied would give critical input to specific domain hypothesis and ligand selectivity (Madabushi et al., 2004). Single-site analysis also contributes to the *in silico* modeling and structural outcome. For example, does selectivity arise from a more restricted general binding pocket as shown by substitution with bulkier amino acids? Is there an alteration of charge in selective areas consistent with altered ligand selectivity? Are some sites of the receptor sequence undergoing more rapid molecular evolution than other sites and, if so, what is the selection pressure at specific amino acid sites?

Finally, through use of cell culture studies, hypotheses concerning the molecular evolution of opioid receptors can be tested. Significant functional differences in transfected amphibian and human MOR are seen when tested at the same time in the same laboratory. For example, the amphibian *mu*-opioid receptor (rpMOR) did show decreased affinity to selective MOR opioid agonists compared to hMOR in transfected cell lines (Brasel et al., 2008). Such studies need to be carried out with a number of cloned opioid receptors from different vertebrate species and correlated with sequence analysis and bioinformatics. While this chapter reviewed the analyses and results using the case example of opioid receptors, efforts using other GPCR families of receptors should be equally as fruitful.

6. CONCLUSIONS

In summary, the degree of opioid receptor sequence divergence within species was correlated with vertebrate evolution. As the primary amino acid sequence (structure) is a determinant of the selective binding (function) of opioid receptors, the correlation of sequence divergence with vertebrate evolution shows that opioid receptor proteins exhibit an evolutionary vector of increased type selectivity. Additionally, the rapid rate of adaptive evolution of vertebrate MOR suggests an evolutionary vector of increased opioid receptor function. The selective pressure underlying the rapid adaptation of MOR is similar in all vertebrate species, an example of convergent evolution. It is of greater interest to note that the target of most clinically used opioids is the *mu*-opioid receptor and selective *mu* opioids are the most potent type of opioid drug in the clinic and in animal models. This provides a striking example of Darwinian fitness at the molecular level should the selection pressure turn out to be a "better" opioid receptor. If confirmed, these findings provide a unique understanding of the pharmacology of vertebrate opioid receptors and the many other families of GPCR proteins encoded by duplicate genes.

ACKNOWLEDGMENTS

The author gratefully acknowledges the past and hopefully future support of the National Institutes of Health, NIDA, through research Grants DA R15-12448 and DA R29-07326, as well as the state of Oklahoma through the Oklahoma Center for the Advancement of Science and Technology (OCAST) Health Research Contracts. Many thanks also to my colleagues at the International Narcotics Research Conference (www.inrcworld.org) for their encouragement and inspiration. This chapter is dedicated to my wife, the "real" doctor, Timmeni L. Harrison, D.O.

REFERENCES

Altschul, S. F., Gish, W., Miller, W., Myers, E. W., & Lipman, D. J. (1990). Basic local alignment search tool. *Journal of Molecular Biology*, *215*, 403–410.

Alvarez, F. A., Rodriguez-Martin, I., Gonzalez-Nunez, V., de Velasco, E. M., Gonzalez, S. R., & Rodriguez, R. E. (2006). New *kappa* opioid receptor from zebrafish *Danio rerio*. *Neuroscience Letters*, *405*, 94–99.

Attwood, T. K. (2001). A compendium of specific motifs for diagnosing GPCR subtypes. *Trends in Pharmacological Sciences*, *22*, 162–165.

Bakalkin, G. Y., Pivovarov, A. S., Kobylyansky, A. G., Nesterenko, P. N., & Yarygin, K. N. (1989). Lateralization of opioid receptors in turtle visual cortex. *Brain Research*, *480*, 268–276.

Bamford, D. H. (2003). Do viruses form lineages across different domains of life? *Research in Microbiology*, *154*, 231–236.

Barrallo, A., González-Sarmiento, R., Alvar, F., & Rodriguez, R. E. (2000). ZFOR2, a new opioid receptor-like gene from the teleost zebrafish (*Danio rerio*). *Brain Research. Molecular Brain Research*, *84*, 1–6.

Barrallo, A., González-Sarmiento, R., Porteros, A., Gracia-Isídoro, M., & Rodríguez, R. E. (1998). Cloning, molecular characterization, and distribution of a gene homologous to δ opioid receptor from zebrafish (*Danio rerio*). *Biochemical and Biophysical Research Communications*, *245*, 544–548.

Beer, B., Erb, R., Pavlic, M., Ulmer, H., Giacomuzzi, S., Riemer, Y., et al. (2013). Association of polymorphisms in pharmacogenetic candidate genes (OPRD1, GAL, ABCB1, OPRM1) with opioid dependence in European population: A case–control study. *PLoS One*, *8*, e75359.

Benyhe, S., Ketevan, A., Simon, J., Hepp, J., Medzihradszky, K., & Borsodi, A. (1997). Affinity labelling of frog brain opioid receptors by dynorphin$_{(1-10)}$ chloromethyl ketone. *Neuropeptides*, *31*, 52–59.

Benyhe, S., Monory, K., Farkas, J., Toth, G., Guerrini, R., Salvadori, S., et al. (1999). Nociceptin binding sites in frog (*Rana esculenta*) brain membranes. *Biochemical and Biophysical Research Communications*, *260*, 592–596.

Benyhe, S., Simon, J., Borsodi, A., Wollemann, M., & Barnard, E. A. (1994). [3H]dynorphin 1-8 binding sites in frog (*Rana esculenta*) brain membranes. *Neuropeptides*, *26*, 359–364.
Benyhe, S., Szucs, M., Borsodi, A., & Wollemann, M. (1992). Species differences in the stereoselectivity of *kappa* opioid binding sites for [3H]U-69593 and [3H] ethylketocyclazocine. *Life Sciences*, *51*, 1647–1655.
Benyhe, S., Varga, E., Hepp, J., Magyar, A., Borsodi, A., & Wollemann, M. (1990). Characterization of *kappa1* and *kappa2* opioid binding sites in frog (*Rana esculenta*) brain membrane. *Neurochemical Research*, *15*, 899–904.
Bird, D. J., Jackson, M., Baker, B. I., & Buckingham, J. C. (1988). Opioid binding sites in the fish brain: An autoradiographic study. *General and Comparative Endocrinology*, *70*, 49–62.
Bockaert, J., & Pin, J. P. (1999). Molecular tinkering of G protein-coupled receptors: An evolutionary success. *The EMBO Journal*, *18*, 1723–1729.
Bond, C., LaForge, K. S., Tian, M., Melia, D., Zhang, S., Borg, L., et al. (1998). Single-nucleotide polymorphism in the human *mu* opioid receptor gene alters β-endorphin binding and activity: Possible implications for opiate addiction. *Proceedings of the National Academy of Sciences of the United States of America*, *95*, 9608–9613.
Bradford, C. S., Walthers, E. A., Searcy, B. T., & Moore, F. L. (2005). Cloning, heterologous expression and pharmacological characterization of a *kappa* opioid receptor from the brain of the rough-skinned newt, *Taricha granulosa*. *Journal of Molecular Endocrinology*, *34*, 809–823.
Bradford, C. S., Walthers, E. A., Stanley, D. J., Baugh, M. M., & Moore, F. L. (2006). *Delta* and *mu* opioid receptors from the brain of a urodele amphibian, the rough-skinned newt, *Taricha granulosa*: Cloning, heterologous expression, and pharmacological characterization. *General and Comparative Endocrinology*, *146*, 275–290.
Brasel, C. M., Sawyer, G. W., & Stevens, C. W. (2008). A pharmacological comparison of the cloned frog and human *mu* opioid receptors reveals differences in opioid affinity and function. *European Journal of Pharmacology*, *599*, 36–43.
Brooks, A. I., Standifer, K. M., Cheng, J., Ciszewska, G. R., & Pasternak, G. W. (1994). Opioid binding in giant toad and goldfish brain. *Receptor*, *4*, 55–62.
Buatti, M. C., & Pasternak, G. W. (1981). Multiple opiate receptors: Phylogenetic differences. *Brain Research*, *218*, 400–405.
Bunzow, J. R., Saez, C., Mortrud, M., Bouvier, C., Williams, J. T., Low, M., et al. (1994). Molecular cloning and tissue distribution of a putative member of the rat opioid receptor gene family that is not a μ, δ, or κ opioid receptor type. *FEBS Letters*, *347*, 284–288.
Campa, D., Gioia, A., Tomei, A., Poli, P., & Barale, R. (2008). Association of ABCB1/MDR1 and OPRM1 gene polymorphisms with morphine pain relief. *Clinical Pharmacology and Therapeutics*, *83*, 559–566.
Cavalier-Smith, T. (2004). Only six kingdoms of life. *Proceedings of the Royal Society of London. Series B, Biological Sciences*, *271*, 1251–1262.
Chaturvedi, K., Christoffers, K. H., Singh, K., & Howells, R. D. (2000). Structure and regulation of opioid receptors. *Biopolymers*, *55*, 334–346.
Chen, Y., Fan, Y., Liu, J., Mestek, A., Tian, M., Kozak, C. A., et al. (1994). Molecular cloning, tissue distribution and chromosomal localization of a novel member of the opioid receptor gene family. *FEBS Letters*, *347*, 279–283.
Chen, X., Liu-Chen, L. Y., Tallarida, R. J., Geller, E. B., de Riel, J. K., & Adler, M. W. (1996). Use of a μ-antisense oligodeoxynucleotide as a μ opioid receptor noncompetitive antagonist *in vivo*. *Neurochemical Research*, *21*, 1363–1368.
Christopoulos, A., & Kenakin, T. (2002). G protein-coupled receptor allosterism and complexing. *Pharmacological Reviews*, *54*, 323–374.
Crist, R. C., Ambrose-Lanci, L. M., Vaswani, M., Clarke, T. K., Zeng, A., Yuan, C., et al. (2013). Case-control association analysis of polymorphisms in the *delta*-opioid receptor,

OPRD1, with cocaine and opioid addicted populations. *Drug and Alcohol Dependence*, *127*, 122–128.

Crist, R. C., Clarke, T. K., Ang, A., Ambrose-Lanci, L. M., Lohoff, F. W., Saxon, A. J., et al. (2013). An intronic variant in OPRD1 predicts treatment outcome for opioid dependence in African-Americans. *Neuropsychopharmacology*, *38*, 2003–2010.

Darlison, M. G., Greten, F. R., Harvey, R. J., Kreienkamp, H., Stuhmer, T., Zwiers, H., et al. (1997). Opioid receptors from a lower vertebrate (*Catostomus commersoni*): Sequence, pharmacology, coupling to a G-protein-gated inward-rectifying potassium channel (GIRK1), and evolution. *Proceedings of the National Academy of Sciences of the United States of America*, *94*, 8214–8219.

Ding, S., Chen, B., Zheng, Y., Lu, Q., Liu, L., & Zhuge, Q. C. (2013). Association study of OPRM1 polymorphisms with schizophrenia in Han Chinese population. *BMC Psychiatry*, *13*, 107.

Dlugos, A. M., Hamidovic, A., Hodgkinson, C., Shen, P. H., Goldman, D., Palmer, A. A., et al. (2011). OPRM1 gene variants modulate amphetamine-induced euphoria in humans. *Genes, Brain, and Behavior*, *10*, 199–209.

Dorus, S., Vallender, E. J., Evans, P. D., Anderson, J. R., Gilbert, S. L., Mahowald, M., et al. (2004). Accelerated evolution of nervous system genes in the origin of *Homo sapiens*. *Cell*, *119*, 1027–1040.

Dreborg, S., Sundstrom, G., Larsson, T. A., & Larhammar, D. (2008). Evolution of vertebrate opioid receptors. *Proceedings of the National Academy of Sciences of the United States of America*, *105*, 15487–15492.

Evans, C. J., Keith, D. E. J., Morrison, H., Magendzo, K., & Edwards, R. H. (1992). Cloning of a *delta* opioid receptor by functional expression. *Science*, *258*, 1952–1955.

Ferguson, D. M., Kramer, S., Metzger, T. G., Law, P. Y., & Portoghese, P. S. (2000). Isosteric replacement of acidic with neutral residues in extracellular loop-2 of the *kappa*-opioid receptor does not affect dynorphin A(1-13) affinity and function. *Journal of Biological Chemistry*, *43*, 1251–1252.

Flower, D. R., & Attwood, T. K. (2004). Integrative bioinformatics for functional genome annotation: Trawling for G protein-coupled receptors. *Seminars in Cell & Developmental Biology*, *15*, 693–701.

Fredriksson, R., Lagerstrom, M. C., Lundin, L. G., & Schiöth, H. B. (2003). The G-protein-coupled receptors in the human genome form five main families: Phylogenetic analysis, paralogon groups, and fingerprints. *Molecular Pharmacology*, *63*, 1256–1272.

Fredriksson, R., & Schiöth, H. B. (2005). The repertoire of G-protein-coupled receptors in fully sequenced genomes. *Molecular Pharmacology*, *67*, 1414–1425.

Fukuda, K., Kato, S., Mori, K., Nishi, M., Takeshima, H., Iwabe, N., et al. (1994). cDNA cloning and regional distribution of a novel member of the opioid receptor family. *FEBS Letters*, *343*, 42–46.

Gaveriaux-Ruff, C., Peluso, J., Befort, K., Simonin, F., Zilliox, C., & Kieffer, B. L. (1997). Detection of opioid receptor mRNA by RT-PCR reveals alternative splicing for the *delta*- and *kappa*-opioid receptors. *Molecular Brain Research*, *48*, 298–304.

Gerra, G., Leonardi, C., Cortese, E., D'Amore, A., Lucchini, A., Strepparola, G., et al. (2007). Human *kappa* opioid receptor gene (OPRK1) polymorphism is associated with opiate addiction. *American Journal of Medical Genetics. Part B, Neuropsychiatric Genetics*, *144b*, 771–775.

Gether, U. (2002). Uncovering molecular mechanisms involved in activation of G protein-coupled receptors. *Endocrine Reviews*, *21*, 90–113.

Gong, X. D., Wang, J. Y., Liu, F., Yuan, H. H., Zhang, W. Y., Guo, Y. H., et al. (2013). Gene polymorphisms of OPRM1 A118G and ABCB1 C3435T may influence opioid requirements in Chinese patients with cancer pain. *Asian Pacific Journal of Cancer Prevention*, *14*, 2937–2943.

Gonzalez-Nunez, V., Barrallo, A., Traynor, J. R., & Rodriguez, R. E. (2006). Characterization of opioid binding sites in zebrafish brain. *The Journal of Pharmacology and Experimental Therapeutics*, *316*, 900–904.

Gu, J., & Gu, X. (2003). Induced gene expression in human brain after the split from chimpanzee. *Trends in Genetics*, *19*, 63–65.

Hastie, B. A., Riley, J. L., 3rd., Kaplan, L., Herrera, D. G., Campbell, C. M., Virtusio, K., et al. (2012). Ethnicity interacts with the OPRM1 gene in experimental pain sensitivity. *Pain*, *153*, 1610–1619.

Hernandez-Avila, C. A., Covault, J., Wand, G., Zhang, H., Gelernter, J., & Kranzler, H. R. (2007). Population-specific effects of the ASN40ASP polymorphism at the *mu*-opioid receptor gene (OPRM1) on HPA-axis activation. *Pharmacogenetics and Genomics*, *17*, 1031–1038.

Hishimoto, A., Cui, H., Mouri, K., Nushida, H., Ueno, Y., Maeda, K., et al. (2008). A functional polymorphism of the μ-opioid receptor gene is associated with completed suicides. *Journal of Neural Transmission*, *115*, 531–536.

Hoehe, M. R. (2003). Haplotypes and the systematic analysis of genetic variation in genes and genomes. *Pharmacogenomics*, *4*, 547–570.

Hoehe, M. R., Kopke, K., Wendel, B., Rohde, K., Flachmeier, C., Kidd, K. K., et al. (2000). Sequence variability and candidate gene analysis in complex disease: Association of *mu* opioid receptor gene variation with substance dependence. *Human Molecular Genetics*, *9*, 2895–2908.

Hokamp, K., McLysaght, A., & Wolfe, K. H. (2003). The 2R hypothesis and the human genome sequence. *Journal of Structural and Functional Genomics*, *3*, 95–110.

Huang, J., Young, B., Pletcher, M. T., Heilig, M., & Wahlestedt, C. (2008). Association between the nociceptin receptor gene (OPRL1) single nucleotide polymorphisms and alcohol dependence. *Addiction Biology*, *13*, 88–94.

Ikeda, K., Ide, S., Han, W., Hayashida, M., Uhl, G. R., & Sora, I. (2005). How individual sensitivity to opiates can be predicted by gene analyses. *Trends in Pharmacological Sciences*, *26*, 311–317.

Ito, E., Xie, G. X., Maruyama, K., & Palmer, P. P. (2000). A core-promoter region functions bi-directionally for human opioid-receptor-like gene ORL1 and its 5′-adjacent gene GAIP. *Journal of Molecular Biology*, *304*, 259–270.

Iwama, H., & Gojobori, T. (2002). Identification of neurotransmitter receptor genes under significantly relaxed selective constraint by orthologous gene comparisons between humans and rodents. *Molecular Biology and Evolution*, *19*, 1891–1901.

Iyer, L. M., Balaji, S., Koonin, E. V., & Aravind, L. (2006). Evolutionary genomics of nucleo-cytoplasmic large DNA viruses. *Virus Research*, *117*, 156–184.

Ji, L. D., Qiu, Y. Q., Xu, J., Irwin, D. M., Tam, S. C., Tang, N. L., et al. (2012). Genetic adaptation of the hypoxia-inducible factor pathway to oxygen pressure among Eurasian human populations. *Molecular Biology and Evolution*, *29*, 3359–3370.

Jordan, I. K., Marino-Ramirez, L., & Koonin, E. V. (2005). Evolutionary significance of gene expression divergence. *Gene*, *345*, 119–126.

Jordan, I. K., Wolf, Y. I., & Koonin, E. V. (2004). Duplicated genes evolve slower than singletons despite the initial rate increase. *BMC Evolutionary Biology*, *4*, 22.

Josefsson, L.-G. (2000). Evidence for kinship between diverse G-protein coupled receptors. *Gene*, *239*, 333–340.

Kieffer, B. L., Befort, K., Gaveriaux-Ruff, C., & Hirth, C. G. (1992). The *delta*-opioid receptor: Isolation of a cDNA by expression cloning and pharmacological characterization. *Proceedings of the National Academy of Sciences of the United States of America*, *89*, 12048–12052.

Klepstad, P., Dale, O., Skorpen, F., Borchgrevink, P. C., & Kaasa, S. (2005). Genetic variability and clinical efficacy of morphine. *Acta Anaesthesiologica Scandinavica*, *49*, 902–908.

Knapp, R. J., Malatynska, E., Fang, L., Li, X., Babin, E., Nguyen, M., et al. (1994). Identification of a human *delta* opioid receptor: Cloning and expression. *Life Sciences, 54*, L463–L469.

Kobayashi, H., Hata, H., Ujike, H., Harano, M., Inada, T., Komiyama, T., et al. (2006). Association analysis of *delta*-opioid receptor gene polymorphisms in methamphetamine dependence/psychosis. *American Journal of Medical Genetics. Part B, Neuropsychiatric Genetics, 141b*, 482–486.

Kondrashhov, F. A., Rogozin, I. B., Wolf, Y. I., & Koonin, E. V. (2002). Selection in the evolution of gene duplications. *Genome Biology, 3*, 8.1–8.9.

Krylov, D. M., Wolf, Y. I., Rogozin, I. B., & Koonin, E. V. (2003). Gene loss, protein sequence divergence, gene dispensability, expression level, and interactivity are correlated in eukaryotic evolution. *Genome Research, 13*, 2229–2235.

Kumar, S., Tamura, K., & Nei, M. (2004). Mega3: Integrated software for molecular evolutionary genetics analysis and sequence alignment. *Briefings in Bioinformatics, 5*, 150–163.

Lachowicz, J. E., Shen, Y., Monsma, F. J. J., & Sibley, D. R. (1995). Molecular cloning of a novel G protein-coupled receptor related to the opiate receptor family. *Journal of Neurochemistry, 64*, 34–40.

LaForge, K. S., Yuferov, V., & Kreek, M. J. (2000). Opioid receptor and peptide gene polymorphisms: Potential implications for addictions. *European Journal of Pharmacology, 410*, 249–268.

Larhammar, D., Dreborg, S., Larsson, T. A., & Sundstrom, G. (2009). Early duplications of opioid receptor and peptide genes in vertebrate evolution. *Annals of the New York Academy of Sciences, 1163*, 451–453.

Law, P.-Y., Wong, Y. H., & Loh, H. H. (1999). Mutational analysis of the structure and function of opioid receptors. *Biopolymers, 51*, 440–455.

Li, X., Keith, D. E., Jr., & Evans, C. J. (1996a). Multiple opioid receptor-like genes are identified in diverse vertebrate phyla. *FEBS Letters, 397*, 25–29.

Li, X., Keith, D. E., Jr., & Evans, C. J. (1996b). *Mu* opioid receptor-like sequences are present throughout vertebrate evolution. *Journal of Molecular Evolution, 43*, 179–184.

Loh, H., & Smith, A. (1990). Molecular characterization of opioid receptors. *Annual Review of Pharmacology and Toxicology, 30*, 123–147.

Lomize, A. L., Pogozheva, I. D., & Mosberg, H. I. (1999). Structural organization of G-protein-coupled receptors. *Journal of Computer-Aided Molecular Design, 13*, 352–353.

Lord, J. A. H., Waterfield, A. A., Hughes, J., & Kosterlitz, H. W. (1977). Endogenous opioid peptides: Multiple agonists and receptors. *Nature, 267*, 495–499.

Lundin, L. G., Larhammar, D., & Hallbook, F. (2003). Numerous groups of chromosomal regional paralogies strongly indicate two genome doublings at the root of the vertebrates. *Journal of Structural and Functional Genomics, 3*, 53–63.

Madabushi, S., Gross, A. K., Philippi, A., Meng, E. C., Wensel, T. G., & Lichtarge, O. (2004). Evolutionary trace of G protein-coupled receptors reveals clusters of residues that determine global and class-specific functions. *The Journal of Biological Chemistry, 279*, 8126–8132.

Makalowski, W. (2001). Are we polyploids? A brief history of one hypothesis. *Genome Research, 11*, 667–670.

Manini, A. F., Jacobs, M. M., Vlahov, D., & Hurd, Y. L. (2013). Opioid receptor polymorphism A118G associated with clinical severity in a drug overdose population. *Journal of Medical Toxicology, 9*, 148–154.

Mansson, E., Bare, L., & Yang, D. (1994). Isolation of a human *kappa* opioid receptor cDNA from placenta. *Biochemical and Biophysical Research Communications, 202*, 1431–1437.

Martin, W. R., Eades, C. G., Thompson, J. A., Huppler, R. E., & Gilbert, P. E. (1976). The effects of morphine-and nalorphine-like drugs in the nondependent and morphine-dependent chronic spinal dog. *Journal of Pharmacology and Experimental Therapeutics*, *197*, 517–532.

Matsunaga, M., Isowa, T., Murakami, H., Kasugai, K., Yoneda, M., Kaneko, H., et al. (2009). Association of polymorphism in the human *mu*-opioid receptor OPRM1 gene with proinflammatory cytokine levels and health perception. *Brain, Behavior, and Immunity*, *23*, 931–935.

Maussang, D., Vischer, H. F., Leurs, R., & Smit, M. J. (2009). Herpesvirus-encoded G protein-coupled receptors as modulators of cellular function. *Molecular Pharmacology*, *76*, 692–701.

Mayer, P., & Hollt, V. (2006). Pharmacogenetics of opioid receptors and addiction. *Pharmacogenetics and Genomics*, *16*, 1–7.

McLysaght, A., Hokamp, K., & Wolfe, K. H. (2002). Extensive genomic duplication during early chordate evolution. *Nature Genetics*, *31*, 200–204.

McPartland, J. M., Norris, R. W., & Kilpatrick, C. W. (2007). Tempo and mode in the endocannaboinoid system. *Journal of Molecular Evolution*, *65*, 267–276.

Menon, S., Lea, R. A., Roy, B., Hanna, M., Wee, S., Haupt, L. M., et al. (2012). The human *mu*-opioid receptor gene polymorphism (A118G) is associated with head pain severity in a clinical cohort of female migraine with aura patients. *Journal of Headache and Pain*, *13*, 513–519.

Merritt, T. J., & Quattro, J. M. (2001). Evidence for a period of directional selection following gene duplication in a neurally expressed locus of triosephosphate isomerase. *Genetics*, *159*, 689–697.

Meunier, J.-C. (1997). Nociceptin/orphanin FQ and the opioid receptor-like ORL1 receptor. *European Journal of Pharmacology*, *340*, 1–15.

Meunier, J.-C., Mollereau, C., Toll, L., Suaudeau, C., Moisand, C., Alvinerie, P., et al. (1995). Isolation and structure of the endogenous agonist of opioid receptor-like ORL1 receptor. *Nature*, *377*, 532–535.

Minami, M., & Satoh, M. (1995). Molecular biology of the opioid receptors: Structures, functions and distributions. *Neuroscience Research*, *23*, 121–145.

Mogil, J. S., & Pasternak, G. (2001). The molecular and behavioral pharmacology of the orphanin FQ/nociceptin peptide and receptor family. *Pharmacological Reviews*, *53*, 381–415.

Moitra, J., Öktem, H. A., & Borsodi, A. (1995). Thermodynamic parameters of frog brain *kappa*-opioid receptors. *Journal of Neurochemistry*, *65*, 798–801.

Mollereau, C., Parmentier, M., Mailleux, P., Butour, J.-L., Moisand, C., Chalon, P., et al. (1994). ORL1, a novel member of the opioid receptor family: Cloning, functional expression and localization. *FEBS Letters*, *341*, 33–38.

Mollereau, C., Pascaud, A., Baillat, G., Mazarguil, H., Puget, A., & Meunier, J. C. (1988). Evidence for a new type of opioid binding site in the brain of the frog *Rana ridibunda*. *European Journal of Pharmacology*, *150*, 75–84.

Nelson, E. C., Lynskey, M. T., Heath, A. C., Wray, N., Agrawal, A., Shand, F. L., et al. (2012). Association of OPRD1 polymorphisms with heroin dependence in a large case–control series. *Addiction Biology*, *19*, 111–121.

Newman, L. C., Sands, S. S., Wallace, D. R., & Stevens, C. W. (2002). Characterization of *mu*, *kappa*, and *delta* opioid binding in amphibian whole brain tissue homogenates. *Journal of Pharmacology and Experimental Therapeutics*, *301*, 364–370.

Newman, L. C., Wallace, D. R., & Stevens, C. W. (2000a). Selective opioid agonist and antagonist displacement of [^{3}H]-naloxone binding in amphibian brain. *European Journal of Pharmacology*, *397*, 255–262.

Newman, L. C., Wallace, D. R., & Stevens, C. W. (2000b). Selective opioid receptor agonist and antagonist displacement of [^{3}H]-naloxone binding in amphibian spinal cord. *Brain Research*, *884*, 184–191.

Ohno, S. (1970). *Evolution by gene duplication*. London: George Allen and Unwin.

Olsen, M. B., Jacobsen, L. M., Schistad, E. I., Pedersen, L. M., Rygh, L. J., Roe, C., et al. (2012). Pain intensity the first year after lumbar disc herniation is associated with the A118G polymorphism in the opioid receptor *mu* 1 gene: Evidence of a sex and genotype interaction. *Journal of Neuroscience*, *32*, 9831–9834.

Ortlund, E. A., Bridgham, J. T., Redinbo, M. R., & Thornton, J. W. (2007). Crystal structure of an ancient protein: Evolution by conformational epistasis. *Science*, *317*, 1544–1548.

Palczewski, K., Kumasaka, T., Hori, T., Behnke, C. A., Motoshima, H., Fox, B. A., et al. (2000). Crystal structure of rhodopsin: A G protein-coupled receptor. *Science*, *289*, 739–746.

Pan, Y.-X., Cheng, J., Xu, J., Rossi, G., Jacobson, E., Ryan-Moro, J., et al. (1995). Cloning and functional characterization through antisense mapping of a κ_3-related opioid receptor. *Molecular Pharmacology*, *47*, 1180–1188.

Pan, Y.-X., Xu, J., Bolan, E., Abbadie, C., Chang, A., Zuckerman, A., et al. (1999). Identification and characterization of three new alternatively spliced μ-opioid receptor isoforms. *Molecular Pharmacology*, *56*, 396–403.

Pan, Y. X., Xu, J., Bolan, E., Moskowitz, H. S., Xu, M., & Pasternak, G. W. (2005). Identification of four novel exon 5 splice variants of the mouse *mu*-opioid receptor gene: Functional consequences of C-terminal splicing. *Molecular Pharmacology*, *68*, 866–875.

Pan, Y. X., Xu, J., Mahurter, L., Xu, M., Gilbert, A. K., & Pasternak, G. W. (2003). Identification and characterization of two new human *mu* opioid receptor splice variants, hMOR-1O and hMOR-1X. *Biochemical and Biophysical Research Communications*, *301*, 1057–1061.

Pan, Y. X., Xu, J., Wan, B. L., Zuckerman, A., & Pasternak, G. W. (1998). Identification and differential regional expression of KOR-3/ORL-1 gene splice variants in mouse brain. *FEBS Letters*, *435*, 65–68.

Pasternak, G. W. (2005). Molecular biology of opioid analgesia. *Journal of Pain and Symptom Management*, *29*, S2–S9.

Pasternak, G. W., & Standifer, K. M. (1995). Mapping of opioid receptors using antisense oligodeoxynucleotides: Correlating their molecular biology and pharmacology. *Trends in Pharmacological Sciences*, *16*, 344–350.

Perez, D. M. (2003). The evolutionarily triumphant G-protein-coupled receptor. *Molecular Pharmacology*, *63*, 1202–1205.

Perez, D. M. (2005). From plants to man: The GPCR "tree of life". *Molecular Pharmacology*, *67*, 1383–1384.

Pert, C. B., Aposhian, D., & Snyder, S. H. (1974). Phylogenetic distribution of opiate binding. *Brain Research*, *75*, 356–361.

Pert, C. B., & Snyder, S. H. (1973). Opiate receptor: Demonstration in nervous tissue. *Science*, *179*, 1011–1014.

Pezalla, P. D., & Dicig, M. (1984). Stress-induced analgesia in frogs: Evidence for the involvement of an opioid system. *Brain Research*, *296*, 356–360.

Pezalla, P. D., & Stevens, C. W. (1984). Behavioral effects of morphine, levorphanol, dextrorphan and naloxone in the frog *Rana pipiens*. *Pharmacology, Biochemistry, and Behavior*, *21*, 213–217.

Porteros, A., Garcia-Isodoro, M., Barrallo, A., González-Sarmiento, R., & Rodriguez, R. E. (1999). Expression of ZFOR1, a delta-opioid receptor, in the central nervous system of the zebrafish (*Danio rerio*). *Journal of Comparative Neurology*, *412*, 429–4358.

Portoghese, P. S., Sultana, M., & Takemori, A. E. (1988). Naltrindole, a highly selective and potent non-peptide *delta* opioid receptor antagonist. *European Journal of Pharmacology*, *146*, 185–186.

Prince, V. E., & Pickett, F. B. (2002). Splitting pairs: The diverging fates of duplicated genes. *Nature Reviews. Genetics*, *3*, 827–837.
Prüll, C.-R. (2003). Part of a scientific master plan? Paul Ehrlich and the origins of his receptor concept. *Medical History*, *47*, 332–356.
Quillen, E. E., Bauchet, M., Bigham, A. W., Delgado-Burbano, M. E., Faust, F. X., Klimentidis, Y. C., et al. (2012). OPRM1 and EGFR contribute to skin pigmentation differences between Indigenous Americans and Europeans. *Human Genetics*, *131*, 1073–1080.
Rana, B. K., Hewett-Emmett, D., Jin, L., Chang, B. H., Sambuughin, N., Lin, M., et al. (1999). High polymorphism at the human melanocortin 1 receptor locus. *Genetics*, *151*, 1547–1557.
Rask-Andersen, M., Almen, M. S., & Schioth, H. B. (2011). Trends in the exploitation of novel drug targets. *Nature Reviews. Drug Discovery*, *10*, 579–590.
Ray, R., Ruparel, K., Newberg, A., Wileyto, E. P., Loughead, J. W., Divgi, C., et al. (2011). Human mu opioid receptor (OPRM1 A118G) polymorphism is associated with brain *mu*-opioid receptor binding potential in smokers. *Proceedings of the National Academy of Sciences of the United States of America*, *108*, 9268–9273.
Reinscheid, R. K., Nothacker, H. P., Bourson, A., Ardati, A., Henningsen, R. A., Bunzow, J. R., et al. (1995). Orphanin FQ: A neuropeptide that activates an opioid-like G protein-coupled receptor. *Science*, *270*, 792–794.
Rodríguez, R. E., Barrallo, A., Garcia-Malvar, F., McFadyen, I. J., González-Sarmiento, R., & Traynor, J. R. (2000). Characterization of ZFOR1, a putative *delta*-opioid receptor from the teleost zebrafish. *Neuroscience Letters*, *288*, 207–210.
Rommelspacher, H., Smolka, M., Schmidt, L. G., Samochowiec, J., & Hoehe, M. R. (2001). Genetic analysis of the *mu*-opioid receptor in alcohol-dependent individuals. *Alcohol*, *24*, 129–135.
Rosenkilde, M. M., Smit, M. J., & Waldhoer, M. (2008). Structure, function and physiological consequences of virally encoded chemokine seven transmembrane receptors. *British Journal of Pharmacology*, *153*(Suppl. 1), S154–S166.
Saito, M., Ehringer, M. A., Toth, R., Oros, M., Szakall, I., Sikela, J. M., et al. (2003). Variants of *kappa*-opioid receptor gene and mRNA in alcohol-preferring and alcohol-avoiding mice. *Alcohol*, *29*, 39–49.
Shahrestanifar, M. S., & Howells, R. D. (1996). Sensitivity of opioid receptor binding to n-substituted maleimides and methanethiosulfonate derivatives. *Neurochemical Research*, *21*, 1295–1299.
Simon, E. J. (1981). Opiate receptors: Some recent developments. In J. W. Lamble (Ed.), *Towards understanding receptors* (pp. 159–165). New York: Elsevier/North-Holland Biomedical Press.
Simon, E. J. (1986). Recent studies on opioid receptors: Heterogeneity and purification. *Annals of the New York Academy of Sciences*, *463*, 31–45.
Simon, J., Benyhe, S., Borsodi, A., Szucs, M., & Wollemann, M. (1985). Separation of κ-opioid receptor subtype from frog brain. *FEBS Letters*, *183*, 395–397.
Simon, J., Benyhe, S., Hepp, J., Khan, A., Borsodi, A., Szucs, M., et al. (1987). Purification of *kappa*-opioid receptor subtype from frog brain. *Neuropeptides*, *10*, 19–28.
Simon, J., Benyhe, S., Hepp, J., Varga, E., Medzihradszky, K., Borsodi, A., et al. (1990). Method for isolation of *kappa*-opioid binding sites by dynorphin affinity chromatography. *Journal of Neuroscience Research*, *25*, 549–555.
Simon, E. J., Hiller, J. M., & Edelman, I. (1973). Stereospecific binding of the potent narcotic analgesic [^{3}H] etorphine to rat-brain homogenate. *Proceedings of the National Academy of Sciences of the United States of America*, *70*, 1947–1949.
Simon, E. J., Hiller, J. M., Groth, J., Itzhak, Y., Holland, M. J., & Beck, S. G. (1982). The nature of opiate receptors in toad brain. *Life Sciences*, *31*, 1367–1370.

Simonin, F., Befort, K., Gaveriaux-Ruff, C., Matthes, H., Nappey, V., Lannes, B., et al. (1994). The human *delta*-opioid receptor: Genomic organization, cDNA cloning, functional expression, and distribution in human brain. *Molecular Pharmacology, 46*, 1015–1021.

Simonin, F., Gaveriaux-Ruff, C., Befort, K., Matthes, H., Lannes, B., Micheletti, G., et al. (1995). Kappa-opioid receptor in humans: cDNA and genomic cloning, chromosomal assignment, functional expression, pharmacology, and expression pattern in the central nervous system. *Proceedings of the National Academy of Sciences of the United States of America, 92*, 7006–7010.

Stevens, C. W. (1992). Alternatives to the use of mammals for pain research. *Life Sciences, 50*, 901–912.

Stevens, C. W. (1995). An amphibian model for pain research. *Lab Animal, 24*, 32–36.

Stevens, C. W. (1996). Relative analgesic potency of *mu*, *delta* and *kappa* opioids after spinal administration in amphibians. *Journal of Pharmacology and Experimental Therapeutics, 276*, 440–448.

Stevens, C. W. (1997). An alternative model for testing opioid analgesics and pain research using amphibians. In L. F. M. van Zutphen & M. Balls (Eds.), *Animal alternatives, welfare and ethics* (pp. 247–251). Amsterdam: Elsevier.

Stevens, C. W. (2004). Opioid research in amphibians: An alternative pain model yielding insights on the evolution of opioid receptors. *Brain Research Reviews, 46*, 204–215.

Stevens, C. W. (2005). Molecular evolution of vertebrate opioid receptor proteins: A preview. In A. Capasso (Ed.), *Recent developments in pain research, 2005* (pp. 13–29). Trivandrum, Kerala, India: Research Signpost.

Stevens, C. W. (2008). Non-mammalian models for the study of pain. In P. M. Conn (Ed.), *Sourcebook of models for biomedical research* (pp. 341–352). Totowa, NJ: Humana Press.

Stevens, C. W., Brasel, C. M., & Mohan, S. (2007). Cloning and bioinformatics of amphibian *mu*, *delta*, *kappa*, and nociceptin opioid receptors expressed in brain tissue: Evidence for opioid receptor divergence in mammals. *Neuroscience Letters, 419*, 189–194.

Stevens, C. W., & Kirkendall, K. (1993). Time course and magnitude of tolerance to the analgesic effects of systemic morphine in amphibians. *Life Sciences, 52*, L111–L116.

Stevens, C. W., Klopp, A. J., & Facello, J. A. (1994). Analgesic potency of *mu* and *kappa* opioids after systemic administration in amphibians. *Journal of Pharmacology and Experimental Therapeutics, 269*, 1086–1093.

Stevens, C. W., Martin, K. K., & Stahlheber, B. W. (2009). Nociceptin produces antinociception after spinal administration in amphibians. *Pharmacology, Biochemistry, and Behavior, 91*, 436–440.

Stevens, C. W., & Newman, L. C. (1999). Spinal administration of selective opioid antagonists in amphibians: Evidence for an opioid unireceptor. *Life Sciences, 64*, PL125–PL130.

Stevens, C. W., & Pezalla, P. D. (1983). A spinal site mediates opiate analgesia in frogs. *Life Sciences, 33*, 2097–2103.

Stevens, C. W., & Pezalla, P. D. (1984). Naloxone blocks the analgesic action of levorphanol but not of dextrorphan in the leopard frog. *Brain Research, 301*, 171–174.

Stevens, C. W., Pezalla, P. D., & Yaksh, T. L. (1987). Spinal antinociceptive action of three representative opioid peptides in frogs. *Brain Research, 402*, 201–203.

Stevens, C. W., & Rothe, K. S. (1997). Supraspinal administration of opioids with selectivity for μ-, δ-, and κ-opioid receptors produces analgesia in amphibians. *European Journal of Pharmacology, 331*, 15–21.

Stevens, C. W., Sangha, S., & Ogg, B. G. (1995). Analgesia produced by immobilization stress and an enkephalinase-inhibitor in amphibians. *Pharmacology, Biochemistry, and Behavior, 51*, 675–680.

Sundstrom, G., Dreborg, S., & Larhammar, D. (2010). Concomitant duplications of opioid peptide and receptor genes before the origin of jawed vertebrates. *PLoS One*, *5*, e10512.

Takemori, A. E., Ho, B. Y., Naeseth, J. S., & Portoghese, P. S. (1988). Nor-binaltorphimine, a highly selective *kappa*-opioid antagonist in analgesic and receptor binding assays. *Journal of Pharmacology and Experimental Therapeutics*, *246*, 255–258.

Taylor, E. W., & Agarwal, A. (1993). Sequence homology between bacteriorhodopsin and G-protein coupled receptors: Exon shuffling or evolution by duplication? *FEBS Letters*, *325*, 161–166.

Terenius, L. (1973). Stereospecific interaction between narcotic analgesics and a synaptic plasma membrane fraction of rat cerebral cortex. *Acta Pharmacologica et Toxicologica*, *32*, 317–320.

Thompson, J. D., Higgins, D. G., & Gibson, T. J. (1994). Clustal W: Improving the sensitivity of progressive multiple sequence alignment through sequence weighting, position-specific gap penalties and weight matrix choice. *Nucleic Acids Research*, *22*, 4673–4680.

Uhl, G. R., Childers, S., & Pasternak, G. (1994). An opiate-receptor gene family reunion. *Trends in Neurosciences*, *17*, 89–93.

Uhl, G. R., Sora, I., & Wang, Z. (1999). The μ opiate receptor as a candidate gene for pain: Polymorphisms, variations in expression, nociception, and opiate responses. *Proceedings of the National Academy of Sciences of the United States of America*, *96*, 7752–7755.

Walthers, E. A., Bradford, C. S., & Moore, F. L. (2005). Cloning, pharmacological characterization and tissue distribution of an ORL1 opioid receptor from an amphibian, the rough-skinned newt, *Taricha granulosa*. *Journal of Molecular Endocrinology*, *34*, 247–256.

Wang, H. L., Chang, W. T., Hsu, C. Y., Huang, P. C., Chow, Y. W., & Li, A. H. (2002). Identification of two c-terminal amino acids, SER(355) and THR(357), required for short-term homologous desensitization of *mu*-opioid receptors. *Biochemical Pharmacology*, *15*, 257–266.

Wang, J. B., Johnson, P. S., Imai, Y., Persico, A. M., Ozenberger, B. A., Eppler, C. M., et al. (1994). cDNA cloning of an orphan opiate receptor gene family member and its splice variant. *FEBS Letters*, *348*, 75–79.

Wang, J. B., Johnson, P. S., Persico, A. M., Hawkins, A. L., Griffin, C. A., & Uhl, G. R. (1994). Human *mu* opiate receptor. cDNA and genomic clones, pharmacologic characterization and chromosomal assignment. *FEBS Letters*, *338*, 217–222.

Wang, H. Q., Kampine, J. P., & Tseng, L.-F. (1996). Antisense oligodeoxynucleotide to a δ-opioid receptor messenger RNA selectively blocks the antinociception induced by intracerebroventricularly administered δ-, but not μ-, ε-, or κ-opioid receptor agonists in the mouse. *Neuroscience*, *75*, 445–452.

Ward, S. J., Portoghese, P. S., & Takemori, A. E. (1982). Pharmacological profiles of *beta*-funaltrexamine (β-FNA) and *beta*-chlornaltrexamine (β-CNA) on the mouse vas deferens preparation. *European Journal of Pharmacology*, *80*, 377–384.

Weerts, E. M., McCaul, M. E., Kuwabara, H., Yang, X., Xu, X., Dannals, R. F., et al. (2013). Influence of OPRM1 ASN40ASP variant (A118G) on [^{11}C]carfentanil binding potential: Preliminary findings in human subjects. *International Journal of Neuropsychopharmacology*, *16*, 47–53.

Wei, L. N., Law, P. Y., & Loh, H. H. (2004). Post-transcriptional regulation of opioid receptors in the nervous system. *Frontiers in Bioscience*, *9*, 1665–1679.

Wick, M. J., Minnerath, S. R., Lin, X., Elde, R., Law, P. Y., & Loh, H. H. (1994). Isolation of a novel cDNA encoding a putative membrane receptor with high homology to the cloned *mu*, *delta*, and *kappa* opioid receptors. *Brain Research*, *27*, 37–44.

Wolfe, K. H., & Li, W. H. (2003). Molecular evolution meets the genomics revolution. *Nature Genetics*, *33*(Suppl), 255–265.

Wollemann, M., Benyhe, S., & Simon, J. (1993). The *kappa* opioid receptor: Evidence for the different subtypes. *Life Sciences*, *52*, 599–611.

Wollemann, M., Farkas, J., Toth, G., & Benyhe, S. (1994). Characterization of [^{3}H] met-enkephalin-arg6-phe7 binding to opioid receptors in frog brain membrane preparations. *Journal of Neurochemistry*, *63*, 1460–1465.

Wollemann, M., Farkas, J., Toth, G., & Benyhe, S. (1999). Comparison of the endogenous heptapeptide met-enkephalin-arg^6-phe^7 binding in amphibian and mammalian brain. *Acta Biologica Hungarica*, *50*, 297–307.

Xia, Y., & Haddad, G. G. (2001). Major difference in the expression of *delta*- and *mu*-opioid receptors between turtle and rat brain. *Journal of Comparative Neurology*, *436*, 202–210.

Xuei, X., Flury-Wetherill, L., Almasy, L., Bierut, L., Tischfield, J., Schuckit, M., et al. (2008). Association analysis of genes encoding the nociceptin receptor (OPRL1) and its endogenous ligand (PNOC) with alcohol or illicit drug dependence. *Addiction Biology*, *13*, 80–87.

Yuferov, V., Fussell, D., LaForge, K. S., Nielsen, D. A., Gordon, D., Ho, A., et al. (2004). Redefinition of the human *kappa* opioid receptor gene (OPRK1) structure and association of haplotypes with opiate addiction. *Pharmacogenetics*, *14*, 793–804.

Zaki, P. A., Bilsky, E. J., Vanderah, T. W., Lai, J., Evans, C. J., & Porreca, F. (1996). Opioid receptor types and subtypes: The *delta* receptor as a model. *Annual Review of Pharmacology and Toxicology*, *36*, 379–401.

Zawilska, J., Lajtha, A., & Borsodi, A. (1988). Selective protection of benzomorphan binding sites against inactivation by n-ethylmaleimide. Evidence for kappa opioid receptors in frog brain. *Journal of Neurochemistry*, *51*, 736–739.

Zhang, J. (2003). Evolution by gene duplication: An update. *Trends in Ecology & Evolution*, *18*, 292–298.

Zhang, H., Gelernter, J., Gruen, J. R., Kranzler, H. R., Herman, A. I., & Simen, A. A. (2010). Functional impact of a single-nucleotide polymorphism in the OPRD1 promoter region. *Journal of Human Genetics*, *55*, 278–284.

Zhu, J., Chen, C., Xue, J. C., Kunapuli, S., DeRiel, J. K., & Liu-Chen, L. Y. (1995). Cloning of a human *kappa* opioid receptor from the brain. *Life Sciences*, *56*, L201–L207.

CHAPTER THREE

Ancestral Vertebrate Complexity of the Opioid System

Dan Larhammar*,[1], Christina Bergqvist*, Görel Sundström*,[2]
*Department of Neuroscience, Unit of Pharmacology, Science for Life Laboratory, Uppsala University, Uppsala, Sweden
[1]Corresponding author: e-mail address: dan.larhammar@neuro.uu.se

Contents

Abstract

The evolution of the opioid peptides and nociceptin/orphanin as well as their receptors has been difficult to resolve due to variable evolutionary rates. By combining sequence comparisons with information on the chromosomal locations of the genes, we have deduced the following evolutionary scenario: The vertebrate predecessor had one opioid precursor gene and one receptor gene. The two genome doublings before the vertebrate radiation resulted in three peptide precursor genes whereupon a fourth copy arose by a local gene duplication. These four precursors diverged to become the prepropeptides for endorphin (POMC), enkephalins, dynorphins, and nociceptin, respectively. The ancestral receptor gene was quadrupled in the genome doublings leading to delta, kappa, and mu and the nociceptin/orphanin receptor. This scenario is corroborated by new data presented here for coelacanth and spotted gar, representing two basal branches in the vertebrate tree. A third genome doubling in the ancestor of teleost fishes generated additional gene copies. These results show that the opioid system was quite complex already in the first vertebrates and that it has more components in teleost fishes than in mammals. From an evolutionary point of view, nociceptin and its receptor can be considered full-fledged members of the opioid system.

[2] Current address: Department of Medical Biochemistry and Microbiology, Science for Life Laboratory, Uppsala University, Box 582, SE-75123 Uppsala, Sweden

Vitamins and Hormones, Volume 97
ISSN 0083-6729
http://dx.doi.org/10.1016/bs.vh.2014.11.001

ABBREVIATIONS

DOP delta opioid receptor
KOP kappa opioid receptor
MOP mu opioid receptor
OPRD1 opioid receptor delta 1 gene
OPRK1 opioid receptor kappa 1 gene
OPRM1 opioid receptor mu 1 gene
OPRL1 opioid receptor-like 1 gene
ORL1 opioid receptor-like 1
PDYN (pre)prodynorphin (protein and gene)
PENK (pre)proenkephalin (protein and gene)
PNOC (pre)pronociceptin (protein and gene)
POMC (pre)proopiomelanocortin (protein and gene)

1. INTRODUCTION

The evolution of biologically active peptides is often challenging to resolve because the mature peptides are usually quite short and well conserved across species, whereas the remainder of their precursors can be quite divergent, sometimes to the extent that the sequences even become difficult to align. These complications concern also the three opioid peptide precursors found in placental mammals, i.e., proenkephalin (PENK), prodynorphin (PDYN), and proopiomelanocortin (POMC), as well as their related precursor pronociceptin (PNOC). One important reason to sort out the evolutionary history of gene and peptide/protein families is that this will increase our understanding of how the functions of related gene products may overlap or differ with regard to interactions with one another and with other components, and also how their regulation of expression may coincide or differ temporally and anatomically. Furthermore, from a medical or veterinary point of view, it is important to be aware of species differences that are of clinical relevance.

After the first report that native peptides from pig brain, named enkephalins, can stimulate opioid receptors (Hughes et al., 1975), several additional peptides were identified and their precursors were cloned as reviewed in Dores, Lecaude, Bauer, and Danielson (2002). The precursors for enkephalins, dynorphins and endorphin all contain the typical opioid peptide core motif YGGF, followed by either M or L for Met-enkephalin and Leu-enkephalin, respectively, or longer extensions for the other opioid peptides, for reviews see Dores et al. (2002), Larhammar, Sundstrom, and Dores

(2013), and Sundstrom, Dreborg, and Larhammar (2010). However, the fourth member of the precursor family that was identified later in mammals did not have the YGGF motif, but rather FGGF. This peptide was named nociceptin and orphanin, respectively, by two different groups of investigators (Meunier et al., 1995; Reinscheid et al., 1995). As this novel peptide did not act on any of the three classical opioid receptors but rather on a fourth related receptor called ORL1 and because the peptide could not be antagonized by the opiate antagonist naloxone, it was not classified as a true opioid peptide, but rather as an opioid-related peptide.

Outside the opioid core motif the four peptide precursors were found to be quite divergent from each other as well as between species. Nevertheless, they all contain a set of conserved cysteine residues in the aminoterminal part of the precursors, six cysteines in PENK, PDYN, and PNOC and four residues in POMC. One additional unifying factor that also supports a common ancestry is the presence of a single intron shortly after the region encoding the signal peptide. Both PENK and PDYN contain multiple copies of the core opioid motif, ranging from 3 to 7 depending on species, whereas POMC has a single core motif (as part of endorphin) and PNOC can have one or two copies depending on class of vertebrates (Sundstrom et al., 2010).

Binding sites for opiates in brain tissue were reported in 1973 (Pert & Snyder, 1973; Simon, Hiller, & Edelman, 1973; Terenius, 1973) and subsequently three distinct categories were described and named delta, mu, and kappa. These three receptor types were molecularly cloned in 1992–1993 (Chen, Mestek, Liu, Hurley, & Yu, 1993; Evans, Keith, Morrison, Magendzo, & Edwards, 1992; Kieffer, Befort, Gaveriaux-Ruff, & Hirth, 1992; Yasuda et al., 1993). The IUPHAR receptor names are DOP, KOP, and MOP and their genes have been named *OPRD1*, *OPRM1*, and *OPRK1*, respectively. They display what we now consider to be typical features of rhodopsin-like G protein-coupled receptors (GPCR) and function by signaling via cytoplasmic G proteins. The three opioid receptors were found to be more closely related to each other than to the other GPCRs that had been cloned at the time. Shortly after, a fourth closely related receptor was cloned from human, mouse, and rat and initially named ORL1 for opioid receptor-like (Bunzow et al., 1994; Mollereau et al., 1994). After identification of its ligand nociceptin, the receptor was named NOP and its gene is called *OPRL1*.

The evolution of these four receptor types has been somewhat difficult to resolve using sequence information, but for other reasons than the evolution of the propeptides, namely, variable evolutionary rates both between the receptor types and between species for receptor orthologs (i.e., species homologs). The NOP receptor sequence has been found to be the most divergent of the four in phylogenetic analyses (Bradford, Walthers, Stanley, Baugh, & Moore, 2006; Larhammar, Dreborg, Larsson, & Sundstrom, 2009; McClendon, Lecaude, Dores, & Dores, 2010; Rivas-Boyero et al., 2011; Waldhoer, Bartlett, & Whistler, 2004). This might be taken as evidence that it was the first to arise by gene duplication whereupon the other three types arose by more recent duplications. However, it is also clear from sequence-based phylogenetic analyses that the NOP receptor has a slightly higher evolutionary rate than the other three types, thus it might have arisen concomitantly with the other three from the ancestral receptor gene. In any event, the slightly greater sequence divergence of the NOP receptor seemed to explain its distinct pharmacological properties and probably contributed to its initial exclusion from the classical or true opioid receptor family. Despite these differences, nociceptin and NOP do influence prominent opioid functions such as reward and nociception, which is why they need to be reconsidered for status as members of the opioid system.

The list of species with identified genes encoding the opioid peptide precursors and receptors is continually growing as a result of the increasing number of more or less completed genome projects. These genomes comprise a rich source of information not only for opioid peptide and receptor sequences but for analysis of the chromosomal locations of the genes in order to investigate conservation of synteny and the mechanisms of the duplication events that generated the peptide and receptor families. In particular, the new genomes serve as tests of the gene evolution scenario that we have previously described for the opioid system, namely that the ancestral peptide and the ancestral receptor gene were located on the same chromosome (Sundstrom et al., 2010), perhaps implying co-evolution. Then followed the two vertebrate tetraploidizations, or genome doublings (Nakatani, Takeda, Kohara, & Morishita, 2007; Putnam et al., 2008), quadrupling the opioid receptor gene (Dreborg, Sundström, Larsson, & Larhammar, 2008). The ancestral propeptide gene was also duplicated in the chromosome doubling events, presumably resulting in four peptide precursor genes, of which one seems to have been lost. Finally, one of the remaining three underwent a local duplication leading to the *PNOC* and *POMC* genes, which are syntenic (colocalized) in both chicken and zebrafish (Sundstrom et al., 2010). All these events took place before the radiation of the vertebrates more than 500 million years ago.

The genes arising by duplication are said to be paralogs. When larger blocks or entire chromosomes are duplicated, they give rise to large paralogous gene regions. Such related paralogous regions or chromosomes have been given the term paralogon (Coulier, Popovici, Villet, & Birnbaum, 2000). In mammals as well as birds, reptiles and most amphibians, a paralogon typically consists of four copies due to the two ancestral vertebrate tetraploidizations (Nakatani et al., 2007; Putnam et al., 2008). Most of the ray-finned fishes have undergone a third tetraploidization, doubling their chromosome quartets to no less than eight paralogon members for many chromosomal regions (Henkel et al., 2012; Jaillon et al., 2004; Ocampo Daza, Sundstrom, Bergqvist, Duan, & Larhammar, 2011; Widmark, Sundstrom, Ocampo Daza, & Larhammar, 2011). The paralogous genes arising from a tetraploidization are called "Ohnologs" in honor of the evolutionary geneticist Susumu Ohno (1928–2000) who emphasized the importance of gene and genome duplications for evolution of new functions in his now classical book "Evolution by gene duplication" (Ohno, 1970).

The evolutionary scenario for the opioid peptide precursor genes and the opioid receptor genes is reviewed here and receives additional support from our analyses of two key species in studies of vertebrate evolution whose genomes have been recently sequenced. One is the African coelacanth *Latimeria chalumnae* (Amemiya et al., 2013), representing the most basally diverging branch among Sarcopterygii (lobe-finned fishes), the group that also contains the tetrapods including, of course, mammals. The other important species is a basal ray-finned fish, the spotted gar, *Lepisosteus oculatus* (Amores, Catchen, Ferrara, Fontenot, & Postlethwait, 2011), which diverged from the ray-finned fish lineage before the teleost ancestor underwent the third tetraploidization. Therefore, it serves as an excellent point of reference for the teleost genomes that have become highly complex after their third tetraploidization. The gar was found to have a quite stable genome (Amores et al., 2011), i.e., it has fewer chromosomal rearrangements than both teleost fishes and many mammals including human, thereby making it exceptionally useful for studies of vertebrate genome evolution. We also present new results for chicken showing that its genome indeed contains a gene for dynorphin, which was missing in previous genome assemblies. This finding brings the chicken opioid system on a par with that of mammals. Taken together, these results confirm the picture of the opioid system as being of ancient vertebrate origin with all of the major components presently found in humans already established more than 500 million years ago.

2. OPIOID PEPTIDE FAMILY

It has been established for a long time that four opioid peptide precursor genes exist in mammals and most other tetrapods (Dores et al., 2002), except polyploid frogs that have additional copies. Cartilaginous fishes have been found to have three of the prepropeptide genes: *POMC*, *PENK*, and *PDYN* have been cloned in the Port Jackson shark *Heterodontus portusjackson* (Dores, Cameron, Lecaude, & Danielson, 2003; Komorowski, Lecaude, Westring, Danielson, & Dores, 2012). Nociceptin has not yet been reported in a shark, but we have identified the NOP receptor in the elephant shark *Callorhinchus milii* (see below), suggesting that nociceptin can be expected to exist also in cartilaginous fishes. The situation is more complicated in teleosts with zebrafish having two copies of both POMC, PENK, and PNOC, resulting from the teleost tetraploidization (Sundstrom et al., 2010), in agreement with previous cloning of duplicate genes in zebrafish (Gonzalez Nunez, Gonzalez Sarmiento, & Rodriguez, 2003; Gonzalez-Nunez, Gonzalez-Sarmiento, & Rodriguez, 2003a, 2003b). PDYN, in contrast, is found as a single copy in zebrafish. Some teleosts have been found to have three copies of the *POMC* gene, for instance, the barfin flounder (Takahashi, Kobayashi, Amano, & Yamanome, 2009), but also other species, some of which have undergone a fourth tetraploidization (Harris, Dijkstra, & Hofmann, 2014).

Although the number of mature opioid or opioid-like peptides may differ between species for each the precursor genes, it is usually easy to determine which precursor is which by examining the sequence. Phylogenetic analysis of the four genes in multiple vertebrates yields a well resolved tree where each of the four clades has maximum bootstrap support (Sundstrom et al., 2010). This reveals that the four genes had already evolved distinct features before the separation of tetrapods and ray-finned fishes (Sundstrom et al., 2010). Figure 1A–D shows alignments for each of the precursors for chicken, the coelacanth *L. chalumnae*, the spotted gar, zebrafish, and human (Table 1). The two new basal species, the coelacanth and the spotted gar, in many instances serve as intermediates between tetrapods and teleosts, as expected.

The PENK precursor contains seven enkephalins, named A–G (Kilpatrick et al., 1981), in coelacanth and gar like in human, chicken, and zebrafish POMCa (Fig. 1A). As reported previously, zebrafish POMCb has mutated peptide F, which is the only Leu-enkephalin YGGFL in

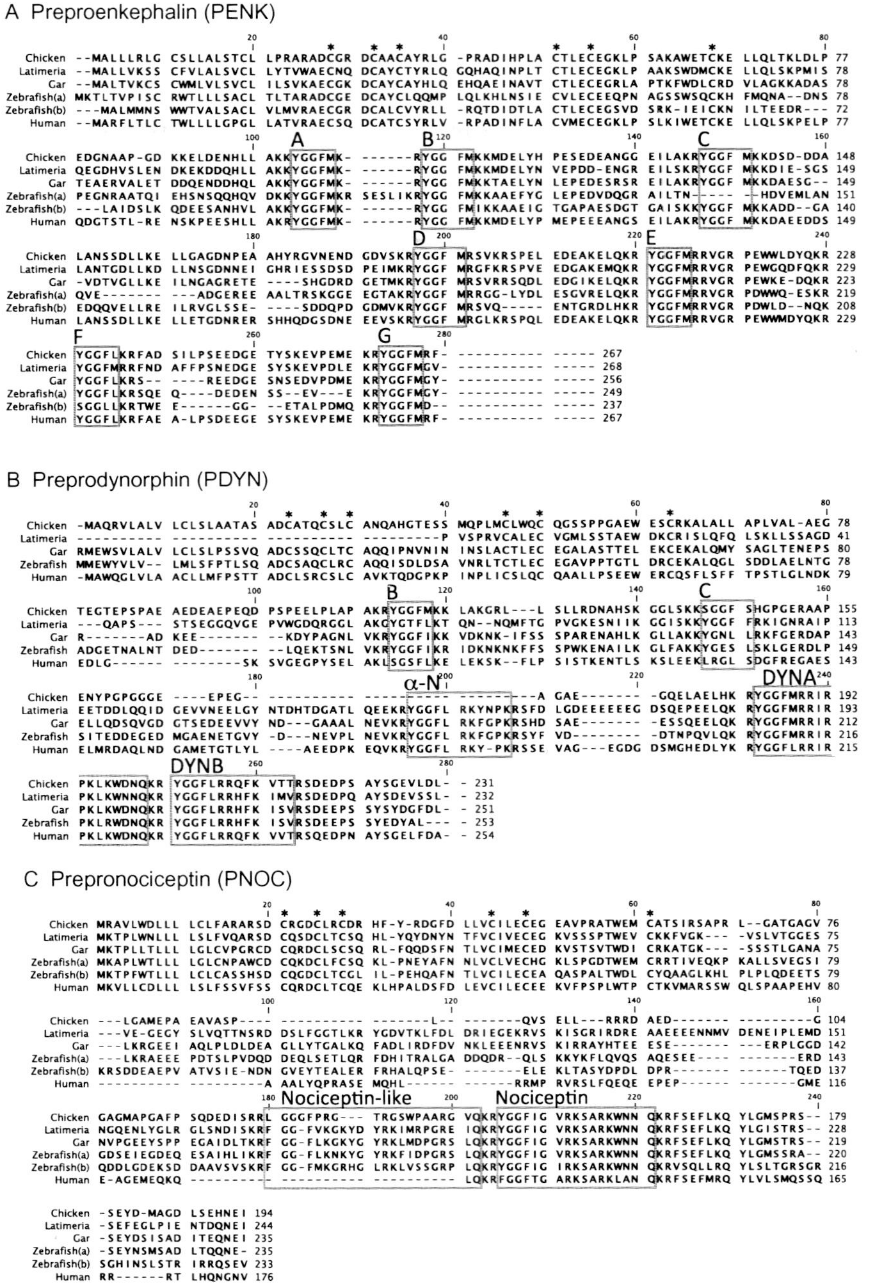

Figure 1 Alignment of prepropeptides for the opioid family precursors. Stars show cysteine residues that are conserved across the precursors. (A) In PENK, seven peptides containing the opioid YGGF core motif are labeled A–G. (B) In PDYN, peptides B and C are named according to the corresponding peptides in PENK. α-N stands for α-neoendorphin, DYNA for dynorphin A, and DYNB for dynorphin B. (C) PNOC has two peptides that contain or resemble the opioid YGGF core motif.

(Continued)

D Preproopiomelanocortin (POMC)

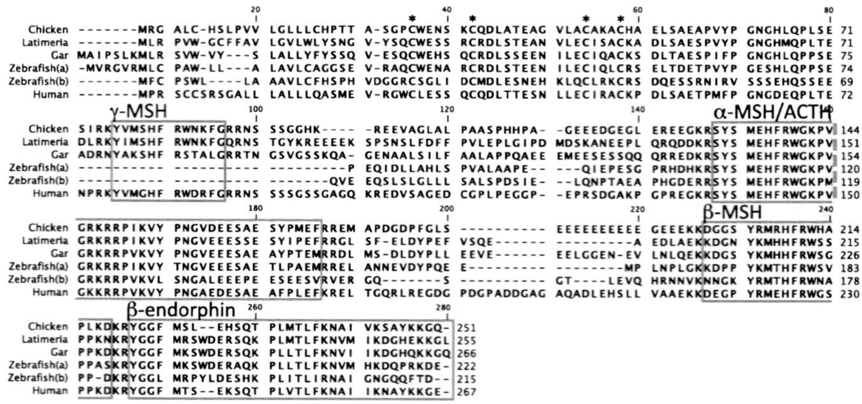

Figure 1—Cont'd (D) POMC has one YGGF core motif at the beginning of β-endorphin. The other marked motifs are variants of MSH (melanocyte-stimulating hormone). The dashed vertical line marks the position where ACTH is cleaved to generate α-MSH (the subsequent G residue is processed to form a carboxyterminal amide). The start of β-MSH is based on the human sequence.

Table 1 Accession codes for opioid prepropeptide sequences from spotted gar, *Lepisosteus oculatus*, and African coelacanth, *Latimeria chalumnae*

Loc, *Lepisosteus oculatus*, spotted gar

	Sequence id	Chromosome: start-stop
pomca	ENSLOCG00000016684	LG1: 43,998,897–44,002,644
penka	ENSLOCG00000005341	LG9: 8,698,989–8,703,661
pdyn	ENSLOCG00000001827	LG18: 2,937,156–2,940,946
pnocb	ENSLOCG00000015967	LG1: 19,306,483–19,327,290

Lch, *Latimeria chalumnae*, African coelacanth

	Sequence id	Chromosome: start-stop
PENK	ENSLACG00000000178	Scaffold JH130364.1: 585–5,576
PDYN	ENSLACG00000000358	Scaffold JH134939.1: 1,581–3,397
POMC	ENSLACG00000011702	Scaffold JH127324.1: 627,671–631,284
PNOC	ENSLACG00000002717	Scaffold JH127452.1: 57,315–75,404

mammals and chicken, to SGGLL (Sundstrom et al., 2010). In the coelacanth, this peptide is YGGFM like all the other enkephalins in the precursor. Thus, the coelacanth cannot produce Leu-enkephalin. Other teleosts, namely, medaka and three-spined stickleback, lack enkephalin C and medaka has mutations in four of the five positions of enkephalin A.

PDYN can give rise to five peptides with the core YGGF motif (Fig. 1B). These are named B, C (because these two motifs correspond to PENK enkephalins B and C), α-neoendorphin (α-N), dynorphin A (DYNA), and dynorphin B (DYNB). Mammals cannot make B and C, but they exist in amphibians and some teleost fishes. As can be seen in Fig. 1B, the PDYN peptide B is unlikely to be cleaved in the coelacanth because it is not flanked by pairs of basic amino acids and furthermore the central amino acid G has mutated to T, but it appears to be cleavable in the gar like in zebrafish and chicken. Peptide C appears to be cleavable in both coelacanth and gar, but both peptides deviate from the common YGGFM or YGGFL sequences: the coelacanth peptide is YGGFF and the gar is YGNLL. Both zebrafish and chicken deviate from the consensus for this peptide, but the species comparison strongly suggests that the ancestral sequence was indeed a cleavable and functional enkephalin, like it is today in frogs (Sundstrom et al., 2010). The third peptide, α-neoendorphin, contains an extension of the core YGGF motif in mammals, with the sequence YGGFRKYPK after cleavage at the following R residue (Kangawa, Minamino, Chino, Sakakibara, & Matsuo, 1981). Both coelacanth and gar seem able to generate to this extended peptide, and both of them have an extra amino acid before the PK at the end, like zebrafish and other teleost fishes (Komorowski et al., 2012). Thus, this suggests that it is the mammalian lineage that has lost this residue. It needs to be confirmed experimentally that coelacanth and gar actually generate this predicted α-neoendorphin variant, and that the carboxyterminal K residue is left uncleaved like in mammals. Finally, the two dynorphin peptides are well conserved with invariant core YGGF motifs and only slight variation in the carboxyterminal parts of the extensions.

Nociceptin has the simplest precursor with just a single YGGF core motif in mammals. Our identification of the *PNOC* gene in coelacanth and gar (Fig. 1C) confirms the high conservation of this peptide reported previously (Sundstrom et al., 2010). Indeed, it is invariant across gar, coelacanth, chicken, and bullfrog, and it has only conservative variation among teleosts. Placental mammals and opossum, in contrast, stand out as clearly divergent, suggesting a distinct route for its evolution in this lineage. Nociceptin is preceded by a similar peptide, called nociceptin-like, in the non-mammalian species whereas mammals have lost this stretch which is also quite divergent in chicken. Nociceptin-like starts with the variant core motif FGGF, like mammalian nociceptin, and has a slightly longer extension and is more variable than nociceptin.

POMC is the most complex of the four precursors due to the insertion of ACTH and other melanocortin-containing motifs. Here, we shall focus on its opioid component, endorphin (endorphin refers to β-endorphin; α- and γ-endorphin are carboxyterminally truncated variants of β-endorphin). Figure 1D shows that the opioid core motif YGGF as well as the following amino acid M are perfectly conserved in coelacanth and gar as well as all other species investigated except some teleost fish POMCb sequences which may even have become nonfunctional (Sundstrom et al., 2010). Immediately after the opioid core motif, the coelacanth, gar, and teleost POMCa-derived sequences have high identity to each other (the sequence KSWD) whereas tetrapods have lost the WD residues, see Fig. 1D and Sundstrom et al. (2010). The ancestrality of WD is shown by its presence in both coelacanth and gar, representing early diverging branches in the lobe-finned and ray-finned lineages, respectively, as well as in the distantly related cartilaginous fishes spiny dogfish, *Squalus acanthias* (Amemiya, Takahashi, Suzuki, Sasayama, & Kawauchi, 1999) and ratfish, *Chimaera phantasma* (Takahashi et al., 2004), whose ancestors diverged from each other approximately 420 million years ago (Inoue et al., 2010). The loss of WD happened in the tetrapod ancestor before the divergence of amphibians and amniotes; see Sundstrom et al. (2010).

The POMC precursor also contains ACTH (adrenocorticotropic hormone), the first part of which constitutes α-MSH (melanocyte-stimulating hormone). POMC also contains additional MSH-like copies, the number of which differs between vertebrate lineages. These peptides are called β, γ, and δ-MSH, the last-mentioned of which is only present in cartilaginous fishes. They are marked in Fig. 1D but they bind to a separate family of GPCRs, the melanocortin receptors (Cortes et al., 2014) why they are beyond the scope of this review. Suffice to say that the complex POMC precursor probably arose when a DNA segment encoding an MSH ancestor was inserted into a preproendorphin gene, as both the aminoterminal part with four conserved cysteines and carboxyterminal part encoding endorphin bear clear resemblance to the other three opioid precursors. Subsequently, the region encoding MSH probably duplicated resulting in the additional MSH variants.

PENK, PDYN, and PNOC all have the six aminoterminal cysteines perfectly conserved also in coelacanth and gar (except that the region encoding the first three cysteines is missing for coelacanth PDYN due to incomplete database entry), and POMC has its four cysteines perfectly conserved also in these species. Presumably, these residues form disulfide bonds that make the precursor form the three-dimensional folds necessary for cleavage and/or transport.

The chromosomal locations of the opioid precursor genes will be discussed below in conjunction with the locations of the opioid receptor genes.

3. OPIOID RECEPTOR FAMILY

Our previous studies showed that the four receptor types arose early in the evolution of the vertebrates, before the divergence of ray-finned and lobe-finned fishes (including tetrapods). The phylogenetic tree shown in Fig. 2 includes novel information from the evolutionarily important and slowly evolving species African coelacanth and spotted gar, along with human, chicken, and western clawed frog and the two teleost fish species zebrafish and three-spined stickleback (Table 2). The tree was calculated with the maximum likelihood method (Guindon et al., 2010). All four receptor types could be identified in coelacanth and gar, thus confirming that all four receptor types existed before the divergence of ray-finned and lobe-finned fishes. This is consistent with origin of the four receptor types before the radiation of gnathostomes.

The tree indicates that NOP has had a higher evolutionary rate, not only basally and in teleost fishes, but also in the lineage leading to humans (Fig. 2), including the basal mammalian lineage (Dreborg et al., 2008). Thus, NOP in mammals may not be representative for this ortholog in other vertebrate lineages. This is consistent with the fact that the opioid core motif of nociceptin is divergent in mammals, FGGF (Fig. 1C) rather than the original YGGF found in enkephalins, dynorphins, and endorphin.

Note that there are two copies of the gene for OPRK1 in stickleback, named a and b, and that OPRD1 is present in two copies in both zebrafish and stickleback (Figs. 2–4) as a result of the teleost tetraploidization (Dreborg et al., 2008). The teleost duplications may explain why the OPRK1 and OPRD1 receptors appear to evolve faster in this lineage than in the other vertebrates. This may also imply ongoing specialization of these receptor duplicates in teleost fishes.

In addition to the sequence similarity among the four opioid receptor types, they also share the same basic exon–intron organization with two introns at corresponding positions in the coding region, the first in the

Table 2 Accession codes for opioid receptor family sequences used for phylogenetic analysis and chromosome comparisons, including the five human somatostatin sequences used as outgroup to root the tree. Species abbreviations as in Figure 2

Species	Assigned sequence name	Gene ID	Chromosome: start position	Name in database
Hsa	Hsa OPRD1	ENSG00000116329	1: 29.14mb	OPRD1
	Hsa OPRM1	ENSG00000112038	6: 154.33mb	OPRM1
	Hsa OPRK1	ENSG00000082556	8: 54.14mb	OPRK1
	Hsa OPRL1	ENSG00000125510	20: 62.71mb	OPRL1
Gga	Gga OPRK1	ENSGALG00000015269	2: 109.85mb	OPRK1
	Gga OPRM1	ENSGALG00000013616	3: 49.32mb	OPRM1
	Gga OPRL1	ENSGALG00000027965	20: 9.60mb	OPRL1
	Gga OPRD1	ENSGALG00000026379	23: 2.11mb	OPRD1
Xtr	Xtr OPRM1	ENSXETG00000008401	GL172710.1: 773.02 kb	oprm1
	Xtr OPRK1	ENSXETG00000008092	GL172719.1: 1711.15 kb	oprk1
	Xtr OPRD1	ENSXETG00000033842	GL173114.1: 852.44 kb	OPRD1
	Xtr OPRL1	ENSXETG00000011546	GL174035.1: 63.99 kb	oprl1
Lch	Lch OPRD1	ENSLACG00000003077	Scaffold JH126714.1: 69.39kb	OPRD1
	Lch OPRL1	ENSLACG00000010572	Scaffold JH127873.1: 515.43kb	OPRL1
	Lch OPRK1	ENSLACG00000005041	Scaffold JH129996.1: 149.28kb	OPRK1
	Lch OPRM1		Scaffold JH132274.1:22.24kb	
Dre	Dre OPRK1	ENSDARG00000006894	2: 30.59mb	oprk1
	Dre OPRM1	ENSDARG00000039434	13: 24.07mb	oprm1
	Dre OPRD1b	ENSDARG00000037159	16: 36.58mb	oprd1b
	Dre OPRD1a	ENSDARG00000041660	19: 14.35mb	oprd1a
	Dre OPRL1	ENSDARG00000071209	23: 8.38mb	oprl1
Loc	Loc OPRM1	ENSLOCG00000015439	LG16: 1052.58kb	oprm1
	Loc OPRL1	ENSLOCG00000005285	LG18: 10.92mb	oprl1
	Loc OPRD1	ENSLOCG00000003690	LG6: 6.67mb	oprd1b
	Loc OPRK1	ENSLOCG00000005766	LG9: 9.50mb	oprk1

Table 2 Accession codes for opioid receptor family sequences used for phylogenetic analysis and chromosome comparisons, including the five human somatostatin sequences used as outgroup to root the tree. Species abbreviations as in Figure 2—cont'd

Species	Assigned sequence name	Gene ID	Chromosome: start position	Name in database
Gac	Gac OPRK1a	ENSGACG00000017114	groupIII: 12.96mb	oprk1
	Gac OPRM1	ENSGACG00000010365	groupVI: 12.88mb	oprm1
	Gac OPRD1a	ENSGACG00000007694	groupX: 11.57mb	oprd1a
	Gac OPRL1	ENSGACG00000010479	groupXII: 13.26mb	oprl1
	Gac OPRD1b	ENSGACG00000007300	groupXX: 8.22mb	oprd1b
	Gac OPRK1b	ENSGACG00000004282	groupXXI: 9.63mb	OPRK1 (1 of 2)
Hsa	Hsa SSTR1	ENSG00000139874	14: 38.68mb	SSTR1
	Hsa SSTR5	ENSG00000162009	16: 1122.76kb	SSTR5
	Hsa SSTR2	ENSG00000180616	17: 71.16mb	SSTR2
	Hsa SSTR4	ENSG00000132671	20: 23.02mb	SSTR4
	Hsa SSTR3	ENSG00000183473	22: 37.60mb	SSTR3

stretch encoding intracellular loop 1 and the second in extracellular loop 2. The *OPRM1* gene in human, rat and mouse has additional exons in the 5′ and 3′ regions that can be used alternately, thereby generating a number of receptor variants differing in the aminoterminal and carboxyterminal parts (Pasternak & Pan, 2013). Some of these variants differ in receptor trafficking whereas others are nonfunctional. It is still unknown to what extent this type of variation can occur for the *OPRM1* gene in other species, or for the other three receptor genes in a broader evolutionary perspective.

The mechanism by which the four opioid receptor genes arose from their common ancestor was found by us to be chromosome duplications (Dreborg et al., 2008) that most likely coincide with the basal vertebrate tetraploidizations. As shown in Fig. 3, this conclusion was based on the observation that all four chromosomes in chicken that encode opioid receptors (shown in yellow) also share members of three adjacent gene families. The most parsimonious explanation for this similarity between chromosomes is that they all arose by duplication of single ancestral chromosome having the ancestors of

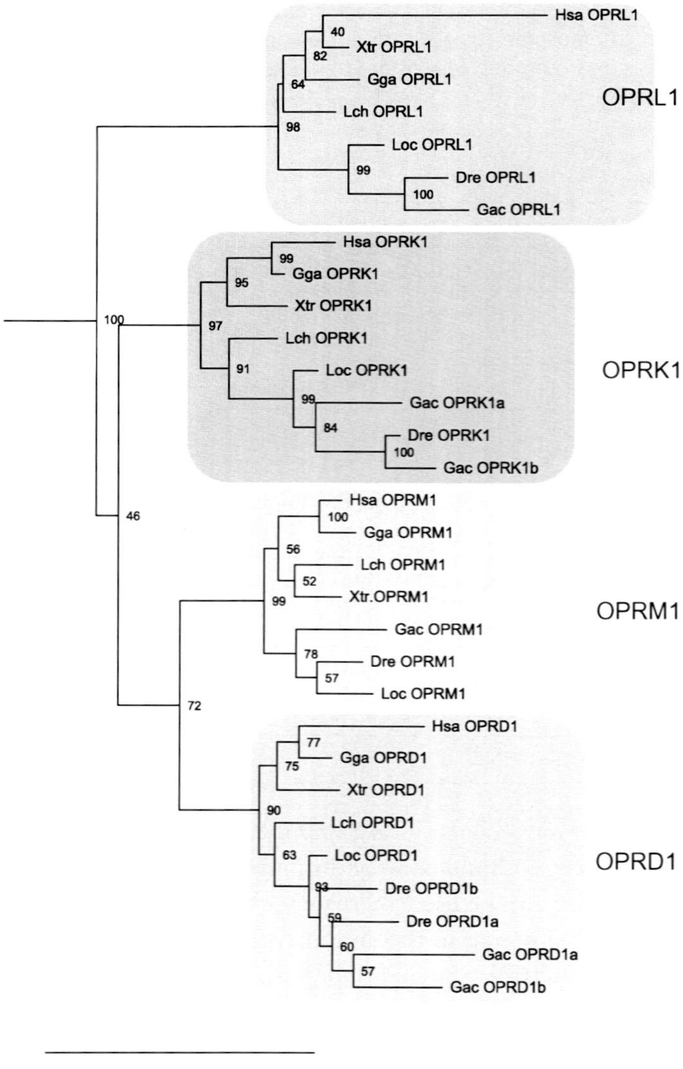

Figure 2 Phylogenetic tree for the opioid receptor family calculated with the maximum likelihood method using the PhyML3.0 algorithm (Guindon et al., 2010). The background colors mark the four receptor clades. The tree was rooted with the five human somatostatin receptors. OPRD1 is the delta receptor (DOP), OPRK1 is kappa (KOP), OPRM1 is mu (MOP), and OPRL1 is the nociceptin (NOP) receptor. Species abbreviations: Hsa, *Homo sapiens*; Gga, *Gallus gallus*, chicken; Xtr, *Xenopus* (Silurana) *tropicalis*, western clawed frog; Lch, *Latimeria chalumnae*, African coelacanth; Dre, *Danio rerio*, zebrafish; Gac, *Gasterosteus aculeatus*, three-spined stickleback; Loc, *Lepisosteus oculatus*, spotted gar. Note that OPRK1 is present in two copies in Gac and that OPRD1 is present in two copies in both Dre and Gac. The alignment of the opioid receptor sequences used to calculate the phylogenetic tree in Figure 2 can be obtained from the corresponding author upon request. (See the color plate.)

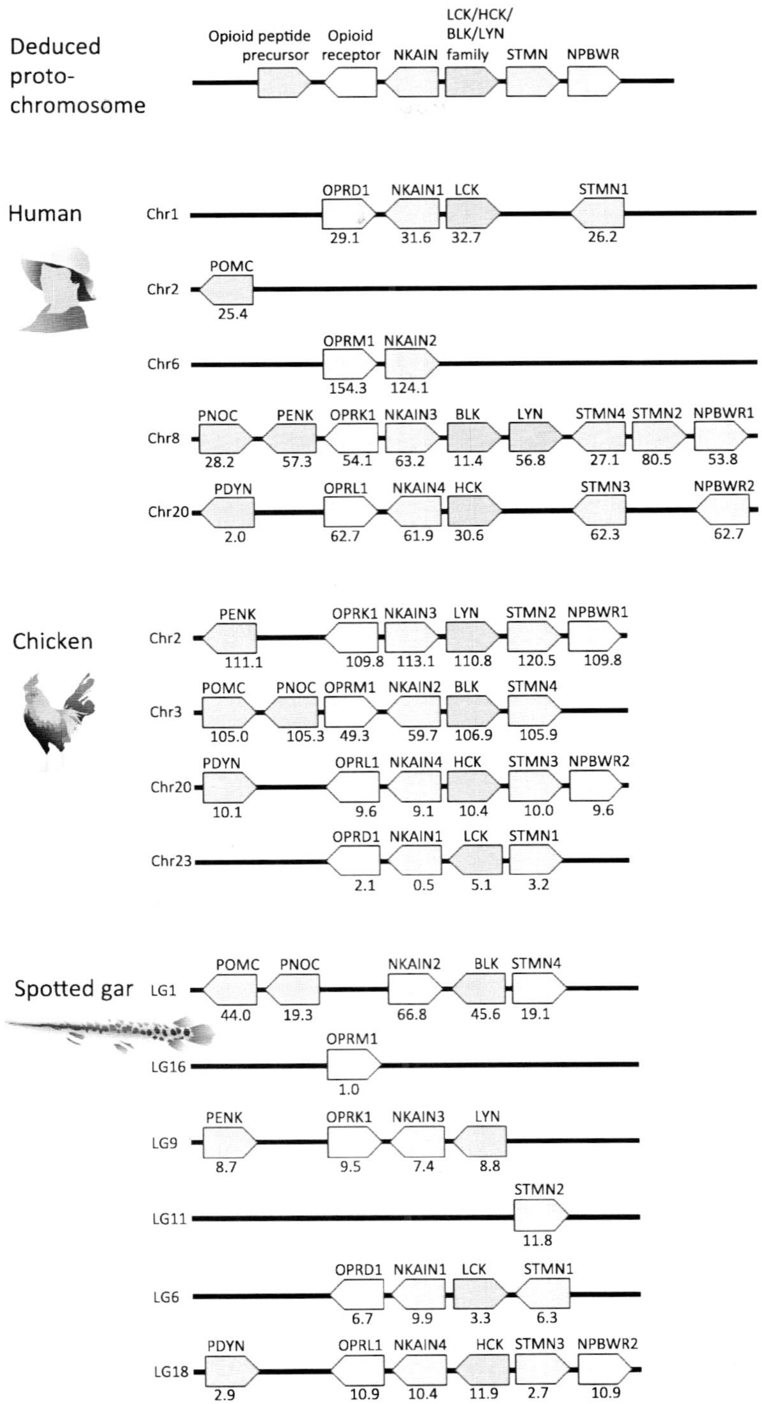

Figure 3 See legend on next page.

these gene families. The deduced ancestral chromosome, in a postulated vertebrate predecessor, is shown at the top of the figure.

In the human genome, however, the chromosome with OPRM1 (chromosome 6) lacks representatives for two of the gene families. We were initially surprised to find that these genes (*BLK* and *STMN4*) are syntenic with OPRK1, suggesting local gene duplications, but after comparing the chromosomal locations of the genes in the genomes of chicken, gray short-tailed opossum (*Monodelphis domestica*) and dog, we concluded that these two genes had been translocated in two steps before they ended up on human chromosome 8 (Dreborg et al., 2008). Nevertheless, with the new more distantly related genomes, we were curious to see the locations of these genes in the spotted gar. Figures 3 and 4 shows that also the spotted gar has a translocation involving this chromosome, but here it seems that it is the OPRM1 receptor gene that has been translocated to LG16 (LG stands for linkage group) whereas all the others are located on LG1. Thus, this chromosome has undergone rearrangements independently in the lineages leading to gar and mammals, while the chromosome has been maintained intact with regard to these genes in chicken. Like the spotted gar, chicken is also known to have a comparatively stable genome, especially concerning interchromosomal rearrangements (Bourque, Zdobnov, Bork, Pevzner, & Tesler, 2005).

We have also identified the four opioid receptors in the genome of a cartilaginous fish, *C. milii*, called elephant shark or Australian ghostshark (data not shown). This species belongs to the subclass *Holocephali* which diverged from the ancestor of *Elasmobranchii* (sharks, rays, and skates) approximately 420 million years ago; see Inoue et al. (2010). The four elephant shark opioid sequences cluster in each of the four opioid receptor clades with high bootstrap support, thus confirming that the four receptor types were already

Figure 3 Chromosomal regions for the four opioid peptide precursor genes and the four opioid family receptor genes in human, chicken, and spotted gar. Horizontal lines symbolize chromosomes. The top line shows the deduced protochromosome prior to the two basal vertebrate tetraploidizations. Four neighboring gene families are shown: NKAIN, the Na^+/K^+ transporting ATPase interacting family; the LCK/HCK/BLK/LYN family of Src-related tyrosine kinases; STMN, stathmin-like proteins; NPBWR, neuropeptide B/W receptors. Each gene family is marked with a color or shade of gray. Chr stands for chromosome, LG stands for linkage group. The numbers below the gene boxes show positions in megabases. The orientation of each gene box is indicated by its pointed end, pointing in the 5′-to-3′ direction, and the relative orientations of the genes within each chromosome have been maintained although the gene order has been modified to highlight the similarities. The order and orientations of the genes on the deduced protochromosome are speculative as these may have shifted multiple times during evolution. (See the color plate.)

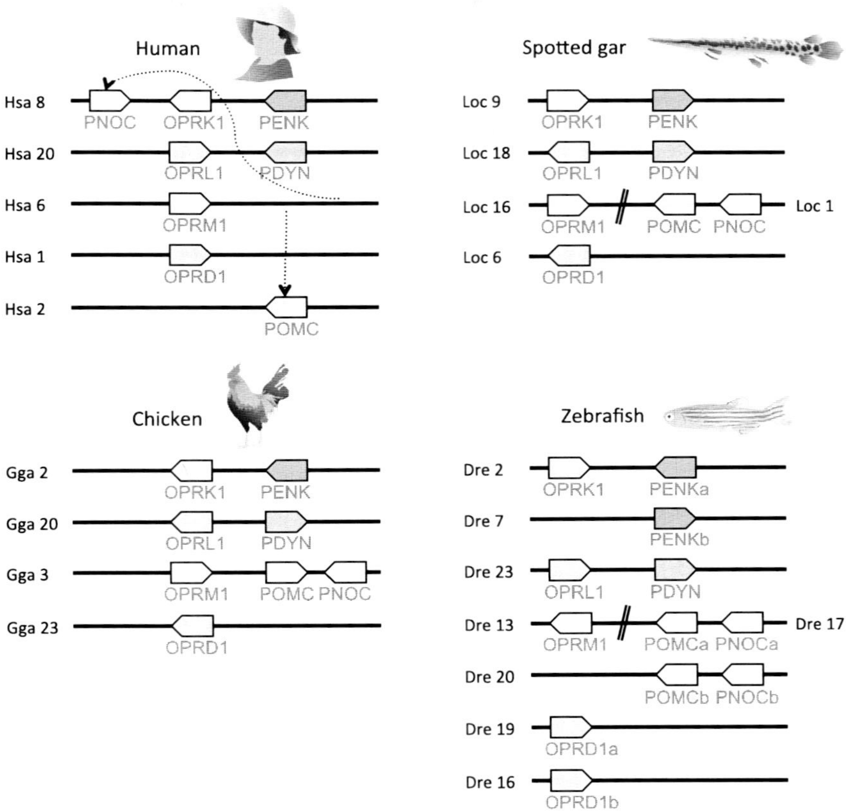

Figure 4 Chromosomal regions for the opioid peptide precursor genes and the opioid family receptor genes in human, chicken, spotted gar, and zebrafish. Horizontal lines symbolize chromosomes. The figure is a modified version of the figure shown by Sundstrom et al. (2010) with addition of PDYN in chicken and the complete repertoire of opioid genes in the spotted gar. (See the color plate.)

established before the radiation of the jawed vertebrates (*Gnathostomata*). As the present assembly of the elephant shark genome consists of fairly short scaffolds (Venkatesh et al., 2014), it has not been possible to investigate conserved synteny for the genes shown in Fig. 3. Therefore, we have not included the elephant shark receptor sequences in Figs. 2–4. Nevertheless, we have been able to confirm conserved synteny for a few genes, thereby supporting the scenario outlined in Fig. 3.

In our initial studies of the opioid receptor family, we found that the genes for two closely related receptors are located in close proximity to two of the opioid family receptor genes in the human genome. These receptors are activated by two peptides that are related to each other, neuropeptide B and neuropeptide W, but seem to be unrelated to the opioid peptides.

The receptor gene *NPBWR1* is adjacent to OPRK1 on human chromosome 8 and the gene for NPBWR2 is close to OPRL1 on chromosome 20. In chicken, only one of the two genes was found: NPBWR1 was likewise found very close to OPRK1. In later chicken genome assemblies, we were able to identify also NPBWR2, and it is indeed close to OPRL1 also in chicken. In the spotted gar genome, we have located NPBWR2 in immediate proximity of PRL1, but no NPBRW1 has been found in any ray-finned fish, strongly suggesting that it was lost early in this lineage.

The high similarity and close synteny of the two NPBW receptor genes with the opioid receptor genes corroborates our gene duplication scenario that they share a common ancestor that underwent a local duplication shortly before the vertebrate tetraploidizations (Dreborg et al., 2008). In the two tetraploidizations, the opioid receptor gene was quadrupled whereas the *NPBWR* gene has only retained two copies in tetrapods and coelacanth, and only one copy in ray-finned fishes. The two *NPBWR* genes lack introns in the coding region, suggesting either that the two opioid receptor introns were inserted after the local gene duplication, or that the ancestral *NPBWR* gene arose by retrotranscription of and opioid receptor mRNA and reinsertion close to the ancestral opioid receptor gene. How the NPBWR found its opioid-unrelated peptide ligand remains to be explored.

Many families of GPCRs for peptides are now known to have undergone duplications in the two basal vertebrate tetraploidizations and then additional duplications in the teleost-specific tetraploidization. GPCR families that have expanded in this way include receptors for neuropeptide Y and its related peptides (Larhammar & Bergqvist, 2013; Larsson et al., 2008, 2009), endothelins (Braasch & Schartl, 2014; Braasch, Volff, & Schartl, 2009), somatostatin and urotensin II (Ocampo Daza, Sundstrom, Bergqvist, & Larhammar, 2012; Tostivint et al., 2014), oxytocin and vasopressin (Lagman et al., 2013; Ocampo Daza, Lewicka, & Larhammar, 2011; Yamaguchi et al., 2012), and several more. The tetraploidizations also led to duplications of many peptide precursor genes as well as signal transduction components. Thus, it is clear that the early stages of vertebrate evolution involved a dramatic increase in the number of genes encoding components that are involved in intercellular as well as intracellular communication. It is tempting to speculate that these events paved the way for evolution of many vertebrate novelties, including more complicated nervous and endocrine signaling systems, by generating duplicates of many genes that could evolve novel or more highly specialized functions.

Recently, all four opioid receptor types have been crystallized and their structures determined at 2.8–3.4 Å resolution. Each one was crystallized in complex with an antagonist. The kappa and NOP receptors were from human (Thompson et al., 2012; Wu et al., 2012) and the delta and mu receptors were from mouse (Granier et al., 2012; Manglik et al., 2012). The outer part of the binding pocket differed substantially between the receptor types which helps explain their ligand selectivities. They have large binding pockets in agreement with ability to bind ligands with widely different structures and, for the NOP receptor, the large NOP peptide. It will now be possible to explore if these structural differences between the receptor types are conserved across species by examining alignments that include sequences for many different species. This will give rise to hypotheses that can be tested by receptor mutagenesis and binding experiments.

4. DISCUSSION: COMPLEXITY, COEVOLUTION, AND DIVERGENCE

The phylogenetic and chromosomal analyses show that the opioid system had already considerable complexity at a very early stage in vertebrate evolution, indeed at the dawn of the vertebrates, and expanded further in the teleost fish lineage while the number of genes has remained constant in most tetrapods. The additional opioid genes in the teleost fishes make these species more complicated than mammals and other tetrapods, at least regarding the number of genes. The independent fourth tetraploidizations that have taken place in for instance the lineages leading to common carp and salmonid fishes predict even greater complexity in these species.

This mode of gene duplication, as a result of chromosome doublings, most likely means that the regulatory regions were initially identical for the duplicates. Subsequent divergence in the regulatory regions of the genes has lead to different expression patterns. Mutations leading to structural changes in the peptides and proteins have resulted in altered binding properties regarding ligand–receptor selectivity and may have also modified the interactions with other components such G proteins. Changes in interactions, regulation, expression level, distribution, and developmental timing all represent possible routes for neofunctionalization (novel functions) and subfunctionalization, i.e., more highly specialized functions in comparison with the ancestral gene. The similarities in expression pattern that are observed today presumably reflect remaining overlap that can be explained by the mode of duplication of the whole chromosomal regions. Areas of future investigation include comparisons across species of anatomical distribution and temporal regulation during development, in addition to the studies already performed in mammals and zebrafish (Gonzalez-Nunez & Rodriguez, 2009).

Three of the peptide precursor genes were clearly present on separate chromosomes after the two tetraploidizations (Fig. 3), but the origin of the fourth propeptide is not that obvious. PNOC and POMC are located close together on chicken chromosome 3, suggesting a local duplication. The time point for this duplication has not been possible to determine from the available genomes why three alternative scenarios were proposed leading to the same end result (Sundstrom et al., 2010). However, these two genes are much further apart in the spotted gar and they are on separate chromosomes in human (Figs. 3 and 4). We have therefore checked the distance in several other species and the two genes are actually close together in several other species including some teleost fishes, and in the softshell turtle they are just over 1 megabase apart. This supports origin by a local duplication event. Another possibility is that PNOC and POMC were on two separate chromosomes after the second tetraploidization, and that they were brought together by a translocation event (unequal crossing-over) while the chromosomes were still very similar to one another after the tetraploidizations.

Our findings also imply that a DNA segment encoding ACTH/melanocortin, or a larger segment encoding multiple melanocortin copies, was inserted into the preproendorphin gene as late as after the two tetraploidizations, because it was only then that a distinct preproendorphin had come into existence, and it was this fusion event that resulted in POMC. Of course a gene encoding ACTH/melanocortin could have arisen separately much earlier, as implied by the fact that the genes encoding the melanocortin receptors also arose as a result of the tetraploidizations from a single ancestral gene that existed before the tetraploidizations (Cortes et al., 2014).

In mammals, the binding properties of the three cloned opioid receptors were quickly determined for a range of native and exogenous compounds (Kieffer, 1995). Among the native opioid peptides, endorphin has a clear preference for MOP while DYNA has a 100-fold preference for KOP. The enkephalins bind with slightly higher affinity to DOP than to MOP and with orders of magnitude lower affinity to KOP. Nociceptin has strong preference for the OPRL1 receptor (Meunier et al., 1995; Reinscheid et al., 1995). For a summary of the binding properties for the native ligands, see fig. 4 in Larhammar et al. (2013).

Because the four receptor types and at least three of the peptide precursors arose in the two tetraploidization events over 500 million years ago, there has been considerable time for them to evolve distinct characteristics. It is therefore possible that they have diverged differentially in the vertebrate classes, evolving ligand-receptor preferences that differ between the lineages. However, to our knowledge, only a single non-mammalian species has so far been thoroughly investigated with regard to peptide–receptor interactions for its native opioid system, namely, the zebrafish. All five receptors, including both copies of OPRD1, have been cloned and expressed by Dr. Raquel E. Rodriguez and coworkers (Alvarez et al., 2006; Barrallo, Gonzalez-Sarmiento, Porteros, Garcia-Isidoro, & Rodriguez, 1998; de Velasco, Law, & Rodriguez, 2009; Pinal-Seoane et al., 2006; Rivas-Boyero et al., 2011), as reviewed in Gonzalez-Nunez and Rodriguez (2009). Zebrafish DYNA has 10-fold higher affinity for KOP than MOP or the two DOP receptors (Gonzalez-Nunez, Marron Fernandez de Velasco, Arsequell, Valencia, & Rodriguez, 2007). Thus, DYNA appears to be less selective for KOP in zebrafish than in human. Met-enkephalin and Leu-enkephalin were investigated for the two DOP receptors and have threefold higher affinity for DOPb (called DOR2 in this chapter) than for DOPa (Gonzalez-Nunez & Rodriguez, 2009), but the statistical significance was not reported nor were the affinities for MOP and KOP. The results for endorphin cannot be easily compared because they are presented as either K_i or EC50 and it is not clear if zebrafish or human endorphin was used. Morphine has approximately 40-fold higher affinity for MOP over DOPa while DOPb and KOP are three orders of magnitude lower than MOP. Thus, the morphine profile resembles the situation in human. The dramatic difference in morphine binding between DOPa and DOPb can probably shed further light on the specific interactions of this agonist with the receptors. Interestingly, the NOP receptor bound not only human nociceptin (which differs from the two zebrafish nociceptins at 4 of the 17 positions, see Fig. 1C) but also DYNA as well as the antagonist naloxone (Rivas-Boyero et al., 2011). DYNA bound with a nanomolar affinity to NOP, thus higher affinity than to the classical opioid receptors (Gonzalez-Nunez & Rodriguez, 2009) and it acted as a partial agonist (Rivas-Boyero et al., 2011). The authors concluded that the zebrafish NOP receptor displays mixed characteristics of mammalian NOP and KOP receptors, which supports the observation in Fig. 1 that it is the human/mammalian NOP receptor (*OPRL1* gene) that has become more divergent due to increased evolutionary rate.

All four receptors have also been cloned and expressed from an amphibian, the rough-skinned newt *Taricha granulosa* (Bradford, Walthers, Searcy, & Moore, 2005; Bradford et al., 2006; Walthers, Bradford, & Moore, 2005), in which opioids had been tested in the 1980s and found to influence locomotion and reproductive behavior and in other amphibians opioids are known to influence nociception. However, as the newt opioid receptors were characterized using human β-endorphin and (presumably) human nociceptin, which can be expected to differ from the native newt peptides (the sequences of these are still unknown), physiological peptide–receptor preferences must be interpreted with caution. Human β-endorphin had approximately fivefold higher affinity for the newt MOP receptor over the DOP (KOP was not investigated). Met-enkephalin had somewhat higher affinity for DOP than for MOP, whereas it was the other way around for Leu-enkephalin (again, KOP was not investigated). DYNA (presumably human) bound to KOP with sub-nanomolar affinity like in human. Human nociceptin, finally, bound to the newt NOP receptor. Like in zebrafish, DYNA (residues 1–13 out of 17, the human peptide) bound with high affinity to newt NOP, again resembling the KOP receptor.

The crystal-based structural model of the human NOP receptor (Thompson) showed that the residues A5.39, Q6.52, and T7.39 have their side chains pointing into the binding pocket and that the corresponding positions in the other three opioid receptors are K, H, and I, respectively. While position 6.52 is indeed a Q also in the zebrafish and newt NOP receptors, like in human NOP, the two other positions differ, leaving human NOP unique in having A6.52 (G in all other NOP except mammals) and T7.39 (I in all other NOP except mammals and chicken, I or V in all other opioid receptors). Thus, there seems to have been a stepwise process where first I7.39 changed to T in an amniote ancestor and then G6.52 changed to an A in a mammalian ancestor. The intermediate chicken NOP would therefore be interesting to investigate with relevant ligands.

Taken together, these binding studies suggest that the properties of NOP in mammals stand out as atypical and that NOP in non-mammals is more like the classical opioid receptors. These results support the classification of the NOP receptor as a member of the opioid receptor family by IUPHAR (http://www.guidetopharmacology.org/GRAC/FamilyIntroduction Forward?familyId=50). Indeed, from an evolutionary temporal point of view nociceptin emerges as a true member of the opioid peptide family, as the time point for its origin coincides with the origin of the other opioid precursors in the same way that the NOP receptor is of equal age as the DOP, KOP, and MOP receptors. Therefore, by widening the perspective beyond the (placental) mammals, there is no doubt that nociceptin and its receptor can be considered to be full-fledged members of the opioid system, albeit the roles in placental mammals may have become somewhat more specialized or divergent.

As the ancestor of the vertebrates, prior to the two tetraploidizations, had one opioid peptide precursor gene and one receptor gene, related sequences have been sought also in invertebrates. However, despite extensive searches no promising candidate genes have yet been found, neither in deuterostomes nor in protostomes. Two extensive searches for peptides and their receptors in multiple invertebrate genomes were reported in 2013. The first study claimed to have found opioid peptides in three protostome species, namely, in an annelid worm and two molluscs, by alignment with vertebrate PENK (Jekely, 2013). The finding of enkephalin in the annelid was also reported by the same author and colleagues in a separate publication (Conzelmann et al., 2013). However, the similarities with the vertebrate opioid core motif YGGF are modest and the peptide extensions quite divergent. Furthermore, the two mollusc sequences do not even display sequence conservation relative to each other in these regions. The annelid sequence contains three YGG motifs but there are also seven YGSL motifs that resemble one another more than they look like any of the vertebrate motifs, therefore probably reflecting an independent expansion by internal duplications. No candidate opioid receptor was found in this study (Jekely, 2013). The second report searched 15 bilaterian genomes, including four invertebrate deuterostomes, for peptides and peptide receptors (Mirabeau & Joly, 2013). Two candidate sequences were shown, obtained from two tunicate genomes. However, the similarity with YGGF was very faint indeed, and furthermore, the two tunicate sequences were quite divergent from each other. The two cysteines at the beginning of one of the sequences appear to be present in the signal peptide. These studies and other extensive searches in invertebrate genomes, without finding any reasonable opioid candidates, lead to the conclusion that the previously reported peptide sequences and POMC precursor sequences for invertebrates (see Sundstrom et al., 2010 for some of the references) are incorrect because no related sequences can be found in any of the searched invertebrate genomes. Thus, until a candidate peptide and a candidate receptor can be found that interact with one another, it remains an open question when the first opioid peptide arose and how it found its GPCR. For the time being, the opioid system components appear to be unique to vertebrates. Also the related NPBW receptor and its peptide ligands seem to lack orthologous sequences in invertebrates.

Regardless when the first opioid peptide and its receptor arose or began to interact with one another, our studies show that both the ancestral opioid peptide gene and its receptor gene were located on the same vertebrate protochromosome before the first tetraploidization. This suggested to us that they may initially have been coselected to interact with each other (Larhammar et al., 2013; Sundstrom et al., 2010). Apparently, any such coselection by synteny was subsequently relieved, because the many family members that exist after the two tetraploidizations are located in many separate chromosomal regions.

5. CONCLUSIONS

The findings that the opioid peptide precursor genes and the opioid receptor genes were duplicated as a result of chromosome duplications means that their origins can be tied to specific events in the evolution of the vertebrate ancestor, namely, the two genome doublings. The time points for these events are still somewhat uncertain, but most likely they took place between 550 and 500 million years ago, after the vertebrate ancestor had diverged from the tunicate ancestor. The absence of clear opioid and receptor orthologs in invertebrates suggests that the system arose *de novo* in the vertebrate lineage, possibly by the appearance of a peptide that could bind to an already existing member of the rhodopsin-like GPCRs. The quadruplication of both the peptide precursor gene and the receptor gene generated opportunities for the components to diverge by subfunctionalization and/or neofunctionalization. Given the origin by chromosome duplication, the functional overlap between the components still seen after eons of time is not surprising. Additional gene duplicates arose in the teleost fish tetraploidization, implying even greater complexity than in mammals.

ACKNOWLEDGEMENT

This work was supported by a grant from the Swedish Research Council.

REFERENCES

Alvarez, F. A., Rodriguez-Martin, I., Gonzalez-Nunez, V., Marron Fernandez de Velasco, E., Gonzalez Sarmiento, R., & Rodriguez, R. E. (2006). New kappa opioid receptor from zebrafish Danio rerio. *Neuroscience Letters*, *405*, 94–99.

Amemiya, C. T., Alfoldi, J., Lee, A. P., Fan, S., Philippe, H., Maccallum, I., et al. (2013). The African coelacanth genome provides insights into tetrapod evolution. *Nature*, *496*, 311–316.

Amemiya, Y., Takahashi, A., Suzuki, N., Sasayama, Y., & Kawauchi, H. (1999). A newly characterized melanotropin in proopiomelanocortin in pituitaries of an elasmobranch, Squalus acanthias. *General and Comparative Endocrinology*, *114*, 387–395.

Amores, A., Catchen, J., Ferrara, A., Fontenot, Q., & Postlethwait, J. H. (2011). Genome evolution and meiotic maps by massively parallel DNA sequencing: Spotted gar, an outgroup for the teleost genome duplication. *Genetics*, *188*, 799–808.

Barrallo, A., Gonzalez-Sarmiento, R., Porteros, A., Garcia-Isidoro, M., & Rodriguez, R. E. (1998). Cloning, molecular characterization, and distribution of a gene homologous to delta opioid receptor from zebrafish (Danio rerio). *Biochemical and Biophysical Research Communications*, *245*, 544–548.

Bourque, G., Zdobnov, E. M., Bork, P., Pevzner, P. A., & Tesler, G. (2005). Comparative architectures of mammalian and chicken genomes reveal highly variable rates of genomic rearrangements across different lineages. *Genome Research*, *15*, 98–110.

Braasch, I., & Schartl, M. (2014). Evolution of endothelin receptors in vertebrates. *General and Comparative Endocrinology* [Epub ahead of print].

Braasch, I., Volff, J. N., & Schartl, M. (2009). The endothelin system: Evolution of vertebrate-specific ligand-receptor interactions by three rounds of genome duplication. *Molecular Biology and Evolution, 26*, 783–799.

Bradford, C. S., Walthers, E. A., Searcy, B. T., & Moore, F. L. (2005). Cloning, heterologous expression and pharmacological characterization of a kappa opioid receptor from the brain of the rough-skinned newt, Taricha granulosa. *Journal of Molecular Endocrinology, 34*, 809–823.

Bradford, C. S., Walthers, E. A., Stanley, D. J., Baugh, M. M., & Moore, F. L. (2006). Delta and mu opioid receptors from the brain of a urodele amphibian, the rough-skinned newt Taricha granulosa: Cloning, heterologous expression, and pharmacological characterization. *General and Comparative Endocrinology, 146*, 275–290.

Bunzow, J. R., Saez, C., Mortrud, M., Bouvier, C., Williams, J. T., Low, M., et al. (1994). Molecular cloning and tissue distribution of a putative member of the rat opioid receptor gene family that is not a mu, delta or kappa opioid receptor type. *FEBS Letters, 347*, 284–288.

Chen, Y., Mestek, A., Liu, J., Hurley, J. A., & Yu, L. (1993). Molecular cloning and functional expression of a mu-opioid receptor from rat brain. *Molecular Pharmacology, 44*, 8–12.

Conzelmann, M., Williams, E. A., Krug, K., Franz-Wachtel, M., Macek, B., & Jekely, G. (2013). The neuropeptide complement of the marine annelid Platynereis dumerilii. *BMC Genomics, 14*, 906.

Cortes, R., Navarro, S., Agulleiro, M. J., Guillot, R., Garcia-Herranz, V., Sanchez, E., et al. (2014). Evolution of the melanocortin system. *General and Comparative Endocrinology*, http://dx.doi.org/10.1016/j.ygcen.2014.04.005. [Epub ahead of print].

Coulier, F., Popovici, C., Villet, R., & Birnbaum, D. (2000). Meta*HOX* gene clusters. *The Journal of Experimental Zoology, 288*, 345–351.

de Velasco, E. M., Law, P. Y., & Rodriguez, R. E. (2009). Mu opioid receptor from the zebrafish exhibits functional characteristics as those of mammalian mu opioid receptor. *Zebrafish, 6*, 259–268.

Dores, R. M., Cameron, E., Lecaude, S., & Danielson, P. B. (2003). Presence of the delta-MSH sequence in a proopiomelanocortin cDNA cloned from the pituitary of the galeoid shark, Heterodontus portusjacksoni. *General and Comparative Endocrinology, 133*, 71–79.

Dores, R. M., Lecaude, S., Bauer, D., & Danielson, P. B. (2002). Analyzing the evolution of the opioid/orphanin gene family. *Mass Spectrometry Reviews, 21*, 220–243.

Dreborg, S., Sundström, G., Larsson, T. A., & Larhammar, D. (2008). Evolution of vertebrate opioid receptors. *Proceedings of the National Academy of Sciences of the United States of America, 105*, 15487–15492.

Evans, C. J., Keith, D. E., Jr., Morrison, H., Magendzo, K., & Edwards, R. H. (1992). Cloning of a delta opioid receptor by functional expression. *Science, 258*, 1952–1955.

Gonzalez Nunez, V., Gonzalez Sarmiento, R., & Rodriguez, R. E. (2003). Characterization of zebrafish proenkephalin reveals novel opioid sequences. *Brain Research. Molecular Brain Research, 114*, 31–39.

Gonzalez-Nunez, V., Gonzalez-Sarmiento, R., & Rodriguez, R. E. (2003a). Cloning and characterization of a full-length pronociceptin in zebrafish: Evidence of the existence of two different nociceptin sequences in the same precursor. *Biochimica et Biophysica Acta, 1629*, 114–118.

Gonzalez-Nunez, V., Gonzalez-Sarmiento, R., & Rodriguez, R. E. (2003b). Identification of two proopiomelanocortin genes in zebrafish (Danio rerio). *Brain Research. Molecular Brain Research, 120*, 1–8.

Gonzalez-Nunez, V., Marron Fernandez de Velasco, E., Arsequell, G., Valencia, G., & Rodriguez, R. E. (2007). Identification of dynorphin a from zebrafish: A comparative study with mammalian dynorphin A. *Neuroscience*, *144*, 675–684.

Gonzalez-Nunez, V., & Rodriguez, R. E. (2009). The zebrafish: A model to study the endogenous mechanisms of pain. *ILAR Journal*, *50*, 373–386.

Granier, S., Manglik, A., Kruse, A. C., Kobilka, T. S., Thian, F. S., Weis, W. I., et al. (2012). Structure of the delta-opioid receptor bound to naltrindole. *Nature*, *485*, 400–404.

Guindon, S., Dufayard, J. F., Lefort, V., Anisimova, M., Hordijk, W., & Gascuel, O. (2010). New algorithms and methods to estimate maximum-likelihood phylogenies: Assessing the performance of PhyML 3.0. *Systematic Biology*, *59*, 307–321.

Harris, R. M., Dijkstra, P. D., & Hofmann, H. A. (2014). Complex structural and regulatory evolution of the pro-opiomelanocortin gene family. *General and Comparative Endocrinology*, *195*, 107–115.

Henkel, C. V., Burgerhout, E., de Wijze, D. L., Dirks, R. P., Minegishi, Y., Jansen, H. J., et al. (2012). Primitive duplicate Hox clusters in the European eel's genome. *PLoS One*, *7*, e32231.

Hughes, J., Smith, T. W., Kosterlitz, H. W., Fothergill, L. A., Morgan, B. A., & Morris, H. R. (1975). Identification of two related pentapeptides from the brain with potent opiate agonist activity. *Nature*, *258*, 577–580.

Inoue, J. G., Miya, M., Lam, K., Tay, B. H., Danks, J. A., Bell, J., et al. (2010). Evolutionary origin and phylogeny of the modern holocephalans (Chondrichthyes: Chimaeriformes): A mitogenomic perspective. *Molecular Biology and Evolution*, *27*, 2576–2586.

Jaillon, O., Aury, J. M., Brunet, F., Petit, J. L., Stange-Thomann, N., Mauceli, E., et al. (2004). Genome duplication in the teleost fish Tetraodon nigroviridis reveals the early vertebrate proto-karyotype. *Nature*, *431*, 946–957.

Jekely, G. (2013). Global view of the evolution and diversity of metazoan neuropeptide signaling. *Proceedings of the National Academy of Sciences of the United States of America*, *110*, 8702–8707.

Kangawa, K., Minamino, N., Chino, N., Sakakibara, S., & Matsuo, H. (1981). The complete amino acid sequence of alpha-neo-endorphin. *Biochemical and Biophysical Research Communications*, *99*, 871–878.

Kieffer, B. L. (1995). Recent advances in molecular recognition and signal transduction of active peptides: Receptors for opioid peptides. *Cellular and Molecular Neurobiology*, *15*, 615–635.

Kieffer, B. L., Befort, K., Gaveriaux-Ruff, C., & Hirth, C. G. (1992). The delta-opioid receptor: Isolation of a cDNA by expression cloning and pharmacological characterization. *Proceedings of the National Academy of Sciences of the United States of America*, *89*, 12048–12052.

Kilpatrick, D. L., Taniguchi, T., Jones, B. N., Stern, A. S., Shively, J. E., Hullihan, J., et al. (1981). A highly potent 3200-dalton adrenal opioid peptide that contains both a [Met]- and [Leu]enkephalin sequence. *Proceedings of the National Academy of Sciences of the United States of America*, *78*, 3265–3268.

Komorowski, L. K., Lecaude, S. G., Westring, C. G., Danielson, P. B., & Dores, R. M. (2012). Evolution of gnathostome prodynorphin and proenkephalin: Characterization of a shark proenkephalin and prodynorphin cDNAs. *General and Comparative Endocrinology*, *177*, 353–364.

Lagman, D., Ocampo Daza, D., Widmark, J., Abalo, X. M., Sundstrom, G., & Larhammar, D. (2013). The vertebrate ancestral repertoire of visual opsins, transducin alpha subunits and oxytocin/vasopressin receptors was established by duplication of their shared genomic region in the two rounds of early vertebrate genome duplications. *BMC Evolutionary Biology*, *13*, 238.

Larhammar, D., & Bergqvist, C. A. (2013). Ancient grandeur of the vertebrate neuropeptide Y system shown by the coelacanth Latimeria chalumnae. *Frontiers in Neuroscience*, *7*, 27.

Larhammar, D., Dreborg, S., Larsson, T. A., & Sundstrom, G. (2009). Early duplications of opioid receptor and Peptide genes in vertebrate evolution. *Annals of the New York Academy of Sciences, 1163*, 451–453.

Larhammar, D., Sundstrom, G., & Dores, R. M. (2013). Evolution of the opioid system. In A. Kastin (Ed.), *Handbook of biologically active peptides* (2nd ed., pp. 1562–1569): San Diego: Elsevier.

Larsson, T. A., Olsson, F., Sundstrom, G., Lundin, L. G., Brenner, S., Venkatesh, B., et al. (2008). Early vertebrate chromosome duplications and the evolution of the neuropeptide Y receptor gene regions. *BMC Evolutionary Biology, 8*, 184.

Larsson, T. A., Tay, B. H., Sundstrom, G., Fredriksson, R., Brenner, S., Larhammar, D., et al. (2009). Neuropeptide Y-family peptides and receptors in the elephant shark, *Callorhinchus milii* confirm gene duplications before the gnathostome radiation. *Genomics, 93*, 254–260.

Manglik, A., Kruse, A. C., Kobilka, T. S., Thian, F. S., Mathiesen, J. M., Sunahara, R. K., et al. (2012). Crystal structure of the micro-opioid receptor bound to a morphinan antagonist. *Nature, 485*, 321–326.

McClendon, J., Lecaude, S., Dores, A. R., & Dores, R. M. (2010). Evolution of the opioid/ ORL-1 receptor gene family. *Annals of the New York Academy of Sciences, 1200*, 85–94.

Meunier, J. C., Mollereau, C., Toll, L., Suaudeau, C., Moisand, C., Alvinerie, P., et al. (1995). Isolation and structure of the endogenous agonist of opioid receptor-like ORL1 receptor. *Nature, 377*, 532–535.

Mirabeau, O., & Joly, J. S. (2013). Molecular evolution of peptidergic signaling systems in bilaterians. *Proceedings of the National Academy of Sciences of the United States of America, 110*, E2028–E2037.

Mollereau, C., Parmentier, M., Mailleux, P., Butour, J. L., Moisand, C., Chalon, P., et al. (1994). ORL1, a novel member of the opioid receptor family. Cloning, functional expression and localization. *FEBS Letters, 341*, 33–38.

Nakatani, Y., Takeda, H., Kohara, Y., & Morishita, S. (2007). Reconstruction of the vertebrate ancestral genome reveals dynamic genome reorganization in early vertebrates. *Genome Research, 17*, 1254–1265.

Ocampo Daza, D., Lewicka, M., & Larhammar, D. (2011). The oxytocin/vasopressin receptor family has at least five members in the gnathostome lineage, including two distinct V2 subtypes. *General and Comparative Endocrinology, 175*, 135–143.

Ocampo Daza, D., Sundstrom, G., Bergqvist, C. A., Duan, C., & Larhammar, D. (2011). Evolution of the insulin-like growth factor binding protein (IGFBP) family. *Endocrinology, 152*, 2278–2289.

Ocampo Daza, D., Sundstrom, G., Bergqvist, C. A., & Larhammar, D. (2012). The evolution of vertebrate somatostatin receptors and their gene regions involves extensive chromosomal rearrangements. *BMC Evolutionary Biology, 12*, 231.

Ohno, S. (1970). *Evolution by gene duplication*. Berlin: Springer-Verlag.

Pasternak, G. W., & Pan, Y. X. (2013). Mu opioids and their receptors: Evolution of a concept. *Pharmacological Reviews, 65*, 1257–1317.

Pert, C. B., & Snyder, S. H. (1973). Opiate receptor: Demonstration in nervous tissue. *Science, 179*, 1011–1014.

Pinal-Seoane, N., Martin, I. R., Gonzalez-Nunez, V., de Velasco, E. M., Alvarez, F. A., Sarmiento, R. G., et al. (2006). Characterization of a new duplicate delta-opioid receptor from zebrafish. *Journal of Molecular Endocrinology, 37*, 391–403.

Putnam, N. H., Butts, T., Ferrier, D. E., Furlong, R. F., Hellsten, U., Kawashima, T., et al. (2008). The amphioxus genome and the evolution of the chordate karyotype. *Nature, 453*, 1064–1071.

Reinscheid, R. K., Nothacker, H. P., Bourson, A., Ardati, A., Henningsen, R. A., Bunzow, J. R., et al. (1995). Orphanin FQ: A neuropeptide that activates an opioid like G protein-coupled receptor. *Science, 270*, 792–794.

Rivas-Boyero, A. A., Herrero-Turrion, M. J., Gonzalez-Nunez, V., Sanchez-Simon, F. M., Barreto-Valer, K., & Rodriguez, R. E. (2011). Pharmacological characterization of a nociceptin receptor from zebrafish (Danio rerio). *Journal of Molecular Endocrinology*, *46*, 111–123.

Simon, E. J., Hiller, J. M., & Edelman, I. (1973). Stereospecific binding of the potent narcotic analgesic (3H) Etorphine to rat-brain homogenate. *Proceedings of the National Academy of Sciences of the United States of America*, *70*, 1947–1949.

Sundstrom, G., Dreborg, S., & Larhammar, D. (2010). Concomitant duplications of opioid peptide and receptor genes before the origin of jawed vertebrates. *PLoS One*, *5*, e10512.

Takahashi, A., Itoh, T., Nakanishi, A., Amemiya, Y., Ida, H., Meguro, H., et al. (2004). Molecular cloning of proopiomelanocortin cDNA in the ratfish, a holocephalan. *General and Comparative Endocrinology*, *135*, 159–165.

Takahashi, A., Kobayashi, Y., Amano, M., & Yamanome, T. (2009). Structural and functional diversity of proopiomelanocortin in fish with special reference to barfin flounder. *Peptides*, *30*, 1374–1382.

Terenius, L. (1973). Stereospecific interaction between narcotic analgesics and a synaptic plasm a membrane fraction of rat cerebral cortex. *Acta Pharmacologica et Toxicologica*, *32*, 317–320.

Thompson, A. A., Liu, W., Chun, E., Katritch, V., Wu, H., Vardy, E., et al. (2012). Structure of the nociceptin/orphanin FQ receptor in complex with a peptide mimetic. *Nature*, *485*, 395–399.

Tostivint, H., Ocampo Daza, D., Bergqvist, C. A., Quan, F. B., Bougerol, M., Lihrmann, I., et al. (2014). Molecular evolution of GPCRs: Somatostatin/urotensin II receptors. *Journal of Molecular Endocrinology*, *52*, T61–T86.

Venkatesh, B., Lee, A. P., Ravi, V., Maurya, A. K., Lian, M. M., Swann, J. B., et al. (2014). Elephant shark genome provides unique insights into gnathostome evolution. *Nature*, *505*, 174–179.

Waldhoer, M., Bartlett, S. E., & Whistler, J. L. (2004). Opioid receptors. *Annual Review of Biochemistry*, *73*, 953–990.

Walthers, E. A., Bradford, C. S., & Moore, F. L. (2005). Cloning, pharmacological characterization and tissue distribution of an ORL1 opioid receptor from an amphibian, the rough-skinned newt Taricha granulosa. *Journal of Molecular Endocrinology*, *34*, 247–256.

Widmark, J., Sundstrom, G., Ocampo Daza, D., & Larhammar, D. (2011). Differential evolution of voltage-gated sodium channels in tetrapods and teleost fishes. *Molecular Biology and Evolution*, *28*, 859–871.

Wu, H., Wacker, D., Mileni, M., Katritch, V., Han, G. W., Vardy, E., et al. (2012). Structure of the human kappa-opioid receptor in complex with JDTic. *Nature*, *485*, 327–332.

Yamaguchi, Y., Kaiya, H., Konno, N., Iwata, E., Miyazato, M., Uchiyama, M., et al. (2012). The fifth neurohypophysial hormone receptor is structurally related to the V2-type receptor but functionally similar to V1-type receptors. *General and Comparative Endocrinology*, *178*, 519–528.

Yasuda, K., Raynor, K., Kong, H., Breder, C. D., Takeda, J., Reisine, T., et al. (1993). Cloning and functional comparison of kappa and delta opioid receptors from mouse brain. *Proceedings of the National Academy of Sciences of the United States of America*, *90*, 6736–6740.

CHAPTER FOUR

Synthesis and Biological Activity of Small Peptides as NOP and Opioid Receptors' Ligands: View on Current Developments

Emilia Naydenova*,[1], Petar Todorov*, Rositza Zamfirova[†]
*Department of Organic Chemistry, University of Chemical Technology and Metallurgy, Sofia, Bulgaria
[†]Institute of Neurobiology, Bulgarian Academy of Sciences, Sofia, Bulgaria
[1]Corresponding author: e-mail address: e_naydenova@abv.bg; pepi_37@abv.bg

Contents

Abstract

The heptadecapeptide nociceptin, also called orphanin FQ (N/OFQ), is the endogenous agonist of the N/OFQ peptide receptor (NOP receptor) and is involved in several central nervous system pathways, such as nociception, reward, tolerance, and feeding. The discovery of small molecule ligands for NOP is being actively pursued for several therapeutic applications. This review presents overview of the several recently reported NOP ligands (agonists and antagonists), with an emphasis of the structural features that may be important for modulating the intrinsic activity of these ligands. In addition, a brief account on the characterization of newly synthesized ligands of NOP receptor with aminophosphonate moiety and β-tryptophan analogues will be presented.

ABBREVIATIONS

SAR structure–activity relationship
NC nociceptin
CG carrageenan
TFA trifluoroacetic acid
ANOVA analysis of variance

Vitamins and Hormones, Volume 97
ISSN 0083-6729
http://dx.doi.org/10.1016/bs.vh.2014.11.005

SPPS solid-phase peptide synthesis
DMF *N,N*-dimethylformamide

1. INTRODUCTION

The endogenous opioid peptides (enkephalins, endomorphins, dinorphyns, etc.) are synthesized and released by neurons in central nervous system (CNS) and peripheral nervous system. They exert their effects by activating specific membrane receptors μ, δ, κ, and ε (Bodnar, 2013; Corbett, Henderson, McKnight, & Paterson, 2006). The opioids are molecules with similar structures, exerting effects on various physiological functions. By this reason, they are named "family." Opioid receptors belong to the large superfamily of seven-transmembrane (7TM)-spanning G protein-coupled receptors (GPCRs). As a class, GPCRs are of fundamental physiological importance mediating the actions of the majority of known neurotransmitters and hormones. The receptors consist of 7TM-spanning α-helices, with an extracellular N-terminus, an intracellular C-terminus, and three interhelical loops on each side of the membrane (Corbett et al., 2006; Henderson & McKnight, 1997; León-Olea et al., 2013; Martin, 1967). Later, in 1995, a new neuropeptide, nociceptin (orphanin FQ), with a sequence similar to that of classical opioids has been isolated from brain tissue (Meunier et al., 1995; Reinscheid et al., 1995; Reinscheid, Ardati, Monsma, & Civelli, 1996; Reinscheid, Higelin, Henningsen, Monsma, & Civelli, 1998). The discovery of nociceptin (nociceptin/orphanin FQ (N/OFQ)) is a prominent example of so-called reverse pharmacology in which a receptor was known before the corresponding ligand could be identified (Civelli, 2008). After the discovery of the δ-opioid receptor by expression, cloning molecular biologists have not only identified the expected μ- and κ-opioid receptors, but also a fourth unperceived receptor, which does not bind classical opioids (Mollereau, Parmentier, & Mailleux, 1994; Mollereau et al., 1996). This new receptor was identified and called opioid receptor like ORL_1 or N/OFQ peptide (NOP) receptor (Henderson & McKnight, 1997). Both the N/OFQ and NOP receptors share a high degree of homology with classical opioid peptides and opioid receptors, respectively, and use similar signal transduction pathways as classical opioids. The NOP receptor has thus been regarded as the fourth member of the opioid receptor family.

It was found that NC–NOP receptor system modulates several biological functions in central and peripheral nerve systems. Agonists of the peptide might be favorable in the treatment of cough, drug dependence, and stress-induced anorexia. Antagonists could be useful in the management of depression, dementia, and inflammatory diseases (rheumatoid arthritis, colitis, myocarditis) and also in the treatment of diseases, accompanied by increased plasma levels of nociception: sepsis (Williams et al., 2008), Parkinsonism (Marthi et al., 2005), liver cirrhosis (Doggrell, 2007), cancer (Horvath et al., 2004; Szalay et al., 2004), and Wilson disease (Hantos et al., 2002).

Since NOP receptor agonists and antagonists both could have therapeutic value, rational approaches for obtaining both within a high-affinity-binding class of compounds are very useful for designing potent and selective NOP ligands with the desired profile of intrinsic efficacy. Development of selective and highly potent peptide and nonpeptide agonist and antagonist ligands is of great importance for the understanding of the physiological and pathological roles of N/OFQ–NOP receptor system, too. Among them are N/OFQ-related peptides (Ambo, Kohara, Kawano, & Sasaki, 2007; Guerrini, Calò, et al., 2000; Guerrini, Rizzi, et al., 2000; Naydenova, Todorov, Mateeva, et al., 2010; Guerrini, Calò, Bigoni, Rizzi, Regoli, & Salvadori, 2001; Guerrini, Calò, Bigoni, Rizzi, Rizzi, et al., 2001), small peptides, identified by screening of peptide combinatorial libraries (Dooley, Chung, Shiller, & Houghten, 1993; Dooley et al., 1997), and nonpeptide ligands such as Ro 64-6198 and JTC-801 (Zaveri, 2003; Zaveri, Jiang, Olsen, Polgar, & Toll, 2005). The synthesis of new NOP receptor ligands with higher affinity and lower enzymatic degradation is a primary goal of many research teams. In recent years, considerable progress has been made in the development of selective NOP receptor peptide ligands having agonist or antagonist properties (Naydenova & Todorov, 2011). The ligands were obtained using various design strategies, including substitution of natural and nonproteinogenic amino acids, conformational restriction, and the bivalent ligand approach (Wollemann, Benyhe, & Simon, 1993). As a part of our research, the synthesis, the characterization, and the biological activity of recently synthesized series of small peptides with aminophosphonates moiety and β-tryptophan analogues as NOP receptor ligands are described (Bocheva et al., 2013; Naydenova, Pavlov, Todorov, Dzhambazova, & Bocheva, 2011; Naydenova, Todorov, Mateeva, et al., 2010; Zamfirova et al., 2013).

2. ENDOGENOUS OPIOID PEPTIDES AND RECEPTORS: NOCICEPTIN AND NOP RECEPTOR LIGANDS

Endogenous opioids are intensively and extensively studied mainly in search for a new powerful analgesic, lacking the adverse effects of most alkaloid opioids. A German chemist, *Friedrich Sertuner*, isolated the active principle morphine (shown in Fig. 1) from opium in 1805, which was then used in therapy. Unfortunately, morphine has just as much potential for abuse as does opium. This prompted medicinal chemists to attempt to develop safer and more efficacious compounds, with the goal of providing analgesia with reduced abuse potential and reduced incidence of side effects (such as respiratory depression). The first pure opioid antagonist, naloxone, was produced in the 1940s, after the synthesis of nalorphine (*N*-allylnormorphine; see Fig. 1), which was previously used to prevent the effects of opioid receptor agonists. However, this action does not concern all opioid receptors, because nalorphine has also been shown to mimic the action of some agonists.

The rigid structural and stereochemical requirements essential for the analgesic actions of morphine and related opioids led to the theory that they produce their effects by interacting with a specific receptor. The concept that there is more than one type of opioid receptor arose to explain the dual

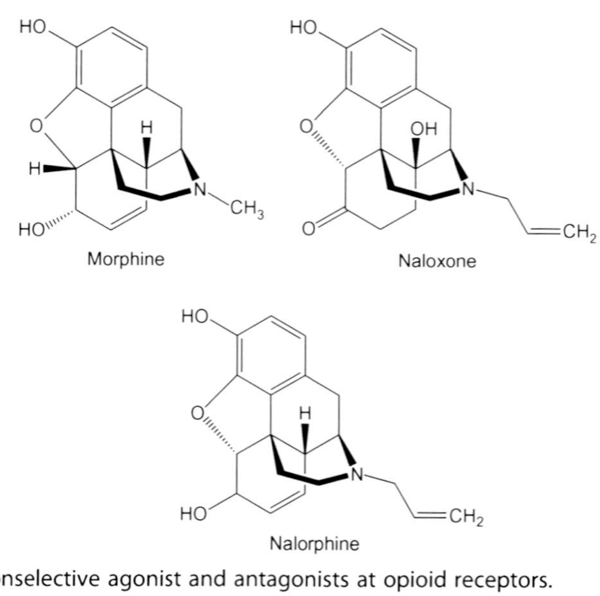

Figure 1 Nonselective agonist and antagonists at opioid receptors.

actions of the synthetic opioid nalorphine, which antagonizes the analgesic effect of morphine in man but also acts as an analgesic in its own right. Martin (1967) concluded that the analgesic action of nalorphine is mediated by a receptor, later called the k-opioid receptor, which is different from the morphine receptor (Calò, Lambert, & Guerrini, 2013; Martin, 1967).

Endogenous opioids exert their physiological and pharmacological effects, respectively, by interacting with opioid receptors belonging to four main types: μ, δ, κ, and ε. Their effectors can be enzymes (like adenylate cyclase) or ion channels (K^+, Ca^{2+}). Opioid receptors are widely distributed in the CNS and the peripheral nervous system (Cesselin, 1995). The wide distribution of the opioid-containing neurons and opioid receptors suggests that the endomorphinergic systems take part in control of various physiological functions. Evidence has been presented for their implication in the control of respiratory, cardiovascular and gastrointestinal functions, pituitary hormonal secretion, motor activity, and the etiology of some pathological states such as epilepsy and psychosis (Olson, Olson, & Kastin, 1995). One of the better-established functional roles of the opioid systems is the control of nociception.

As it was mentioned before, nociceptin (Phe–Gly–Gly–Phe–Thr–Gly–Ala–Arg–Lys–Ser–Ala–Arg–Lys–Leu–Ala–Asn–Gln) is a neuropeptide, structurally related to opioid peptides, particularly dynorphin A (Meunier et al., 1995; Reinscheid et al., 1995), but despite the primary sequence similarity, it does not bind to classical opioid receptors (Henderson & McKnight, 1997; Nicholson, Paterson, Menzies, Corbett, & McKnight, 1998) N-terminal fragment of N/OFQ (Phe–Gly–Gly–Phe) is similar to N-terminal "quartet" Tyr–Gly–Gly–Phe of dynorphins, endorphins, and enkephalins. The main difference between them and the other opioid peptides is the replacement of Tyr^1 with Phe^1. It is so-called "message domain," which is involved in receptor binding and activation (Butour, Moisant, Mazarguil, Mollereau, & Meunier, 1997; Dooley & Houghten, 1996; Guerinni, Tramontana, & Maggi, 1997). A crucial role for the selective receptor occupation by N/OFQ is "address domain," including positively charged amino acids Arg^8–Lys^{12} and Arg^9–Lys^{13} (Dooley et al., 1997; Rizzi et al., 2002). Like opioid agonists, N/OFQ inhibits adenylate cyclase and calcium currents and hyperpolarizes neurons by opening potassium channels (Hawes, Graziano, & Lambert, 2000; Moran, Abdulla, & Smith, 2000). In contrast, the physiological and behavioral functions of N/OFQ are agreed to be quite different from those of classical opioids (Calò, Bigoni, Rizzi, et al., 2000; Calò, Guerrini, Bigoni, et al.,

2000; Calò, Guerrini, Rizzi, Salvadori, & Regoli, 2000; Calò et al., 2002; Ciccocioppo, Angeletti, Panocka, & Massi, 2000; Ciccocioppo, Panocka, Polidori, Regoli, & Massi, 1999; Heinricher, 2003; Mogil & Pasternak, 2001; Reinscheid & Civelli, 2002; Reinscheid et al., 1995; Tariq, Nurulain, Tekes, & Adeghate, 2013). The specific location of the NOP receptor and nociceptin in the brain (Fukada et al., 1994; Neal et al., 1999; Nothacker et al., 1996; Robinson & Becker, 1986) shows that they take part in the modulation of multiple processes, controlled by CNS. The N/OFQ–NOP receptor system has been reported to modulate certain biological functions, including pain transmission, stress and anxiety, learning and memory, locomotor activity, and food intake. Concerning peripheral nervous system, the activation of nociceptin receptors influence cardiac, digestive, secretory, renal, respiratory, and immune systems.

However, native opioid peptides are unable to bind to NOP receptor. The ligand nociceptin, despite its primary sequence similarity with that of some opioid peptides, does not bind to opioid receptors. Nociceptin was found to be inactive in several preparations of smooth muscles, whereas it inhibited the contractions induced by electrical field stimulation in the mouse *vas deferens* (MVD), the guinea pig ileum (GPI), and other preparations. It is worthy to note that the MVD and the GPI, besides the classical opioid receptors δ and μ, respectively, contain NOP receptors. Despite its structural similarity to opioid peptides, nociceptin exerts a potent hyperalgesic effect at supraspinal level while it produces analgesia at spinal level. In addition, nociceptin is able to block supraspinal antinociception produced either by morphine or by selective opioid receptor agonists (Bodnar, 2011, 2013). There is a considerable interest in the development of peptides retaining high activity and receptor selectivity after peripheral administration and several strategies have been used to enhance peptide delivery to the CNS. For this purpose, the scientists have performed structure–activity relationship (SAR) studies using nociceptin (NC) (1–13)NH_2, the minimal fragment of NC retaining the same potency and efficacy as the natural peptide (Bigoni et al., 1999; Calò et al., 2013; Guerinni et al., 1997; Guerrini, Calò, Bigoni, Rizzi, Regoli, & Salvadori, 2001; Guerrini, Calò, Bigoni, Rizzi, Rizzi, et al., 2001; Varani et al., 1999). Previous studies have led to the identification of several interesting molecules, such as [Tyr^1]NC(1–13)NH_2, a nonselective agonist, which is able to activate both NOP and classical opioid receptors (Varani et al., 1999), [$Phe^1\psi(CH_2NH)Gly^2$]NC(1–13)NH_2 (Guerrini et al., 1998), a low-efficacy agonist that behaves as an antagonist, a partial agonist, or even a full agonist, depending on the preparation under study (Calò, Bigoni, Rizzi, et al., 2000; Calò, Guerrini, Bigoni, et al., 2000; Calò et al., 2005; Calò, Guerrini, Rizzi, et al., 2000), and [$Nphe^1$]NC(1–13)NH_2, the first pure and selective NOP receptor antagonist of peptide structure (Calò, Bigoni, Rizzi, et al., 2000; Calò, Guerrini, Bigoni, et al., 2000; Calò, Guerrini, Rizzi, et al., 2000). These analogues were identified using a classical SAR approach performed on the Phe^1 residue. The Phe^4 in the nociceptin sequence has been identified as the most critical residue for receptor interaction. Bigoni et al. (2002) investigated the pharmacological activity of a series of NC(1–13)NH_2 analogues, in which the hydrogen atom in the *para* position of Phe^4 was substituted with –F, $-NO_2$, $-CN$, $-Cl$, $-Br$, $-I$, $-CH_3$, $-OH$, or $-NH_2$.

In receptor-binding studies, performed using Chinese hamster ovary (CHO) cells expressing the recombinant human NOP receptor (CHO_{hNOP}) and in rat cerebral cortex membranes, [(pF)Phe^4]NC(1–13)NH_2, [(pNO_2)Phe^4]NC(1–13)NH_2, and [(pCN)Phe^4]NC(1–13)NH_2 displayed higher affinity than NC(1–13)NH_2. The affinity of [(pCl)Phe^4]NC(1–13)NH_2 was essentially identical to that of NC(1–13)NH_2, while the remaining compounds displayed reduced affinity. In a series of functional assays, the *para*-substituted analogues behaved as full agonists (with the exception of [(pOH)Phe^4]NC(1–13)NH_2, which acted as a partial agonist in the GTPγS-binding assays) with the following rank order potency: $pF = pNO_2 \geq pCN \geq NC = NC(1\text{–}13)NH_2 = pCl \geq pBr > pI = pCH_3 > pOH > pNH_2$. The effects of all analogues were not modified by naloxone. The selective NOP receptor antagonist [$Nphe^1$]NC(1–13)NH_2, tested in all preparations against one or both of the highly potent derivatives [(pF)Phe^4] NC(1–13)NH_2 and [(pNO_2)Phe^4]NC(1–13)NH_2, showed pA2 values similar to those found against NC. This study further supports the notion that Phe^4 of NC is the critical residue for receptor occupation and activation.

An interesting idea for modifying the structure of a biologically active peptide was the proposal to shift the side chain of a certain amino acid residue, from the α-carbon to the α-nitrogen atom of the peptide bond. The resulting N-substituted glycine is achiral and can be used for the synthesis of peptide derivatives, called peptoids or peptide–peptoid hybrids, possessing proteolytic stability and high flexibility (Kessler, 1993; Simon et al., 1992). Glycosylation of peptides is another promising approach in rational drug design and the synthesis of glycopeptides and neoglycopeptides has been extensively reviewed (Davies, 2002; Negri et al., 1999; Schweizer, 2002) and described the synthesis of some and reported their opioid receptor affinity and their analgesic potency after subcutaneous injection in mice. Biondi et al. (2006) described the synthesis and some preliminary pharmacological experiments of new NC(1–13)NH_2 analogues in which either one or both the Phe^1 and Phe^4, the Arg^8 and Arg^{12}, and the Ala^7 and Ala^{11} have been replaced by *N*-benzyl-glycine, *N*-(3-guanidino-propyl)glycine, and β-alanine, respectively. The preparation is also described of NC(1–13) NH_2 analogues in which either galactose or *N*-acetyl-galactosamine is β-*O*-glycosidically linked to Thr^5 and/or to Ser^{10}.

Structure–activity and nuclear magnetic resonance (NMR) studies on N/OFQ demonstrated that Aib substitution of Ala^7 and/or Ala^{11} increases the peptide potency through an alpha helix structure induction mechanism. On these bases Arduin et al. (2007) synthesized and evaluated pharmacologically in the MVD assay a series of N/OFQ-NH_2 analogues substituted in positions 7 and 11 with Cα,α-disubstituted cyclic, linear, and branched amino acids. None of the 20 novel N/OFQ analogues produced better results than [Aib^7] N/OFQ-NH_2. Thus, this substitution was combined with other chemical modifications known to modulate peptide potency- and/or efficacy-generating compound [Nphe1$Aib^7Arg^{14}Lys^{15}$]N/OFQ-NH_2 (coded as UFP-111), compound [(pF)Phe4$Aib^7Arg^{14}Lys^{15}$]N/OFQ-NH_2 (UFP-112), and compound [Phe^1W(CH_2–NH)Gly^2(pF)$Phe^4Aib^7Arg^{14}Lys^{15}$] N/OFQ$NH_2$ (UFP-113). These novel peptides behaved as highly potent NOP receptor ligands showing full (UFP-112) and partial (UFP-113) agonists and pure antagonist (UFP-111) activities in a series of *in vitro* functional assays performed on pharmacological preparations expressing native as well as recombinant NOP receptors.

During the last few years, new shortened analogues of nociceptin by Fmoc solid-phase method have been performed by our team (Naydenova, Todorov, Mateeva, et al., 2010; Vezenkov, Zhivkova, & Naydenova, 2006). The influence of Gly and Ala residues has been investigated through their consecutive substitution with βAla in H–Phe–Gly–Gly–Phe–Thr–βAla–NH_2 and H–Phe–Gly–Gly–Phe–Thr–Gly–βAla–NH_2 peptides. In the H–Phe–Gly–Gly–Phe–Thr–Gly–Ala–Arg–Orn–NH_2 peptides, the Lys^9 residue has been substituted with ornithine (Orn). The biological activity of the newly synthesized compounds has been established *in vitro* via electrostimulated *rats. vas deferens* (RVD). Contrary to nociceptin, the shorter fragments show lower or no activity. These findings are consistent with the hypothesis that the sequence length is an important factor for the peptide binding to the receptor. The biological activity tests also show that the substitution of Ala with βAla does not affect the activity of the shortened analogues. On the other hand, the inhibitory effect has been enhanced by the substitution of lysine (Lys) by Orn in the new NC(1–9)NH_2 analogue (Hirschmann et al., 1994; Vezenkov et al., 2006).

Many analogues were also examined in search of variations of their functions, compared to the "original" molecule (Chen et al., 2002; Rizzi et al., 2001). Guerinni et al. (1997) and Guerrini et al. (1998) proposed that the N-terminal tridecapeptide sequence of the nociceptin molecule suffices for its full biological activities. It has been reported that $Arg^{8,12}$ and $Lys^{9,13}$ seem to be crucial for receptor occupation, most probably interacting with the acidic amino acids present in the second extracellular loop of the NOP receptor.

Following the "message/address" concept in the frame of the SAR (Guerinni et al., 1997), we have investigated the importance of the side-chain amino group from the peptide backbone on the biological activity. We have replaced the Lys^9 and Lys^{13} by their structural analogues Orn, diaminobutanoic (Dab) acid, and diaminopropanoic (Dap) acid (Naydenova et al., 2006).

The results obtained suggested that replacement of Lys by Orn at position 9 statistically significantly enhanced the biological activity, while the substitution at position 13 exerted an opposite effect, decreasing the biological activity of N/OFQ(1–13)NH_2. These data revealed the importance of the side-chain length of the amino acid at positions 9 and 13 for the biological activity of nociceptin derivatives.

Further shortening of the side chain of N/OFQ(1–13)NH_2 at position 9 [Dab^9]N/OFQ(1–13)NH_2, [Dap^9]N/OFQ(1–13)NH_2 decreased the biological activity of the tested compounds. The newly synthesized peptides have been also studied regarding potential toxicity and antioxidant capacity on SH-SY5Y neuroblastoma cells. The N/OFQ(1–13)NH_2 and its structural analogue [Orn^9]N/OFQ(1–13)NH_2 possess more favorable profiles than the other two N/OFQ analogues. The results suggest that shortening of the side chain of the Lys might increase cell damage and reduce the viability of the cells. Moreover, such alterations may lead to changes in free-oxygen-generating systems and in antioxidant status in rat brain and liver tissues (Kirkova et al., 2009). It was found also that NC (1–13)NH_2 depresses the carrageenan-induced acute inflammation, compared to the controls (Zamfirova et al., 2009). Based on the fact that carrageenan (CG) injected in the rat paw produces acute inflammation and pain via enhancement of free radical formation, we also studied the changes in the antioxidant status of the inflammated paw tissues as a possible mechanism of the peptide action. The results showed that needier N/OFQ(1–13)NH_2, nor [Orn^9]

N/OFQ(1–13)NH_2, posses pro- or antioxidant activity. The same has been found in the brain of kainic acid-treated rats. N/OFQ(1–13)NH_2 and [Orn^9] N/OFQ(1–13)NH_2 did not affect the antioxidant cell defense systems (Tzvetanova et al., 2009).

We investigated the effect of N/OFQ(1–13)NH_2 on the fast oscillations of interpulse interval (IPI) on conscious normotensive *Wistar* rats and spontaneously hypertensive rats (SHR). N/OFQ(1–13)NH_2 inhibits the fast oscillations in IPI mediated mainly by the sympathetic nerve activity in *Wistar* rats. The lack of changes in SHR may be a result of existing high sympathetic drive to the heart, which probably leads to reduced capacity of the heart rate control mechanisms to respond to OP4 stimulation (Markova, Girchev, Naydenova, & Vezenkov, 2007).

The effects of nociceptin analogues N/OFQ(1–13)NH_2 or [Orn^9] N/OFQ(1–13)NH_2 on the blood pressure variability were studied in conscious *Wistar* rats. The replacement of Lys with Orn in the ninth position abolished the effects of nociceptin analogue N/OFQ(1–13)NH_2 on blood pressure variability in experimental animals (Girchev, Markova, Naydenova, & Vezenkov, 2009). Girchev, Markova, Naydenova, and Vezenkov (2010) indicated functional role of amino acid placed in the ninth position in the structure of nociceptin for the realization of its effects on fast oscillations of arterial blood pressure in both normotensive and SHR (Girchev et al., 2009, 2010).

Phosphonopeptides are phosphorus analogues of naturally occurring peptides containing a tetrahedral phosphorus atom. Their importance is obvious from the fact that they have been widely used as enzyme inhibitors and, as haptens in catalytic antibody research, because they can be considered as stable mimetics of tetrahedral transition states in ester and amide hydrolysis and formation (Cunningham, Drag, Kafarski, & Bell, 2008; Hirschmann et al., 1994). To date, several efficient synthetic routes have been developed for the synthesis of phosphonopeptides and phosphinopeptides, containing C-terminal α-aminoalkylphosphinic acids (Kukhar, Soloshonok, & Solodenko, 1994; Meng & Xu, 2010).

Recently have been reported (Naydenova, Todorov, Mateeva, et al., 2010), SAR of new N-modified analogues of N/OFQ(1–13)NH_2 with aminophosphonate moiety containing five-, seven-, and eight-membered cycloalkane rings with the following sequences (Fig. 2).

Our research has shown that the selectivity of the peptides containing 1-[(methoxyphosphono) methylamino]cycloalkane carboxylic acids to the N-side of Phe is not changed—they remain selective agonists of

X—Gly—Gly—Phe—Thr—Gly—Ala—Arg—Lys—Ser—Ala—Arg—Lys—NH_2

X—Phe—Gly—Gly—Phe—Thr—Gly—Ala—Arg—Lys—Ser—Ala—Arg—Lys—NH_2

Where **X** is :

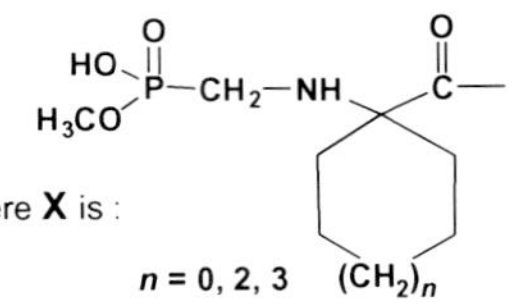

Figure 2 N-modified analogues of N/OFQ(1–13)NH_2 with aminophosphonate moiety.

NOP receptors. The derivative with the largest ring demonstrated efficacy similar to that of N/OFQ(1–13)NH_2, but in a 10-fold higher concentration (Naydenova et al., 2011; Naydenova, Todorov, Mateeva, et al., 2010; Naydenova, Todorov, & Troev, 2010).

3. HEXAPEPTIDES WITH NOP RECEPTOR AFFINITY

The synthesis of new NOP receptor ligands with higher affinity and lower enzymatic degradation is a primary goal of much research. Using combinatorial chemistry analysis, several hexapeptides with a general formula Ac–Arg–Tyr–Tyr–Arg/Lys–Trp/Ile–Arg/Lys–NH_2 have been identified as the least peptide sequence with high NOP receptor affinity and selectivity (Dooley et al., 1997; Mason, Ho, Nicholson, & McKnight, 2001). The hexapeptides Ac–Arg–Tyr–Tyr–Arg–Trp–Lys–NH_2 and Ac–Arg–Tyr–Tyr–Arg–Ile–Lys–NH_2 were found to act as partial or full agonists or antagonists of NOP receptor in different *in vivo* and *in vitro* systems (Ambo et al., 2007; Dooley et al., 1997). The peptide Ac–Arg–Tyr–Tyr–Arg–Ile–Lys–NH_2 blocks the effects of N/OFQ in rat cortex (stimulation of GTPg^{35}S binding) or heart (positive chronotropic effect in isolated myocytes; Berger, Albrecht, Wallukat, & Bienert, 1999).

These hexapeptides were used as chemical templates in SAR studies (Ambo et al., 2007; Judd et al., 2004). New ligands were synthesized, substituting Orn, Dab, and Dap acids for Lys based on the above-mentioned templates (Vezenkov et al., 2006). We reported the effect of newly synthesized NOP receptor ligands on the neurogenic contraction of isolated RVD. The study reveals that substitution of Lys with shorter amino acids could increase agonist or antagonist properties of the peptide (Dzambazova,

Nocheva, & Bocheva, 2008; Kasakov, Nashar, Naydenova, Vezenkov, & Vlaskovska, 2010).

Based on the Ac–Arg–Tyr–Tyr–Arg–Trp–Lys–NH_2 as a chemical template, we prepared and evaluated for agonistic activity *in vitro* new N-modified analogues with α-aminocycloalkane carboxylic acid, bearing (methoxyphosphonyl) methyl group. The 1-[(dimethoxyphosphono) methylamino]cycloalkane carboxylic acids were previously prepared by our group following Kabachnik–Fields reaction (Naydenova et al., 2008; Naydenova, Todorov, & Troev, 2010). The peptides were prepared with good yield by solid phase synthesis using 2-(1*H*-benzotriazole-1-yl)-1,1,3,3-tetramethyluronium tetrafluoroborate (TBTU), an efficient peptide-coupling reagent. In view of the low solubility of α-aminophosphonates, the coupling reaction was performed in *N,N*-dimethylformamide (DMF) at high excess of *N,N*-diisopropylethylamine (DIEA) and the coupling reaction time was 15 h. During the trifluoroacetic acid (TFA) cleavage, one of the methoxy groups from aminophosphonic residue was removed. The new hexapeptides have the following sequences (Fig. 3).

In our study, we tested the newly synthesized ligands *in vitro* on electrically stimulated smooth-muscle preparations from rat *vasa deferentia*. The effects of newly synthesized short-chain peptides were compared with Ac–Arg–Tyr–Tyr–Arg–Trp–Lys–NH_2, which is known to be a very potent and selective partial agonist of NOP receptors. In the hexapeptide derivatives, Arg^1 was substituted by 1-[(methoxyphosphono)methylamino] cycloalkane carboxylic acids with five-, seven-, and eight-membered rings. As a whole, the N-terminal modification with aminophosphonates moiety decreased the affinity of the ligands to NOP receptor.

This newly synthesized heptapeptide has the following sequence (Fig. 4).

The removal of the acetyl group in the parent peptide dramatically reduced both the potency and efficacy. Furthermore, the replacement of the acetyl group by aminophosphonates moiety (Fig. 4) also reduced the

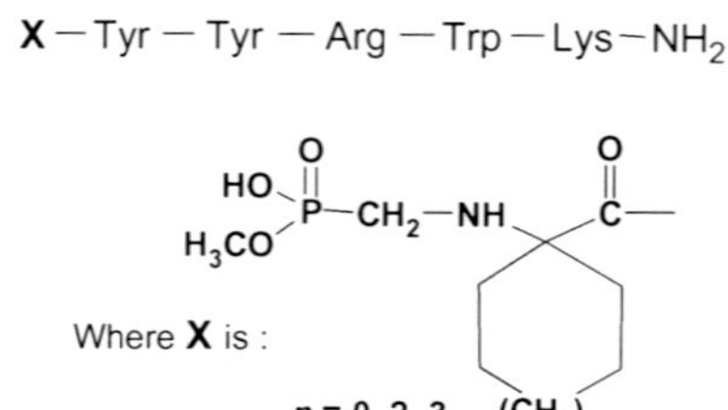

Figure 3 N-modified hexapeptide analogues with α-aminocycloalkane carboxylic acid.

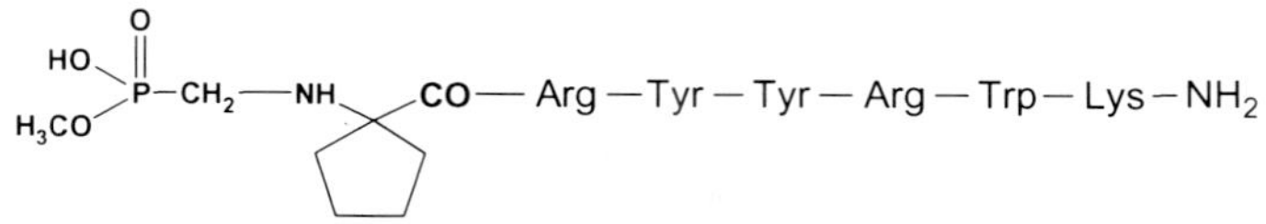

Figure 4 Newly synthesized heptapeptide.

potency of the peptide. Nevertheless, its action remains specific for NOP receptors: applied after opioid receptor blockade by naloxone, the peptide does not change its effect. The blockade of NOP receptors with naloxone benzoylhydrazone prevented completely the inhibitory effect of this compound. Taken together, our results showed that Arg^1 could be decisive for selectivity and biological activity of hexapeptides and that N-terminal acetyl group is required for their high affinity to NOP receptors.

Because pain is frequently observed symptom of various diseases and development of analgesic drugs is one of the greatest achievements in medicine, we investigated N-modified analogues of the H–Arg–Tyr–Tyr–Arg–Trp–Lys–NH_2 with aminophosphonates moiety during acute pain in male Wistar rats. Antinociceptive effects were evaluated by two nociceptive tests—paw pressure (PP) and hot plate and statistically accessed by analysis of variance (ANOVA) (Naydenova et al., 2011).

The obtained results showed that incorporation of 1-[(methoxyphosphono) methylamino]cycloalkane carboxylic acid in position 1 of hexapeptide molecule increases the analgesic effects of the newly synthesized peptide analogues with five carbon atoms, while enlargement of the cycle with eight carbon atoms diminished its analgesic effect. The analgesic effect of newly synthesized small peptides was more pronounced in compound where the Arg^1 was preserved.

JTC-801 (NOP receptor antagonist) inhibited the analgesic effects of compounds in PP test, while naloxone (Nal) reversed them to hyperalgesic. These data suggest the involvement of nociceptin and opioid neuromediatory systems in their mechanisms of action during acute pain.

With a view to developing ligands with more potent analgesic activity and lower enzymatic degradation, a new series of hexapeptides containing β^2-tryptophan analogues in positions 4 and 5, based on the templates Ac–Arg–Phe–Met–Trp–Met–Lys–NH_2 (opioid receptor antagonist) and Ac–Arg–Tyr–Tyr–Arg–Trp–Lys–NH_2 (highly potent and selective NOP receptor agonist), have been synthesized and biological evaluated (Bocheva et al., 2013; Zamfirova et al., 2013). The mechanisms of peptides' action were attempted using naloxone (an opioid antagonist) and JTC-801

Ac–Arg–Tyr–Tyr–Arg–**X**–Lys–NH_2

Ac–Arg–Phe–Met–**X**–Met–Lys–NH_2

Where **X** is:

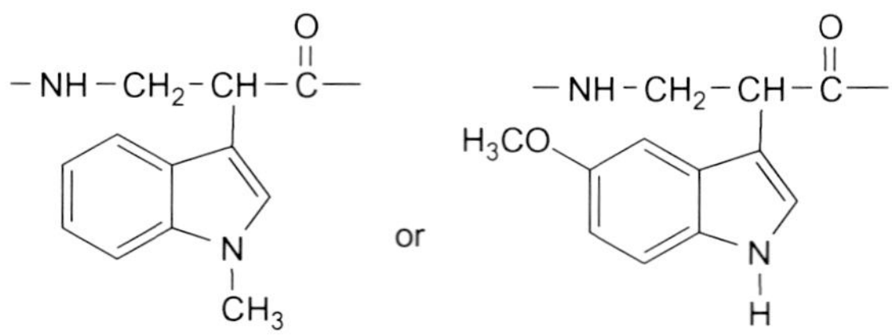

Figure 5 Hexapeptides-containing β^2-tryptophan analogues.

(a NOP receptor antagonist). The novel hexapeptides containing β^2-tryptophan analogues, as NOP receptor ligands, have the following sequences (Fig. 5).

These hexapeptides containing β^2-tryptophan analogues expressed analgesic activity through both NOP and opioid receptors. Similar results have been obtained in structure–activity studies on isolated smooth-muscle contractions. These data give us ground to suggest that in this case not only the position of modification, but also the nature of the incorporated group leads to significant changes in peptide's selectivity and affinity. Based on the fact that activation of the both types of receptors influences the process of inflammation, we examine the effect of one of the compounds on acute carrageenan-induced inflammation. We found that it slows up the increase of edema volume compared to the controls, more pronounced in the first 120 min after application realized mainly by the activation of opioid receptors (unpublished data).

4. SOLID-PHASE PEPTIDE SYNTHESIS

The basis of the solid-phase peptide synthesis (SPPS) assumes that the peptide sequence can be build up on a solid support—a specially treated polyvinyl resin, which connects to the C-terminal of the first amino acid and, after obtaining the peptide sequence, this bond can be cleaved. Depending on the type of the resin, two types of protection groups for –NH_2 are widely used in SPPS: the Fmoc- and Boc-protection groups. Depending on the choice of the protecting group, two types of strategies are employed—Boc strategy (Merrifield synthesis) and Fmoc strategy (synthesis of Shepard).

The consecutive linkage of N-α-protected amino acids is currently done by a few major methods in the SPPS. The Boc strategy requires a permanent unblocking of the protective group with TFA, which may result in the break of peptide bonds or catalyzation of unwanted secondary reactions, while this step is omitted in the Fmoc strategy. This is indeed the main advantage of the Fmoc over Boc strategy. In the Fmoc strategy, the unblocking of the peptide is done by a weak alkaline treatment with piperidine.

The use of TFA is limited to the final cleavage of the peptide from the resin. The cleavage of the final Boc peptide from the resin is done by treatment with hydrogen fluoride (HF)—even more reactive than TFA. Those facts and the slightly more expensive equipment required for Boc strategy limits its overall use.

In both cases, the main employed solvents in the condensation and dissolution reactions are dichloromethane (DCM) and DMF.

The SPPS main steps include (see Scheme 1):

1. Linking the C-terminal amino acid to the resin.
2. The next step requires the unblocking of the N-α-amino group; in Fmoc strategy, 20% piperidin in DMF is used.
3. The third step is the peptide bond formation. For these purpose, condensation reagents like DCC/1-hydroxy-benzotriazole (1-HOBt), 2-(1*H*-benzotriazole-1-yl)-1,1,3,3-tetramethyluronium hexafluorophosphate (HBTU), and TBTU are needed. The use of *N,N'*-dicyclohexylcarbodiimide (DCC) as condensation reagent if Gln and Asn residues are present is not recommended as dehydration of the side-chain amide group is observed. The completeness of the reaction is monitored by the standard Kaiser test. This step may be repeated a few times in order to have a complete reaction.
4. The next very important stage is the cleavage of the peptide from the resin. In the Fmoc strategy, the most suitable conditions for unblocking the peptide from the resin depend on the individual peculiarities of the sequence, the present side chains, the trifunctional side chains, and the reactivity of the linker of the resin. The typical procedure in Fmoc strategy is represented by a treatment of the peptide resin with TFA (1–90%) in solution with different scavengers that help the subsequent alkylation during the unblocking and also interact with the secondary products resulting from this process (anisol, thioanisol, ethanditiol, ethylmethylsulfide, triisopropylsilane, etc.). During the unblocking from the resin, the protective groups from the side chains are also removed.

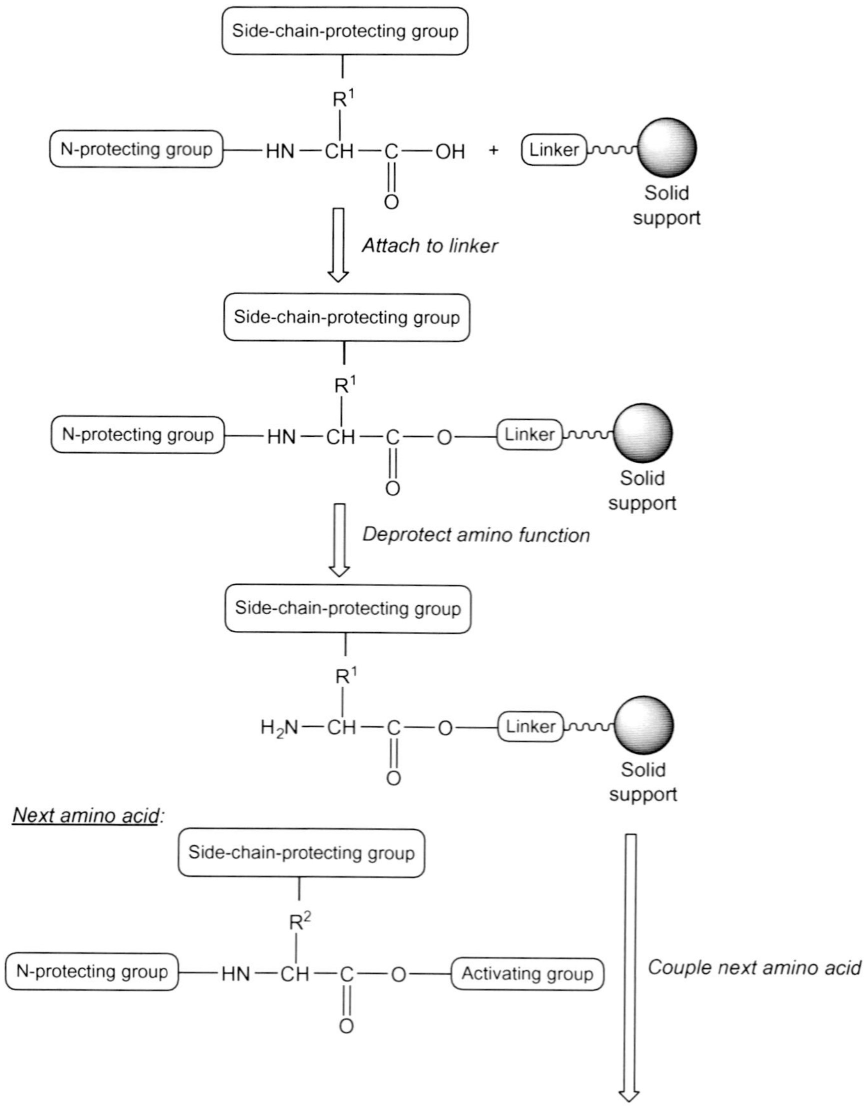

Scheme 1 See legend on the opposite page. *(Continued)*

5. The last step encompasses all the methods for purification of the obtained peptides from the resins, enantiomers (dimers), and shortened analogues. The methods currently employed for the purification and analysis of synthetic peptides include reversed-phase chromatography (RP-HPLC), liquid chromatography with classical colons with silica gel, electrophoresis, gel filtration, ion exchange, and thin layer chromatography. The

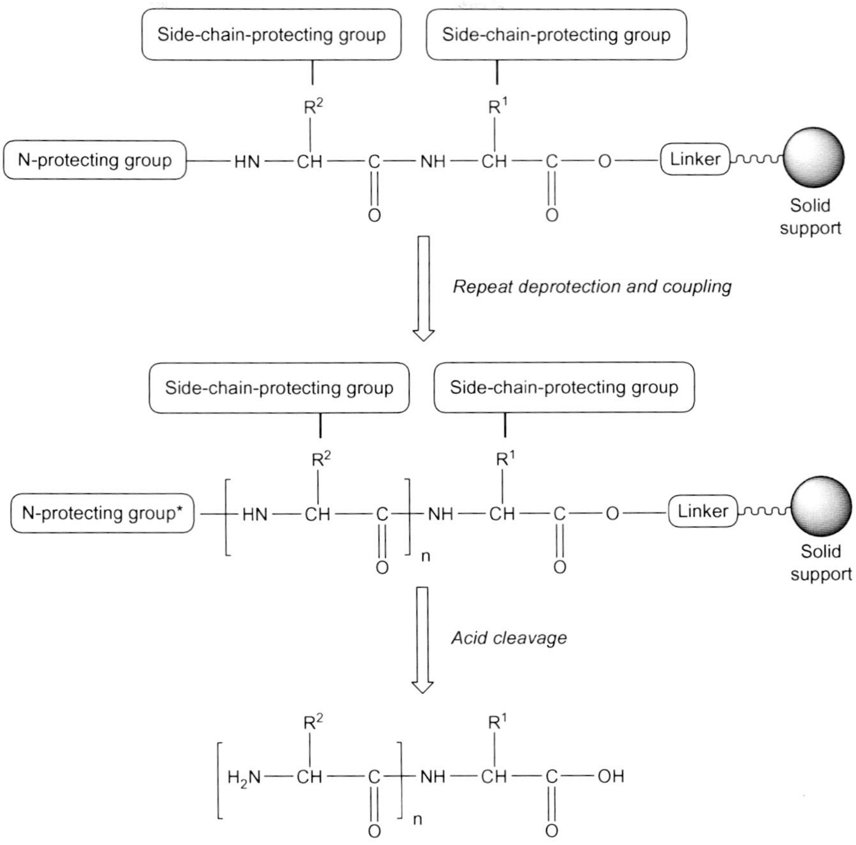

Scheme 1 Basic steps in SPPS.

instrumental methods for analysis are ^{1}H and ^{13}C NMR, circular dichroism, mass spectroscopy, etc.

The main advantages of the SPPS method are: no purification of the intermediate products is required; a possibility for automatization of the process; no loss of products (during the different manipulations); no toxicity; and combinatory chemistry (break/dissolution, sorting/combining).

Our modified peptide analogues were synthesized by SPPS method using Fmoc strategy, because of the advantages of this method. As a solid-state support, the Rink Amide 4-methylbenzhydrylamine (MBHA) resin was employed. The condensation reagents were TBTU, HBTU, DIEA, and DIC (1,3-diisopropylcarbodiimide). As antiracemization, additive 1-HOBt was used.

As the nociceptin analogues with a C-terminal amide group exhibit superior activity than those with C-terminal carboxylic group, all of the newly synthesized peptide was C-terminal amides. It has also been determined that the peptides with C-terminal amide group are more resistant to enzyme degradation and that their conformation suits better the interaction with NOP receptor.

The protective group of the trifunctional amino acids has been chosen to allow an easy cleavage from the resin, while they remain stable during the peptide bond formation. The guanidyl moiety of the Arg was protected by *N*-(2,2,5,7,8-pentamethylchroman-6-sulfonyl) (Pmc) group and/or by 2,2,4,6,7-pentamethyldihydrobenzofuran-5-sulfonyl (Pbf) group, the hydroxyl group of the Thr—*tert*-butyl group in order to inhibit the acylation. The condensation reaction was carried out using a mixture of amino acid/TBTU(HBTU)/HOBt/DIEA/ in 3/2.9/3/9/1 ratio. The condensation reaction time is typically 3.5 h. However, for some amino acids like Phe and Thr, a second condensation with prolongation of the reaction time to 5–6 h may be necessary. A 20% piperidine solution in DMF was used to remove the Fmoc group at every step. The coupling and deprotection reactions were checked by the Kaiser test—a free amino group shows a blue coloration, while in case that all the amino groups have reacted no coloration occurs.

An important decision for the SPPS is the choice of suitable solvent for the condensation reactions. In order to increase the solubility of the Fmoc-protected amino acids, a polar solvent (DMF) was employed. Since in the SPPS, the solvent and the resin are in contact, in the condensation reaction all the reagents (resin, amino acids, solvent, etc.) must be of high purity in order to avoid secondary reaction. The dissolution of the reagents from the resin condensation reagent and HOBt was effectuated with 1:1 ratio of DMF:DCM.

One of the most important steps of the SPPS is the cleavage of the peptide from the resin and the removal of the side-chain-protective groups for the trifunctional amino acids. A mixture of TFA (95%), water (2.5%), and triisopropylsilane (TIS) (2.5%) was used in the deblocking of the peptide from the resin. TIS is used as a scavenger-blocking secondary reaction. The required time for the last procedure is typically around 4.5 h.

For precipitation of the peptides, diethyl ether was employed. After filtration of the precipitate, it is dissolved in ($AcOH/H_2O$) water buffer and is lyophilized. The last step is the purification of the peptides from the salts by high-performance liquid chromatography (HPLC). The peptide was characterized by electrospray ionization mass spectrometry and through optical angle of rotation. Amino acid analysis of the peptide acid hydrolysates gave the expected amino acid ratios.

5. CONCLUSIONS

N/OFQ is an endogenous ligand of the nociceptin opioid peptide (NOP) receptor, structurally and functionally related to the classical opioid receptors. The specific location of the NOP receptor and nociceptin in the brain shows that they take part in the modulation of multiple processes, controlled by CNS and peripheral nervous system. This heptadecapeptide has been found to play a direct role on pain perception. The opportunity to affect nociceptin action by highly potent and selective ligands for its receptors makes much reliable the investigation of neurobiology of NC/NOP system. For that reason, the synthesis of new nociceptin analogues with some modifications is interesting for chemists and physicians. The present review extended our previous research about the synthesis, the characterization and the biological activity of recently synthesized peptides with aminophosphonates moiety and β-tryptophan analogues as NOP and opioid receptors ligands.

The representative newly synthesized peptide analogues and their biological activity shown in this review indicate at least their usefulness as the lead compounds for the design and preparation of new drugs. Especially successful seems to be the approaches for the design of analgesic agents.

ACKNOWLEDGMENT

We gratefully acknowledge the financial support by Grant DTK 02/61 of the Ministry of Education and Science (Bulgaria).

REFERENCES

Ambo, A., Kohara, H., Kawano, S., & Sasaki, Y. (2007). Opioid receptor-like 1 (ORL1) receptor binding and the biological properties of Ac–Arg–Tyr–Tyr–Arg–Ile–Arg–NH_2 and its analogs. *Journal of Peptide Science*, *13*, 672.

Arduin, M., Spagnolo, B., Calò, G., Guerrini, R., Carrà, G., Fischetti, C., et al. (2007). Synthesis and biological activity of nociceptin/orphanin FQ analogues substituted in position 7 or 11 with Cα, α-dialkylated amino acids. *Bioorganic & Medicinal Chemistry*, *15*, 4434–4443.

Berger, H., Albrecht, E., Wallukat, G., & Bienert, M. (1999). Antagonism by acetyl-RYYRIK-NH_2 of G protein activation in rat brain preparations and of chronotropic effect on rat cardiomyocytes evoked by nociceptin/orphanin FQ. *British Journal of Pharmacology*, *126*, 555–558.

Bigoni, R., Giuliani, S., Calò, G., Rizzi, A., Guerrini, R., Salvadori, S., et al. (1999). Characterization of nociceptin receptors in the periphery: In vitro and in vivo studies. *Naunyn-Schmiedeberg's Archives of Pharmacology*, *359*, 160–167.

Bigoni, R., Rizzi, A., Rizzi, D., Camarda, V., Guerrini, R., Lambert, D. G., et al. (2002). Pharmacological characterisation of [(pX)Phe4]nociceptin(1–13)amide analogues. 1. In vitro studies. *Naunyn-Schmiedeberg's Archives of Pharmacology, 365*, 442–449.

Biondi, B., Goldin, D., Giannini, E., Lattanzi, R., Negri, L., Melchiorri, P., et al. (2006). Novel nociceptin analogues: Synthesis and biological activity. *International Journal of Peptide Research and Therapeutics, 12*, 139–144.

Bocheva, A., Nocheva, H., Pavlov, N., Todorov, P., Calmès, M., Martinez, J., et al. (2013). Synthesis and analgesic effects of novel β^2-tryptophan hexapeptide analogs. *Amino Acids, 45*, 983–988.

Bodnar, R. J. (2011). Review, endogenous opiates and behaviour: 2010. *Peptides, 32*, 2522–2552.

Bodnar, R. J. (2013). Endogenous opiates and behavior: 2012. *Peptides, 50*, 55–95.

Butour, J., Moisant, C., Mazarguil, H., Mollereau, C., & Meunier, J. (1997). Recognition and activation of the opioid receptor-like ORL1 receptor by nociceptin, nociceptin analogs and opioids. *European Journal of Pharmacology, 321*, 97–103.

Calò, G., Bigoni, R., Rizzi, A., Guerrini, R., Salvadori, S., & Regoli, D. (2000). Nociceptin/orphanin FQ receptor ligands. *Peptides, 21*, 935–947.

Calò, G., Guerrini, R., Bigoni, R., Rizzi, A., Marzola, G., Okawa, H., et al. (2000). Characterization of [nphe(1)]nociceptin(1–13)NH(2), a new selective nociceptin receptor antagonist. *British Journal of Pharmacology, 129*, 1183–1193.

Calò, G., Guerrini, R., Rizzi, A., Salvadori, S., Burmeister, M., Kapusta, D. R., et al. (2005). UFP-101, a peptide antagonist selective for the nociceptin_orphanin FQ receptor. *CNS Drug Reviews, 11*, 97–112.

Calò, G., Guerrini, R., Rizzi, A., Salvadori, S., & Regoli, D. (2000). Pharmacology of nociceptin and its receptor: A novel therapeutic target. *British Journal of Pharmacology, 129*, 1261–1283.

Calò, G., Lambert, D. G., & Guerrini, R. (2013). Nociceptin/orphanin FQ. In A. Kastin (Ed.), *Handbook of biologically active peptides* (2nd ed., pp. 1577–1585). Academic press. ISBN: 978-0-12-385095-9 (Chapter 215).

Calò, G., Rizzi, A., Bigoni, R., Guerrini, R., Salvadori, S., & Regoli, D. (2002). Pharmacological profile of nociceptin/orphanin FQ receptors. *Clinical and Experimental Pharmacology and Physiology, 29*, 223–228.

Cesselin, F. (1995). Opioid and anti-opioid peptides. *Fundamental & Clinical Pharmacology, 9*, 409–433.

Chen, L. X., Wang, Z. Z., Wu, H., Fang, Q., Chen, Y., & Wang, R. (2002). Effects of nociceptin (13–17) in pain modulation at supraspinal level in mice. *Neuroscience Letters, 331*, 95–98.

Ciccocioppo, R., Angeletti, S., Panocka, I., & Massi, M. (2000). Nociceptin/orphanin FQ and drugs of abuse. *Peptides, 21*, 1071–1080.

Ciccocioppo, R., Panocka, I., Polidori, C., Regoli, D., & Massi, M. (1999). Effect of nociceptin on alcohol intake in alcoholpreferring rats. *Psychopharmacology, 141*, 220–224.

Civelli, O. (2008). The orphanin FQ/nociceptin (OFQ/N) system. *Results and Problems in Cell Differentiation, 46*, 1–25.

Corbett, A. D., Henderson, G., McKnight, A. T., & Paterson, S. J. (2006). 75 years of opioid research: The exciting but vain quest for the Holy Grail. *British Journal of Pharmacology, 147*, S153–S162.

Cunningham, E., Drag, M., Kafarski, P., & Bell, A. (2008). Chemical target validation studies of aminopeptidase in malaria parasites using alpha-aminoalkylphosphonate and phosphonopeptide inhibitors. *Antimicrobial Agents and Chemotherapy, 52*, 3221–3228.

Davies, B. G. (2002). Synthesis of glycoproteins. *Chemical Reviews, 102*, 579–601.

Doggrell, S. A. (2007). Cardiovascular and renal effects of nociceptin/orphanin FQ: A new mediator to target? *Current Opinion in Investigational Drugs, 8*, 742–749.

Dooley, C. T., Chung, N. N., Shiller, P. W., & Houghten, R. A. (1993). Acetalins: Opioid receptor antagonists determined through the use of synthetic peptide combinatorial libraries. *Proceedings of the National Academy of Sciences of the United States of America*, *90*, 10911.

Dooley, C. T., & Houghten, R. A. (1996). Orphanin FQ: Receptor binding and analog structure activity relationships in rat brain. *Life Sciences*, *59*, L23–PL29.

Dooley, C. T., Spaeth, C. G., Berzetei-Gurske, I. P., Craymer, K., Adapa, I. D., Brandt, S. R., et al. (1997). Binding and in vitro activities of peptides with high affinity for the nociceptin/orphanin FQ receptor, ORL1. *The Journal of Pharmacology and Experimental Therapeutics*, *283*, 735–741.

Dzambazova, E., Nocheva, H., & Bocheva, A. (2008). Analgesic effects of some newly synthesized nociceptin analogues in rats. *Pharmacy*, *55*, 30–34.

Fukada, K., Kato, S., Mori, K., Nishi, M., Takeshima, H., Iwabe, N., et al. (1994). cDNA cloning and regional distribution of novel member of the opioid receptor family. *FEBS Letters*, *343*, 42–46.

Girchev, R. A., Markova, P. P., Naydenova, E. D., & Vezenkov, L. T. (2009). Fast oscillations of arterial blood pressure during nociceptin analogues application in Wistar rats. *Bulgarian Chemical Communications*, *41*, 127–132.

Girchev, R., Markova, P., Naydenova, E., & Vezenkov, L. (2010). The influence of nociceptin analogues on the modulation of blood pressure oscillation in spontaneously hypertensive rats. *Trakia Journal of Sciences*, *8*, 107–113.

Guerrini, R., Calò, G., Bigoni, R., Rizzi, A., Regoli, D., & Salvadori, S. (2001). Structure–activity relationship of [$Nphe^1$]-NC-(1–13)-NH_2, a pure and selective nociceptin/orphanin FQ receptor antagonist. *The Journal of Peptide Research*, *57*, 215–222.

Guerrini, R., Calò, G., Bigoni, R., Rizzi, D., Rizzi, A., Zucchini, M., et al. (2001). Structure–activity studies of the Phe(4) residue of nociceptin(1–13)-NH(2): Identification of highly potent agonists of the nociceptin/orphanin FQ receptor. *Journal of Medicinal Chemistry*, *44*, 3956–3964.

Guerrini, R., Calò, G., Rizzi, A., Bigoni, R., Bianchi, C., Salvadori, S., et al. (1998). A new selective antagonist of the nociceptin receptor. *British Journal of Pharmacology*, *123*, 163–165.

Guerrini, R., Calò, G., Rizzi, A., Bigoni, R., Rizzi, D., Regoli, D., et al. (2000). Structure–activity relationships of nociceptin and related peptides: Comparison with dynorphin A. *Peptides*, 7, 923–933.

Guerrini, R., Rizzi, A., Varani, K., Toth, G., Gessi, S., Hashiba, E., et al. (2000). Further studies on nociceptin-related peptides: Discovery of a new chemical template with antagonist activity on the nociceptin receptor. *Journal of Medicinal Chemistry*, *43*, 2805–2813.

Guerinni, S., Tramontana, M., & Maggi, C. A. (1997). Effects of nocicepton on heart rate and blood pressure in anesthetized rats. *European Journal of Pharmacology*, *333*, 177–179.

Hantos, M. B., Szalay, F., Lakatos, P. L., Hegedûs, D., Firneisz, G., Reiczigel, J., et al. (2002). Elevated plasma nociceptin level in patients with Wilson disease. *Brain Research Bulletin*, *58*, 311–313.

Hawes, B. E., Graziano, M. P., & Lambert, D. G. (2000). Cellular actions of nociceptin: Transduction mechanisms. *Peptides*, *21*, 961–967.

Heinricher, M. M. (2003). Orphanin FQ/nociceptin: From neural circuitry to behaviour. *Life Sciences*, *73*, 813–822.

Henderson, G., & McKnight, A. T. (1997). The orphan opioid receptor and its endogenous ligand—Nociceptin/orphanin FQ. *Trends in Pharmacological Sciences*, *18*, 293–300.

Hirschmann, R., Taylor, C. M., Benkovic, P. A., Taylor, S. D., Yager, K. M., Sprengeler, P. A., et al. (1994). Peptide synthesis catalyzed by an antibody containing a binding site for variable amino acids. *Science*, *265*, 234–237.

Horvath, A., Folhoffer, A., Lakatos, P. L., Halasz, J., Hylies, G., & Schaff, Z. (2004). Rising plasma nociceptin level during development of HCC: A case report. *World Journal of Gastroenterology*, *10*, 152–154.

Judd, A. K., Tuttle, D. J., Jones, R. W., Sanchez, A., Polgar, W., Berzetei-Gurske, I., et al. (2004). Structure–activity studies on high affinity NOP-active hexapeptides. *The Journal of Peptide Research*, *64*, 87–94.

Kasakov, L., Nashar, M., Naydenova, E., Vezenkov, L., & Vlaskovska, M. (2010). In vitro studies of the activity of newly synthesized nociceptin/orphanin FQ receptor ligand analogues. *Protein and Peptide Letters*, *17*, 616–620.

Kessler, H. (1993). Peptoids—A new approach to the development of pharmaceuticals. *Angewandte Chemie International Edition in English*, *32*, 543–544.

Kirkova, M., Zamfirova, R., Leoekiewicz, M., Kubera, M., Lason, W., & Todorov, S. (2009). Study of the cytotoxicity and antioxidant capacity of N/OFQ(1–13)NH_2 and its structural analogues. *Pharmacological Reports*, *61*, 1163–1172.

Kukhar, V. P., Soloshonok, V. A., & Solodenko, V. A. (1994). Asymmetric synthesis of phosphorus analogs of amino acids. *Phosphorus, Sulfur, and Silicon and the Related Elements*, *92*, 239–264.

León-Olea, M., Miller-Pérez, C., Sánchez-Islas, E., Mendoza-Sotelo, J., Garduño-Gutiérrez, R., de Gortari, P., et al. (2013). The nociceptin/orphanin FQ-like opioid peptide in nervous periesophageal ganglia of land snail Helix aspersa. *Brain Research*, *1505*, 22–46.

Markova, P., Girchev, R., Naydenova, E., & Vezenkov, L. (2007). Heart rate variability after application of nociceptin analog N/OFQ(1–13)NH_2 in spontaneously hypertensive rats. *Trakia Journal of Sciences*, *5*, 1–6.

Marthi, M., Mela, F., Fartin, M., Zucchini, S., Brown, J. M., Witta, J., et al. (2005). Blockade of nociceptin/orphanin FQ transmission attenuates symptoms and neurodegeneration associated with Parkinson's disease. *The Journal of Neuroscience*, *25*, 9591–9601.

Martin, W. R. (1967). Opioid antagonists. *Pharmacological Reviews*, *19*, 463–521.

Mason, S. L., Ho, M., Nicholson, J., & McKnight, A. T. (2001). In vitro characterization of Ac-RYYRWK-NH_2, Ac-RYYRIK-NH_2 and [Phe$^1\Psi$(CH_2-NH)Gly^2] nociceptin (l–13)NH_2, at rat native and recombinant ORLr receptors. *Neuropeptides*, *35*, 244–256.

Meng, F., & Xu, J. (2010). Direct synthesis of phosphinopeptides containing C-terminal alpha-aminoalkylphosphinic acids. *Amino Acids*, *39*, 533–538.

Meunier, J. C., Mollereau, C., Toll, L., Suaudeau, C., Moisand, C., Alvinerie, P., et al. (1995). Isolation and structure of the endogenous agonist of opioid receptor-like ORL1 receptor. *Nature*, *377*, 532–535.

Mogil, J. S., & Pasternak, G. W. (2001). The molecular and behavioral pharmacology of the orphanin FQ/nociceptin peptide and receptor family. *Pharmacological Reviews*, *53*, 381–415.

Mollereau, C., Parmentier, M., & Mailleux, P. (1994). ORL1, a novel member of the opioid receptor family. Cloning, functional expression and localization. *FEBS Letters*, *341*, 33–38.

Mollereau, C., Simons, M. J., Soularue, P., Liners, F., Vassart, G., Meunier, J. C., et al. (1996). Structure, tissue distribution, and chromosomal localization of the prepronociceptin gene. *Proceedings of the National Academy of Sciences of the United States of America*, *93*, 8666–8670.

Moran, T. D., Abdulla, F. A., & Smith, P. A. (2000). Cellular neurophysiological actions of nociceptin/orphanin FQ. *Peptides*, *21*, 969–976.

Naydenova, E., Pavlov, N., Todorov, P., Dzhambazova, E., & Bocheva, A. (2011). Analgesic effects of novel hexapeptide analogues as nociceptin receptor ligands. In *Proceedings of the 22nd American Peptide Symposium on American Peptide Society* (pp. 356–357).

Naydenova, E., & Todorov, P. (2011). Nociceptin/orphanin (N/OFQ) analogues: Chemistry and biology. *Journal of the University of Chemical Technology and Metallurgy*, *46*, 333–348.

Naydenova, E., Todorov, P., Mateeva, P., Zamfirova, R., Pavlov, N., & Todorov, S. (2010). Synthesis and biological activity of novel small peptides with aminophosphonates moiety as NOP receptor ligands. *Amino Acids*, *39*, 1537–1543.

Naydenova, E., Todorov, P., Topashka-Ancheva, M., Momekov, G., Yordanova, Tz, Konstantinov, S., et al. (2008). Novel N-(phosphonomethyl)glycine derivatives. Design, characterization and biological activity. *European Journal of Medicinal Chemistry*, *43*(6), 1199–1205.

Naydenova, E., Todorov, P., & Troev, K. (2010). Resent synthesis of aminophosphonic acids as potential biological importance. *Amino Acids*, *38*, 23–30.

Naydenova, E., Zhivkova, V., Zamfirova, R., Vezenkov, L., Dobrinova, Y., & Mateeva, P. (2006). Synthesis and biological activity of nociceptin/orphanin FQ(1–13)NH_2 analogues modified in 9 and/or 13 position. *Bioorganic & Medicinal Chemistry Letters*, *16*, 4071–4074.

Neal, C. R., Mansour, A., Reinsheid, R., Nothacker, H. P., Givelli, O., & Watson, S. J. (1999). Localization of orphanin FQ (nociceptin) peptide and messenger RNA in the central nervous system of the rat. *The Journal of Comparative Neurology*, *406*, 503–547.

Negri, L., Lattanzi, R., Tabacco, F., Orrù, I., Severini, C., Scolaro, B., et al. (1999). Dermorphin and deltorphin glycosylated analogues: Synthesis and antinociceptive activity after systemic administration. *Journal of Medicinal Chemistry*, *42*, 400–404.

Nicholson, J. R., Paterson, S. J., Menzies, J. R., Corbett, A. D., & McKnight, A. T. (1998). Pharmacological studies on the "orphan" opioid receptor in central and peripheral sites. *Canadian Journal of Physiology and Pharmacology*, *76*, 304–313.

Nothacker, H. P., Reinshield, R. K., Monsour, A., Henningsen, R. A., Ardati, A., Monsma, F. J., Jr., et al. (1996). Primary structure and tissue distribution of the orphan FQ precursor. *Proceedings of the National Academy of Sciences of the United States of America*, *93*, 8677–8682.

Olson, G. A., Olson, R. D., & Kastin, A. J. (1995). Endogenous opiates: 1994. *Peptides*, *16*, 1517–1555.

Reinscheid, R. K., Ardati, A., Monsma, F. J., & Civelli, O. (1996). Structure–activity relationship studies on the novel neuropeptide orphanin FQ. *The Journal of Biological Chemistry*, *271*, 14163–14168.

Reinscheid, R. K., & Civelli, O. (2002). The orphanin FQ/nociceptin knockout mouse: A behavioral model for stress responses. *Neuropeptides*, *36*, 72–76.

Reinscheid, R., Higelin, J., Henningsen, R., Monsma, F. J., & Civelli, O. (1998). Structures and that delineate orphanin FQ and dinorphin A pharmacological selectivities. *The Journal of Biological Chemistry*, *273*, 1490–1495.

Reinscheid, R. K., Nothacker, H. P., Bourson, A., Ardati, A., Henningsen, R. A., Bunzow, J. R., et al. (1995). Orphanin FQ: A neuropeptide that activates an opioidlike G protein-coupled receptor. *Science*, *270*, 792–794.

Rizzi, A., Bigoni, R., Kamarda, V., Marzola, G., Guerrini, R., Risi, C., et al. (2002). [Arg14, Lys15] nociceptin/orphanin FQ receptor: In vitro and in vivo studies. *The Journal of Pharmacology and Experimental Therapeutics*, *300*, 57–63.

Rizzi, A., Marzola, G., Bigoni, R., Guerrini, R., Salvadori, S., Mogil, J. S., et al. (2001). Endogenous nociceptin signaling and stress-induced analgesia. *Neuroreport*, *12*, 3009–3013.

Robinson, T. E., & Becker, J. B. (1986). Enduring changes in brain and behavior produced by chronic amphetamine administration: A review and evaluation of animal models of amphetamine psychosis. *Brain Research*, *396*, 157–198.

Schweizer, F. (2002). Glycosamino acids: Building blocks for combinatorial synthesis—Implications for drug discovery. *Angewandte Chemie International Edition in English*, *41*, 230–253.

Simon, R. J., Kania, R. S., Zuckermann, R. N., Huebner, V. D., Jewell, D. A., Banville, S., et al. (1992). Peptoids: A modular approach to drug discovery. *Proceedings of the National Academy of Sciences of the United States of America*, *89*, 9367–9371.

Szalay, F., Hantos, M. B., Horvath, A., Lakatos, P. L., Folhoffer, A., & Dunkel, K. (2004). Increased nociceptin/orphanin FQ plasma levels in hepatocellular carcinoma. *World Journal of Gastroenterology*, *10*, 42–45.

Tariq, S., Nurulain, S., Tekes, K., & Adeghate, E. (2013). Deciphering intracellular localization and physiological role of nociceptin and nocistatin. *Peptides*, *43*, 174–183.

Tzvetanova, E., Pavlova, A., Alexandrova, A., Nenkova, G., Petrov, L., Kirkova, M., et al. (2009). Are nociceptin(1–13)NH_2 and its structural analogue [Orn^9] nociceptin(1–13) NH_2 able to affect brain antioxidant status in control and kainic acid-treated rats? *Cell Biochemistry and Function*, *27*, 243–250.

Varani, K., Rizzi, A., Calò, G., Bigoni, R., Toth, G., Guerrini, R., et al. (1999). Pharmacology of [Tyr1]nociceptin analogs: Receptor binding and bioassay studies. *Naunyn-Schmiedeberg's Archives of Pharmacology*, *360*, 270–277.

Vezenkov, L., Zhivkova, V., & Naydenova, E. (2006). Design and synthesis of new analogues of the nociceptin/orphanin FQ receptor ligands Ac–Arg–Tyr–Tyr–Arg–Trp–Lys–NH_2 and Ac–Arg–Tyr–Tyr–Arg–Ile–Lys–NH_2. *Bulgarian Chemical Communications*, *38*, 64–66.

Williams, J. P., Thompson, J. P., Young, S. P., Gold, S. J., McDonald, J., Rowbotham, D. J., et al. (2008). Nociceptin and urotensin-II concentrations in critically ill patients with sepsis. *British Journal of Anaesthesia*, *100*, 810–814.

Wollemann, M., Benyhe, S., & Simon, J. (1993). The kappa-opioid receptor: Evidence for the different subtypes. *Life Sciences*, *52*, 599–611.

Zamfirova, R., Pavlov, N., Todorov, P., Mateeva, P., Martinez, J., Calmès, M., et al. (2013). Synthesis and changes in affinity for NOP and opioid receptors of novel hexapeptides containing β^2-tryptophan analogues. *Bioorganic & Medicinal Chemistry Letters*, *23*, 4052–4055.

Zamfirova, R., Tzvetanova, E., Alexandrova, A., Petrov, L., Mateeva, P., Pavlova, A., et al. (2009). In vivo effects of N/OFQ(1–13)NH_2 and its structural analogue [Orn^9]N/OFQ (1–13)NH_2 on carrageenan-induced inflammation: Rat-paw oedema and antioxidant status. *Central European Journal of Biology*, *4*, 170–178.

Zaveri, N. (2003). Peptide and nonpeptide ligands for the nociceptin/orphanin FQ receptor ORL1: Research tools and potential therapeutic agents. *Life Sciences*, *73*, 663–678.

Zaveri, N., Jiang, F., Olsen, C., Polgar, W., & Toll, L. (2005). Small-molecule agonists and antagonists of the opioid receptor-like receptor (ORL1, NOP): Ligand-based analysis of structural factors influencing intrinsic activity at NOP. *The AAPS Journal*, 7, E345–E352. Article 34.

CHAPTER FIVE

Pain Regulation by Nocistatin-Targeting Molecules: G Protein-Coupled-Receptor and Nocistatin-Interacting Protein

Emiko Okuda-Ashitaka*,[1], Seiji Ito[†]
*Department of Biomedical Engineering, Osaka Institute of Technology, Osaka, Japan
[†]Department of Medical Chemistry, Kansai Medical University, Hirakata, Japan
[1]Corresponding author: e-mail address: ashitaka@bme.oit.ac.jp

Contents

Abstract

Nociceptin/orphanin FQ (N/OFQ) and nocistatin (NST) are neuropeptides produced from the same precursor protein. N/OFQ is involved in a broad range of central functions including pain, learning, memory, anxiety, and feeding. However, NST has opposite effects on various central functions evoked by N/OFQ. The regulation of their receptors may be important for these opposite functions of NST and N/OFQ. Although N/OFQ binds to a specific N/OFQ receptor, the target molecule of NST remains unclear. Some biological effects of NST are mediated by a G protein-coupled receptor. Furthermore, using high-performance affinity nanobeads, we recently identified a

Vitamins and Hormones, Volume 97
ISSN 0083-6729
http://dx.doi.org/10.1016/bs.vh.2014.12.001

4-nitrophenylphosphatase domain and nonneuronal SNAP25-like protein homolog 1 (NIPSNAP1) as a protein that interacts with NST in the mouse spinal cord. The inhibition of N/OFQ-evoked tactile pain allodynia by NST is mediated by NIPSNAP1. This review focuses on the molecular mechanisms of pain regulation by the target molecules of NST including a G protein-coupled receptor and NIPSNAP1.

1. INTRODUCTION

Nociceptin/orphanin FQ (N/OFQ) is generated from the precursor protein preproN/OFQ, by proteolytic processing, which is post-translationally processed into mature bioactive peptides (Fig. 1A; Meunier et al., 1995; Okuda-Ashitaka et al., 1996; Reinscheid et al., 1995). N/OFQ is composed of 17 amino acid residues that are conserved among species and is structurally similar to the opioid peptide dynorphin A. N/OFQ selectively binds to a specific N/OFQ receptor (NOP), which is highly homologous to the classical opioid receptors μ, δ, and κ. The N/OFQ and NOP systems are involved in a broad range of biological functions in the central and peripheral tissues, such as the nervous system, cardiovascular system, airways, gastrointestinal tract, urogenital tract, and immune system (Lambert, 2008). N/OFQ and NOP are widely distributed

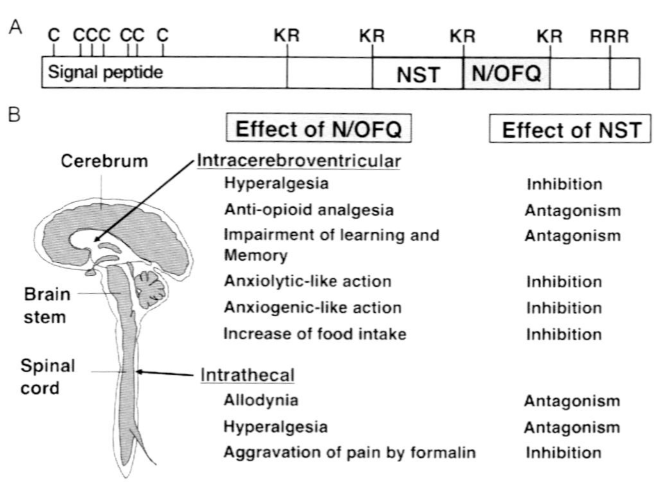

Figure 1 (A) Schematic representation of the structure of N/OFQ and NST peptide precursors. (B) Effects of NST on the N/OFQ-induced biological functions.

in the part of the central nervous system associated with pain transmission (Boom et al., 1999; Houtani et al., 2000; Ikeda et al., 1998). Although the classical opioids and their receptor systems have been implicated in analgesic effects, N/OFQ has been reported to produce both pronociceptive and antinociceptive responses depending on the testing paradigms, injection sites, and doses (Ito, Okuda-Ashitaka, & Minami, 2001; Mogil & Pasternak, 2001; Zeilhofer & Calo', 2003). Intracerebroventricular (i.c.v.) injections of N/OFQ induce pronociceptive effects such as hyperalgesia, which is defined as increased response to stimuli that is physiologically painful and reverses opioid-induced analgesia. Intrathecal (i.t.) administration of N/OFQ induces, in addition to analgesia, hyperalgesia, and touch-evoked allodynia, which is pain due to innocuous stimuli that do not normally provoke pain (Hara et al., 1997; Okuda-Ashitaka et al., 1998).

Nocistatin (NST) is a neuropeptide generated from the same precursor protein as N/OFQ (Fig. 1A; Okuda-Ashitaka et al., 1998). NST differs in the number of amino acid residues among various species: 17 amino acid residues in bovine, 30 in human, 35 in rat, and 41 in mouse. In fact, these various lengths of NST amino acid residues have been detected in the brain of these species and human cerebrospinal fluid (Joseph et al., 2006; Lee et al., 1999; Okuda-Ashitaka et al., 1998). We first found that the i.t. administration of bovine NST blocks the N/OFQ-induced nociceptive transmission such as hyperalgesia and allodynia (Fig. 1B; Okuda-Ashitaka et al., 1998). All of the various species of NST have inhibitory activity for N/OFQ-induced allodynia (Minami et al., 1998; Okuda-Ashitaka & Ito, 2000; Okuda-Ashitaka et al., 1998). The i.t. administration of NST also inhibits the N/OFQ-induced aggravation of pain by formalin (Nakano et al., 2000). Furthermore, the i.c.v. administration of NST reverses N/OFQ-induced biological activities such as hyperalgesia, anti-opioid analgesia (Liu, Nishiuchi, Kimura, & Tachibana, 2006; Scoto, Santangelo, & Parenti, 2005; Zhao et al., 1996), impairment of learning and memory (Hiramatsu & Inoue, 1999), anxiolytic and anxiogenic actions (Gavioli, Rae, Calo, Guerrini, & de Lima, 2002; Kamei, Matsunawa, Miyata, Tanaka, & Saitoh, 2004), and an increase in food intake (Olszewski, Shaw, Grace, Billington, & Levine, 2000). NST has no effects under physiologically normal conditions, but NST itself has inhibitory effects on inflammatory pain responses induced by formalin and carrageenan/kaolin (Nakagawa, Kaneko, Inamura, & Satoh, 1999; Nakano et al., 2000; Yamamoto & Sakashita, 1999), hyperalgesia in chronic constriction injury

(CCI) to the sciatic nerve, and morphine tolerance (Ge et al., 2007; Sun et al., 2001). NST at high doses (nmol) has also been reported to induce pronociceptive effects on the inflammatory pain response by formalin (Ahmadi et al., 2003; Zeilhofer, Selbach, Gühring, Erb, & Ahmadi, 2000), hyperalgesia in CCI, and nociceptive flexor reflexes (Inoue, Kawashima, Allen, & Ueda, 2003).

The pharmacological characterization of N/OFQ and NST has been studied extensively *in vivo*, and subsequent characterization of the cellular actions is required for understanding the biological functions. The NOP receptor is a member of G protein-coupled receptor superfamily. N/OFQ, which is sensitive to a $G_{i/o}$ inhibitor, pertussis toxin (PTX), is implicated in the inhibition of cyclic AMP production, the opening of inwardly rectifying K^+ channels, and the closing of voltage-gated Ca^{2+} channels (Lambert, 2008). The βγ subunits of $G_{i/o}$ also activate phospholipase C (PLC) pathway (New & Wong, 2002). Furthermore, N/OFQ is implicated in the activation of PLC pathway through PTX-insensitive G_{14} or G_{16}, and in the activation of mitogen-activated protein (MAP) kinase through PTX-insensitive $G_{z,}$, $G_{12,}$, $G_{14,}$, or G_{16}. In contrast, NST does not interfere with the binding of N/OFQ to the NOP receptor and the N/OFQ-induced inhibition of cAMP formation in the NOP cDNA-transfected cells, although it affects various biological responses induced by N/OFQ (Okuda-Ashitaka & Ito, 2000; Okuda-Ashitaka et al., 1996). Several biological activities of NST are inhibited by PTX, and therefore NST may bind an as yet unidentified $G_{i/o}$ protein-coupled receptor (Table 1). NST suppresses inhibitory neurotransmission such as glycine in the rat spinal cord *in situ* (Zeilhofer et al., 2000), inhibits 5-hydroxytryptamine release in the mouse neocortex synaptosome *in vitro* (Fantin, Fischetti, Trapella, & Morari, 2007), and induces nociception *in vivo* through PTX-sensitive $G_{i/o}$ (Inoue et al., 2003). NST has been reported to depolarize the projection neurons of periaqueductal gray (PAG) through PTX-insensitive $G_{q/11}$ (Chen, Li, Yeh, Chou, & Wang, 2009; Chen et al., 2010). Identification of the NST receptor will contribute to a better understanding of biological responses evoked by NST.

The G protein-coupled receptors hitherto have been identified by using expression cloning, whereas radioligand binding and signal transduction served as screening assays. The first successful cloning of an opioid receptor, the δ opioid receptor, resulted from the expression screening methods using radioligand binding (Evans, Keith, Morrison, Magendzo, & Edwards, 1992).

Table 1 Putative signaling pathway of NST through G protein

G protein	Signal transduction	Effective dose	Effect	Reference
Neurotransmission				
$G_{i/o}$ (PTX-sensitive)		1–10 μM	Inhibition of inhibitory synaptic transmission	Zeilhofer et al. (2000)
$G_{i/o}$ (PTX-sensitive)		0.001–30 μM	Inhibition of 5-hydroxytryptamine release	Fantin et al. (2007)
$G_{q/11}$	PLCβ–PKC–TRPC1	0.01–3 μM	Depolarization of neuron	Chen et al. (2010)
Pain response				
$G_{i/o}$ (PTX-sensitive)	PLCβ-Substance P release	0.01–10 pmol, i.pl.	Nociceptive flexor response	Inoue et al. (2003)

Subsequently, other homologous-receptor cDNA clones were obtained by hybridization screening and polymerase chain reaction. In contrast, affinity purification is a useful technique for the identification of ligand-binding proteins. The approach for purifying solubilized receptor protein from tissue was unsuccessful, primarily because of the low efficiency of affinity purification and nonspecific binding. High-performance affinity nanobeads (SG beads) have excellent features such as low nonspecific protein interaction and high purification efficiency (Ohtsu et al., 2005; Shimizu et al., 2000). The SG beads that conjugate with naltrindole, a selective antagonist against the δ-opioid receptor, can purify the δ-opioid receptor from the expressing cells (Hasegawa et al., 2006). Therefore, we tried affinity purification using the NST-conjugated SG beads from synaptosomal membranes of the mouse spinal cord and identified a 4-nitrophenylphosphatase domain and non-neuronal SNAP25-like protein homolog 1 (NIPSNAP1) as a protein that interacts with NST (Okuda-Ashitaka et al., 2012). The inhibition of N/OFQ-evoked allodynia by NST was lacking in NIPSNAP1-deficient mice. This review focuses on pain responses regulated by target molecules of NST including a G protein-coupled receptor and NIPSNAP1.

2. BIOLOGICAL ACTIVITY BY NST THROUGH G PROTEIN-COUPLED RECEPTOR

2.1 Regulation of presynaptic neurotransmitter release through putative $G_{i/o}$-coupled NST receptor

Effects of opioids, including N/OFQ, on the inhibition of cyclic AMP formation and the opening of inwardly rectifying K^+ channels through the α-subunit of the $G_{i/o}$ protein reduce neuronal excitability and their effects on the closing of voltage-gated Ca^{2+} channel through the βγ subunit of $G_{i/o}$ protein reduce the neurotransmitter release. In the dorsal horn of the spinal cord which is an important site for pain transmission, the excitatory neurotransmitters such as glutamate and substance P released from the primary afferent terminals enhance the excitability of secondary spinal neurons (Basbaum, Bautista, Scherrer, & Hulius, 2009; Costigan, Scholz, & Woolf, 2009; Dray, Urban, & Dickenson, 1994; Gold & Gebhart, 2010; Kuner, 2010). The i.t. N/OFQ-induced analgesic activity results from the inhibition of the nociceptive transmission through $G_{i/o}$-coupled receptor. Electrophysiological studies *in situ* indicate that N/OFQ inhibits the glutamatergic neurotransmission in the superficial dorsal horn of the spinal cord (Chaplan, Malmberg, & Yaksh, 1997; Lutfy, Cai, Woodward, & Weber, 1997; Nakano et al., 2000). Similar to N/OFQ, NST has been reported to be involved in neurotransmitter release through a PTX-sensitive $G_{i/o}$ protein in the dorsal horn of the spinal cord (Zeilhofer et al., 2000). NST (1–10 μM) does not directly interfere with the release of glutamate from the presynaptic terminal and inhibits the release of glycine from inhibitory interneurons in the dorsal horn of the spinal cord. Inhibitory neurotransmitters released from interneurons suppress the spinal nociceptive transmission (Lu & Peri, 2003; Lyadomi, Lyadomi, Kumamto, Tomokuni, & Yoshimura, 2000; Yoshimura & Nishi, 1995; Zeilfofer, 2005; Zeilfofer, Winder, & Yevebes, 2012), and blocking of glycinergic neurotransmission leads to the hypersensitivity of the spinal dorsal horn neurons (Cronin, Bradbury, & Lidierth, 2004; Sivilotti & Woolf, 1994; Yaksh, 1989). The i.t. administration of NST at 1–10 nmol increases the formalin-evoked inflammatory pain and neuropathic pain of CCI model (Ahmadi et al., 2003; Ma, Xie, Dong, Wang, & Wu, 2003; Xu et al., 1996). The pronociceptive effects of i.t. NST are due to the suppression of inhibitory neurotransmission through the blocking of glycine release in the spinal cord. A recent report indicated that NST (1 μM) inhibits the high K^+-induced

5-hydroxytryptamine release from cerebrocortical synaptosomes through $G_{i/o}$ protein (Fantin et al., 2007). The i.c.v. administration of NST has been shown to be a potent anxiogenic-like peptide, and it inhibits both the anxiolytic and anxiogenic effects evoked by N/OFQ through the regulation of serotonergic transmission. Serotonergic transmission is also implicated in the descending pathway from the brainstem and it exerts a strong inhibitory effect on the spinal dorsal horn transmission. In addition to glycine release, the suppression of serotonin release by NST may enhance nociceptive responses in the spinal cord.

The $\beta\gamma$ subunits of $G_{i/o}$ activate the PLC pathway, and activated PLC cleaves phosphatidylinositol 4,5-bisphosphate (PIP_2) to generate diacylglycerol (DG) and inositol 1,4,5-triphosphate (IP_3) for Ca^{2+}-release and the activation of protein kinase C (PKC). The intraplantal (i.pl.) administration of NST (0.01–1 pmol) induces a pronociceptive response in the allogenic-induced nociceptive flexion test (Inoue et al., 2003). The NST-induced nociceptive flexor response is blocked by the i.pl. injection of PTX and U-73122, a PLC inhibitor. The nociceptive response of NST is mediated by the polymodal C-fibers primary afferent neuron-containing substance P. NST may induce the release of substance P from the peripheral nerve endings of the nociceptive primary afferent neurons by the PLC activation through $\beta\gamma$ subunits of $G_{i/o}$.

2.2 Regulation of postsynaptic transmission through putative $G_{i/o}$-coupled NST receptor

NST may have different biological effects depending on doses; low doses (fmol–pmol) of NST induce antinociception and high doses (nmol), pronociception. The i.t. administration of NST significantly inhibits the formalin-evoked pain behaviors at fmol–pmol order (Ahmadi et al., 2003; Nakano et al., 2000), but NST at 1–10 nmol significantly increases it (Ahmadi et al., 2003; Xu et al., 1999). The suppression of glycine release by NST inhibits not only the inhibitory neurotransmission through the glycine receptor but also the activation of the N-methyl-D-aspartate (NMDA) receptor (Ahmadi et al., 2003). Glycine is known to interact with a single binding site of the NR1 subunit of the glutamate receptor NMDA type (Dingledine, Borges, Bowie, & Traynellis, 1999; MacBain, Kleckner, Wyrick, & Dingledine, 1989; Zeilfofer et al., 2012). The activation of NMDA by glycine is more sensitive than the activation of the glycine receptor because the ED_{50} of glycine is approximately 100 μM for the glycine receptor and 0.1–1 μM for the glycine-binding site of the NMDA receptor. The suppression of glycine release by a low dose of NST via $G_{i/o}$ may reduce nociceptive response through the inhibition of the postsynaptic NMDA receptor activation (Ahmadi et al., 2003).

2.3 Depolarization of projection neurons by a putative $G_{q/11}$-coupled NST receptor

NST has been reported to link a receptor coupled with PTX-insensitive $G_{q/11}$. NST depolarizes the projection neurons of the central nucleus of the amygdala (CeA) to PAG (Chen et al., 2009). PAG and rostral ventromedial medulla (RVM) are the components of descending antinociceptive circuitry and can inhibit sensory processing in the spinal dorsal horn (Porrenca, Ossipov, & Gebhart, 2000). The CeA–PAG–RVM descending pain-modulating pathway is the supraspinal action site of opioid-induced analgesia. PAG is interconnected with the limbic forebrain areas, including the dorsomedial nucleus of the hypothalamus, amygdala, medial prefrontal cortex, anterior cingulate cortex, and CeA, the output nuclei in amygdala (Heinricher, Tavares, Leith, & Lumb, 2009). Although NST failed to affect the membrane excitability of PAG and RVM neurons directly (Vaughan et al., 2001), it induced inward currents through a non-selective cation channel, the classical transient receptor potential (TRPC) channels, in medial CeA neurons. The NST-induced TRPC cation current is completely inhibited by the anti-$G_{q/11}$ antibody, an inhibitor of PLC U-73122, and a potent inhibitor of PKC inhibitor GF109203, but not by a Ca^{2+} chelator BAPTA. Although PLC activation induced both PKC activation and IP_3-medicated Ca^{2+} release, the opening of TRPC cation channels is mediated by the PLC–PKC pathway through the $G_{q/11}$-coupled receptor (Ramsey, Delling, & Clapham, 2006). NST also depolarizes the projection neurons of rostral agranular insular cortex (RAIC)-PAG (Chen et al., 2009, 2010). RAIC-PAG projection neurons are an essential component of brainstem descending antinociceptive circuitry, and RAIC in the insular cortex is also involved in supraspinal antinociceptive effects produced by opioid, dopamine, and GABA. The i.c.v. administration of NST at pmol–nmol order inhibited the N/OFQ-induced hyperalgesia (Liu et al., 2006; Scoto et al., 2005) and the carrageenan/kaolin-induced inflammatory pain (Nakagawa et al., 1999). Therefore, i.c.v. NST leads to supraspinal antinociception by the activation of descending antinociceptive pathways.

3. PAIN REGULATION THROUGH AN NST-INTERACTING PROTEIN

3.1 Purification of an NST-interacting protein using high-performance affinity latex nanobeads

Affinity purification is a useful technique for the identification of ligand-binding proteins. The use of conventional matrices such as agarose gels and sepharose gels has several problems such as low efficiency of affinity purification and nonspecific binding of proteins to matrices. Therefore, various kinds of matrices related to affinity purifications have been developed. In particular, high-performance affinity latex nanobeads (called SG beads), which are glycidylmethacrylate-covered glycidylmethacrylate-styrene copolymer core beads, were found to have excellent features such as low nonspecific protein interaction and high purification efficiency (Shimizu et al., 2000). The copolymer core of SG beads is made up of nonporous particles to reduce nonspecific binding, and the covered surface possesses moderate hydrophilicity and adequate functional groups to allow surface chemical modification. The SG beads have been successfully used to purify various proteins including drug receptors and transcription factors (Ohtsu et al., 2005; Sakamoto, Kabe, Hatakeyama, Yamaguchi, & Handa, 2009; Shimizu et al., 2000). With regard to the purification of membrane proteins such as receptors, SG beads conjugated with naltrindole, a selective antagonist against δ-opioid receptors, were reported to purify the δ-opioid receptor from the membrane fractions of cells transfected with δ-opioid receptor cDNA (Hasegawa et al., 2006).

To identify molecules that bound to NST, we performed affinity purification using the NST-conjugated SG beads (Okuda-Ashitaka et al., 2012). NST was immobilized on the surface of SG beads through hydrophilic ethylene glycol diglycidyl ether spacer (Fig. 2A). We tried to isolate NST-interacting proteins from the synaptosomal membranes of the mouse spinal cord. Solubilization conditions for membrane-associated molecules are quite important. Therefore, we assessed the activity of [^{3}H]NST binding to the solubilized membranes fractions with several detergents. One of them, *n*-octyl-β-D-thioglucoside, exhibited the highest binding activity. Thereafter, we used the extract of synaptosomal membranes in mouse spinal cord solubilized with n-octyl-β-D-thioglucoside, and the bound proteins were subjected to sodium dodecyl sulfate-polyacrylamide gel electrophoresis (SDS-PAGE) and silver staining (Fig. 2B). Three proteins with apparent

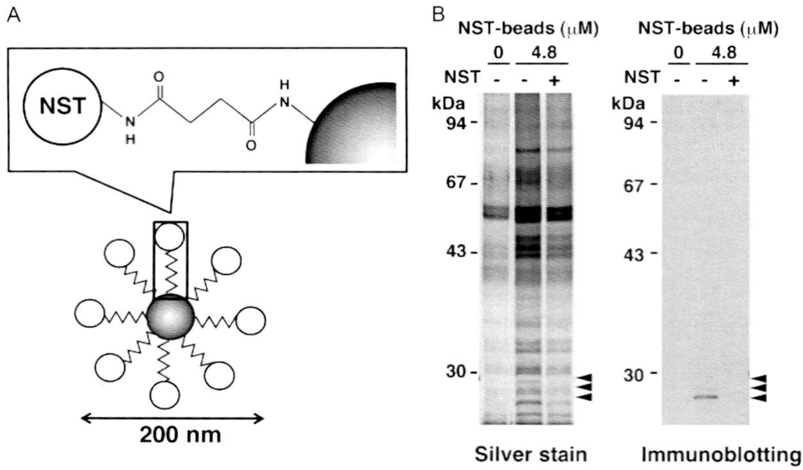

Figure 2 (A) Schematic representation of the NST-conjugated SG beads. (B) Purification of NST-interacting proteins from synaptosomal membrane extracts of the spinal cord using NST-conjugated SG beads. Eluates of bound proteins were subjected to SDS-PAGE followed by silver staining and immunoblotting with anti-NIPSNAP1 antibody that recognized the carboxyl-terminal region. Arrowheads indicate the positions of the interacting proteins. This research was originally published in J Biol Chem (Okuda-Ashitaka et al., 2012), © the American Society from Biochemistry and Molecular Biology.

molecular masses of approximately 29, 28, and 27 kDa specifically interacted with the NST-conjugated SG beads. The bands eluates were analyzed by matrix-assisted laser desorption/ionization time-of-flight mass spectrometry and which identified NIPSNAP1.

3.2 Identification of NIPSNAP1 as an NST-interacting protein

NIPSNAP1 is a member of the NIPSNAP family which comprises NIPSNAP1, NIPSNAP2, NIPSNAP3, and NIPSNAP4 (also called NIPSNAP3A). In *C. elegans*, the gene for NIPSNAP1 is regulated by a polycistronic operon containing 4-nitrophenylphosphatase and nonneuronal SNAP25-like proteins, after which NIPSNAP1 was named (Seroussi, Pan, Kedra, Roe, & Dumanski, 1998). NIPSNAP1 is a membrane protein-containing a single transmembrane helix instead of seven transmembrane segments as seen with G protein-coupled receptors. The predicted protein from NIPSNAP1 cDNA consists of 284 amino acid residues with a calculated molecular weight of 33,363. However, we purified three proteins with apparent molecular masses of approximately 29, 28, and 27 kDa using NST-conjugated SG beads (Okuda-Ashitaka et al., 2012). These protein bands were detected by the same affinity purification procedure using NST-conjugated SG beads with extracts from the membranes of Cos7 cells expressing NIPSNAP1. The amino-terminal and carboxyl-terminal regions of NIPSNAP1 were not obtained by mass spectrometry analysis, and the purified protein bands of NIPSNAP1 were detected by an anti-NIPSNAP1 antibody that recognized the carboxyl-terminal region but not by an antibody that recognized the amino-terminal region (Fig. 2B). Theses results indicate that NST binds the mature form of NIPSNAP1 truncated at the amino-terminal region. Radiolabeled NST, $[^3H]$NST, also bound to the membrane fraction of NIPSNAP1-transfected Cos7 cells. $[^3H]$NST binding to the synaptosomal membrane of the spinal cord was significantly reduced in the NIPSNAP1-deficient mice (NIPSNAP1$^{-/-}$). Theses evidence led us to conclude that the mature form of NIPSNAP1 is an NST-interacting protein.

3.3 Pain regulation induced by NIPSNAP1

To address whether NIPSNAP1 affected pain behaviors, we prepared the NIPSNAP1$^{-/-}$ mice. These mice were fertile and obtained at the expected Mendelian frequency. The 29-kDa protein of NIPSNAP1 was expressed in the brain, spinal cord, liver, and kidney in wild-type mice, but the protein band disappeared in NIPSNAP1$^{-/-}$ mice. Further, specific $[^3H]$NST binding was reduced by approximately 30% in their spinal cord membrane fraction. NIPSNAP1$^{-/-}$ mice exhibited no gross histological abnormalities of the brain or spinal cord, and the expression of NST was not affected in the spinal cord. The mature form of NIPSNAP1 is localized in the synaptic membranes and mitochondria in the brain. It has also been reported that NIPSNAP1 is localized in the postsynaptic density (PSD) fraction of synapses (Satoh et al., 2002). The expression levels of PSD proteins related to pain transmission, such as glutamate receptors (NR1, NR2A, NR2B, GluR1, and GluR2) and neuronal nitric oxide synthase (nNOS), remained unchanged in the homogenates from the hippocampus and the spinal cord of NIPSNAP1$^{-/-}$ mice compared with those in the wild-type mice. Further, no significant change in HSP60 (a major molecular chaperone found in the mitochondrial matrix) and voltage-dependent anion channel (a pore-forming protein in the outer mitochondrial membrane) was noticed in the hippocampus and spinal cord.

The i.t. injection of N/OFQ induces touch-evoked allodynia, a pain response to tactile stimuli applied to the flank with a paint brush, in conscious mice (Hara et al., 1997; Okuda-Ashitaka et al., 1996). The dose–response curves of N/OFQ for allodynia is bell-shaped pattern from 0.55 to 55 fmol. Although the i.t. administration of NST (up to 250 fmol) by itself does not affect nociception in tactile stimuli, the allodynia evoked by 25 fmol N/OFQ is dose-dependently blocked by NST, with a half-maximal inhibitory dose of 0.37 fmol and 95% confidence limits of 0.018–2.2 fmol (Okuda-Ashitaka et al., 1998). NST alone did not induce allodynia in either NIPSNAP1$^{-/-}$ or wild-type mice. The N/OFQ (25 fmol)-induced allodynia was inhibited by 25 fmol NST in wild-type mice, whereas the inhibition by NST disappeared in NIPSNAP1$^{-/-}$ mice (Fig. 3; Okuda-Ashitaka et al., 2012). Taken together, these results suggest that the inhibition of N/OFQ-evoked allodynia by NST is mediated by NIPSNAP1, with no difference in NST-immunoreactivity in the dorsal horn of spinal cord between wild-type and NIPSNAP1$^{-/-}$ mice (Okuda-Ashitaka et al., 2012).

The i.t. N/OFQ-induced allodynia is blocked by glycine, but not affected by the $GABA_A$ and $GABA_B$ receptor agonists (Hara et al., 1997). The N/OFQ-induced allodynia may involve in a neurochemical event beginning with the disinhibition of the inhibitory glycinergic responses, and which leads to the release of glutamate from the presynaptic terminal. Furthermore, the N/OFQ-induced allodynia disappears in NMDA receptor NR2A subunit knockout mice (Minami et al., 2000). The

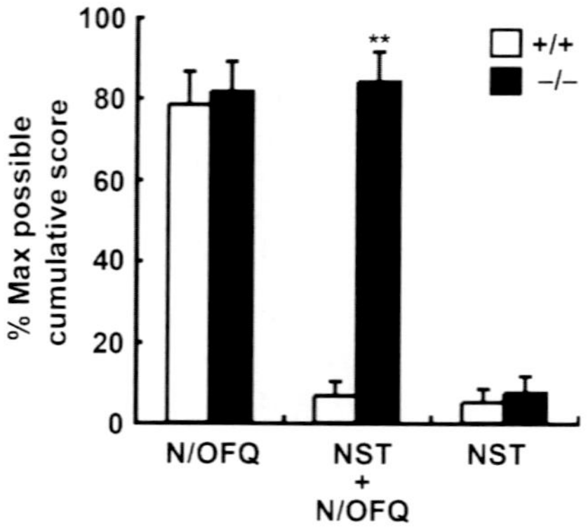

Figure 3 Inhibition of N/OFQ-induced tactile allodynia induced by NST. NST (50 pg) and/or N/OFQ (50 pg) were injected i.t. into the wild-type (+/+) and NIPSNAP1$^{-/-}$ (−/−) mice. The values shown are expressed as a percent of the maximum possible cumulative score for allodynia over the 50-min observation period. **$p < 0.01$ versus wild-type value. This research was originally published in J Biol Chem (Okuda-Ashitaka et al., 2012), © the American Society from Biochemistry and Molecular Biology.

N/OFQ-evoked allodynia is mediated by the activation of the NMDA receptor comprising the NR2A subunit by glutamate in the postsynaptic neurons. The increases in intracellular Ca^{2+} following the activation of NMDA receptor enhances nNOS activity. The N/OFQ-induced allodynia is blocked by methylene blue, a soluble guanylate cyclase inhibitor, and hemoglobin, a NO scavenger (Minami et al., 2000). N/OFQ at low dose of 1–100 nM also enhances nNOS activity significantly in the superficial layer of the spinal cord (Xu et al., 2007). The N/OFQ-stimulated nNOS activity is reduced by glycine and NMDA receptor antagonists and disappeared in NR2A subtype deficient mice. From these results, we consider that low dose N/OFQ-induced allodynia develops through NO production due to the activation of NMDA receptors including NR2A, by increased release of glutamate from the synaptic terminal, which is due to the disinhibition induced by the inhibition of glycine release. On the other hand, NST attenuates the N/OFQ-enhanced nNOS activity in the spinal cord (Xu et al., 2007), correlating with the inhibition of allodynia induced by low dose of N/OFQ. NST also attenuates the NMDA-enhanced nNOS activity in the spinal cord, suggesting that postsynapse is considered to have the site of action of NST. Additional studies focusing on the relationship between NST and NIPSANP1 are needed to define the NST-evoked pain regulation.

3.4 Other functions of NIPSNAP1

NIPSNAP1 has been reported to interact with several proteins such as transient receptor potential vanilloid channel 6 (TRPV6) (Schoeber et al., 2007), amyloid precursor protein (APP) (Tummala, Li, & Homayouni, 2010), and branched-chain α-ketoacid dehydrogenase enzyme complex (BCKDC) (Nautilya et al., 2010). TRPV6 is a highly selective Ca^{2+} channel, and it is implicated in intracellular Ca^{2+} homeostasis in the kidney and ileum. NIPSNAP1 protein is localized in the apical part of cells of the inner medullary collecting ducts of the kidney, and it inhibits the Ca^{2+}-selective channel activity through TRPV6 at cell-surface membranes (Schoeber et al., 2007). In addition to a membrane protein, the interaction of NIPSNAP1 with APP and BCKDC is observed in the mitochondria. NIPSNAP1 is targeted to mitochondria through the interaction of the amino-terminal NIPSNAP1 with the outer mitochondria membrane protein TOM 22 (Tummala et al., 2010), which is present in the matrix or inner membrane (Nautilya et al., 2010). Overexpression of APP inhibits mitochondrial localization and decreases in protein level of NIPSNAP1. Recent reports have shown that overexpression of APP induces an abnormal mitochondrial distribution, mitochondrial fragmentation, a decrease in mitochondrial fusion proteins, and increase in mitochondrial fission proteins (Cho et al., 2009; Wang et al., 2008), suggesting that the lack of NIPSNAP1 may implicate mitochondrial dysfunction. Further, NIPSNAP1 is associated with the E2 subunit of BCKDC attached in the inner membrane of the mitochondria (Nautilya et al., 2010). However, NIPSNAP1 does not affect the enzyme activity of BCKDC component proteins. Thus, in *in vitro* conditions, NIPSNAP1 appears to have multifunctions acting through protein–protein interactions. Furthermore, NIPSNAP1 localized in the PSD fraction of synapses is increased during generalized seizures caused by kainate (Satoh et al., 2002). The level of NIPSNAP1 mRNA is reduced in the brain of a mouse model for phenylketonuria, an inborn error of amino acid metabolism caused by phenylalanine hydroxylase deficiency (Surendran, Tyring, & Matalon, 2005). NIPSNAP1 may thus be involved in several central functions other than pain transmission.

4. CONCLUSIONS

NST regulates several pathological pains in different pain transmission sites through putative G protein-coupled receptors. Similar to opioids including N/OFQ, NST is involved in the neurotransmitter release through PTX-sensitive $G_{i/o}$ protein in the spinal dorsal horn. The inhibition of cyclic AMP formation, the opening of inwardly rectifying K^+ channels, and the closing of voltage-gated Ca^{2+} channels reduce neurotransmitter release. N/OFQ reduces the release of glutamate from the presynaptic terminals, whereas NST reduces the release of glycine from the inhibitory interneurons. Blocking of glycinergic neurotransmission leads to the aggravation of inflammatory pain and neuropathic pain. However, the decrease of glycine by NST inhibits the postsynaptic NMDA activation through the binding of glycine, which implicates the antinociceptive response in inflammatory pain. NST also activates PLC through $G_{i/o}$, and it induces the nociceptive flexor response through the release of substance P from the peripheral nerve endings of nociceptive primary afferent neurons. Supraspinal NST leads to antinociception by the activation of descending antinociceptive pathways through PTX-insensitive $G_{q/11}$. NST depolarizes CeA and RAIC neurons projecting to PAG through the opening of TRPC channels mediated by PLC–PKC pathway. Furthermore, NIPSNAP1, a novel NST-interacting protein, is implicated in the inhibition of tactile allodynia by NST independent of G protein-coupled receptor signal pathways. Further studies focusing on the G protein-coupled NST receptor and pain in addition to allodynia regulated by NIPSANP1 will contribute to our better understanding of pain transmission evoked by NST.

ACKNOWLEDGMENT

Our works in this article were supported in part by grants-in-aid for Scientific Research (C) from Japan Society for the Promotion of Science.

REFERENCES

Ahmadi, S., Muth-Selbach, U., Lauterbach, A., Lipfert, P., Neuhuber, W. L., & Zeilhofer, H. U. (2003). Facilitation of spinal NMDA receptor currents by spillover of synaptically release glycine. *Science*, *300*, 2094–2097.

Basbaum, A. I., Bautista, D. M., Scherrer, G., & Hulius, D. (2009). Cellular and molecular mechanisms of pain. *Cell*, *139*, 267–284.

Boom, A., Mollereau, C., Meunier, J. C., Vassart, G., Parmentier, M., Vanderhaeghen, J. J., et al. (1999). Distribution of the nociceptin and nocistatin precursor transcript in the mouse central nervous system. *Neuroscience*, *91*, 991–1007.

Chaplan, S. R., Malmberg, A. B., & Yaksh, T. L. (1997). Efficacy of spinal NMDA receptor antagonism in formalin hyperalgesia and nerve injury evoked allodynia in the rat. *The Journal of Pharmacology and Experimental Therapeutics*, *280*, 829–838.

Chen, Y. L., Li, A. H., Yeh, T. H., Chou, A. H., & Wang, H. L. (2009). Nocistatin and nociceptin exert opposite effects on the excitability of central amygdala nucleus periaqueductal gray projection neurons. *Molecular and Cellular Neurosciences*, *40*, 76–88.

Chen, Y. L., Li, A. H., Yen, T. H., Chou, A. H., Weng, Y. S., & Wang, H. L. (2010). Nocistain excites rostral agranular insular cortex-periaqueductal gray projection neurons by enhancing transient receptor potential cation conductance via Gα/11-PLC-protein kinase C pathway. *Neuroscience*, *168*, 226–239.

Cho, D. H., Nakamura, T., Fang, J., Cieplak, P., Godzik, A., Gu, Z., et al. (2009). S-nitrosylation of Drp1 mediates β amyloid-related mitochondrial fission and neuronal injury. *Science*, *324*, 102–105.

Costigan, M., Scholz, J., & Woolf, C. J. (2009). Neuropathic pain: A maladaptive response of the nervous system to damage. *Annual Review of Neuroscience*, *32*, 1–32.

Cronin, J. N., Bradbury, E. J., & Lidierth, M. (2004). Laminar distribution of GABAA and glycine-receptor mediated tonic inhibition in the dorsal horn of the lumbar spinal cord: Effects of picrotoxin and strychnine on expression of Fos-like immunoreactivity. *Pain*, *112*, 156–163.

Dingledine, R., Borges, K., Bowie, D., & Traynellis, S. F. (1999). The glutamate receptor ion channels. *Pharmacological Reviews*, *51*, 7–61.

Dray, A., Urban, L., & Dickenson, A. (1994). Pharmacology of chronic pain. *Trends in Pharmacological Sciences*, *15*, 190–197.

Evans, C. J., Keith, D. E., Jr., Morrison, H., Magendzo, K., & Edwards, R. H. (1992). Cloning of a δ receptor by functional expression. *Science*, *258*, 1952–1955.

Fantin, M., Fischetti, C., Trapella, C., & Morari, M. (2007). Nocistatin inhibits 5-hydroxytryptamine release in the mouse neocortex via presynaptic $G_{i/o}$ protein linked pathways. *British Journal of Pharmacology*, *152*, 549–555.

Gavioli, E. C., Rae, G. A., Calo, G., Guerrini, R., & de Lima, T. C. (2002). Central injections of nocistatin or its C-terminal hexapeptide exert anxiogenic-like effect on behaviour of mice in the plus-maze test. *British Journal of Pharmacology, 136*, 764–772.

Ge, Z. J., Zhang, L. C., Zeng, Y. M., Dai, T. J., Chang, L., Wang, J. K., et al. (2007). Involvement of local orphanin FQ in the tolerance induced by repeated microinjections of morphine into ventrolateral periaqueductal gray in rats. *Pharamacology, 80*, 261–268.

Gold, M. S., & Gebhart, G. F. (2010). Nociceptor sensitization in pain pathogenesis. *Nature Medicine, 16*, 1248–1257.

Hara, N., Minami, T., Okuda-Ashitaka, E., Sugimoto, T., Sakai, M., Onaka, M., et al. (1997). Characterization of nociceptin hyperalgesia and allodynia in conscious mice. *British Journal of Pharmacology, 121*, 401–408.

Hasegawa, M., Ohno, H., Tanaka, H., Hatakeyama, M., Kawaguchi, H., Takahashi, T., et al. (2006). Affinity identification of δ-opioid receptors using latex nanoparticles. *Bioorganic & Medicinal Chemistry Letters, 16*, 158–161.

Heinricher, M. M., Tavares, I., Leith, J. L., & Lumb, B. M. (2009). Descending control of nociception: Specificity, recruitment and plasticity. *Brain Research Reviews, 60*, 214–225.

Hiramatsu, M., & Inoue, K. (1999). Nociceptin/orphanin FQ and nocistatin on learning and memory impairment induced by scopolamine in mice. *British Journal of Pharmacology, 127*, 655–660.

Houtani, T., Nishi, M., Takeshima, H., Sato, K., Sakuma, S., Kakimoto, S., et al. (2000). Distribution of nociceptin/orphanin FQ precursor protein and receptor in brain and spinal cord: A study using in situ hybridization and X-gal histochemistry in receptor-deficient mice. *The Journal of Comparative Neurology, 424*, 489–508.

Ikeda, K., Watanabe, M., Ichikawa, T., Kobayashi, T., Yano, R., & Kumanishi, T. (1998). Distribution of prepro-nociceptin/orphanin FQ mRNA and its receptor mRNA in developing and adult mouse central nervous systems. *The Journal of Comparative Neurology, 14*, 139–151.

Inoue, M., Kawashima, T., Allen, R. G., & Ueda, H. (2003). Nocistatin and prepronociceptin/orphanin FQ 160-187 cause nociception through activation of $G_{i/o}$ in capsaicin-sensitive and G_s in capsaicin-insensitive nociceptors, respectively. *The Journal of Pharmacology and Experimental Therapeutics, 306*, 141–146.

Ito, S., Okuda-Ashitaka, E., & Minami, T. (2001). Central and peripheral roles of prostaglandins in pain and their interactions with novel neuropeptides nociceptin and nocistatin. *Neuroscience Research, 41*, 299–332.

Joseph, T., Lee, T. L., Ning, C., Nishiuchi, Y., Kimura, T., Jikuya, H., et al. (2006). Identification of mature nocistatin and nociceptin in human brain and cerebrospinal fluid by mass spectrometry combined with affinity chromatography and HPLC. *Peptides, 27*, 122–130.

Kamei, J., Matsunawa, Y., Miyata, S., Tanaka, S., & Saitoh, A. (2004). Effects of nociceptin on the exploratory behavior of mice in the hole-board test. *European Journal of Pharmacology, 489*, 77–87.

Kuner, R. (2010). Central mechanisms of pathological pain. *Nature Medicine, 16*, 1258–1266.

Lambert, D. G. (2008). The nocicepton/orphanin FQ receptor: A target with broad therapeutic potential. *Nature Reviews. Drug Discovery, 7*, 694–710.

Lee, T.-L., Fung, M. Y., Chen, F. G., Chou, N., Okuda-Ashitaka, E., Ito, S., et al. (1999). Identification of human, rat and mouse nocistatin in brain and human cerebrospinal fluid. *Neuroreport, 10*, 1537–1541.

Liu, E. H., Nishiuchi, Y., Kimura, T., & Tachibana, S. (2006). Supraspinal nocistatin and its amide derivative antagonize the hyperalgesic effects oh nociceptin in mice. *Neuroscience Letters, 397*, 59–63.

Lu, Y., & Peri, E. R. (2003). A specific inhibitory pathway between substantia gelatinosa neurons receiving direct C-fiber input. *The Journal of Neuroscience, 23*, 8752–8758.

Lutfy, K., Cai, S. X., Woodward, R. M., & Weber, E. (1997). Antinociceptive effects of NMDA and non-NMDA receptor antagonists in the tail flick test in mice. *Pain, 70*, 31–40.

Lyadomi, M., Lyadomi, I., Kumamto, E., Tomokuni, K., & Yoshimura, M. (2000). Presynaptic inhibition by baclofen of miniature EPSCs and IPSCs in substantia gelatinosa neurons of the adult rat spinal dorsal horn. *Pain, 85*, 385–393.

Ma, F., Xie, H., Dong, Z. Q., Wang, Y. Q., & Wu, G. C. (2003). Effect of intrathecal nocistatin on nociceptin/orphanin FQ analgesia in chronic constriction injury rat. *Brain Research, 988*, 189–192.

MacBain, C. J., Kleckner, N. W., Wyrick, S., & Dingledine, R. (1989). Structural requirements for activation of the glycine coagonist site of N-methly-D-asparatate receptor expresses in Xenopus oocytes. *Molecular Pharmacology, 36*, 556–565.

Meunier, J. C., Mollereau, C., Toll, L., Suaudeau, C., Moisand, C., Alvinerie, P., et al. (1995). Isolation and structure of the endogenous agonist of opioid receptor-like ORL1 receptor. *Nature, 377*, 532–535.

Minami, T., Okuda-Ashitaka, E., Mori, H., Sakimura, K., Watanabe, M., Mishina, M., et al. (2000). Characterization of nociceptin/orphanin FQ-induced pain responses in conscious mice: Neonatal capsaicin treatment and N-methyl-D-aspartate receptor GluRε subunit knockout mice. *Neuroscience, 97*, 133–142.

Minami, T., Okuda-Ashitaka, E., Nishiuchi, Y., Kimura, T., Tachibana, S., Mori, H., et al. (1998). Anti-nociceptive responses produced by human putative counterpart of nocistatin. *British Journal of Pharmacology, 124*, 1016–1018.

Mogil, J. S., & Pasternak, W. (2001). The molecular and behavioral pharmacology of the orphanin FQ/nociceptin peptide and receptor family. *Pharmacological Reviews, 53*, 381–415.

Nakagawa, T., Kaneko, M., Inamura, S., & Satoh, M. (1999). Intracerebroventricular administration of nocistatin reduces inflammatory hyperalgesia in rats. *Neuroscience Letters, 265*, 64–66.

Nakano, H., Minami, T., Abe, K., Arai, T., Tokumura, M., Ibii, N., et al. (2000). Effect of intrathecal nocistatin on the formalin-induced pain in mice versus that of nociceptin/orphanin FQ. *The Journal of Pharmacology and Experimental Therapeutics, 292*, 331–336.

Nautilya, M., Sweatt, A. J., MacKenzie, J. A., Mark Payne, R., Szucs, S., Matalon, R., et al. (2010). Neuronal localization of the mitochondrial protein NIPSANP1 in rat nervous systems. *The European Journal of Neuroscience, 32*, 560–569.

New, D. C., & Wong, Y. H. (2002). The ORL1 receptor: Molecular pharmacology and signaling mechanisms. *Neurosignals, 11*, 197–212.

Ohtsu, Y., Ohba, R., Imamura, Y., Kobayashi, M., Hatori, H., Zenkoh, T., et al. (2005). Selective ligand purification using high-performance affinity beads. *Analytical Biochemistry, 338*, 245–252.

Okuda-Ashitaka, E., & Ito, S. (2000). Nocistatin: A novel neuropeptide encoded by the gene for the nociceptin/orphanin FQ precursor. *Peptide, 21*, 1101–1109.

Okuda-Ashitaka, E., Minami, T., Tsubouchi, S., Kiyonari, H., Iwamatsu, A., Noda, T., et al. (2012). Identification of NIPSANP1 as a nocistatin-interacting protein involving pain transmission. *The Journal of Biological Chemistry, 287*, 10403–10413.

Okuda-Ashitaka, E., Minmi, T., Tachibana, S., Yoshihara, Y., Nishiuchi, Y., Kimura, T., et al. (1998). Nocistatin, a peptide that blocks nociceptin action in pain transmission. *Nature, 392*, 286–289.

Okuda-Ashitaka, E., Tachibana, S., Houtani, T., Minami, T., Masu, Y., Nishi, M., et al. (1996). Identification and characterization of an endogenous ligand for opioid receptor homologue ROR-C: Its involvement in allodynic response to innocuous stimulus. *Molecular Brain Research*, *43*, 96–104.

Olszewski, P. K., Shaw, T. J., Grace, M. K., Billington, C. J., & Levine, A. S. (2000). Nocistatin inhibits food intake in rats. *Brain Research*, *872*, 181–187.

Porrenca, F., Ossipov, M. H., & Gebhart, G. F. (2000). Chronic pain and medullary descending facilitation. *Trends in Neurosciences*, *25*, 319–325.

Ramsey, I. S., Delling, M., & Clapham, D. E. (2006). An introduction to TRP channels. *Annual Review of Physiology*, *68*, 619–647.

Reinscheid, R. K., Nothacker, H. P., Bourson, A., Ardati, A., Henningsen, R. A., Bunzow, J. R., et al. (1995). Ophanin FQ: A neuropeptide that activates an opioid like G protein-coupled receptor. *Science*, *270*, 792–794.

Sakamoto, S., Kabe, Y., Hatakeyama, M., Yamaguchi, Y., & Handa, H. (2009). Development and application of high-performance affinity beads: Toward chemical biology and drug discovery. *Chemical Record*, *9*, 66–85.

Satoh, K., Takeuchi, M., Oda, Y., Deguchi-Tawarada, M., Sakamoto, Y., Matsubara, K., et al. (2002). Identification of activity-regulated proteins in the postsynaptic density fraction. *Genes to Cells: Devoted to Molecular & Cellular Mechanisms*, 7, 187–197.

Schoeber, J. P., Topala, C. N., Lee, K. P., Lambers, T. T., Ricard, G., van der Kemp, A. W., et al. (2007). Identification of NIPSANP1 as a novel auxiliary protein inhibiting TRPV6 activity. *European Journal of Pharmacology*, *152*, 549–555.

Scoto, G. M., Santangelo, N., & Parenti, C. (2005). Effect of supraspinal nocistatin on nociceptin/orphanin FQ antagonism of selective opioid analgesia. *Neuroscience Letters*, *387*, 126–129.

Seroussi, E., Pan, H. Q., Kedra, D., Roe, B. A., & Dumanski, J. P. (1998). Characterization of the human NIPSNAP1 gene fro 22qqw. A member of a novel gene family. *Gene*, *212*, 13–20.

Shimizu, N., Sugimoto, K., Tang, J., Nishi, T., Sato, I., Hiramoto, M., et al. (2000). High-performance affinity beads for identifying drug receptors. *Nature Biotechnology*, *18*, 877–881.

Sivilotti, L., & Woolf, C. J. (1994). The contribution of $GABA_A$ and glycine receptors to central sensitization: Distribution and touch-evoked allodynia in the spinal cord. *Journal of Neurophysiology*, *72*, 169–179.

Sun, R. Q., Zhao, C. S., Wang, H. J., Jing, Z., Wang, W., Yang, K., et al. (2001). Nocistatin, a peptide reversing acute and chronic morphine tolerance. *Neuroreport*, *12*, 1789–1792.

Surendran, S., Tyring, S. K., & Matalon, R. (2005). Expression of calpastatin, minopontin, NIPSNAP1, rabaptin-5 and neuronatin in the phenylketonuria (PKU) mouse brain: Possible role on cognitive defect seen in PKU. *Neurochemistry International*, *48*, 595–599.

Tummala, H., Li, X., & Homayouni, R. (2010). Interaction of a novel mitochondrial protein, 4-nitrophenylphosphatase domain and non-neuronal SNAP25-like protein homolog 1 (NIPSNAP1), with the amyloid precursor protein family. *The European Journal of Neuroscience*, *31*, 1926–1934.

Vaughan, C. W., Connor, M., Jennings, E. A., Marinelli, S., Allen, R. G., & Christine, M. J. (2001). Actions of nociceptin/orphanin FQ and other prepronociceptin products on rat rostral ventromedial medulla neurons in vitro. *The Journal of Physiology*, *534*, 849–859.

Wang, X., Su, B., Siedlak, S. L., Moreira, P. I., Fujioka, H., & Wang, Y. (2008). Amyloid-β overexpression causes abnormal mitochondrial dynamics via differential modulation of mitochondrial fission/fusion proteins. *Proceedings of the National Academy of Sciences of the United States of America*, *105*, 19318–19323.

Xu, I. S., Hashemi, M., Calo, G., Regoli, D., Wiesenfeld-Hallin, Z., & Xu, X. J. (1999). Effects of intrathecal nocistatin on the flexor reflex and its interaction with orphanin FQ nociceptin. *Neuroreport*, *10*, 3681–3684.

Xu, L., Okuda-Ashitaka, E., Matsmura, S., Mabuchi, T., Okamoto, S., Sakimura, K., et al. (2007). Signal pathways coupled to activation of neuronal nitric oxide synthase in the spinal cord by nociceptin/orphanin FQ. *Neruopharamacology*, *52*, 1318–1325.

Yaksh, T. L. (1989). Behavioral and autonomic correlates of the tactile evoked allodynia produced by spinal glycine inhibition: Effects of modulatory receptor systems and excitatory amino acid antagonists. *Pain*, *37*, 111–123.

Yamamoto, T., & Sakashita, Y. (1999). Effect of nocistatin and its interaction with nociceptin/orphanin FQ on the rat formalin test. *Neuroscience Letters*, *262*, 179–182.

Yoshimura, M., & Nishi, S. (1995). Primary afferent-evoked glycine- and GABA-mediated IPSPs in substantia gelatinosa neurons in the rat spinal cord in vitro. *The Journal of Physiology*, *482*, 29–38.

Zeilfofer, H. U. (2005). Synaptic modulation in pain pathways. *Reviews of Physiology, Biochemistry and Pharmacology*, *154*, 73–100.

Zeilfofer, H. U., Winder, H., & Yevebes, G. E. (2012). Fast synaptic inhibition in spinal sensory processing and pain control. *Physiological Reviews*, *92*, 193–235.

Zeilhofer, H. U., & Calo', G. (2003). Nociceptin/orphanin FQ and its receptor—Potential targets for pain therapy? *The Journal of Pharmacology and Experimental Therapeutics*, *306*, 423–429.

Zeilhofer, H. U., Selbach, U. M., Gühring, H., Erb, K., & Ahmadi, S. (2000). Selective suppression of inhibitory transmission by nocistatin in the rat spinal cord dorsal horn. *The Journal of Neuroscience*, *20*, 4922–4929.

Zhao, C. S., Li, B. S., Zhao, G. Y., Liu, H. X., Luo, F., Wang, Y., et al. (1996). Nocistatin reverses the effect of orphanin FQ/nociceptin in antagonizing morphine analgesia. *Neuroreport*, *10*, 297–299.

CHAPTER SIX

Nociceptin and Meiosis during Spermatogenesis in Postnatal Testes

Ko Eto[1]

Department of Biological Sciences, Graduate School of Science and Technology, Kumamoto University, Kumamoto City, Kumamoto, Japan

[1]Corresponding author: e-mail address: etoko@gpo.kumamoto-u.ac.jp

Contents

Vitamins and Hormones, Volume 97
ISSN 0083-6729
http://dx.doi.org/10.1016/bs.vh.2014.10.003

Abstract

Phosphorylated Rec8, a key component of cohesin, mediates the association and dis-association, "dynamics," of chromosomes occurring in synaptonemal complex formation, crossover recombination, and sister chromatid cohesion during meiosis in germ cells. Yet, the extrinsic factors triggering meiotic chromosome dynamics remained unclear. In postnatal testes, follicle-stimulating hormone (FSH) acts directly on somatic Sertoli cells to activate gene expression via an intracellular signaling pathway composed of cAMP, cAMP-dependent protein kinase (PKA), and cAMP-response element-binding protein (CREB), and promotes germ cell development and spermatogenesis indirectly. Yet, the paracrine factors mediating the FSH effects to germ cells remained elusive. We have shown that nociceptin, known as a neuropeptide, is upregulated by FSH signaling through cAMP/PKA/CREB pathway in Sertoli cells of postnatal murine testes. Chromatin immunoprecipitation from Sertoli cells demonstrated that CREB phosphorylated at Ser133 associates with prepronociceptin gene encoding nociceptin. Analyses with Sertoli cells and testes revealed that both prepronociceptin mRNA and the nociceptin peptide are induced after FSH signaling is activated. In addition, the nociceptin peptide is induced in testes after 9 days post partum following FSH surge. Thus, our findings may identify nociceptin as a novel paracrine mediator of the FSH effects in the regulation of spermatogenesis; however, very little has known about the functional role of nociceptin in spermatogenesis. We have shown that nociceptin induces Rec8 phosphorylation, triggering chromosome dynamics, during meiosis in spermatocytes of postnatal murine testes. The nociceptin receptor Oprl-1 is exclusively expressed in the plasma membrane of testicular germ cells, mostly spermatocytes. Treatment of testes with nociceptin resulted in a rapid phosphorylation of Rec8. Injection of nociceptin into mice stimulated Rec8 phosphorylation and meiotic chromosome dynamics in testes, whereas injection of nocistatin, a specific inhibitor for nociceptin, abolished them. Therefore, our findings suggest that nociceptin is a novel extrinsic factor that plays a crucial role in the progress of meiosis during spermatogenesis.

1. INTRODUCTION

1.1 Meiotic chromosome dynamics during spermatogenesis

Spermatogenesis is a highly organized process dependent on the local microenvironment within testes, culminating in the production of spermatozoa. Spermatogonial stem cells undergo a series of mitotic divisions, producing not only the self-renewed stem cells but also differentiated type A spermatogonia; the spermatogonia differentiate into diploid primary spermatocytes. The spermatocytes undergo two meiotic divisions, producing haploid round spermatids; the spermatids undergo spermiogenesis, generating mature sperm.

The first meiotic division (meiosis I) begins with a long prophase when each chromosome consists of two parallel chromatids at the leptotene stage; chromosomes align and pair with the homologous partners in a step termed synapsis through synaptonemal complex formation at the zygotene stage; two homologous chromosomes undergo crossover recombination at the pachytene and diplotene stage (Page & Hawley, 2003; Petronczki, Siomos, & Nasmyth, 2003). During meiosis I, sister chromatid cohesion is maintained until the metaphase to anaphase transition, but cohesion between chromosome arms is lost from sister chromatids at anaphase I, permitting homologous chromosomes to separate. Until the second meiotic division, sister chromatids remain attached at the centromeres, but centromeric cohesion is lost from sister chromatids at anaphase II, allowing the disjunction of individual sister chromosomes to each haploid gamete.

Rec8 is a key component of the meiotic cohesin complex that is required for chromosome dynamics in synaptonemal complex formation, crossover recombination, and sister chromatid cohesion (Guacci, Koshland, & Strunnikov, 1997; Klein et al., 1999; Losada, Hirano, & Hirano, 1998; Michaelis, Ciosk, & Nasmyth, 1997; Molnar, Bahler, Sipiczki, & Kohli, 1995; Nasmyth, 2001; Pasierbek et al., 2001; Watanabe & Nurse, 1999; Xu, Beasley, Warren, van der Horst, & McKay, 2005). The loss of cohesion is caused by the dissociation of cohesin, which is triggered by the proteolytic cleavage of phosphorylated Rec8, resulting in meiotic divisions (Buonomo et al., 2000; Eijpe, Offenberg, Jessberger, Revenkova, & Heyting, 2003; Lee, Iwai, Yokota, & Yamashita, 2003). However, these raise the question as to how cohesin functions and chromosome dynamics are regulated in meiosis.

1.2 A role in spermatogenesis of FSH and the mechanism of its action

Follicle-stimulating hormone (FSH), a glycoprotein produced in the pituitary, plays an essential role in pubertal spermatogenesis (Simoni, Gromoll, & Nieschlag, 1997). The total levels of FSH in serum increase in mice from 8 days post partum (dpp) onward (Barakat, O'Connor, Gold, de Kretser, & Loveland, 2008). FSH acts directly on somatic Sertoli cells to activate gene expression via an intracellular signaling pathway comprising cAMP, cAMP-dependent protein kinase (PKA), and cAMP-response element-binding protein (CREB) (Kangasniemi et al., 1990; Rannikki, Zhang, & Huhtaniemi, 1995; Simoni et al., 1997;

Sprengel, Braun, Nikolics, Segaloff, & Seeburg, 1990), and germ cell development indirectly (de Kretser et al., 2001; Mather, Moore, & Li, 1997). Nevertheless, few of the critical players secreted from Sertoli cells to germ cells have been identified to mediate the FSH effects, particularly the players specific to meiosis.

It is generally known that when FSH binds to the Gs protein-coupled seven-transmembrane receptor (FSH receptor), adenylyl cyclase is activated, resulting in an increased concentration of intracellular cAMP, a second messenger; subsequently PKA is activated and phosphorylates multiple substrates such as CREB, a transcription factor, at Ser133, leading to the activation of gene expression (Simoni et al., 1997). Because FSH receptor is exclusively expressed in Sertoli cells (Kangasniemi et al., 1990; Rannikki et al., 1995), it is therefore considered that the ultimate effects of FSH on germ cells are mediated through not only intimate interactions of Sertoli cells with germ cells but also paracrine interactions within seminiferous tubules at multiple steps of germ cell differentiation (Wright, 1993).

Recently we have identified nociceptin, also called orphanin FQ, that is upregulated by FSH signaling in Sertoli cells of postnatal murine testes (Eto, Shiotsuki, Sakai, & Abe, 2012), suggesting that nociceptin is a novel paracrine factor mediating the FSH effects from Sertoli cells to germ cells on spermatogenesis. Nociceptin is traditionally known as a neuropeptide belonging to the opioid peptide family and the endogenous agonist for the opioid receptor-like receptor-1 (Oprl-1), a Gi protein-coupled seven-transmembrane receptor (Childers, 1991; Reisine & Bell, 1993). This gene encodes a precursor protein of prepronociceptin, from which the mature peptide nociceptin consisting of 17 amino acid residues is produced. We found that nociceptin is an extrinsic factor directing the progress of meiosis by inducing Rec8 phosphorylation in spermatocytes of postnatal murine testes (Eto, Shiotsuki, & Abe, 2013). Thus, nociceptin may be responsible for the paracrine regulation of cohesin-mediated chromosome dynamics during meiosis in spermatogenesis.

2. REGULATION OF NOCICEPTIN EXPRESSION BY FSH SIGNALING IN SERTOLI CELLS

2.1 Phosphorylation of CREB following cAMP stimulation in a Sertoli cell line

CREB is expressed and phosphorylated in germ and Sertoli cells during the early phase of murine spermatogenesis (Kim et al., 2009). To verify Sertoli cell-specific phosphorylation of CREB by cAMP/PKA pathway, we used Sertoli B cells, a Sertoli cell line. The time course of CREB phosphorylation was examined in cells after dibutyryl cAMP (db-cAMP), a membrane-permeable analog of cAMP, was added to the cultures by Western blotting for CREB phosphorylated on Ser133. The phosphorylation reached the maximum level at 10 min after the addition of db-cAMP, and then returned to the basal level by 60 min; however, this was suppressed to the basal level by an inhibitor for PKA (H-89).

We validated the presence of the FSH-operated intracellular signaling pathway mediating CREB phosphorylation in Sertoli B cells. Because endogenous FSH receptor was undetectable in the cells, we transfected them with an expression plasmid for murine FSH receptor and detected its exogenous expression by Western blotting. In cells overexpressing FSH receptor, CREB phosphorylation was enhanced after FSH was added to the cultures; however, this was suppressed to the basal level by H-89. These results indicated that CREB is a target for FSH signaling comprising cAMP/PKA pathway in Sertoli B cells (Fig. 1).

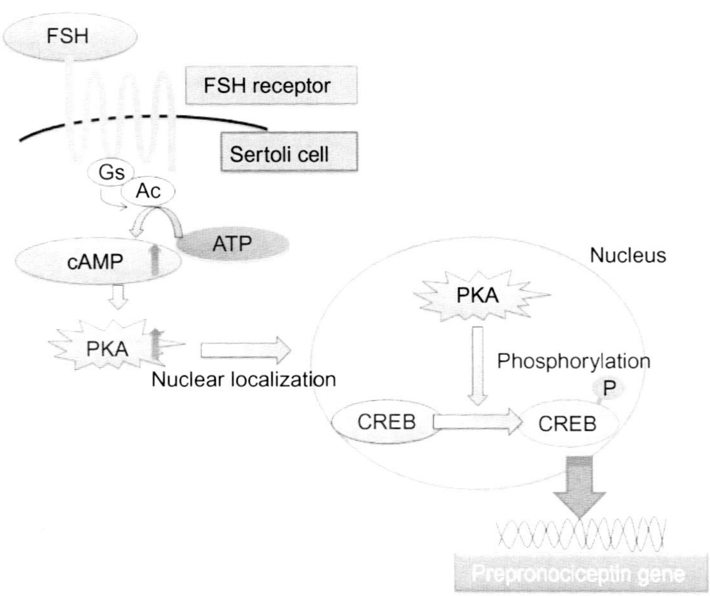

Figure 1 The FSH intracellular signal transduction inducing prepronociceptin gene transcription in Sertoli cells. Gs, Gs protein coupled to the seven-transmembrane receptor for FSH; Ac, adenylyl cyclase; PKA, cAMP-dependent protein kinase, or protein kinase A; CREB, cAMP-response element-binding protein; P, phosphorylated. Upward arrows indicate increases in the concentration and the activity. (See the color plate.)

2.2 Identification of prepronociceptin gene associating cAMP-dependently with phosphorylated CREB

To identify novel genes that are regulated by CREB phosphorylated at Ser133 in Sertoli cells, we performed chromatin immunoprecipitation (ChIP) from Sertoli B cells. After cells were stimulated for 10 min with db-cAMP, extracts were prepared and processed for ChIP with the same antibody to phosphorylated CREB. We screened by PCR several genes, whose proximal promoter regions associate with phosphorylated CREB, and investigated murine prepronociceptin gene (Zaveri, Waleh, & Toll, 2006). The proximal promoter of murine prepronociceptin gene has one functional CRE site in a different location from the human promoter (Zaveri, Green, Polgar, Huynh, & Toll, 2002; Zaveri et al., 2006). The DNA fragment from the putative transcription start site to the ATG translation start codon (252 bp) was detected only in cells treated with db-cAMP but not in untreated cells. None could be detected from immunoprecipitates with an unrelated antibody. Nucleotide sequencing of the detected DNA fragment confirmed the presence of a consensus CRE sequence (CGTCA) at 30 bp upstream of the ATG translation start codon in the proximal promoter of murine prepronociceptin gene as reported (Zaveri et al., 2006). These results indicated that phosphorylated CREB associates with the proximal promoter region of prepronociceptin gene in Sertoli B cells (Fig. 1). This gene encodes a precursor protein of prepronociceptin, from which the mature nociceptin peptide consisting of 17 amino acid residues is produced. Nociceptin, also known as orphanin FQ, is a neuropeptide belonging to the opioid peptide family and shares the identical amino acid sequence between mice and other species.

2.3 Effects of cAMP and FSH on the expressions of prepronociceptin mRNA and the nociceptin peptide in Sertoli cells and testes

To determine whether the association of phosphorylated CREB with the proximal promoter region of prepronociceptin gene activates the transcription in Sertoli cells, we cultured Sertoli B cells with db-cAMP for 12, 24, or 48 h and analyzed by RT-PCR. The prepronociceptin mRNA was transiently induced 24 h after db-cAMP addition to the cultures; however, this was completely suppressed by H-89. Interestingly, none was seen in cells exposed for 12 or 48 h to db-cAMP.

We determined by RT-PCR whether FSH activates the expression of prepronociceptin mRNA in Sertoli cells. Here we used for organ cultures the 7-dpp testes just before total levels of serum FSH rise in mice (Barakat et al., 2008), because we wanted to ignore the effect of endogenous FSH on the mRNA expression. The mRNA was clearly induced when testes were cultured for 24 h following FSH addition; however, this was suppressed by H-89.

We then examined the effect of FSH on the expression of the nociceptin peptide in organ cultures of testes. After the 7-dpp testes were exposed for 24 h to FSH, their extracts containing endogenous peptides were prepared and analyzed by Western blotting. The peptide was induced by FSH stimulation; however, this was completely suppressed by H-89. To evaluate the specificity of the nociceptin antibody reactivity, the antibody was immunodepleted by incubating with the synthetic nociceptin peptide, followed by Western blotting of the same samples as above. No immunopositive signal was detected, revealing that the antibody recognized the nociceptin peptide specifically. These results indicated that FSH activates the transcription of prepronociceptin gene through cAMP/PKA/CREB pathway, resulting in an increased production of the nociceptin peptide, in Sertoli cells (Fig. 1).

2.4 Expression of the endogenous nociceptin peptide in testes

We finally examined the expression of the endogenous nociceptin peptide in testes. The 7- to 20-dpp testes were used because a drastic FSH surge, differentiation of premeiotic spermatogonia into meiotic spermatocytes, and their differentiation into postmeiotic spermatids occur during this age range. Endogenous peptides were prepared from testes at each age and analyzed by Western blotting. Strikingly, the nociceptin peptide started to increase drastically after 9 dpp and the amounts in the extracts were estimated as less than 20 pmol.

3. FUNCTION OF NOCICEPTIN DURING MEIOSIS IN SPERMATOCYTES

3.1 The expression of endogenous Oprl-1 and the phosphorylation of endogenous Rec8 in testes

In mice the first wave of spermatogenesis is completed in 35 dpp (Qin, Richardson, Jasin, Handel, & Arnheim, 2004; Russell, Ettlin, Sinha Hikim, & Clegg, 1990): by 6 dpp seminiferous tubules contain spermatogonia and Sertoli cells; in 9 dpp spermatogonia differentiate into primary spermatocytes; around 11–16 dpp premeiotic spermatocytes develop; after 17 dpp meiotic cells are generated; by 18 dpp postmeiotic haploid spermatids appear; and after 28 dpp mature sperm are formed.

To survey possible functions of nociceptin in spermatogenesis including Rec8 phosphorylation, the cellular and subcellular localization of the receptor Oprl-1 in testes were analyzed by immunohistochemistry. We used the 9- and 10-dpp testes containing spermatogonia and spermatocytes because the nociceptin-stimulated Rec8 phosphorylation could not occur in testes younger than 9 dpp but did at 10 dpp in both *in vitro* and *in vivo* analyses as described below. Similar results were obtained in testes of both 9 and 10 dpp. The immunoreactivity for Oprl-1 was exclusively detected in the regular ridge of large, rounded germ cells located near the center of seminiferous tubules but less at their periphery. Staining with DAPI or hematoxylin and eosin demonstrated that Oprl-1 was predominantly expressed in spermatocytes and localized to the plasma membrane, but barely in spermatogonia locating at the basal compartment of seminiferous tubules and Sertoli cells having cytological features such as irregular shaped nucleus, abundant clear cytoplasm, and irregular demarcation of the cellular limits. Sections incubated with normal rabbit immunoglobulins showed no positive staining.

Phosphorylation of endogenous Rec8 in testes at various ages was examined by Western blotting. When Rec8 undergoes phosphorylation, two additional bands with slower electrophoretic mobility (phosphorylated proteins) and one original band with faster electrophoretic mobility (a nonphosphorylated protein) appear at the range of 80–100 kDa. The phosphorylation gradually elevated in testes after 10 dpp, reaching the maximum level around 16 dpp, although Rec8 was constantly expressed from 8 to 20 dpp. These *in vivo* analyses indicated that the elevation of Rec8 phosphorylation in testes after 10 dpp seems to follow the increase of nociceptin production in testes after 9 dpp (Eto et al., 2012), and this is supported by the *in vitro* analyses described below, suggesting the involvement of nociceptin/Oprl-1 signaling in Rec8 phosphorylation.

3.2 Effect of nociceptin on the phosphorylation of Rec8 in testes

To determine whether nociceptin induces Rec8 phosphorylation in testes, we cultured testes of several ages for various periods with nociceptin and analyzed by Western blotting for Rec8. First, we determined that the 10-dpp testes were optimal for observing Rec8 phosphorylation because the testes contained phosphorylated Rec8 at the lower level than the older testes and the younger testes were negative. The phosphorylation was significantly detected by 15 min after nociceptin was added to the cultures and increased with longer incubation periods; however, this was suppressed to the basal level by pertussis toxin, a selective inhibitor for the Gi protein coupled to Oprl-1, or nocistatin, a specific inhibitor for nociceptin.

To determine whether nociceptin stimulates Rec8 phosphorylation *in vivo*, mice of 10, 12, or 14 dpp were intraperitoneally injected with the vehicle (water), nociceptin, or nocistatin and 12 or 24 h later the testes were analyzed by Western blotting for Rec8. The phosphorylation faintly increased 24 h after mice were injected with water (control). Expectedly, nociceptin injection into 10-dpp mice caused a significant enhancement of Rec8 phosphorylation in the testes 12 and 24 h later compared to the control. On the other hand, an increased Rec8 phosphorylation was hardly observed in the testes 24 h after nociceptin injection into 12- or 14-dpp mice because the endogenous nociceptin-induced Rec8 phosphorylation became higher and we could not validate the effect of exogenous nociceptin on Rec8 phosphorylation easily. Conversely, nocistatin administration completely suppressed Rec8 phosphorylation in testes of any ages. These results indicated that nociceptin directly and specifically stimulates Rec8 phosphorylation through Oprl-1 in testes *in vitro* and *in vivo* (Fig. 2).

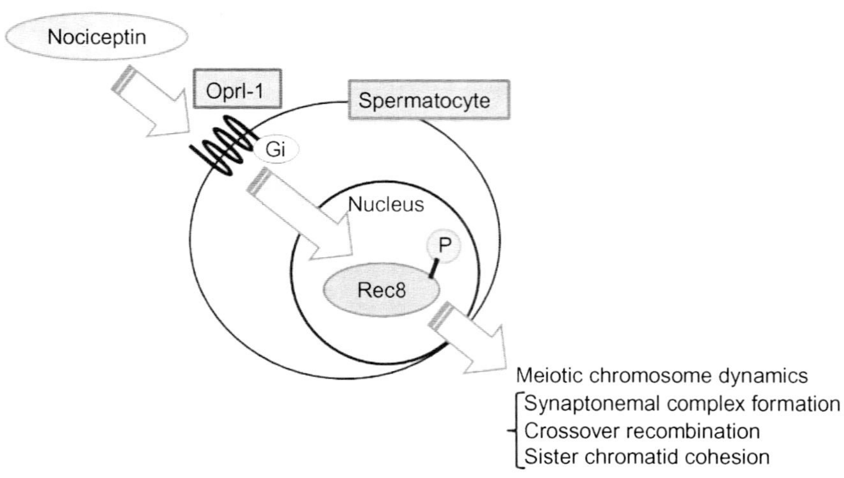

Figure 2 The nociceptin intracellular signal transduction inducing Rec8 phosphorylation and meiotic chromosome dynamics in spermatocytes. Gi, Gi protein coupled to the seven-transmembrane receptor Oprl-1 for nociceptin; P, phosphorylated. (See the color plate.)

3.3 Effect of nociceptin on the progress of meiosis during spermatogenesis

To determine whether the nociceptin-induced Rec8 phosphorylation affects meiosis and spermatogenesis *in vivo*, we injected 10-dpp mice intraperitoneally daily with the vehicle (water), nociceptin, or nocistatin for 10 days and then analyzed sections of the testes at 20 dpp by immunohistochemistry for spermatogenic cell proliferation and histology for spermatogenic cell differentiation. Immunohistochemistry was performed on proliferating cells with an antibody to proliferating cell nuclear antigen (PCNA) that labels strongly nuclei of spermatogonia and early spermatocytes but labels weakly late stages of spermatocytes. Histology was performed on postmeiotic spermatids with hematoxylin and eosin staining. Normally, the 10-dpp testes contain spermatogonia and early spermatocytes, but the 20-dpp testes contain late spermatocytes and spermatids in addition to them. In testes of mice injected with water (control) or nociceptin, PCNA labeling persisted weakly in a subpopulation of cells in the basal layer of seminiferous tubules, demonstrating normal progress of germ cell development including the progress of meiosis. In testes of control mice, endogenous nociceptin may be effective on Rec8 phosphorylation and spermatogenensis. In contrast, the majority of cells in testes of mice injected with nocistatin exhibited strong PCNA staining near the center of seminiferous tubules even at 20 dpp, demonstrating abnormal existence of proliferating spermatogonia and early spermatocytes and prominent lack of late spermatocytes and spermatids, which corresponds to the contents of germ cells in the 10-dpp testes. Furthermore, histological analyses clearly showed that spermatids were present near the center of seminiferous tubules in testes of control mice or nociceptin-injected mice, whereas the tubules derived from nocistatin-injected mice contained few spermatids but not a few spermatogonia and spermatocytes. These results indicated that the nociceptin-induced Rec8 phosphorylation is likely responsible for the progress of meiosis during spermatogenesis (Fig. 2).

To clarify which stage and how nociceptin modulates meiosis through Rec8 phosphorylation during prophase of meiosis I *in vivo*, we injected 12- or 14-dpp mice intraperitoneally once with water, nociceptin, or nocistatin and 24 h later analyzed meiotic chromosome dynamics with chromosome spreads prepared from the testes of 13 or 15 dpp by immunostaining for Rec8. Our immunostaining analyses on the localization in spread chromosomes of normal spermatocytes were consistent with previous reports (Bolcun-Filas et al., 2009; Buonomo et al., 2000; Eijpe et al., 2003; Pasierbek et al., 2001; Xu et al., 2005). Based on the observations, spermatocytes were staged during prophase I: spermatocytes with short and fine chromosomes are at the leptotene stage; spermatocytes with longer and more concentrated chromosomes are at the zygotene stage when synaptonemal complex formation and synapsis occur; spermatocytes with thickened, or synapsed, and desynapsed chromosomes are at the pachytene and diplotene stage, respectively, when crossover recombination occurs. Immunostaining of chromosome spreads for Rec8 showed that spermatocytes at the zygotene and pachytene stage were present in addition to at the leptotene stage at 13 dpp in testes of control mice or nociceptin-injected mice; however, only spermatocytes at the leptotene stage were observed in testes of nocistatin-injected mice. These results revealed that nociceptin promotes the transition from the leptotene to the zygotene stage during prophase I, and that nocistatin suppresses spermatocyte development by inhibiting nociceptin signaling. We could detect with great difficulty the immunoreactive signals for Rec8 on spread chromosomes prepared from testes of nocistatin-injected mice by microscopic observation under the high background following the long exposure because the leptotene chromosomes are very thin. Testes of control mice or nociceptin-injected mice contained spermatocytes at the pachytene and diplotene stage in addition to at the leptotene and zygotene stage at 15 dpp, whereas spermatocytes at the leptotene and zygotene stage were observed in testes of nocistatin-injected mice. These results demonstrated that nociceptin promotes the transition from the zygotene to the pachytene stage during prophase I. In testes of control mice, endogenous nociceptin may be effective on spermatocyte development. Thus, nociceptin triggers chromosome dynamics through Oprl-1 during meiosis in spermatocytes (Fig. 2).

3.4 Effect of FSH on the phosphorylation of Rec8 in testes

To determine whether FSH stimulates Rec8 phosphorylation via nociceptin/Oprl-1 signaling in testes, we cultured the 9-dpp testes for various periods with FSH and analyzed by Western blotting for Rec8 because it takes at least 24 h for FSH to induce nociceptin transiently (Eto et al., 2012) and nociceptin stimulates Rec8 phosphorylation significantly in the 10-dpp testes *in vivo* and *in vitro*. The phosphorylation was significantly enhanced after 25 h of FSH addition to the cultures; however, this was suppressed by H-89, a potent inhibitor for PKA, or nocistatin. These results indicated

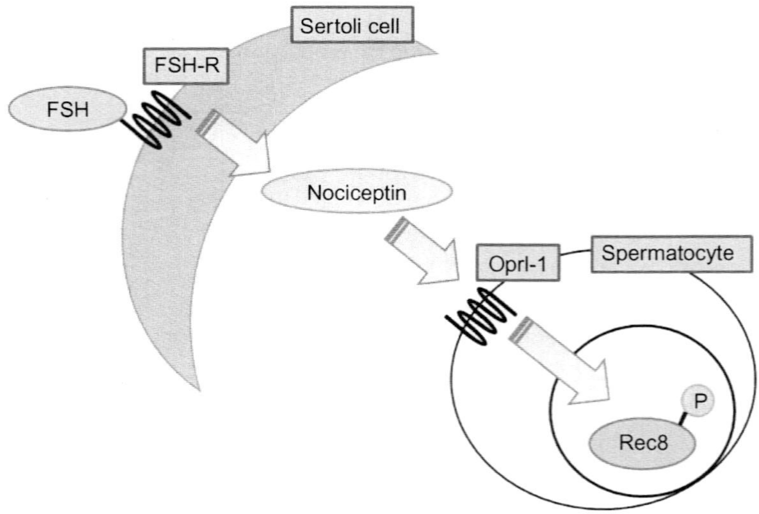

Figure 3 The nociceptin-mediated intercellular signal transduction communicating between Sertoli cells and spermatocytes within testicular tubules. (See the color plate.)

that FSH signaling in Sertoli cells stimulates Rec8 phosphorylation in spermatocytes, suggesting that this FSH effect is mediated by nociceptin signaling between Sertoli cells and spermatocytes in testes (Fig. 3).

4. NOCICEPTIN IS A NOVEL PARACRINE FACTOR THAT IS INDUCED IN SERTOLI CELLS AND MEDIATES TO GERM CELLS THE EFFECT OF FSH ON SPERMATOGENESIS

4.1 Prepronociceptin gene is transcriptionally regulated by FSH signaling in Sertoli cells

In response to FSH, its receptor may trigger multiple signaling pathways that regulate Sertoli cell functions including gene expression directly and lead to germ cell development indirectly (de Kretser et al., 2001; Mather et al., 1997; Simoni et al., 1997; Walker, Fucci, & Harbener, 1995). However, which signaling pathways are activated in Sertoli cells and how the signals affect germ cells are poorly understood. CREB is highly expressed and phosphorylated in Sertoli cells during the early phase of spermatogenesis (Kim et al., 2009). In addition, cAMP/PKA system phosphorylating CREB exists in mammalian Sertoli cells (Simoni et al., 1997; Walker et al., 1995). That is why we used Sertoli B cells (Fujino et al., 2006) to gain insights into Sertoli cell-specific signal transduction and gene expression. We revealed that Sertoli B cells exhibit the same physiological responses to FSH via cAMP in at least PKA activation and CREB phosphorylation/activation as testicular Sertoli cells excluding the undetectable level of FSH receptor expression. Our ChIP assay demonstrated that cAMP/PKA pathway may regulate the association of phosphorylated CREB with a putative consensus CRE sequence in the proximal promoter of murine prepronociceptin gene, which seems to be highly conserved between Sertoli cells in testes and neurons in brains (Zaveri et al., 2002, 2006). Moreover, we showed for the first time by RT-PCR analyses of Sertoli B cells and testes and Western blot analyses of peptide extracts from testes that the expressions of prepronociceptin mRNA and the nociceptin peptide may be upregulated by FSH signaling via cAMP/PKA/CREB pathway in Sertoli cells. We found that it takes 24 h for FSH to induce nociceptin, consistent with previous reports showing that gene and/or protein induction for cellular responses to FSH is generally delayed by at least 24 h (Gnozalez-Robayna, Alliston, Buse, Firestone, & Richards, 1999; Pei, Dodson, Schoderbek, Maurer, & Mayo, 1991), although the reason remains unknown.

Interestingly, this is the first report on the expression of nociceptin in testes. The opioid peptide family members such as proopiomelanocortin, preprodynorphin, and preproenkephalin are transcriptionally regulated in a similar fashion by raising the level of intracellular cAMP in brains (Loeffler, Kley, Pittius, & Hollt, 1986; Simpson & McGinty, 1995), consistent with our result for the FSH-activated transcription of prepronociceptin gene in Sertoli cells. Other peptides than nociceptin that might be expressed and function in testes are the focus of our future investigations. Overall, these data suggested that the FSH-operated cAMP/PKA/CREB pathway transcriptionally upregulates prepronociceptin gene expression, leading to the production of the nociceptin peptide, in Sertoli cells (Fig. 1).

4.2 Nociceptin is a paracrine factor mediating the FSH-regulated germ cell development

The functional roles of FSH in spermatogenesis have been addressed in mammals (Simoni et al., 1997), but more information is required. The FSH effects on germ cells are mediated through the receptor expressed restrictedly in Sertoli cells (Kangasniemi et al., 1990; Rannikki et al., 1995; Simoni et al., 1997), whereby paracrine factors are produced and secreted locally to act on germ cells. In mammalian testes, stem cell factor (SCF) is expressed and upregulated by FSH stimulation in Sertoli cells and is critical for proliferation of differentiating type A spermatogonia (Orth, Qiu, Jester, & Pilder, 1997; Sandlow, Feng, Cohen, & Sandra, 1996; Yoshinaga et al., 1991). However, few paracrine factors are known beyond SCF, which are essential for spermatogenesis. There is a dramatic surge of FSH in serum after 8 dpp in mice (Barakat et al., 2008) and we found that the nociceptin peptide is hardly detected in the 7- and 8-dpp testes; therefore, we considered that these are correlated with a constant induction of nociceptin in testes after 9 dpp following FSH surge, consistent with our *in vitro* results showing that FSH signaling is capable of inducing nociceptin in testes (Fig. 4).

Northern blot analyses for the distribution of prepronociceptin mRNA in rat tissues demonstrated that it is abundantly present in spinal cords and brains but barely in testes (Mollereau et al., 1996). However, our analyses with Sertoli B cells and testes demonstrated a transient and local increase in the expression of nociceptin in Sertoli cells following the activation of FSH signaling. This may be necessary for determining the temporal and spatial specificity for nociceptin-mediated paracrine functions during

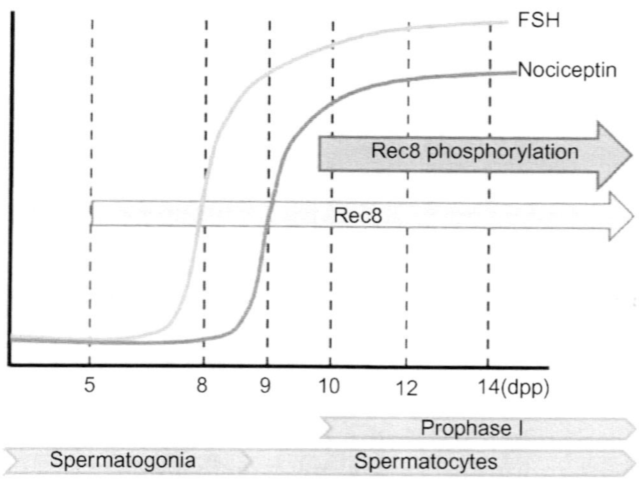

Figure 4 The temporal and spatial relationships among serum FSH levels, testicular nociceptin levels, germ Rec8 levels and phosphorylation, germ cell development, and meiosis initiation during the early phase of spermatogenesis in postnatal murine testes. After birth, spermatogonia exist as the only germ cells in testes until 9 dpp; serum FSH levels increase from 8 dpp onward; after further 24 h testicular nociceptin levels increase and spermatocytes appear in addition to spermatogonia; and Rec8 phosphorylation begins to happen around 10 dpp when meiosis I begins with a prophase (prophase I) although Rec8 expression is already observed from 5 dpp. (See the color plate.)

spermatogenesis within seminiferous tubules in postnatal testes. Overall, nociceptin has a potential role in the FSH-regulated spermatogenesis. Identification of nociceptin described here provides new cue for understanding the molecular mechanisms underlying spermatogenesis. The question concerning the physiological significance and functional roles of nociceptin still remains unanswered and is now under investigation with *in vitro* and *in vivo* analyses.

5. NOCICEPTIN IS A NOVEL EXTRINSIC FACTOR INDUCING REC8 PHOSPHORYLATION AND CHROMOSOME DYNAMICS DURING MEIOSIS IN SPERMATOCYTES

Our study reports the paracrine mechanism for nociceptin to stimulate Rec8 phosphorylation directing chromosome dynamics including synaptonemal complex formation, crossover recombination, and sister chromatid cohesion during meiosis in spermatocytes (Eto et al., 2013). Therefore, we concluded that nociceptin may be a novel extrinsic factor inducing Rec8 phosphorylation and subsequently meiotic chromosome dynamics, consequently maintaining spermatogenesis in murine testes (Fig. 2).

5.1 Nociceptin is an extrinsic regulator for Rec8 phosphorylation during meiosis in spermatocytes

With immunohistochemical analyses for Oprl-1, we identified spermatocytes as one of the target cells for the action of nociceptin in the 10-dpp testes, where spermatocytes appear and prophase of meiosis I is initiated (Fig. 4). Rec8 phosphorylation is a crucial step for initiating chromosome dynamics in premeiotic spermatocytes (Buonomo et al., 2000; Eijpe et al., 2003; Lee et al., 2003; Nasmyth, 2001), but which extrinsic factors trigger the phosphorylation remained unclear. We showed that the expression of endogenous nociceptin in testes after 9 dpp (Eto et al., 2012) is involved in the phosphorylation of endogenous Rec8 after 10 dpp (Eto et al., 2013), raising the possibility that nociceptin is secreted from Sertoli cells and targets spermatocytes to stimulate Rec8 phosphorylation within seminiferous tubules during meiosis. Our *in vitro* and *in vivo* analyses provided novel evidence for that possibility. We also demonstrated *in vivo* that inhibition of the endogenous nociceptin-stimulated Rec8 phosphorylation by exogenous nocistatin resulted in defective meiosis and ultimately disturbed spermatogenesis, suggesting that nociceptin regulates meiosis by modulating Rec8 phosphorylation. This is supported by the previous finding that Rec8 phosphorylation induces the removal of cohesins by proteolytic cleavage and subsequent loss of cohesion, initiating two steps of chromosome segregation in spermatocytes during meiotic divisions to produce haploid spermatids (Buonomo et al., 2000; Eijpe et al., 2003; Lee et al., 2003; Nasmyth, 2001). In addition, our *in vivo* analyses revealed that nociceptin triggers the progress of meiosis by inducing Rec8 phosphorylation at the early phase of meiosis. We showed by immunostaining analyses with chromosome spreads for Rec8 that the dysfunction of nociceptin by nocistatin delayed or/and suppressed the transition from the leptotene to the zygotene stage and the transition from the zygotene to the pachytene stage in spermatocytes, suggesting that nociceptin regulates synapsis via synaptonemal complex formation occurring at the zygotene stage by modulating chromosome dynamics at the leptotene stage and meiotic crossover recombination occurring at the pachytene stage by modulating chromosome dynamics at the zygotene stage, respectively. Thus, we could define nociceptin as an extrinsic factor responsible for inducing Rec8 phosphorylation in spermatocytes, eventually leading to the progress of meiosis during germ cell development (Fig. 2).

We showed that nociceptin is able to induce Rec8 phosphorylation apparently in 10-dpp testes but less in testes younger than 9 dpp or older than 11 dpp (Rec8 phosphorylation stimulated by exogenous nociceptin was difficult to be detected because of Rec8 phosphorylation elevated by endogenous nociceptin) in both *in vitro* and *in vivo* systems, indicating that the extrinsic signal must not be sufficient for Rec8 phosphorylation. This suggests the requirement of intrinsic regulations for tuning the timing of Rec8 phosphorylation. In addition, Rec8 phosphorylation was stimulated when the 9-dpp testes were exposed for over 25 h to FSH because nociceptin was induced by FSH signaling 24 h after culture, suggesting that testes spontaneously acquire the characters and contents of the 10-dpp testes, which are at least required for Rec8 phosphorylation (Fig. 4). The intrinsic mechanisms might comprise the induction of a kinase phosphorylating Rec8 such as aurora kinase AIR-2 in *Caenorhabditis elegans* (Rogers, Bishop, Waddle, Schumacher, & Lin, 2002) or polo-like kinase CDC5 in budding yeast (Clyne et al., 2003; Lee & Amon, 2003), which have not yet been identified in mammals, although nociceptin is probably capable of activating such a kinase and stimulating Rec8 phosphorylation directly. Hence, meiotic chromosome dynamics might be precisely accompanied by restricting the timing for Rec8 phosphorylation through both the nociceptin-mediated extrinsic mechanisms and the unknown intrinsic mechanisms in spermatocytes.

5.2 Nociceptin is a testicular peptide, "testipeptide," that is expressed and functions locally within testes

Interestingly, this is the first report on the function of nociceptin in spermatogenesis in murine testes. The localization of Oprl-1 in the plasma membrane of spermatocytes suggested that the receptor is able to respond to nociceptin, which is expressed and secreted from Sertoli cells (Eto et al., 2012), and function in spermatocytes for inducing Rec8 phosphorylation. In addition, our *in vitro* analyses indicated that FSH stimulates Rec8 phosphorylation, suggesting that nociceptin signaling is a paracrine mechanism mediating the FSH-stimulated Rec8 phosphorylation and subsequent meiotic chromosome dynamics in spermatocytes (Fig. 3). This may be necessary for determining the temporal and spatial specificity for Rec8 phosphorylation in spermatocytes, thereby ensuring proper chromosome dynamics during meiosis. We therefore propose a novel idea that nociceptin is a testicular peptide, "testipeptide," that normally is expressed and functions locally within testes. However, how nociceptin stimulates Rec8 phosphorylation in spermatocytes remains to be unexplained.

6. CONCLUSIONS

Our study reports that nociceptin is induced in Sertoli cells through the FSH-activated intracellular signaling pathway and provides insights into a novel paracrine regulation of spermatogenesis in postnatal murine testes (Eto et al., 2012). In addition, we have presented new findings that nociceptin is a critical factor for Rec8 phosphorylation and has an important impact on the phosphorylated Rec8-mediated chromosome dynamics during meiosis in spermatocytes (Eto et al., 2013). We have also provided insight into the functional role of nociceptin/Oprl-1 signaling in testicular germ cells in mice (Eto et al., 2013). Our data may demonstrate the physiological mechanism that the FSH surge-induced temporal expression of nociceptin in Sertoli cells after 9 dpp (Eto et al., 2012) continuously stimulates and maintains Rec8 phosphorylation responsible for meiotic chromosome dynamics in spermatocytes after 10 dpp during germ cell development (Eto et al., 2013) (Fig. 4). This study should contribute to a better understanding of the molecular mechanisms underlying meiosis in testes.

ACKNOWLEDGMENTS

I am grateful to Mr. Masahiro Shiotsuki, Miss Tomomi Sakai, and Dr. Shin-ichi Abe for the contribution to this work.

REFERENCES

Barakat, B., O'Connor, A. E., Gold, E., de Kretser, D. M., & Loveland, K. L. (2008). Inhibin, activin, follistatin and FSH serum levels and testicular production are highly modulated during the first spermatogenic wave in mice. *Reproduction, 136*, 345–359.

Bolcun-Filas, E., Speed, R., Taggart, M., Grey, C., de Massay, B., Benavente, R., et al. (2009). Mutation of mouse Syce1 gene disrupts synapsis and suggests a link between synaptonemal complex structural components and DNA repair. *PLoS Genetics, 5*, e1000393.

Buonomo, S. B. C., Clyne, R. K., Fuchs, J., Loidl, J., Uhlmann, F., & Nasmyth, K. (2000). Disjunction of homologous chromosomes in meiosis I depends on proteolytic cleavage of the meiotic cohesin REC8 by separin. *Cell, 103*, 387–398.

Childers, S. R. (1991). Opioid receptor-coupled second messenger systems. *Life Sciences, 48*, 1991–2003.

Clyne, R. K., Katis, V. L., Jessop, L., Benjamin, K. R., Herskowitz, I., Lichten, M., et al. (2003). Polo-like kinase Cdc5 promotes chiasmata formation and cosegregation of sister centromeres at meiosis I. *Nature Cell Biology, 5*, 480–485.

de Kretser, D. M., Loveland, K. L., Meehan, T., O'Bryan, M. K., Phillips, D. J., & Wreford, N. G. (2001). Inhibins, activins and follistatin: Actions on the testis. *Molecular and Cellular Endocrinology, 180*, 87–92.

Eijpe, M., Offenberg, H., Jessberger, R., Revenkova, E., & Heyting, C. (2003). Meiotic cohesin REC8 marks the axial elements of rat synaptonemal complexes before cohesions SMC1β and SMC3. *The Journal of Cell Biology, 160*, 657–670.

Eto, K., Shiotsuki, M., & Abe, S. (2013). Nociceptin induces Rec8 phosphorylation and meiosis in postnatal murine testes. *Endocrinology, 154*, 2891–2899.

Eto, K., Shiotsuki, M., Sakai, T., & Abe, S. (2012). Nociceptin is upregulated by FSH signaling in Sertoli cells in murine testes. *Biochemical and Biophysical Research Communications, 421*, 678–683.

Fujino, R., Tanaka, K., Morimatsu, M., Tamura, K., Kogo, H., & Hara, T. (2006). Spermatogonial cell-mediated activation of IκBζ-independent nuclear factor-κB pathway in Sertoli cells induces transcription of the lipocalin-2 gene. *Molecular Endocrinology, 20*, 904–915.

Gnozalez-Robayna, I., Alliston, T. N., Buse, P., Firestone, G. L., & Richards, J. S. (1999). Functional and subcellular changes in the A-kinase-signaling pathway: Relation to aromatase and Sgk expression during the transition of granulosa cells to luteal cells. *Molecular Endocrinology, 13*, 1318–1337.

Guacci, V., Koshland, D., & Strunnikov, A. (1997). A direct link between sister chromatid cohesion and chromosome condensation revealed through the analysis of MCD1 in *S. cerevisiae*. *Cell, 91*, 47–58.

Kangasniemi, M., Kaipia, A., Toppari, J., Perheentupa, A., Huhtaniemi, I., & Parvinen, M. (1990). Cellular regulation of follicle-stimulating hormone (FSH) binding in rat seminiferous tubules. *Journal of Andrology, 11*, 336–343.

Kim, J. S., Song, M. S., Seo, H. S., Yang, M., Kim, S. H., Kim, J. C., et al. (2009). Immunohistochemical analysis of cAMP response element-binding protein in mouse testis during postnatal development and spermatogenesis. *Histochemistry and Cell Biology, 131*, 501–507.

Klein, F., Mahr, P., Galova, M., Buonomo, S. B., Michaelis, C., Nairz, K., et al. (1999). A central role for cohesins in sister chromatid cohesion, formation of axial elements, and recombination during yeast meiosis. *Cell, 98*, 91–103.

Lee, B. H., & Amon, A. (2003). Role of polo-like kinase CDC5 in programming meiosis I chromosome segregation. *Science, 300*, 482–486.

Lee, J., Iwai, T., Yokota, T., & Yamashita, M. (2003). Temporally and spatially selective loss of Rec8 protein from meiotic chromosomes during mammalian meiosis. *Journal of Cell Science, 116*, 2781–2790.

Loeffler, J. P., Kley, N., Pittius, C. W., & Hollt, V. (1986). Regulation of pro-opiomelanocortin (POMC) mRNA levels in primary pituitary cultures. *NIDA Research Monograph, 75*, 397–400.

Losada, A., Hirano, M., & Hirano, T. (1998). Identification of *Xenopus* SMC protein complexes required for sister chromatid cohesion. *Genes and Development, 12*, 1986–1997.

Mather, J. P., Moore, A., & Li, R. H. (1997). Activins, inhibins, and follistatins: Further thoughts on a growing family of regulators. *Proceedings of the Society for Experimental Biology and Medicine, 215*, 209–222.

Michaelis, C., Ciosk, R., & Nasmyth, K. (1997). Cohesins: Chromosomal proteins that prevent premature separation of sister chromatids. *Cell, 91*, 35–46.

Mollereau, C., Simons, M. J., Soularue, P., Liners, F., Vassart, G., Meunier, J. C., et al. (1996). Structure, tissue distribution, and chromosomal localization of the prepronociceptin gene. *Proceedings of the National Academy of Sciences of the United States of America, 93*, 8666–8670.

Molnar, M., Bahler, J., Sipiczki, M., & Kohli, J. (1995). The rec8 gene of *Schizosaccharomyces pombe* is involved in linear element formation, chromosome pairing and sister-chromatid cohesion during meiosis. *Genetics, 141*, 61–73.

Nasmyth, K. (2001). Disseminating the genome: Joining, resolving, and separating sister chromatids during mitosis and meiosis. *Annual Review of Genetics, 35*, 673–745.

Orth, J. M., Qiu, J., Jester, W. F., Jr., & Pilder, S. (1997). Expression of the c-kit gene is critical for migration of neonatal rat gonocytes in vitro. *Biology of Reproduction, 57*, 676–683.

Page, S. L., & Hawley, R. S. (2003). Chromosome choreography: The meiotic ballet. *Science, 301*, 785–789.

Pasierbek, P., Jantsch, M., Melcher, M., Schleiffer, A., Schweizer, D., & Loidl, J. (2001). A *Caenorhabditis elegans* cohesion protein with functions in meiotic chromosome pairing and disjunction. *Genes and Development, 15*, 1349–1360.

Pei, L., Dodson, R., Schoderbek, W. E., Maurer, R. A., & Mayo, K. E. (1991). Regulation of the α inhibin gene by cyclic adenosine 3′,5′-monophosphate after transfection into rat granulosa cells. *Molecular Endocrinology, 5*, 521–534.

Petronczki, M., Siomos, M. F., & Nasmyth, K. (2003). Un ménage à quatre: The molecular biology of chromosome segregation in meiosis. *Cell, 112*, 423–440.

Qin, J., Richardson, L. L., Jasin, M., Handel, M. A., & Arnheim, N. (2004). Mouse strains with an active H2-Ea meiotic recombination hot spot exhibit increased levels of H2-Ea-specific DNA breaks in testicular germ cells. *Molecular and Cellular Biology, 24*, 1655–1666.

Rannikki, A. S., Zhang, F. P., & Huhtaniemi, I. T. (1995). Ontogeny of follicle-stimulating hormone receptor gene expression in the rat testis and ovary. *Molecular and Cellular Endocrinology, 107*, 199–208.

Reisine, T., & Bell, G. I. (1993). Molecular biology of opioid receptors. *Trends in Neurosciences, 16*, 506–510.

Rogers, E., Bishop, J. D., Waddle, J. A., Schumacher, J. M., & Lin, R. (2002). The aurora kinase AIR-2 functions in the release of chromosome cohesion in *Caenorhabditis elegans* meiosis. *The Journal of Cell Biology, 157*, 219–229.

Russell, L. D., Ettlin, R. A., Sinha Hikim, A. P., & Clegg, E. D. (1990). *Histological and histopathological evaluation of the testis* (1st ed.). Cleanwater, FL: Cache River Press.

Sandlow, J. I., Feng, H. L., Cohen, M. B., & Sandra, A. (1996). Expression of c-kit and its ligand, stem cell factor, in normal and subfertile human testicular tissue. *Journal of Andrology, 17*, 403–408.

Simoni, M., Gromoll, J., & Nieschlag, E. (1997). The follicle-stimulating hormone receptor: Biochemistry, molecular biology, physiology, and pathophysiology. *Endocrine Reviews, 18*, 739–773.

Simpson, J. N., & McGinty, J. F. (1995). Forskolin induces preproenkephalin and preprodynorphin mRNA in rat striatum as demonstrated by in situ hybridization histochemistry. *Synapse, 19*, 151–159.

Sprengel, R., Braun, T., Nikolics, K., Segaloff, D., & Seeburg, P. H. (1990). The testicular receptor for follicle stimulating hormone: Structure and functional expression of cloned cDNA. *Molecular Endocrinology, 4*, 525–530.

Walker, W. H., Fucci, L., & Harbener, J. F. (1995). Expression of the gene encoding transcription factor cyclic adenosine 3′,5′-monophosphate (cAMP) response element-binding protein (CREB): Regulation by follicle-stimulating hormone-induced cAMP signaling in primary rat Sertoli cells. *Endocrinology, 136*, 3534–3545.

Watanabe, Y., & Nurse, P. (1999). Cohesin Rec8 is required for reductional chromosome segregation at meiosis. *Nature, 400*, 461–464.

Wright, W. (1993). Cellular interactions in the seminiferous epithelium. In C. Desjardins & L. Ewing (Eds.), *Cell and molecular biology of the testis* (pp. 377–399). New York, NY: Oxford University Press.

Xu, H., Beasley, M. D., Warren, W. D., van der Horst, G. T. J., & McKay, M. J. (2005). Absence of mouse REC8 cohesin promotes synapsis of sister chromatids in meiosis. *Developmental Cell, 8*, 949–961.

Yoshinaga, K., Nishikawa, S., Ogawa, M., Hayashi, S., Kunisada, T., Fujimoto, T., et al. (1991). Role of c-kit in mouse spermatogenesis: Identification of spermatogonia as a specific site of c-kit expression and function. *Development, 113*, 689–699.

Zaveri, N. T., Green, C. J., Polgar, W. E., Huynh, N., & Toll, L. (2002). Regulation of transcription of the human prepronociceptin gene by Sp1. *Gene, 290*, 45–52.

Zaveri, N. T., Waleh, N., & Toll, L. (2006). Regulation of the prepronociceptin gene and its effect on neuronal differentiation. *Gene, 384*, 27–36.

CHAPTER SEVEN

Orphanin FQ-ORL-1 Regulation of Reproduction and Reproductive Behavior in the Female

Kevin Sinchak*[,1], **Lauren Dalhousay***, **Nayna Sanathara**†
*Department of Biological Sciences, California State University, Long Beach, California, USA
†Department of Pharmacological Sciences, University of California, Irvine, California, USA
[1]Corresponding author: e-mail address: kevin.sinchak@csulb.edu

Contents

Abstract

Orphanin FQ (OFQ/N) and its receptor, opioid receptor-like receptor-1 (ORL-1), are expressed throughout steroid-responsive limbic and hypothalamic circuits that regulate female ovarian hormone feedback and reproductive behavior circuits. The arcuate nucleus of the hypothalamus (ARH) is a brain region that expresses OFQ/N and ORL-1 important for both sexual behavior and modulating estradiol feedback loops. Within the ARH, the activation of the OFQ/N-ORL-1 system facilitates sexual receptivity (lordosis) through the inhibition of β-endorphin neuronal activity. Estradiol initially activates ARH β-endorphin neurons to inhibit lordosis. Simultaneously, estradiol upregulates coexpression of OFQ/N and progesterone receptors and ORL-1 in ARH β-endorphin neurons. Ovarian hormones regulate pre- and postsynaptic coupling of ORL-1 to its G protein-coupled signaling pathways. When the steroid-primed rat is nonreceptive, estradiol acts pre- and postsynaptically to decrease the ability of the OFQ/N-ORL-1 system to inhibit ARH β-endorphin neurotransmission. Conversely, when sexually receptive, ORL-1 signaling is restored to inhibit β-endorphin neurotransmission. Although steroid signaling that facilitates lordosis converges to deactivate ARH

Vitamins and Hormones, Volume 97
ISSN 0083-6729
http://dx.doi.org/10.1016/bs.vh.2014.11.002

β-endorphin neurons, estradiol-only facilitation of lordosis requires the activation of ORL-1, but estradiol + progesterone does not, indicating that multiple circuits mediate ovarian hormone signaling to deactivate ARH β-endorphin neurons. Research on the role of OFQ/N-ORL-1 in ovarian hormone feedback loops is just beginning. In the rat, OFQ/N may act to terminate gonadotropin-releasing hormone and luteinizing hormone release under positive and negative feedbacks. In the ewe, it appears to directly inhibit gonadotropin-releasing hormone release to mediate progesterone-negative feedback. As a whole, the localization and actions of OFQ/N-ORL-1 system indicate that it may mediate the actions of estradiol and progesterone to synchronize reproductive behavior and ovarian hormone feedback loops.

ABBREVIATIONS

β-END β-endorphin
ARH arcuate nucleus of the hypothalamus
AVPV anteroventral periventricular nucleus of the hypothalamus
Ca^{2+} calcium
DBB diagonal band of Broca
E2 17β-estradiol
EB 17β-estradiol benzoate
EB + P 2 μg 17β-estradiol benzoate + 500 μg progesterone
E-biotin 17β-estradiol conjugated to biotin molecule
EB-only 5–50 μg 17β-estradiol benzoate
EN-RET endoplasmic reticulum
ER estrogen receptor
FSH follicle-stimulating hormone
G-1 GPER30 agonist
G-15 GPER30 antagonist
GIRK-1 G protein-gated inwardly rectifying K^+ channel
GnRH gonadotropin-releasing hormone
GPER30 G protein-coupled estrogen receptor-1
HPO hypothalamic–pituitary–ovarian
LH luteinizing hormone
MBH mediobasal hypothalamic
MeApd posterodorsal medial amygdala
MEPO median preoptic nucleus
mER membrane estrogen receptor
mGluR1a metabotropic glutamate receptor type-1a
MOP μ-opioid receptors
MPN medial preoptic nucleus of the hypothalamus
NPY neuropeptide Y
OFQ/N orphanin FQ
ORL-1 opioid receptor-like receptor-1
OVX ovariectomized
P progesterone
POMC proopiomelanocortin

PR progesterone receptor
PVN paraventricular nucleus
RP3V rostral periventricular region of the third ventricle
VGCC voltage-gated calcium channels
VMH ventromedial hypothalamus
Y1 neuropeptide Y-Y1 receptor

1. INTRODUCTION

Successful reproduction in the female is dependent on the proper synchronization of hormonal feedback loops of the hypothalamic–pituitary–ovarian (HPO) axis. These feedback loops control ovarian hormone release to coordinate reproductive behavior with reproductive physiology associated with ovulation and uterine development to increase the likelihood of fertilization and pregnancy following copulation. The ovarian hormones, estradiol and progesterone, are important signaling molecules that act on neural circuits to regulate reproductive behavior and physiology. Opioid neuropeptide systems are important mediators of ovarian steroid signaling that regulate reproduction in the female (Devorshak-Harvey, Bona-Gallo, & Gallo, 1987; Genazzani & Petraglia, 1989; Goodman & Inskeep, 2006; Lado-Abeal et al., 2002; Micevych & Sinchak, 2013; Pfaus & Gorzalka, 1987; Sinchak & Micevych, 2003). Approximately, 20 years ago, cloning studies revealed the existence of a novel opioid receptor, the opioid receptor-like receptor-1 (ORL-1) (Bunzow et al., 1994; Chen et al., 1994; Fukuda et al., 1994; Lachowicz, Shen, Monsma, & Sibley, 1995; Marchese et al., 1994; Mollereau et al., 1994; Wang et al., 1994; Wick et al., 1994). Shortly afterward, the heptadecapeptide, orphanin FQ (OFQ/N; aka nociceptin) was determined to be the endogenous ligand for the orphan receptor, ORL-1 (Meunier et al., 1995; Reinscheid et al., 1995; Saito et al., 1996; Saito, Maruyama, Saido, & Kawashima, 1995). Both OFQ/N and ORL-1 are expressed in steroid hormone-responsive brain regions that are important for the integrative control of the HPO axis and reproductive behavior in females (Neal, Mansour, Reinscheid, Nothacker, Civelli, Akil, et al., 1999; Neal, Mansour, Reinscheid, Nothacker, Civelli, & Watson, 1999; Sanathara, Moraes, Mahavongtrakul, & Sinchak, 2014; Sinchak, Romeo, & Micevych, 2006). The initial ORL-1 mapping studies were performed in the male rat and demonstrated that ORL-1 was highly expressed in the ventromedial hypothalamus (VMH) (Neal, Mansour, Reinscheid, Nothacker, Civelli, et al., 1999; Neal, Mansour, Reinscheid, Nothacker, Civelli, & Watson, 1999), indicating the potential for regulation of reproductive behavior. In the female rat, the VMH is an important nucleus that is responsive to estradiol and progesterone and contains motor output pathways that regulate sexual receptivity (Pfaff & Sakuma, 1979a, 1979b). Thus, the initial study that demonstrated a role for OFQ/N-ORL-1 system in reproduction utilized OFQ/N infusion into the VMH region. In estradiol-primed sexually nonreceptive rats, OFQ/N infusion facilitated sexual receptivity, demonstrating that indeed the OFQ/N-ORL-1 system could regulate aspects of reproduction (Sinchak, Hendricks, Baroudi, & Micevych, 1997). However, since the discovery of the OFQ/N-ORL-1 opioid system, besides investigations by me and my collaborators, very few groups have investigated the roles of the OFQ/N-ORL-1 system in regulating female reproduction and reproductive behavior. This chapter reviews the role of the OFQ/N-ORL-1 system in the regulation of sexual receptivity in the rat and the studies that have investigated the role of OFQ/N in positive and negative feedback systems in the female rat and ewe (An, Yu, Feng, Chen, & Zhang, 2007; Foradori et al., 2007; Nestor et al., 2013).

2. OVARIAN HORMONE REGULATION OF REPRODUCTIVE BEHAVIOR AND NEUROENDOCRINE FEEDBACK LOOPS

2.1 Neuroendocrine feedback loops

The release of ovarian hormones is regulated by negative and positive feedback mechanisms of the HPO axis. In the rat, the estrous cycle is normally 4–5 days in length. On diestrous days 1 and 2 of the cycle the pituitary hormones, follicle stimulating hormone (FSH) and luteinizing hormone (LH), induce ovarian follicle maturation and synthesis of ovarian steroids. Initially, circulating estradiol levels slowly rise and are regulated by estradiol-induced negative feedback. However, this is followed by estrogen-positive feedback that produces a rapid rise in circulating estradiol through the morning of proestrus which peaks on that afternoon. This estrogen-positive feedback induces the LH surge that induces ovulation and luteinization of the follicular cells for progesterone production. The sequential release of estradiol and progesterone from the ovary acts simultaneously in reproductive behavioral neurocircuits, neuroendocrine feedback neurocircuits, and reproductive organs to coordinate the onset of reproductive behavior with the development of the uterine endometrium and ovulation to increase the probability that copulation results in successful fertilization and implantation of the embryo.

2.2 Reproductive behavior

Two main behavioral components of female sexual behavior in the rat are sexual proceptivity and receptivity (Beach, 1948). Proceptive behaviors are appetitive behaviors that are associated with her motivational state to seek a male and solicit copulation (reviewed in Cummings & Becker, 2012). These include ear wiggling, hopping and darting, and pacing behaviors. The female rat, if given the opportunity, will pace the rate at which she receives copulations from the male. Her copulatory rate is lower than the rate the male would set if he were allowed to set the pace (Adler, 1969, 1978; Erskine, Kornberg, & Cherry, 1989; McClintock, 1984; Mendelson & Gorzalka, 1987; Pfaus, Smith, & Coopersmith, 1999; Yang & Clemens, 1997). This female-induced pacing of the copulatory rate increases sperm transport into the uterus and induces a neuroendocrine reflex that elicits a progestational state increasing the likelihood of pregnancy (Chester & Zucker, 1970; Erskine, 1987; Kornberg & Erskine, 1994; Matthews & Adler, 1977; McClintock, Toner, Adler, & Anisko, 1882; Terkel & Sawyer, 1978). Estradiol-only priming that induces maximal sexual receptivity produces very low levels of proceptivity, whereas maximal levels of proceptivity and receptivity are produced by estradiol and progesterone exposure as seen during the estrous cycle (Edwards & Pfeifle, 1983; Hardy & Debold, 1971; Hlinak & Madlafousek, 1981; Tennent, Smith, & Davidson, 1980; Whalen, 1974). To this date, the role of the OFQ/N-ORL-1 system in the regulation of proceptivity in the female has not been investigated.

Sexual receptivity is marked by consummatory behaviors associated with the ability to copulate in response to stimuli received from a male rat. In the female rat, exposure of neural circuits to the proper timing and levels of estradiol or estradiol and progesterone induces sexual receptivity. If the female rat is sexually receptive, then when a male mounts and palpates her flanks, she will respond reflexively with a stereotypic lordosis posture associated with copulation. Lordosis is the arching of the back with extension of the fore and hind legs, and deflection of the tail. This positions the perineum and the pelvic bones so that the male can successfully achieve intromission and ejaculation (Pfaff, Diakow, Montgomery, & Jenkins, 1978).

In ovariectomized (OVX) rats, sexual receptivity can be induced by mimicking the sequential release of estradiol and progesterone from the ovary. A subcutaneous priming dose of estradiol (2 μg 17β-estradiol conjugated to a benzoate molecule; EB) produces circulating levels of estradiol that are equivalent to peak levels on proestrus that act on lordosis behavioral circuits, but does not facilitate lordosis on its own (Jones, Farrell, Gregory, & Pfaus, 2013; Priest, Eckersell, & Micevych, 1995; Sinchak & Micevych, 2001). However, subsequent progesterone (P; 500 μg) treatment can be given 26–56 h after EB priming to induce maximal levels of sexual receptivity (Clemens & Weaver, 1985; Jones et al., 2013; Priest et al., 1995; Sinchak & Micevych, 2001). Reproductively mature, cycling female rats are spontaneous ovulators that come into behavioral estrus and ovulate with each cycle (reviewed in Clemens & Weaver, 1985). However, as the rat ages and enters early menopause, the rising levels of estradiol no longer induce the LH surge and ovulation, but the brain is exposed to higher levels of estradiol for a prolonged period of time (Lu, Hopper, Vargo, & Yen, 1979; Micevych, Soma, & Sinchak, 2008). This prolonged exposure to estradiol induces a persistent state of sexual receptivity, and copulation with a male will now reflexively trigger the LH surge and ovulation (Day, Morales, & Lu, 1988). Thus, at this stage of her reproductive life, the rat is a reflex ovulator (Hoffman et al., 2013). The induction of sexual receptivity can be mimicked in the OVX rat by longer exposures (48 h+) to higher levels of only estradiol (5–50 μg EB) (Blaustein, Finkbohner, & Delville, 1987; Clemens & Weaver, 1985; Jones et al., 2013; Pfaff, 1970; Priest et al., 1995; Quadagno, McCullough, & Langan, 1972; Sinchak & Micevych, 2001; Sodersten & Eneroth, 1981). Estradiol-only facilitation of sexual receptivity can be induced by either a single high dose of EB (5–50 μg) or repeated injections of smaller doses over consecutive days (Blaustein et al., 1987; Clemens & Weaver, 1985; Pfaff, 1970; Quadagno et al., 1972; Sinchak & Micevych, 2001; Sodersten & Eneroth, 1981). The onset of sexual receptivity in estradiol-only treated animals is delayed and begins about 48 h after initial estradiol treatment, whereas progesterone can facilitate lordosis if given as early as 20–24 h after EB priming (Boling & Blandau, 1939; Sinchak & Micevych, 2001). A third, less used steroid priming paradigm that also induces maximal levels of sexual receptivity primes first with 2 μg EB followed by 17β-estradiol not conjugated to a benzoate molecule (non-esterfied-estradiol) in the place of progesterone (Long, Serey,

Welborn, & Sinchak, 2013; Long, Serey, & Sinchak, 2014; Parsons, Rainbow, Snyder, & McEwen, 1984; Pfaff, Schwartz-Giblin, McCarthy, & Kow, 1994). Interestingly, in this last paradigm, lordosis is facilitated within 30 min of non-esterfied-estradiol treatment, and this non-esterfied-estradiol has been shown to be activating an extranuclear estrogen receptor (ER), G protein-coupled estrogen receptor-1 (GPR30; aka GPER-1; Fig. 2) (Filardo et al., 2006; Kuo et al., 2007; Long et al., 2013, 2014; Smith et al., 2009, 2007). Thus, the ability of these different steroid treatment paradigms to facilitate sexual receptivity indicates that multiple neural pathways may mediate both the induction of sexual receptivity and the LH surge (ovulation) such that the patterns are synchronized to optimize the chances that copulation results in pregnancy. Our studies support the idea that OFQ/N-ORL-1 circuits in the arcuate nucleus of the hypothalamus (ARH) are required for estrogen-only facilitation of lordosis, but not EB + P (Sanathara, Moraes, Kanjiya, & Sinchak, 2011; Sinchak, Dewing, Cook, & Micevych, 2007).

Whether the animal is gonadally intact or OVX, facilitation of sexual receptivity requires the behavioral neurocircuits to be initially exposed to estradiol. This initial estradiol treatment has numerous effects on reproductive circuits that include: (1) priming or upregulating lordosis-facilitative neurocircuits through transcriptional or extranuclear signaling mechanisms; (2) activating lordosis-inhibitory pathways to prevent copulation from occurring prior to ovulation and allow for the steroid-induced development of the uterine endometrium; and (3) modulating functional coupling to and signaling of opioid receptors through their G protein signaling pathways. An example of a priming effect of estradiol is the upregulation of progesterone receptor (PR) in the VMH, ARH, and medial preoptic nucleus of the hypothalamus (MPN) (Alves et al., 2000; Blaustein & Feder, 1979; MacLusky & McEwen, 1978, 1980; Moguilewsky & Raynaud, 1979; Sanathara et al., 2014; Shughrue, Lubahn, NegroVilar, Korach, & Merchenthaler, 1997). Estradiol-induction of PR synthesis takes about 20 h and is why progesterone treatment must be given at least 20–24 h after EB priming for facilitation of lordosis and why progesterone follows the rise in estradiol in the rat estrous cycle (Alves et al., 2000; Shughrue et al., 1997; Sinchak & Micevych, 2001). Estradiol also induces OFQ/N and ORL-1 mRNA and protein expression in several brain regions that are important for sexual receptivity (Sanathara et al., 2014; Sinchak et al., 2006). In addition to upregulating enzymes, ligands and receptors (Dohanich, Witcher, Weaver, & Clemens, 1982; Luine, Khylchevskaya, & McEwen, 1975;

Luine, Park, Joh, Reis, & McEwen, 1980; Rainbow, Degroff, Luine, & McEwen, 1980), estradiol concurrently modulates receptor signaling pathways (Kellert, Nguyen, Nguyen, Nguyen, & Wagner, 2009; Kelly, Loose, & Ronnekleiv, 1992; Lagrange, Ronnekleiv, & Kelly, 1994; Qiu et al., 2003; Tang, Tran, & Wagner, 2005). For example, estradiol priming decouples ORL-1 from the G protein-gated inwardly rectifying K^+ (GIRK-1) channel in ARH neurons (Borgquist, Kachani, Tavitian, Sinchak, & Wagner, 2013; Borgquist, Mela Rivas, Kachani, Sinchak, & Wagner, 2014).

Much of our understanding of role of the OFQ/N-ORL-1 system in the regulation of sexual receptivity has been determined in a model multisynaptic opioid neural circuit that originates in the ARH and projects to the MPN (Fig. 1; reviewed in Micevych & Sinchak, 2013; Sinchak & Wagner, 2012). All of the steroid priming paradigms that facilitate lordosis converge to regulate the output of ARH β-endorphin (β-END) neurons that project to the MPN (Borgquist et al., 2013, 2014; Cheung & Hammer, 1995; Christensen et al., 2012; Christensen, Dewing, & Micevych, 2011; Christensen & Micevych, 2012, 2013; Dewing et al., 2007; Dewing, Christensen, Bondar, & Micevych, 2008; Long, Chhorvann, & Sinchak, 2012; Long et al., 2013, 2014; Micevych, Rissman, Gustafsson, & Sinchak, 2003; Mills, Sohn, & Micevych, 2004; Sanathara et al., 2011, 2014; Sinchak et al., 2013; Sinchak & Micevych, 2001; Sinchak, Shahedi, Dewing, & Micevych, 2005; Fig. 1). β-END is one of the posttranslational products released by proopiomelanocortin (POMC) neurons and an endogenous ligand for μ-opioid receptors (MOP). MPN MOP activation inhibits lordosis in maximally receptive females (Acosta-Martinez & Etgen, 2002; Pfaus & Pfaff, 1992; Sinchak & Micevych, 2001; Sirinathsinghji, 1986). The MOP, being a G protein-coupled receptor, are internalized rapidly into early endosomes upon activation by endogenous ligands (Arden, Segredo, Wang, Lameh, & Sadaee, 1995; Keith et al., 1998, 1995, 1996Mantyh et al., 1995; Mestek et al., 1995; Sharma, Klee, & Nirenberg, 1977; von Zastrow, Keith, Zaki, & Evans, 1994). Using immunohistochemistry, MOP internalization is easily visualized and is positively associated with MOP immunoreactive staining intensity (Fig. 1; Dewing et al., 2007; Eckersell, Popper, & Micevych, 1998; Mills et al., 2004; Sinchak & Micevych, 2001). We have used this technique as a measure of MOP activation to determine whether our various hormonal and pharmacological manipulations turn on or off the ARH β-END neurons that project to the MPN and then compare the MPN MOP activational state to the sexual receptive state (Fig. 1; Dewing

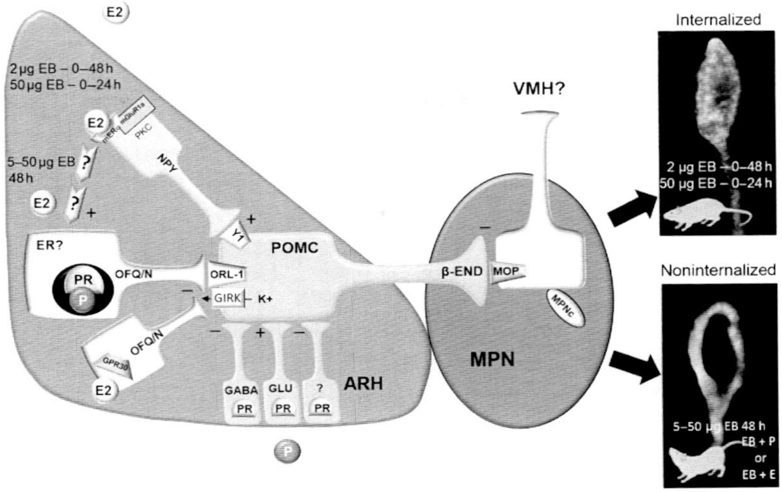

Figure 1 Proposed model circuit for estradiol (E2) and progesterone (P) regulation or OFQ/N-ORL-1 lordosis circuit within the arcuate nucleus of the hypothalamus (ARH) and medial preoptic nucleus (MPN). Estradiol binds to membrane estrogen receptor-α (mERα) complexed with metabotropic glutamate receptor-1a (mGluR1a) that signals through PKC to induce the release of neuropeptide Y (NPY) that binds to the NPY-Y1 receptor (Y1) on a subpopulation of POMC neurons that project to the MPN. This rapidly activates and maintains activation to induce β-END release in the MPN. β-END activates and internalizes μ-opioid receptors (MOP) to actively inhibit lordosis (Internalized photomicrograph). A priming dose of estradiol upregulates progesterone receptors (PR) and OFQ/N coexpression and ORL-1 expression in ARH POMC neurons that project to the MPN while maintaining MOP for up to 48 h. Lordosis is facilitated when this circuit is deactivated by 48 h of exposure to high doses of estradiol (5–50 μg; EB-only high dose; Noninternalized photomicrograph) by (1) the downregulation of mERα-mGluR1a and (2) increasing OFQ/N release to activate ORL-1 and decrease β-END release through increasing outward GIRK-1 K^+ currents (−). This recycles MPN MOP back to the plasma membrane and reduces MPN MOP activation/internalization to facilitate lordosis (Noninternalized photomicrograph). Following estradiol priming, progesterone also deactivates this circuit to facilitate lordosis. Although PR are expressed in OFQ/N neuron, OFQ/N is not necessary for deactivating MPN MOP and progesterone may induce the release of other inhibitory neurotransmitters such as GABA to inhibit β-END release to facilitate lordosis (Borgquist et al., 2013, 2014; Dewing et al., 2007, 2008; Eckersell et al., 1998; Mills et al., 2004; Sanathara et al., 2011, 2014; Sinchak et al., 2007; Sinchak & Micevych, 2001; Wagner et al., 1998).

et al., 2007; Eckersell et al., 1998; Mills et al., 2004; Sinchak & Micevych, 2001). By combining immunohistochemical and behavioral data, we have shown that lordosis is actively inhibited when MPN MOP are activated/internalized in steroid-primed rodents, whereas following estradiol priming, the deactivation of MPN MOP is associated with facilitation of sexual receptivity (Fig. 1; Dewing et al., 2007; Eckersell et al., 1998; Micevych et al., 2003; Mills et al., 2004; Sinchak & Micevych, 2001; Sinchak et al., 2005). Thus, we have optimized methods for determining whether hormonal or pharmacological manipulations that regulate sexual receptivity also regulate the activity of this ARH-MPN lordosis circuit (Dewing et al., 2007; Sanathara et al., 2011; Sinchak et al., 2013; Sinchak & Micevych, 2001). Although we have a good understanding that MPN MOP activity mediates hormonal regulation of the sexual receptive state of the rat, we know little about these MPN MOP neurons. We have begun to study the projection and phenotype of the MPN MOP neurons. Using the retrograde tract tracer, Fluorogold, we have seen that a subpopulation of these MPN MOP neurons project to the VMH region (Sinchak et al., 2010). Further, a subpopulation of MPN MOP neurons expresses either ERα or ORL-1 (Polovin et al., 2012). Since MOP activation inhibits the activity of the MOP neuron, the actions of the neurotransmitters released by MOP neurons projecting to the VMH should be facilitative to lordosis given that MOP activation is inhibitory to lordosis (Calizo & Flanagan-Cato, 2002, 2003; Cottingham, Femano, & Pfaff, 1987; Pfeifle & Edwards, 1983; Pfeifle, Shivers, & Edwards, 1980).

Although OVX rats are not sexually receptive, the ARH-MPN lordosis-inhibitory circuit is deactivated (Eckersell et al., 1998), suggesting that activation of lordosis-inhibitory pathways is not required when ovarian hormones are absent. However, EB treatment activates MPN MOP within 20 min, presumably through the release of β-END (Eckersell et al., 1998; Sirinathsinghji, 1986). This rapid estradiol-induced MPN MOP activation that inhibits lordosis is produced by estradiol binding to extranuclear ERα trafficked to the plasma membrane (mERα) that complex to and signal through metabotropic glutamate receptors-type 1a (mGluR1a; mERα–mGluR1a; Fig. 1; Boulware, Kordasiewicz, & Mermelstein, 2007; Christensen et al., 2011; Dewing et al., 2007, 2008; Dominguez, Dewing, Kuo, & Micevych, 2013; Mills et al., 2004; Sanathara et al., 2014). The mERα–mGluR1a signaling in the ARH stimulates the release of neuropeptide Y (NPY) that acts on NPY-Y1 receptors in ARH β-END neurons to stimulate the release of β-END in MPN (Dewing et al., 2007; Mills et al., 2004; Sanathara et al., 2014). This activates and internalizes MPN MOP and actively inhibits lordosis to allow for priming effects of estradiol in lordosis-facilitative circuits (Eckersell et al., 1998; Sanathara et al., 2014; Sinchak & Micevych, 2001; Sinchak et al., 2006). Lordosis is facilitated when ARH β-END neurons are deactivated following EB priming by either subsequent progesterone (EB + P), or non-esterfied-estradiol, or after exposure to a high dose of estradiol for 48 h (Fig. 1; Eckersell et al., 1998; Long et al., 2013, 2014; Sinchak & Micevych, 2001). MPN MOP deactivation after estradiol priming is an important step in the facilitation of lordosis (Acosta-Martinez & Etgen, 2002; Christensen et al., 2011; Dewing et al., 2007; Eckersell et al., 1998; Kelly, Lagrange, Wagner, & Ronnekleiv, 1999; Sinchak & Micevych, 2001). Further, MPN MOP activation fluctuates in association with the state of sexual receptivity throughout the estrous cycle (Sinchak & Micevych, 2003). For example, MPN MOP are activated during diestrous days 1 and 2 when she is nonreceptive and estradiol levels are starting to slowly rise. On the evening of proestrus when sexually receptive, MOP are deactivated. Then on the morning of estrus, when she is no longer sexually receptive, MPN MOP are reactivated (Sinchak & Micevych, 2003). Thus, the timing of the onset and termination of sexual receptivity appears to be mediated by activity of this ARH-MPN lordosis-inhibitory circuit. In maximally receptive, EB + P-primed rats, MPN MOP activation inhibits sexual receptivity, indicating that the MPN regulates output of downstream lordosis motor output pathways of the VMH (Calizo & Flanagan-Cato, 2002, 2003; Cottingham et al., 1987; Eckersell et al., 1998; Flanagan-Cato, Lee, & Calizo, 2006; Flanagan, Lee, & Calizo, 2006; Pfeifle & Edwards, 1983; Pfeifle et al., 1980; Sinchak & Micevych, 2001). Thus, hormone signaling that regulates sexual receptivity converges on the ARH β-END neuron to regulate its neurotransmission (Fig. 1). The initial actions of estradiol are to activate the β-END ARH-MPN lordosis-inhibitory circuit and prime lordosis-facilitative circuits and subsequent hormone actions that facilitate lordosis deactivate this circuit. Collectively, our studies into the ovarian hormone regulation of OFQ/N-ORL-1 circuits and the facilitation of lordosis have revealed that the different steroid paradigms converge to inhibit the activity the ARH-β-END neuron and deactivate MPN MOP through multiple neuronal pathways.

3. OFQ/N-ORL-1 REGULATION OF SEXUAL RECEPTIVITY

Because of studies demonstrating the expression of ORL-1 mRNA in the VMH and OFQ/N mRNA in regions that regulate sexual receptivity and project to the VMH (Neal, Mansour, Reinscheid, Nothacker, Civelli, et al., 1999; Neal, Mansour, Reinscheid, Nothacker, Civelli, & Watson, 1999b), our first studies infused OFQ/N into the VMH region of estradiol-primed nonreceptive female rats (Sinchak et al., 1997). Within 10 min of infusion, OFQ/N dose-dependently facilitated lordosis (Sinchak et al., 1997). Originally, we concluded that OFQ/N was acting through ORL-1 expressed in descending motor output pathways of the VMH to facilitate lordosis (Calizo & Flanagan-Cato, 2002, 2003; Cottingham et al., 1987; Pfeifle et al., 1980; Sinchak et al., 1997). However, our interpretation of where OFQ/N is acting has evolved as we performed more studies. Our first insight that EB-only and EB + P activate different neurocircuits to facilitate lordosis was when we tested whether OFQ/N release is required for facilitation of lordosis (Sinchak et al., 2007). At the time, an antagonist to ORL-1 was not available, therefore we infused OFQ/N antibodies into the VMH to bind to and inactivate released OFQ/N by immunoneutralization. OVX rats were treated with either EB + P or EB-only (5 μg) to facilitate lordosis. Interestingly, OFQ/N immunoneutralization blocked sexual receptivity in EB-only-treated but not the EB + P-treated OVX rats (Sinchak et al., 2007). These data indicate that EB-only facilitation of lordosis requires the release of OFQ/N and activation of ORL-1 in the region of the VMH and the MPN to facilitate sexual receptivity. However, EB + P may have either induced the release of more OFQ/N than could be immunoneutralized or activates neurocircuits other than the OFQ/N-ORL-1 in the region to facilitate lordosis (Sinchak et al., 2007). We saw similar results in a later experiment using the ORL-1 antagonist, UFP-101, in which we measured sexual receptivity and MPN MOP activation (Sanathara et al., 2011). Blocking ORL-1 increased MPN MOP activation and inhibited sexual receptivity in the EB-only-treated OVX rats, whereas the EB + P-treated rats were maximally receptive and MPN MOP were deactivated (Sanathara et al., 2011). The results of the OFQ/N immunoneutralization and ORL-1 inhibition studies indicate that EB-only and EB + P steroid treatments regulate the deactivation of ARH β-END neurons that project to the MPN through different pathways. These results pose an interesting question: how does the high dose of estradiol treatments switch their signaling from activating the ARH-MPN circuit and inhibiting lordosis to inhibiting the output of the β-END neuron to facilitate lordosis? Besides activating OFQ/N neurons through an unknown mechanism, we have shown that EB-only facilitation of lordosis also reduces excitatory input to β-END neurons after 48 h (Mahavongtrakul, Kanjiya, Maciel, Kanjiya, & Sinchak, 2013). The high dose of EB downregulates the levels of mERα–mGluR1a complexes in the ARH to reduce the estradiol-induced activation of MPN MOP 48 h after treatment compared to the 2 μg EB priming dose (Mahavongtrakul et al., 2013).

Like EB-only, we have also found that facilitation of lordosis by the EB + non-esterfied-estradiol requires the activation of ORL-1 (Long et al., 2013, 2014). If EB-primed nonreceptive rats are infused with UFP-101 prior to the non-esterfied-estradiol infusion, then MPN MOP remain activated and sexual receptivity is not facilitated (Long et al., 2013, 2014). Thus, EB-only and EB + non-esterfied-estradiol facilitation of lordosis induced the release of OFQ/N and requires the activation of ORL-1 to reduce the activity of ARH β-END neurons to deactivate MPN MOP. The mechanisms through which these steroid priming paradigms activate the release of OFQ/N in the ARH are not clear. However, it appears that each uses different ER signaling pathways. Antagonizing GPR30 inhibited facilitation of lordosis by subsequent non-esterfied-estradiol; however, it did not inhibit facilitation of lordosis by EB-only treatment (Fig. 2; Long et al., 2013, 2014). GPR30 may be located in the membrane of the endoplasmic reticulum and alter neuronal activity through the release of internal calcium,

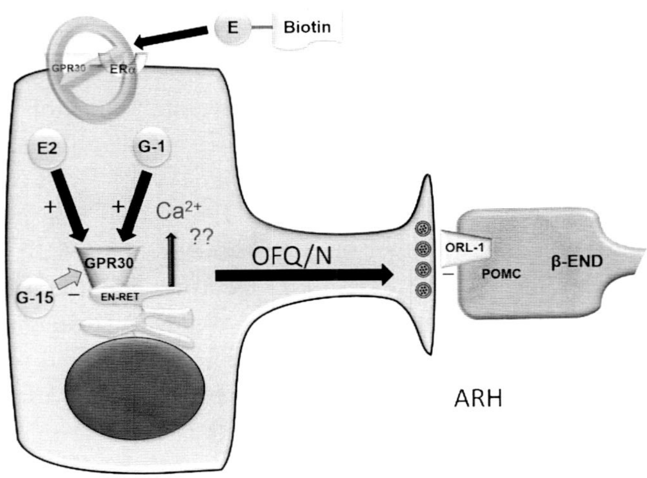

Figure 2 Model lordosis circuit in the arcuate nucleus (ARH) for rapid GPR30 facilitation of sexual receptivity in estradiol-primed nonreceptive rats. Based on our results, EB (2 μg) priming increases OFQ/N expression in ARH neurons. Subsequent non-esterfied-estradiol (E2) activates an extranuclear GPR30 located in the endoplasmic reticulum (EN-RET) to induce the release of intracellular stores of calcium (Ca^{2+}) as part of its signaling mechanism to induce the release of OFQ/N. The OFQ/N activates ORL-1 in ARH POMC neurons to inhibit the release of β-endorphin (β-END) in the medial preoptic nucleus to rapidly facilitate lordosis within 30 min. The GPR30 agonist, G-1, facilitated lordosis while antagonism of GPR30 with G-15 blocked E2 facilitation of lordosis. Since membrane impermeable E-biotin failed to facilitate lordosis, E2 is likely acting on an intracellular ER and not ERα or GPR30 located in the plasma membrane.

since 17β-estradiol conjugated to biotin (E-biotin) a plasma membrane impermeable construct, did not facilitate lordosis as seen when either non-esterfied-estradiol or G1 (GPR30 agonist) was given (Fig. 2; Long et al., 2013, 2014). This is supported by the lack of GPR30 detection in the plasma membrane of neurons and astrocytes (Bondar, Kuo, Hamid, & Micevych, 2009; Gorosito, Lorenzo, & Cambiasso, 2008; Kuo, Hamid, Bondar, Prossnitz, & Micevych, 2010). GPR30 has been associated with the endoplasmic reticulum (Filardo, Quinn, Bland, & Frackelton, 2000; Filardo et al., 2007; Funakoshi, Yanai, Shinoda, Kawano, & Mizukami, 2006; Otto et al., 2008; Revankar, Cimino, Sklar, Arterburn, & Prossnitz, 2005). In nonneural cellular preparations, the GPR30 has been shown to be trafficked between the plasma membrane and perinuclear locations such as the endoplasmic reticulum (Cheng, Graeber, Quinn, & Filardo, 2011). Since GPR30 can rapidly mobilize intracellular Ca^{2+} stores, estradiol's ability to rapidly facilitate lordosis may be through GPR30 located in the endoplasmic reticulum that regulates intracellular calcium levels (Noel, Keen, Baumann, Filardo, & Terasawa, 2009; Revankar et al., 2005; Terasawa, Noel, & Keen, 2009).

Although PR are expressed in ARH OFQ/N neurons, we expected that progesterone facilitation of lordosis would require activation of ORL-1 (Sanathara et al., 2014), However, EB + P does not require the activation of ORL-1 to deactivate MPN MOP and facilitate lordosis. It is likely that subsequent progesterone induces release of OFQ/N but that progesterone also regulates the release of other neurotransmitters like GABA and glutamate to reduce the activity of β-END neurons and facilitate lordosis (Fig. 1). Thus, estradiol initially activates the MPN MOP through excitation of the ARH β-END neurons, and then lordosis is facilitated when MPN MOP are deactivated through the actions of progesterone, a high dose of estradiol, or non-esterfied-estradiol that inhibits β-END release, which use different neurocircuits and signaling pathways that regulate the output of ARH β-END neurons that project to the MPN (Eckersell et al., 1998; Sinchak & Micevych, 2001).

Although OFQ/N-ORL-1 facilitation of lordosis appears to require the deactivation of MPN MOP, ORL-1 is expressed throughout the VMH which contain motor output pathways. We were not sure whether OFQ/N facilitated lordosis through the ARH, VMH, or maximal levels of sexual receptivity were induced through the activation of both ARH and VMH. In this set of experiments, we attempted to determine whether OFQ/N facilitated lordosis by acting in the ARH to deactivate MPN

MOP, VMH descending motor output pathways or through activation of both the ARH ascending pathways and VMH descending pathways to facilitate lordosis. OFQ/N was infused into four different sites that started at the ARH and moved dorsolaterally through and then outside of the VMH, and then sexual receptivity and levels of MPN MOP activation were measured. We found that OFQ/N infusions facilitated lordosis to the same level, whether they were aimed at the ARH or where we expected OFQ/N to be acting only in the VMH. Further, if the OFQ/N infusion induced lordosis, then MPN MOP was deactivated, suggesting that OFQ/N facilitation of lordosis requires the inhibition of the ARH β-END neurons that project to the MPN (Sanathara et al., 2011). Since ARH β-END neurons that project to the MPN express ORL-1 and OFQ/N hyperpolarizes these neurons via increasing outward GIRK-1 currents (Borgquist et al., 2013, 2014; Sanathara et al., 2014), all of the OFQ/N infusions may have diffused into the ARH to directly inhibit the activation of ARH β-END neurons to deactivate MPN MOP. It is possible that OFQ/N infusions were contained within the VMH and still only activated ARH neurons directly, since Golgi staining revealed that some ARH processes extend into the VMH (Millhouse, 1973; van den Pol & Cassidy, 1982). Tract tracing studies have not clearly defined neurocircuits between the VMH and ARH (Bouret et al., 2008; Canteras, Simerly, & Swanson, 1994; Saper, Swanson, & Cowan, 1976). However, laser scanning photostimulation studies suggest that there are excitatory inputs to POMC neuron from the VMH as well as inhibitory inputs to NPY ARH neurons (Sternson, Shepherd, & Friedman, 2005). Thus, potential VHM–ARH pathways exist, and OFQ/N may act in the VMH to indirectly regulate ARH β-END output to the MPN.

Only one other region has been explored for OFQ/N-ORL-1 regulation of sexual receptivity and that is the MPN (Sinchak et al., 2007). Like the ARH region, OFQ/N infusion into the MPN facilitates lordosis in EB-primed OVX rats (Sinchak et al., 2007). Further, it also appears the EB-only facilitation of lordosis requires ORL-1 activation in this region. Immunoneutralization of OFQ/N was able to block EB-only, but not EB + P facilitation of lordosis (Sinchak et al., 2007). It is unclear what type of neurons OFQ/N acts on to facilitate lordosis. It is possible that OFQ/N may be acting at the terminals of the ARH β-END neurons to inhibit neurotransmission which would reduce MPN MOP activation and facilitate lordosis. Further, activation of ORL-1 may inhibit the output of MPN lordosis-inhibitory neurons that project to the VMH motor output pathway or may deactivate an inhibitory interneuron within the MPN to facilitate lordosis. We do know that EB + P treatment and not EB-only upregulates OFQ/N and ORL-1 mRNA expression in the medial MPN and OFQ/N binding and ORL-1 signaling through its G protein (Quesada & Micevych, 2008; Sinchak et al., 2006). However, at this point little is known about the content and projection of OFQ/N and ORL-1 neurons within the MPN.

4. OVARIAN STEROID REGULATION OF OFQ/N AND ORL-1 EXPRESSION AND SIGNALING

The distributions of OFQ/N- and ORL-1-expressing neurons overlap extensively with the distribution of ER and PR expression in the brain. Although both expression of OFQ/N and ORL-1 is regulated by ovarian hormones (Quesada & Micevych, 2008; Sanathara et al., 2014; Sinchak et al., 2006), only OFQ/N has been shown to be expressed in ER and PR neurons due to the technical limitation of identifying ORL-1 *in vivo* (Nestor et al., 2013; Sanathara et al., 2014). In the ewe, double-label immunohistochemistry revealed that subpopulations of OFQ/N neurons in the preoptic area, anterior hypothalamic area, and ARH expressed ERα and PR (Nestor et al., 2013). In the ARH, over 85% of the OFQ/N neurons expressed ERα, and over 70% expressed PR (Nestor et al., 2013). In the rat, we also observed colocalization of OFQ/N and PR in the ARH, MPN, posterodorsal medial amygdala (MeApd) and red nucleus (Mana, Haase, Franklin, Noboru, & Sinchak, 2008; Sanathara et al., 2014). In the ARH, estradiol treatment increased the number of immunopositive neurons that expressed either OFQ/N or PR as well as the number of neurons that coexpressed PR and OFQ/N (Sanathara et al., 2014). In the rat, approximately 40% of OFQ/N neurons expressed PR which is much lower than in the ewe (Nestor et al., 2013). However, in the rat about 70% of the PR neurons expressed OFQ/N. Measuring OFQ/N mRNA expression demonstrated that estradiol priming increased expression in the caudal region of the MeApd, and subsequent progesterone treatment increased OFQ/N mRNA expression in the medial portion of the MPN (Sinchak et al., 2006). Although immunohistochemical studies demonstrate the presence of OFQ/N fibers and processes in the VMH, there is a distinct absence of both OFQ/N mRNA and immunohistochemical staining the neuronal somas in the VMH, indicating that the release of OFQ/N in the VMH is from other brain regions that project to the VMH and not local circuits within the VMH (Sanathara et al., 2014; Sinchak et al., 2006). These data demonstrate that both estradiol and progesterone can directly modulate the OFQ/N neurons through either expression or release of OFQ/N. Because of the abundance of OFQ/N-PR expression in the ARH neurons, we expected that progesterone facilitation of lordosis would be dependent on progesterone inducing the release of OFQ/N to activate ORL-1 (Sanathara et al., 2011). We suspect that progesterone does induce the release of OFQ/N in the ARH region to inhibit β-END release in the MPN; however, progesterone likely also induces release of other neurotransmitters that inhibit ARH β-END neurons and therefore are not obligate to solely the actions of ORL-1 activity in the ARH to facilitate lordosis.

Anatomical studies are lacking that demonstrate ORL-1 neurons express ER or PR. However, ORL-1 mRNA expression, OFQ/N-binding studies, and electrophysiological studies indicate that multiple types of ER may be acting directly on ORL-1 neurons to regulate expression and function. In the female rat, ORL-1 is expressed in numerous steroid-responsive circuits that express ER and PR (Quesada & Micevych, 2008; Sanathara et al., 2014; Sinchak et al., 2006). These include the ARH, VMH, anteroventral periventricular nucleus of the hypothalamus (AVPV; aka rostral periventricular region of the third ventricle, RP3V in other species), MeApd, and paraventricular nucleus (PVN). Further, estradiol has been shown to increase ORL-1 mRNA expression in the ARH, VMH, AVPV (RP3V), and the median preoptic nucleus (MEPO) (Sanathara et al., 2014; Sinchak et al., 2006). These are brain regions that regulate sexual receptivity and neuroendocrine feedback loops. ORL-1 mRNA expression and OFQ/N binding is increased in the MPN by sequential EB + P treatment but not EB-only (Quesada & Micevych, 2008; Sinchak et al., 2006). Infusing the retrograde tract tracer, Fluorogold, into the MPN and performing double-labeled fluorescent *in situ* hybridization, we demonstrated that ORL-1 is expressed in ARH POMC (β-END) neurons that project to the MPN (Sanathara et al., 2014). Further, estradiol priming increased the coexpression of ORL-1 and POMC in ARH neurons that project to the MPN. Thus, one of the estradiol priming effects is the upregulation of ORL-1 in the β-END neurons. This estradiol-induced ORL-1 expression was also observed in OFQ/N-binding studies. Estradiol treatment increase OFQ/N binding in membranes taken from the MBH, which includes the ARH and VMH (Quesada & Micevych, 2008). These results paralleled mRNA expression studies that demonstrated EB-increased ORL-1 expression in the ARH and throughout the VMH (Sanathara et al., 2014; Sinchak et al., 2006). Thus, ORL-1 expression is under site-specific regulation by ovarian hormones, estradiol upregulates ORL-1 in the ARH and VMH, whereas in the MPN both estradiol and progesterone are required to increase expression. However, it is unclear which ER or PR is activated to induce ORL-1 expression.

Electrophysiological studies and ORL-1 mRNA expression studies indicate that estradiol may be acting over time through multiple types of ER to modulate the activity of ARH ORL-1 neurons that project to the MPN (Farhang, Pietruszewski, Lutfy, & Wagner, 2010; Lagrange, Ronnekleiv, & Kelly, 1997; Qiu et al., 2003, 2006). First, estradiol and progesterone regulate the release of OFQ/N that can act on both presynaptic neurons that synapse on ARH β-END neurons, and postsynaptically to modulate the output of ARH β-END neurons through modulation of ORL-1 expression and signaling efficiency (Fig. 3; Borgquist et al., 2013, 2014; Quesada & Micevych, 2008). Second, as a $G_{i/o}$-coupled receptor, ORL-1 directly modulates cell excitability by increasing postsynaptic K^+ currents to decrease β–END excitability (Borgquist et al., 2013, 2014; Farhang et al., 2010; Wagner, Ronnekleiv, Grandy, & Kelly, 1998).

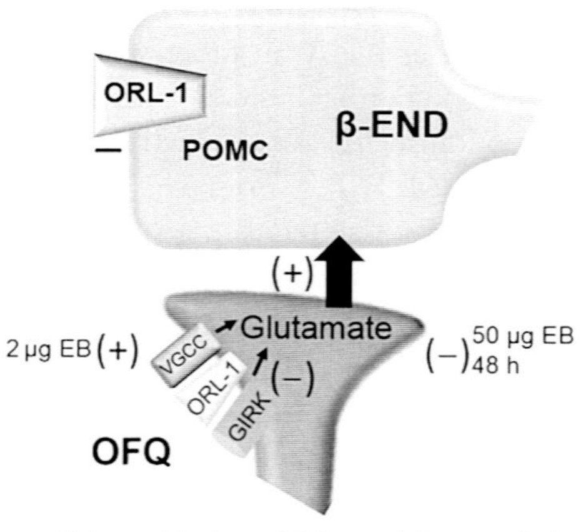

Figure 3 Model of estradiol modulation of ORL-1 inhibition of glutamate neurons that synapse on ARH β-endorphin (β-END) neurons that project to the MPN. Activation of ORL-1 at the presynaptic terminal inhibits glutamate release by producing hyperpolarizing outward K^+ currents through GIRK-1 that reduce the opening of voltage-gated calcium channels (VGCC). A priming dose of estradiol reduces the ability of OFQ/N to inhibit glutamate neurotransmission. This effectively maintains activation of ARH β-END neurons and MPN MOP to inhibit lordosis. However, 48 h after exposure to high doses of estradiol, OFQ/N inhibition of glutamate release is restored. This reduces excitatory input to β-END neurons and decreases MPN MOP activation as part of the mechanisms that inhibit β-END release to facilitate lordosis (Borgquist et al., 2013, 2014; Farhang et al., 2010). The ER mechanism(s) involved that modulate presynaptic ORL-1 signaling over time are not known at the moment.

ORL-1 can act at the presynaptic terminal to inhibit neurotransmitter release to the β-END neuron through reducing the activity of voltage-gated calcium channels (Farhang, Diaz, Tang, & Wagner, 2009). For example, OFQ/N presynaptically inhibits excitatory glutamatergic input to ARH neurons (Emmerson & Miller, 1999; Wagner et al., 1998). Our studies indicate that estradiol and progesterone regulate the efficacy of ORL-1 to signal through these pathways. Postsynaptically, although estradiol upregulates the expression of ORL-1 in ARH β-END neurons that project to the MPN, estradiol appears to decouple ORL-1 from the GIRK-1 (Borgquist et al., 2013, 2014; Wagner et al., 1998). Estradiol priming that inhibits lordosis reduces OFQ/N-induced K^+ current in the ARH β-END neurons that project to the MPN (Fig. 3; Borgquist et al., 2013, 2014). This in effect reduces the ability of OFQ/N to inhibit β-END release as part of a mechanism to maintain MPN MOP activation and inhibit lordosis. ORL-1 is recoupled to the GIRK-1 48 h after exposure to a high dose of EB (50 μg) when sexual receptivity is facilitated and MPN MOP is deactivated (Borgquist et al., 2014). In EB + P-treated animals, it appears that ORL-1 recoupling is only partially restored compared to the EB high-dose animals (Borgquist et al., 2014). This pattern of ORL-1 modulation is also seen in regulating presynaptic glutamatergic excitatory input onto ARH β-END neurons. The priming dose of estradiol decreases OFQ/N presynaptic inhibition of glutamatergic neurotransmission (Fig. 2; Borgquist et al., 2013, 2014). This reduces ORL-1 inhibitory tone and allows for greater release of glutamate and increased excitatory input to ARH β-END neurons at a time when lordosis should be inhibited. In contrast, 48 h after exposure to 50 μg EB, the OFQ/N-induced presynaptic inhibition of glutamate release is restored (Borgquist et al., 2014). This reduces the excitatory input to the β-END neurons at a time when the animal is sexually receptive and deactivation of MPN MOP should be maintained. As seen with ORL-1 recoupling to GIRK-1 channels, EB + P only partially restores OFQ/N inhibition of glutamate neurotransmission. Thus, ovarian hormonal signaling converges to modulate the activity of the ARH β-END neurons through regulating OFQ/N-ORL-1 system. Estradiol initially upregulates OFQ/N and ORL-1 but does not release OFQ/N and decreases ORL-1 signaling presynaptically to reduce inhibition glutamate release and postsynaptically to reduce GIRK-1 channel K^+ current that would decrease ARH β-END neuron excitation. These actions decrease the inhibitory tone to ARH β-END neurons to increase β-END neurotransmission that activates MPN MOP and inhibits lordosis. After 48 h of exposure to a high dose of EB, OFQ/N is released within the ARH, and OFQ/N-induced inhibition of glutamate neurotransmission is increased, as is ORL-1 signaling through GIRK-1 channels in ARH β-END neurons. This inhibits the release of β-END and reduces MPN MOP activation to facilitate lordosis. EB + P partially restores these effects of the ORL-1 system; however, maximizing ORl-1 signaling is not necessary since it appears that progesterone deactivation of ARH β-END neurons is mediated by another neurotransmitter system in addition to the OFQ/N-ORL-1 system.

At this time, it is unclear how the high dose of EB initially inhibits the release of OFQ/N and then 48 h later induces the release of OFQ/N to facilitate lordosis. It will be interesting to determine the types of ER(s) and signaling pathways being used over time to upregulate expression and modulate ORL-1 signaling and OFQ/N release to deactivate MPN MOP to facilitate lordosis. Since progesterone does not require ORL-1 activation to facilitate lordosis, why is there such a high population of OFQ/N neurons that express PR? Further, we know very little about the OFQ/N and ORL-1 neurons within the MPN. At the level of steroid regulation ORL-1 signaling, it would be interesting to determine whether ORL-1 can signal through multiple pathway within a given neuron and whether steroid priming alters which signaling pathway is used. Finally, although we know OFQ/N in the MPN facilitate lordosis, we know neither the phenotype of these neurons nor where they project.

5. OFQ/N-ORL-1 REGULATION OF GnRH AND LH RELEASE DURING POSITIVE AND NEGATIVE FEEDBACK

Gonadotropin-releasing hormone (GnRH) neurons lack ERα and PR (Herbison, Robinson, & Skinner, 1993; Herbison & Theodosis, 1992; Lehman & Karsch, 1993; Skinner, Caraty, & Allingham, 2001), indicating that ovarian steroids must act through intermediary neurons to regulate GnRH and LH release. Although the neuropeptide kisspeptin has been implicated to mediate estrogen-positive and -negative feedback regulation of GnRH neurons and subsequent LH release from the pituitary (reviewed in Clarkson & Herbison, 2009; Oakley, Clifton, & Steiner, 2009; Przekop & Ciechanowska, 2012), these feedback loops are also modulated by classical endogenous opioid peptides (Goodman, Gibson, Skinner, & Lehman, 2002;

Goodman, Parfitt, Evans, Dahl, & Karsch, 1995; Kalra, 1993; Richter et al., 2001; Taylor, Goubillon, Broad, & Robinson, 2007). In rodents, kisspeptin is expressed in two main regions of the hypothalamus to regulate feedback. Kisspeptin neurons in the RP3V region (aka, AVPV) project to the diagonal band of Broca (DBB) to activate GnRH neurons at the soma to induce the luteinizing hormone surge via positive feedback (Clarkson & Herbison, 2006; Liu & Herbison, 2008; Smith, Popa, Clifton, Hoffman, & Steiner, 2006). The negative feedback effects of estradiol appear to be mediated through kisspeptin neurons located in the ARH that project to the median eminence to GnRH nerve terminals (reviewed in Roa, Navarro, & Tena-Sempere, 2011).

Since OFQ/N and ORL-1 are expressed in these regions as well as other brain regions that express ER and PR, investigations into whether OFQ/N modulates steroid feedback systems have been initiated. *In vitro*, OFQ/N was shown to inhibit GnRH release from sections of the rat hypothalamus (Dhandapani & Brann, 2002), and mediobasal hypothalamic (MBH) sections from OVX rats (An et al., 2005), indicating the potential for OFQ/N to regulate GnRH release. In female rats treated with estradiol and progesterone to induce an LH surge, infusion of OFQ/N into the third ventricle reduced the magnitude of the LH surge but did not eliminate the surge (An et al., 2007). Pretreatment with an ORL-1 antagonist blocked the OFQ/N inhibition of LH release (An et al., 2007). Interestingly, this same group demonstrated that OFQ/N also reduced GnRH levels in the median eminence of untreated OVX rats and that an ORL-1 antagonist increased GnRH levels (An, He, Feng, Feng, & Yu, 2009), indicating that OFQ/N is acting within the hypothalamus or median eminence to reduce GnRH release and not at the level of the pituitary to reduce LH release. In both cases, the effects of OFQ/N were delayed 40 min to 2 h after treatment, suggesting that OFQ/N actions may be slow to develop and may act to mediate the onset or termination of LH release during both positive and negative feedback. However, given the route of OFQ/N administration and limited hormonal manipulations, it is difficult to make any firm conclusions on the role of the OFQ/N-ORL-1 system in negative and positive feedback regulation of LH release in the female rat. However, since circulating progesterone levels rise concurrently or shortly after the onset of the LH surge (Feder, Brown-Grant, & Corker, 1971; Kalra & Kalra, 1974; Smith, Freeman, & Neill, 1975), progesterone may act to stimulate the release of OFQ/N to terminate the surge release of GnRH and LH.

Work in the ewe has helped define the actions of OFQ/N in positive and negative feedback mechanisms regulating GnRH and LH release (Foradori, Goodman, & Lehman, 2003; Goodman et al., 2013; Nestor et al., 2013). OFQ/N is easily visualized in the soma of neurons by immunocytochemistry in sheep and has made the ability to determine the coexpression of OFQ/N more manageable, but still costly given the model system. In the ewe, OFQ/N is expressed in both ERα- and PR-positive neurons in areas that regulate reproduction: the preoptic area, anterior hypothalamic area, and ARH (Nestor et al., 2013). In the ARH, over 85% of the OFQ/N neurons expressed ERα, and over 70% expressed PR (Nestor et al., 2013). Further, in the ARH, OFQ/N neurons also colocalized with POMC, but not kisspeptin, agouti-related peptide or tyrosine hydroxylase (Nestor et al., 2013). Interestingly, over 90% of the GnRH neurons throughout the hypothalamus express OFQ/N (Foradori et al., 2007). Thus, OFQ/N system is poised to mediate the actions of estradiol and progesterone in GnRH regulatory circuits. These regions include the DBB and ventromedial preoptic area, anterior hypothalamus, and MBH. The coexpression of GnRH and OFQ/N was also observed within the processes and terminals of neurons in the brain regions; however, OFQ/N and GnRH immunostaining were distinct from each other, indicating that these peptides are in separate secretory vesicles. This allows for the ability to differentially regulate the release of OFQ/N and GnRH. Like the rodent, infusion of OFQ/N into the lateral ventricle of the ewe reduced LH secretion. In an OVX ewe, both LH pulse frequency and amplitude were reduced by OFQ/N treatment, whereas in the ewe in the luteal phase only pulse frequency was reduced (Foradori et al., 2007). In the OVX ewe that has been treated with progesterone and estradiol to mimic the luteal phase levels of hormones, ORL-1 antagonist was infused in the lateral ventricle to determine whether LH levels increased (Nestor et al., 2013). LH pulse frequency was increased; however, levels of LH and pulse amplitude were not affected. In OVX ewes primed with only estradiol which produces high frequency but low LH pulse amplitude, blocking ORL-1 did not increase LH pulse frequency, which is extremely low in these animals and had no effect on any other aspect of LH release (Nestor et al., 2013). Thus, the ability to increase the LH pulse frequency by blocking ORL-1 in progesterone and estradiol-primed animals indicates that ORL-1 mediates progesterone-negative feedback effects. Although anatomical confirmation of ORL-1 expression in GnRH neurons is lacking, OFQ/N appears to be able to directly inhibit GnRH neurons in the ARH through increasing potassium currents via GIRK-1 channels

(Wagner et al., 1998). Thus, OFQ/N could autoregulate GnRH release and/or modulate the activity of an interneuron to modulate GnRH release. In addition to its actions at GnRH neurons, OFQ/N immunoreactivity was localized to the median eminence. Therefore, it is possible that OFQ/N could act directly in the pituitary to modulate LH release (Foradori et al., 2007; Foradori & Handa, 2004; Wagner et al., 1998). Further, it is possible that OFQ/N originating from non-GnRH neurons may synapse on GnRH neurons to regulate GnRH release within the hypothalamus or at GnRH terminals in the median eminence (Foradori et al., 2007; Foradori & Handa, 2004; Wagner et al., 1998). Moreover, OFQ/N release may be compartmentalized within the ARH, for example, to regulate the activity of β-END neurons (Sanathara et al., 2011) that regulates GnRH release (Goodman & Inskeep, 2006). Thus, our understanding of the OFQ/N-ORL-1 system's role in regulating estradiol-positive and -negative feedback is just beginning.

6. CONCLUSIONS

OFQ/N and ORL1 are expressed in estradiol and progesterone responsive neurocircuits. Accordingly, the OFQ/N-ORL-1 system has been shown to mediate ovarian hormone regulation of sexual receptivity and inhibit the release of LH associated with neuroendocrine feedback loops within the hypothalamus. Other systems in the female, like energy balance and nociception, are regulated by ovarian hormones and their actions are also mediated through the OFQ/N-ORL-1 system (reviewed in Gupta et al., 2001; Sinchak & Wagner, 2012; Zhang, Donica, & Standifer, 2012). Thus, understanding interactions between ovarian hormones and OFQ/N-ORL-1 systems regulating neural systems function is of fundamental importance. Basic research has been slowed by the lack of tools and technical difficulties that are associated with anatomical studies that would allow for tract tracing and phenotyping of OFQ/N and ORL-1 neurons. Additionally, an ORL-1 antagonist has only been recently available. Thus, understanding the importance and role of the OFQ/N-ORL-1 system in mediating ovarian hormonal information is just underway.

ACKNOWLEDGMENTS

Supported by NIH Grant RO1HD058638.

REFERENCES

Acosta-Martinez, M., & Etgen, A. M. (2002). Activation of mu-opioid receptors inhibits lordosis behavior in estrogen and progesterone-primed female rats. *Hormones and Behavior*, *41*(1), 88–100.

Adler, N. T. (1969). Effects of the male's copulatory behavior on successful pregnancy of the female rat. *Journal of Comparative and Physiological Psychology*, *69*(4), 613–622.

Adler, N. T. (1978). On the mechanisms of sexual behavior and their evolutionary constraints. In J. B. Hutchison (Ed.), *Biological determinants of sexual behavior* (pp. 657–694). New York: Wiley.

Alves, S. E., McEwen, B. S., Hayashi, S., Korach, K. S., Pfaff, D. W., & Ogawa, S. (2000). Estrogen-regulated progestin receptors are found in the midbrain raphe but not hippocampus of estrogen receptor alpha (ER alpha) gene-disrupted mice. *Journal of Comparative Neurology*, *427*(2), 185–195.

An, X. F., Chen, H. P., Ma, S. L., Feng, Y., Hao, J. W., & Chen, B. Y. (2005). Involvement of nociceptin/orphanin FQ in release of hypothalamic GnRH mediated by ORL1 receptor in ovariectomized rats. *Acta Pharmacologica Sinica*, *26*(9), 1039–1044. http://dx.doi.org/10.1111/j.1745-7254.2005.00172.x.

An, X. F., He, M., Feng, Y., Feng, H., & Yu, J. Y. (2009). Central administration of Orphanin FQ inhibits GnRH secretion by ORL1 receptor in the median eminence of freely moving ovariectomized rats. *Neuroscience Bulletin*, *25*(1), 1–6. http://dx.doi.org/10.1007/s12264-009-1119-5.

An, X. F., Yu, J. Y., Feng, Y., Chen, B. Y., & Zhang, S. L. (2007). Role of hypothalamus nociceptin/orphanin FQ in pre-ovulatory luteinizing hormone surge of estrogen and progesterone-primed, ovariectomized rats. *Acta Pharmacologica Sinica*, *28*(8), 1189–1197. http://dx.doi.org/10.1111/j.1745-7254.2007.00587.x.

Arden, J. R., Segredo, V., Wang, Z., Lameh, J., & Sadaee, W. (1995). Phosphorylation and agonist-specific intracellular trafficking of an epitope-tagged mu-opioid receptor expressed in HEK 293 cells. *Journal of Neurochemistry*, *65*(4), 1636–1645.

Beach, F. A. (1948). *Hormones and behavior*. New York: Paul B Hoeber.

Blaustein, J. D., & Feder, H. H. (1979). Cytoplasmic progestin receptors in female guinea pig brain and their relationship to refractoriness in expression of female sexual behavior. *Brain Research*, *177*(3), 489–498.

Blaustein, J. D., Finkbohner, R., & Delville, Y. (1987). Estrogen-induced and estrogen-facilitated female rat sexual behavior is not mediated by progestin receptors. *Neuroendocrinology*, *45*(2), 152–159.

Boling, F. L., & Blandau, R. J. (1939). The estrogen-progesterone induction of mating responses in the spayed female rat. *Endocrinology*, *25*, 359–364.

Bondar, G., Kuo, J., Hamid, N., & Micevych, P. (2009). Estradiol-induced estrogen receptor-alpha trafficking. *The Journal of Neuroscience*, *29*(48), 15323–15330. http://dx.doi.org/10.1523/JNEUROSCI.2107-09.2009.

Borgquist, A., Kachani, M., Tavitian, N., Sinchak, K., & Wagner, E. J. (2013). Estradiol negatively modulates the pleiotropic actions of orphanin FQ/nociceptin at proopiomelanocortin synapses. *Neuroendocrinology*, *98*(1), 60–72. http://dx.doi.org/10.1159/000351868.

Borgquist, A., Mela Rivas, V., Kachani, M., Sinchak, K., & Wagner, E. (2014). Gonadal steroids differentially modulate the actions of orphanin FQ/nociceptin at a physiologically relevant circuit controlling female sexual receptivity. *Journal of Neuroendocrinology*, *26*(5), 329–340.

Boulware, M. I., Kordasiewicz, H., & Mermelstein, P. G. (2007). Caveolin proteins are essential for distinct effects of membrane estrogen receptors in neurons. *Journal of Neuroscience*, *27*(37), 9941–9950. http://dx.doi.org/10.1523/JNEUROSCI.1647-07.2007.

Bouret, S. G., Gorski, J. N., Patterson, C. M., Chen, S., Levin, B. E., & Simerly, R. B. (2008). Hypothalamic neural projections are permanently disrupted in diet-induced obese rats. *Cell Metabolism*, 7(2), 179–185. http://dx.doi.org/10.1016/j.cmet.2007.12.001.

Bunzow, J. R., Saez, C., Mortrud, M., Bouvier, C., Williams, J. T., Low, M., et al. (1994). Molecular cloning and tissue distribution of a putative member of the rat opioid receptor gene family that is not a mu, delta or kappa opioid receptor type. *FEBS Letters*, *347*(2–3), 284–288.

Calizo, L. H., & Flanagan-Cato, L. M. (2002). Estrogen-induced dendritic spine elimination on female rat ventromedial hypothalamic neurons that project to the periaqueductal gray. *The Journal of Comparative Neurology*, *447*(3), 234–248. http://dx.doi.org/10.1002/cne.10223.

Calizo, L. H., & Flanagan-Cato, L. M. (2003). Hormonal-neural integration in the female rat ventromedial hypothalamus: Triple labeling for estrogen receptor-alpha, retrograde tract tracing from the periaqueductal gray, and mating-induced Fos expression. *Endocrinology*, *144*(12), 5430–5440. http://dx.doi.org/10.1210/en.2003-0331.

Canteras, N. S., Simerly, R. B., & Swanson, L. W. (1994). Organization of projections from the ventromedial nucleus of the hypothalamus: A phaseolus vulgaris-leucoagglutinin study in the rat. *The Journal of Comparative Neurology*, *348*(1), 41–79.

Chen, Y., Fan, Y., Liu, J., Mestek, A., Tian, M., Kozak, C. A., et al. (1994). Molecular cloning, tissue distribution and chromosomal localization of a novel member of the opioid receptor gene family. *FEBS Letters*, *347*(2–3), 279–283.

Cheng, S. B., Graeber, C. T., Quinn, J. A., & Filardo, E. J. (2011). Retrograde transport of the transmembrane estrogen receptor, G-protein-coupled-receptor-30 (GPR30/GPER) from the plasma membrane towards the nucleus. *Steroids*, *76*(9), 892–896. http://dx.doi.org/10.1016/j.steroids.2011.02.018.

Chester, R. V., & Zucker, I. (1970). Influence of male copulatory behavior on sperm transport, pregnancy and pseudopregnancy in female rats. *Physiology & Behavior*, *5*(1), 35–43.

Cheung, S., & Hammer, R. (1995). Gonadal steroid hormone regulation of proopiomelanocortin gene expression in the arcuate neurons that innervate the medial preoptic are of the rat. *Neuroendocrinology*, *62*, 283–292.

Christensen, A., Bentley, G. E., Cabrera, R., Ortega, H. H., Perfito, N., Wu, T. J., et al. (2012). Hormonal regulation of female reproduction. *Hormone and Metabolic Research*, *44*(8), 587–591. http://dx.doi.org/10.1055/s-0032-1306301.

Christensen, A., Dewing, P., & Micevych, P. (2011). Membrane-initiated estradiol signaling induces spinogenesis required for female sexual receptivity. *The Journal of Neuroscience*, *31*(48), 17583–17589. http://dx.doi.org/10.1523/JNEUROSCI.3030-11.2011.

Christensen, A., & Micevych, P. (2012). CAV1 siRNA reduces membrane estrogen receptor-alpha levels and attenuates sexual receptivity. *Endocrinology*, *153*(8), 3872–3877. http://dx.doi.org/10.1210/en.2012-1312.

Christensen, A., & Micevych, P. (2013). A novel membrane estrogen receptor activated by STX induces female sexual receptivity through an interacation with mGluR1a. *Neuroendocrinology*, *97*(4), 363–368.

Clarkson, J., & Herbison, A. E. (2006). Postnatal development of kisspeptin neurons in mouse hypothalamus; sexual dimorphism and projections to gonadotropin-releasing hormone neurons. *Endocrinology*, *147*(12), 5817–5825. http://dx.doi.org/10.1210/en.2006-0787.

Clarkson, J., & Herbison, A. E. (2009). Oestrogen, kisspeptin, GPR54 and the pre-ovulatory luteinising hormone surge. *Journal of Neuroendocrinology*, *21*(4), 305–311. http://dx.doi.org/10.1111/j.1365-2826.2009.01835.x.

Clemens, L. G., & Weaver, D. R. (1985). The role of gonadal hormone in the activation of feminine sexual behavior. In N. Adler, D. Pfaff, & R. W. Goy (Eds.), *Handbook of behavioral neurobiology: Vol. 7*. (pp. 183–227). New York: Plenum Press.

Cottingham, S. L., Femano, P. A., & Pfaff, D. W. (1987). Electrical stimulation of the midbrain central gray facilitates reticulospinal activation of axial muscle EMG. *Experimental Neurology*, *97*(3), 704–724.

Cummings, J. A., & Becker, J. B. (2012). Quantitative assessment of female sexual motivation in the rat: Hormonal control of motivation. *Journal of Neuroscience Methods*, *204*(2), 227–233. http://dx.doi.org/10.1016/j.jneumeth.2011.11.017.

Day, J. R., Morales, T. H., & Lu, J. K. (1988). Male stimulation of luteinizing hormone surge, progesterone secretion and ovulation in spontaneously persistent-estrous, aging rats. *Biology of Reproduction*, *38*(5), 1019–1026.

Devorshak-Harvey, E., Bona-Gallo, A., & Gallo, R. V. (1987). Endogenous opioid peptide regulation of pulsatile luteinizing hormone secretion during pregnancy in the rat. *Neuroendocrinology*, *46*(5), 369–378.

Dewing, P., Boulware, M. I., Sinchak, K., Christensen, A., Mermelstein, P. G., & Micevych, P. (2007). Membrane estrogen receptor-alpha interactions with metabotropic glutamate receptor 1a modulate female sexual receptivity in rats. *The Journal of Neuroscience*, *27*(35), 9294–9300. http://dx.doi.org/10.1523/JNEUROSCI.0592-07.2007.

Dewing, P., Christensen, A., Bondar, G., & Micevych, P. (2008). Protein kinase C signaling in the hypothalamic arcuate nucleus regulates sexual receptivity in female rats. *Endocrinology*, *149*(12), 5934–5942.

Dhandapani, K. M., & Brann, D. W. (2002). Orphanin FQ inhibits GnRH secretion from rat hypothalamic fragments but not GT1-7 neurons. *Neuroreport*, *13*(10), 1247–1249.

Dohanich, G. P., Witcher, J. A., Weaver, D. R., & Clemens, L. G. (1982). Alteration of muscarinic binding in specific brain areas following estrogen treatment. *Brain Research*, *241*(2), 347–350.

Dominguez, R., Dewing, P., Kuo, J., & Micevych, P. (2013). Membrane-initiated estradiol signaling in immortalized hypothalamic N-38 neurons. *Steroids*, *78*(6), 607–613. http://dx.doi.org/10.1016/j.steroids.2012.12.008.

Eckersell, C. B., Popper, P., & Micevych, P. E. (1998). Estrogen-induced alteration of mu-opioid receptor immunoreactivity in the medial preoptic nucleus and medial amygdala. *The Journal of Neuroscience*, *18*, 3967–3976.

Edwards, D. A., & Pfeifle, J. K. (1983). Hormonal control of receptivity, proceptivity and sexual motivation. *Physiology & Behavior*, *30*(3), 437–443.

Emmerson, P. J., & Miller, R. J. (1999). Pre- and postsynaptic actions of opioid and orphan opioid agonists in the rat arcuate nucleus and ventromedial hypothalamus in vitro. *Journal of Physiology (London)*, *517*(2), 431–445.

Erskine, M. S. (1987). Serum 5 alpha-androstane-3 alpha, 17 beta-diol increases in response to paced coital stimulation in cycling female rats. *Biology of Reproduction*, *37*(5), 1139–1148.

Erskine, M. S., Kornberg, E., & Cherry, J. A. (1989). Paced copulation in rats: Effects of intromission frequency and duration on luteal activation and estrus length. *Physiology & Behavior*, *45*(1), 33–39.

Farhang, B., Diaz, S., Tang, S. L., & Wagner, E. J. (2009). Sex differences in the cannabinoid regulation of energy homeostasis. *Psychoneuroendocrinology*, *34*(Suppl. 1), S237–S246. http://dx.doi.org/10.1016/j.psyneuen.2009.04.007.

Farhang, B., Pietruszewski, L., Lutfy, K., & Wagner, E. J. (2010). The role of the NOP receptor in regulating food intake, meal pattern, and the excitability of proopiomelanocortin neurons. *Neuropharmacology*, *59*(3), 190–200. http://dx.doi.org/10.1016/j.neuropharm.2010.05.007.

Feder, H. H., Brown-Grant, K., & Corker, C. S. (1971). Pre-ovulatory progesterone, the adrenal cortex and the 'critical period' for luteinizing hormone release in rats. *The Journal of Endocrinology*, *50*(1), 29–39.

Filardo, E. J., Graeber, C. T., Quinn, J. A., Resnick, M. B., Giri, D., DeLellis, R. A., et al. (2006). Distribution of GPR30, a seven membrane-spanning estrogen receptor, in primary breast cancer and its association with clinicopathologic determinants of tumor progression. *Clinical Cancer Research: An Official Journal of the American Association for Cancer Research*, *12*(21), 6359–6366. http://dx.doi.org/10.1158/1078-0432.CCR-06-0860.

Filardo, E. J., Quinn, J. A., Bland, K. I., & Frackelton, A. R., Jr. (2000). Estrogen-induced activation of Erk-1 and Erk-2 requires the G protein-coupled receptor homolog, GPR30, and occurs via trans-activation of the epidermal growth factor receptor through release of HB-EGF. *Molecular Endocrinology*, *14*(10), 1649–1660.

Filardo, E., Quinn, J., Pang, Y., Graeber, C., Shaw, S., Dong, J., et al. (2007). Activation of the novel estrogen receptor G protein-coupled receptor 30 (GPR30) at the plasma membrane. *Endocrinology*, *148*(7), 3236–3245. http://dx.doi.org/10.1210/en.2006-1605.

Flanagan, L. M., Lee, B., & Calizo, L. (2006). Co-localization of midbrain projections, progestin receptors, and mating-induced fos in the hypothalamic ventromedial nucleus of the female rat. *Hormones and Behavior*, *50*(1), 52–60.

Flanagan-Cato, L. M., Lee, B. J., & Calizo, L. H. (2006). Co-localization of midbrain projections, progestin receptors, and mating-induced fos in the hypothalamic ventromedial nucleus of the female rat. *Hormones and Behavior*, *50*(1), 52–60. http://dx.doi.org/10.1016/j.yhbeh.2006.01.012.

Foradori, C. D., Amstalden, M., Coolen, L. M., Singh, S. R., McManus, C. J., Handa, R. J., et al. (2007). Orphanin FQ: Evidence for a role in the control of the reproductive neuroendocrine system. *Endocrinology*, *148*(10), 4993–5001. http://dx.doi.org/10.1210/en.2007-0011.

Foradori, C. D., Goodman, R. L., & Lehman, M. N. (2003). Colocalization of orphanin FQ peptide in GnRH neurons of the ovine preoptic area and hypothalamus. In *Paper presented at the Society for Neuroscience, New Orleans, LA, USA*.

Foradori, C. D., & Handa, R. J. (2004). The orphanin FQ receptor, opioid receptor like-1, is expressed in GT1-7 cells and pituitary. In *Paper presented at the Society for Neuroscience, San Diego, CA, USA*.

Fukuda, K., Kato, S., Mori, K., Nishi, M., Takeshima, H., Iwabe, N., et al. (1994). cDNA cloning and regional distribution of a novel member of the opioid receptor family. *FEBS Letters*, *343*(1), 42–46.

Funakoshi, T., Yanai, A., Shinoda, K., Kawano, M. M., & Mizukami, Y. (2006). G protein-coupled receptor 30 is an estrogen receptor in the plasma membrane. *Biochemical and Biophysical Research Communications*, *346*(3), 904–910. http://dx.doi.org/10.1016/j.bbrc.2006.05.191.

Genazzani, A. R., & Petraglia, F. (1989). Opioid control of luteinizing hormone secretion in humans. *Journal of Steroid Biochemistry*, *33*(4B), 751–755.

Goodman, R. L., Gibson, M., Skinner, D. C., & Lehman, M. N. (2002). Neuroendocrine control of pulsatile GnRH secretion during the ovarian cycle: Evidence from the ewe. *Reproduction Supplement*, *59*, 41–56.

Goodman, R. L., Hileman, S. M., Nestor, C. C., Porter, K. L., Connors, J. M., Hardy, S. L., et al. (2013). Kisspeptin, neurokinin B, and dynorphin act in the arcuate nucleus to control activity of the GnRH pulse generator in ewes. *Endocrinology*, *154*(11), 4259–4269. http://dx.doi.org/10.1210/en.2013-1331.

Goodman, R. L., & Inskeep, E. K. (2006). Control of the ovarian cycle in the sheep. In J. D. Neill (Ed.), *Knobil's and Neill's Physiology of Reproduction* (3rd ed., pp. 2389–2447). St. Louis: Elsevier.

Goodman, R. L., Parfitt, D. B., Evans, N. P., Dahl, G. E., & Karsch, F. J. (1995). Endogenous opioid peptides control the amplitude and shape of gonadotropin-releasing hormone pulses in the ewe. *Endocrinology*, *136*(6), 2412–2420.

Gorosito, S. V., Lorenzo, A. G., & Cambiasso, M. J. (2008). Estrogen receptor alpha is expressed on the cell-surface of embryonic hypothalamic neurons. *Neuroscience*, *154*(4), 1173–1177. http://dx.doi.org/10.1016/j.neuroscience.2008.05.001.

Gupta, D. S., Kelson, A. B., Polgar, W. E., Toll, L., Szucs, M., & Gintzler, A. R. (2001). Ovarian sex steroid-dependent plasticity of nociceptin/orphanin FQ and opioid modulation of spinal dynorphin release. *The Journal of Pharmacology and Experimental Therapeutics*, *298*(3), 1213–1220.

Hardy, D. F., & Debold, J. F. (1971). Effects of mounts without intromission upon the behavior of female rats during the onset of estrogen-induced heat. *Physiology & Behavior*, *7*(4), 643–645.

Herbison, A. E., Robinson, J. E., & Skinner, D. C. (1993). Distribution of estrogen receptor-immunoreactive cells in the preoptic area of the ewe: Co-localization with glutamic acid decarboxylase but not luteinizing hormone-releasing hormone. *Neuroendocrinology*, *57*(4), 751–759.

Herbison, A. E., & Theodosis, D. T. (1992). Localization of oestrogen receptors in preoptic neurons containing neurotensin but not tyrosine hydroxylase, cholecystokinin or luteinizing hormone-releasing hormone in the male and female rat. *Neuroscience*, *50*(2), 283–298.

Hlinak, Z., & Madlafousek, J. (1981). Estradiol treatment and precopulatory behavior in ovariectomized female rats. *Physiology & Behavior*, *26*(2), 171–176.

Hoffman, G. E., Galor, T., Szabo, F. K., Adetola, A., Best, C., Subedi, K., et al. (2013). *Paths of activation for reflex vaginocervical stimulation of luteinizing hormone (LH) secretion in female rats*, In *Paper presented at the Society for Neuroscience, San Diego, CA, USA*.

Jones, S. L., Farrell, S., Gregory, J. G., & Pfaus, J. G. (2013). Sensitization of sexual behavior in ovariectomized rats by chronic estradiol treatment. *Hormones and Behavior*, *64*(1), 8–18. http://dx.doi.org/10.1016/j.yhbeh.2013.04.012.

Kalra, S. P. (1993). Mandatory neuropeptide-steroid signaling for the preovulatory luteinizing hormone-releasing hormone discharge. *Endocrine Reviews*, *14*(5), 507–538.

Kalra, S. P., & Kalra, P. S. (1974). Temporal interrelationships among circulating levels of estradiol, progesterone and LH during the rat estrous cycle: Effects of exogenous progesterone. *Endocrinology*, *95*(6), 1711–1718.

Keith, D. E., Anton, B., Murray, S. R., Zaki, P. A., Chu, P. C., Lissin, D. V., et al. (1998). mu-Opioid receptor internalization: Opiate drugs have differential effects on a conserved endocytic mechanism in vitro and in the mammalian brain. *Molecular Pharmacology*, *53*(3), 377–384.

Keith, D., Murray, S., Zaki, P., Chu, P., Lissin, D., Kang, L., et al. (1995). Rapid endocytosis of opioid receptors: Differential regulation by opioid peptide and morphine. *Society for Neuroscience – Abstracts*, *21*, 1353.

Keith, D. E., Murray, S. R., Zaki, P. A., Chu, P. C., Lissin, D. V., Kang, L., et al. (1996). Morphine activates opioid receptors without causing their rapid internalization. *The Journal of Biological Chemistry*, *271*(32), 19021–19024.

Kellert, B. A., Nguyen, M. C., Nguyen, C., Nguyen, Q. H., & Wagner, E. J. (2009). Estrogen rapidly attenuates cannabinoid-induced changes in energy homeostasis. *European Journal of Pharmacology*, *622*(1–3), 15–24. http://dx.doi.org/10.1016/j.ejphar. 2009.09.001.

Kelly, M. J., Lagrange, A. H., Wagner, E. J., & Ronnekleiv, O. K. (1999). Rapid effects of estrogen to modulate G protein-coupled receptors via activation of protein kinase A and protein kinase C pathways. *Steroids*, *64*(1–2), 64–75.

Kelly, M. J., Loose, M. D., & Ronnekleiv, O. K. (1992). Estrogen suppresses mu-opioid- and GABAB-mediated hyperpolarization of hypothalamic arcuate neurons. *The Journal of Neuroscience*, *12*(7), 2745–2750.

Kornberg, E., & Erskine, M. S. (1994). Effects of differential mating stimulation on the onset of prolactin surges in pseudopregnant rats. *Psychoneuroendocrinology*, *19*(4), 357–371.

Kuo, W. H., Chang, L. Y., Liu, D. L., Hwa, H. L., Lin, J. J., Lee, P. H., et al. (2007). The interactions between GPR30 and the major biomarkers in infiltrating ductal carcinoma of the breast in an Asian population. *Taiwanese Journal of Obstetrics & Gynecology*, *46*(2), 135–145. http://dx.doi.org/10.1016/S1028-4559(07)60007-2.
Kuo, J., Hamid, N., Bondar, G., Prossnitz, E. R., & Micevych, P. (2010). Membrane estrogen receptors stimulate intracellular calcium release and progesterone synthesis in hypothalamic astrocytes. *Journal of Neuroscience*, *30*(39), 12950–12957. http://dx.doi.org/10.1523/JNEUROSCI.1158-10.2010.
Lachowicz, J. E., Shen, Y., Monsma, F. J., Jr., & Sibley, D. R. (1995). Molecular cloning of a novel G protein-coupled receptor related to the opiate receptor family. *Journal of Neurochemistry*, *64*(1), 34–40.
Lado-Abeal, J., Clapper, J. A., Chen Zhu, B., Hough, C. M., Syapin, P. J., & Norman, R. L. (2002). Hypoglycemia-induced suppression of luteinizing hormone (LH) secretion in intact female rhesus macaques: Role of vasopressin and endogenous opioids. *Stress*, *5*(2), 113–119. http://dx.doi.org/10.1080/10253890290027886.
Lagrange, A. H., Ronnekleiv, O. K., & Kelly, M. J. (1994). The potency of mu-opioid hyperpolarization of hypothalamic arcuate neurons is rapidly attenuated by 17 beta-estradiol. *The Journal of Neuroscience*, *14*(10), 6196–6204.
Lagrange, A. H., Ronnekleiv, O. K., & Kelly, M. J. (1997). Modulation of G protein-coupled receptors by an estrogen receptor that activates protein kinase A. *Molecular Pharmacology*, *51*(4), 605–612.
Lehman, M. N., & Karsch, F. J. (1993). Do gonadotropin-releasing hormone, tyrosine hydroxylase-, and beta-endorphin-immunoreactive neurons contain estrogen receptors? A double-label immunocytochemical study in the Suffolk ewe. *Endocrinology*, *133*(2), 887–895.
Liu, X., & Herbison, A. E. (2008). Small-conductance calcium-activated potassium channels control excitability and firing dynamics in gonadotropin-releasing hormone (GnRH) neurons. *Endocrinology*, *149*(7), 3598–3604. http://dx.doi.org/10.1210/en.2007-1631.
Long, N. P., Chhorvann, S., & Sinchak, K. (2012). In estradiol primed rats subsequent free estradiol rapidly facilitates lordosis through G-protein coupled receptor 30 (GPR30). In *Paper presented at the Society for Neuroscience, New Orleans*.
Long, N., Serey, C., Welborn, A., & Sinchak, K. (2013). Free estradiol signals through G-protein coupled receptor 30 (GPER) to rapidly facilitate lordosis via deactivation of medial preoptic nucleus-μ-opioid receptors in estradiol primed rats. In *Paper presented at the Society for Neuroscience, San Diego, CA, USA*.
Long, N., Serey, C., & Sinchak, K. (2014). 17β-estradiol rapidly facilitates lordosis through G protein-coupled estrogen receptor 1 (GPER) via deactivation of medial preoptic nucleus μ-opioid receptors in estradiol primed female rats. *Hormones and Behavior*, *66*(4), 663–666.
Lu, K. H., Hopper, B. R., Vargo, T. M., & Yen, S. S. (1979). Chronological changes in sex steroid, gonadotropin and prolactin secretions in aging female rats displaying different reproductive states. *Biology of Reproduction*, *21*(1), 193–203.
Luine, V. N., Khylchevskaya, R. I., & McEwen, B. S. (1975). Effect of gonadal steroids on activities of monoamine oxidase and choline acetylase in rat brain. *Brain Research*, *86*(2), 293–306.
Luine, V., Park, D., Joh, T., Reis, D., & McEwen, B. (1980). Immunochemical demonstration of increased choline acetyltransferase concentration in rat preoptic area after estradiol administration. *Brain Research*, *191*(1), 273–277.
MacLusky, N. J., & McEwen, B. S. (1978). Oestrogen modulates progestin receptor concentrations in some rat brain regions but not in others. *Nature*, *274*(5668), 276–278.
MacLusky, N. J., & McEwen, B. S. (1980). Progestin receptors in the rat brain: Distribution and properties of cytoplasmic progestin binding sites. *Endocrinology*, *106*, 192–202.

Mahavongtrakul, M., Kanjiya, M. P., Maciel, M., Kanjiya, S., & Sinchak, K. (2013). Estradiol dose-dependent regulation of membrane estrogen receptor-alpha, metabotropic glutamate receptor-1a, and their complexes in the arcuate nucleus of the hypothalamus in female rats. *Endocrinology, 154*(9), 3251–3260. http://dx.doi.org/10.1210/en.2013-1235.

Mana, A., Haase, J., Franklin, K., Noboru, K., & Sinchak, K. (2008). Colocalization progesterone receptor and orphanin FQ/nociceptin immunocytochemical colocalization in the red nucleus. In *Paper presented at the Society for Neuroscience, Washington, DC.*

Mantyh, P. W., Allen, C. J., Ghilardi, J. R., Rogers, S. D., Mantyh, C. R., Liu, H., et al. (1995). Rapid endocytosis of a G protein-coupled receptor: Substance P evoked internalization of its receptor in the rat striatum in vivo. *Proceedings of the National Academy of Sciences of the United States of America, 92*(7), 2622–2626.

Marchese, A., Docherty, J. M., Nguyen, T., Heiber, M., Cheng, R., Heng, H. H., et al. (1994). Cloning of human genes encoding novel G protein-coupled receptors. *Genomics, 23*(3), 609–618. http://dx.doi.org/10.1006/geno.1994.1549.

Matthews, M., & Adler, N. T. (1977). Facilitative and inhibitory influences of reproductive behavior on sperm transport in rats. *Journal of Comparative and Physiological Psychology, 91*(4), 727–741.

McClintock, M. K. (1984). Estrous synchrony: Modulation of ovarian cycle length by female pheromones. *Physiology & Behavior, 32*(5), 701–705.

McClintock, M. K., Toner, J. P., Adler, N. T., & Anisko, J. J. (1882). Postejaculatory quiescence in female and male rats: Consequences for sperm transport during group mating. *Journal of Comparative and Physiological Psychology, 96*, 268–277.

Mendelson, S. D., & Gorzalka, B. B. (1987). An improved chamber for the observation and analysis of the sexual behavior of the female rat. *Physiology & Behavior, 39*(1), 67–71.

Mestek, A., Hurley, J. H., Bye, L. S., Campbell, A. D., Chen, Y., Tian, M., et al. (1995). The human mu opioid receptor. *Journal of Neuroscience, 15*(3 Pt 2), 2396–2406.

Meunier, J. C., Mollereau, C., Toll, L., Suaudeau, C., Moisand, C., Alvinerie, P., et al. (1995). Isolation and structure of the endogenous agonist of opioid receptor-like ORL1 receptor. *Nature, 377*(6549), 532–535.

Micevych, P. E., Rissman, E. F., Gustafsson, J. A., & Sinchak, K. (2003). Estrogen receptor-alpha is required for estrogen-induced mu-opioid receptor internalization. *Journal of Neuroscience Research, 71*(6), 802–810.

Micevych, P., & Sinchak, K. (2013). Temporal and concentration dependent estradiol effects on neural pathways mediating sexual receptivity. *Journal of Neuroendocrinology, 25*(11), 1012–1023. http://dx.doi.org/10.1111/jne.12103.

Micevych, P., Soma, K. K., & Sinchak, K. (2008). Neuroprogesterone: Key to estrogen positive feedback? *Brain Research Reviews, 57*(2), 470–480. http://dx.doi.org/10.1016/j.brainresrev.2007.06.009.

Millhouse, O. E. (1973). The organization of the ventromedial hypothalamic nucleus. *Brain Research, 55*(1), 71–87.

Mills, R. H., Sohn, R. K., & Micevych, P. E. (2004). Estrogen-induced mu-opioid receptor internalization in the medial preoptic nucleus is mediated via neuropeptide Y-Y1 receptor activation in the arcuate nucleus of female rats. *The Journal of Neuroscience, 24*(4), 947–955.

Moguilewsky, M., & Raynaud, J. P. (1979). The relevance of hypothalamic and hypophyseal progestin receptor regulation in the induction and inhibtion of sexual behavior in the female rat. *Endocrinology, 105*, 516–522.

Mollereau, C., Parmentier, M., Mailleux, P., Butour, J. L., Moisand, C., Chalon, P., et al. (1994). ORL1, a novel member of the opioid receptor family. Cloning, functional expression and localization. *FEBS Letters, 341*(1), 33–38.

Neal, C. R., Mansour, A., Reinscheid, R., Nothacker, H. P., Civelli, O., Akil, H., et al. (1999). Opioid receptor-like (ORL1) receptor distribution in the rat central nervous system: Comparison of ORL1 receptor mRNA expression with I-125-[(14)Tyr]-orphanin FQ binding. *Journal of Comparative Neurology, 412*(4), 563–605.

Neal, C. R., Mansour, A., Reinscheid, R., Nothacker, H. P., Civelli, O., & Watson, S. J. (1999b). Localization of orphanin FQ (nociceptin) peptide and messenger RNA in the central nervous system of the rat. *Journal of Comparative Neurology, 406*(4), 503–547.

Nestor, C. C., Coolen, L. M., Nesselrod, G. L., Valent, M., Connors, J. M., Hileman, S. M., et al. (2013). Evidence that orphanin FQ mediates progesterone negative feedback in the Ewe. *Endocrinology, 154*(11), 4249–4258. http://dx.doi.org/10.1210/en.2013-1274.

Noel, S. D., Keen, K. L., Baumann, D. I., Filardo, E. J., & Terasawa, E. (2009). Involvement of G protein-coupled receptor 30 (GPR30) in rapid action of estrogen in primate LHRH neurons. *Molecular Endocrinology, 23*(3), 349–359. http://dx.doi.org/10.1210/me.2008-0299.

Oakley, A. E., Clifton, D. K., & Steiner, R. A. (2009). Kisspeptin signaling in the brain. *Endocrine Reviews, 30*(6), 713–743. http://dx.doi.org/10.1210/er.2009-0005.

Otto, C., Rohde-Schulz, B., Schwarz, G., Fuchs, I., Klewer, M., Brittain, D., et al. (2008). G protein-coupled receptor 30 localizes to the endoplasmic reticulum and is not activated by estradiol. *Endocrinology, 149*(10), 4846–4856. http://dx.doi.org/10.1210/en.2008-0269.

Parsons, B., Rainbow, T. C., Snyder, L., & McEwen, B. S. (1984). Progesterone-like effects of estradiol on reproductive behavior and hypothalamic progestin receptors in the female rat. *Neuroendocrinology, 39*(1), 25–30.

Pfaff, D. W. (1970). Nature of sex hormone effects on rat sex behavior: Specificity of effects and individual patterns of response. *Journal of Comparative and Physiological Psychology, 73*, 349–358.

Pfaff, D. W., Diakow, C., Montgomery, M., & Jenkins, F. A. (1978). X-ray cinematographic analysis of lordosis in female rats. *Journal of Comparative and Physiological Psychology, 92*(5), 937–941.

Pfaff, D. W., & Sakuma, Y. (1979a). Deficit in the lordosis reflex of female rats caused by lesions in the ventromedial nucleus of the hypothalamus. *The Journal of Physiology, 288*, 203–210.

Pfaff, D. W., & Sakuma, Y. (1979b). Facilitation of the lordosis reflex of female rats from the ventromedial nucleus of the hypothalamus. *The Journal of Physiology, 288*, 189–202.

Pfaff, D. W., Schwartz-Giblin, S., McCarthy, M., & Kow, L. M. (1994). Cellular and molecular mechanisms of female reproductive behaviors. In E. Knobil & J. D. Neill (Eds.), *The physiology of reproduction: Vol. 2.* (pp. 107–220). New York: Raven Press, Ltd.

Pfaus, J. G., & Gorzalka, B. B. (1987). Opioids and sexual behavior. *Neuroscience and Biobehavioral Reviews, 11*(1), 1–34.

Pfaus, J. G., & Pfaff, D. W. (1992). Mu-, delta-, and kappa-opioid receptor agonists selectively modulate sexual behaviors in the female rat: Differential dependence on progesterone. *Hormones and Behavior, 26*(4), 457–473.

Pfaus, J. G., Smith, W. J., & Coopersmith, C. B. (1999). Appetitive and consummatory sexual behaviors of female rats in bilevel chambers. I. A correlational and factor analysis and the effects of ovarian hormones. *Hormones and Behavior, 35*(3), 224–240.

Pfeifle, J. K., & Edwards, D. A. (1983). Midbrain lesions eliminate sexual receptivity but spare sexual motivation in female rats. *Physiology & Behavior, 31*(3), 385–389.

Pfeifle, J. K., Shivers, M., & Edwards, D. A. (1980). Parasagittal hypothalamic knife cuts and sexual receptivity in the female rat. *Physiology & Behavior, 24*(1), 145–150.

Polovin, G., Bowlby, R., Garcia, B. L., Thach, V., Tea, P., Seng, H., et al. (2012). Subpopulation of μ-opioid receptor neurons in the medial preoptic nucleus express estrogen receptor-a and opioid receptor-like receptor-1. In *Paper presented at the Society for Neuroscience, New Orleans, LA, USA*.

Priest, C. A., Eckersell, C. B., & Micevych, P. E. (1995). Estrogen regulates preproenkephalin-a mRNA levels in the rat ventromedial nucleus: Temporal and cellular aspects. *Brain Research. Molecular Brain Research*, *28*(2), 251–262.

Przekop, F., & Ciechanowska, M. (2012). Kisspeptin (kiss 1) network signaling of hypothalamic gonadotropin-releasing hormone (GnRH) neurons. *Journal of Animal and Feed Sciences*, *21*(3), 397–424.

Qiu, J., Bosch, M. A., Tobias, S. C., Grandy, D. K., Scanlan, T. S., Ronnekleiv, O. K., et al. (2003). Rapid signaling of estrogen in hypothalamic neurons involves a novel G-protein-coupled estrogen receptor that activates protein kinase C. *The Journal of Neuroscience*, *23*(29), 9529–9540.

Qiu, J., Bosch, M. A., Tobias, S. C., Krust, A., Graham, S. M., Murphy, S. J., et al. (2006). A G-protein-coupled estrogen receptor is involved in hypothalamic control of energy homeostasis. *The Journal of Neuroscience*, *26*(21), 5649–5655. http://dx.doi.org/10.1523/JNEUROSCI.0327-06.2006.

Quadagno, D. M., McCullough, J., & Langan, R. (1972). The effect of varying amounts of exogenous estradiol benzoate on estrous behavior in the rat. *Hormones and Behavior*, *3*(3), 175–179.

Quesada, A., & Micevych, P. (2008). Estrogen and progesterone modulate [S]GTPgammaS binding to nociceptin receptors. *Neuroendocrinology*, *88*(1), 35–42. http://dx.doi.org/10.1159/000113933.

Rainbow, T. C., Degroff, V., Luine, V. N., & McEwen, B. S. (1980). Estradiol 17 beta increases the number of muscarinic receptors in hypothalamic nuclei. *Brain Research*, *198*(1), 239–243.

Reinscheid, R. K., Nothacker, H. P., Bourson, A., Ardati, A., Henningsen, R. A., Bunzow, J. R., et al. (1995). Orphanin FQ: A neuropeptide that activates an opioid-like G protein-coupled receptor. *Science*, *270*(5237), 792–794.

Revankar, C. M., Cimino, D. F., Sklar, L. A., Arterburn, J. B., & Prossnitz, E. R. (2005). A transmembrane intracellular estrogen receptor mediates rapid cell signaling. *Science*, *307*(5715), 1625–1630. http://dx.doi.org/10.1126/science.1106943.

Richter, T. A., Spackman, D. S., Robinson, J. E., Dye, S., Harris, T. G., Skinner, D. C., et al. (2001). Role of endogenous opioid peptides in mediating progesterone-induced disruption of the activation and transmission stages of the GnRH surge induction process. *Endocrinology*, *142*(12), 5212–5219.

Roa, J., Navarro, V. M., & Tena-Sempere, M. (2011). Kisspeptins in reproductive biology: Consensus knowledge and recent developments. *Biology of Reproduction*, *85*(4), 650–660. http://dx.doi.org/10.1095/biolreprod.111.091538.

Saito, Y., Maruyama, K., Kawano, H., Hagino-Yamagishi, K., Kawamura, K., Saido, T. C., et al. (1996). Molecular cloning and characterization of a novel form of neuropeptide gene as a developmentally regulated molecule. *The Journal of Biological Chemistry*, *271*(26), 15615–15622.

Saito, Y., Maruyama, K., Saido, T. C., & Kawashima, S. (1995). N23K, a gene transiently up-regulated during neural differentiation, encodes a precursor protein for a newly identified neuropeptide nociceptin. *Biochemical and Biophysical Research Communications*, *217*(2), 539–545.

Sanathara, N. M., Moraes, J., Kanjiya, S., & Sinchak, K. (2011). Orphanin FQ in the mediobasal hypothalamus facilitates sexual receptivity through the deactivation of medial preoptic nucleus mu-opioid receptors. *Hormones and Behavior*, *60*(5), 540–548. http://dx.doi.org/10.1016/j.yhbeh.2011.08.002.

Sanathara, N. M., Moraes, J., Mahavongtrakul, M., & Sinchak, K. (2014). Estradiol upregulates progesterone receptor and orphanin FQ colocalization in arcuate nucleus neurons and opioid receptor-like receptor-1 expression in proopiomelanocortin neurons that project to the medial preoptic nucleus. *Neuroendocrinology,* May 9. [Epub ahead of print].

Saper, C. B., Swanson, L. W., & Cowan, W. M. (1976). The efferent connections of the ventromedial nucleus of the hypothalamus of the rat. *The Journal of Comparative Neurology, 169*(4), 409–442.

Sharma, S. K., Klee, W. A., & Nirenberg, M. (1977). Opiate-dependent modulation of adenylate cyclase. *Proceedings of the National Academy of Sciences of the United States of America, 74*(8), 3365–3369.

Shughrue, P. J., Lubahn, D. B., NegroVilar, A., Korach, K. S., & Merchenthaler, I. (1997). Responses in the brain of estrogen receptor alpha-disrupted mice. *Proceedings of the National Academy of Sciences of the United States of America, 94*(20), 11008–11012.

Sinchak, K., Dewing, P., Cook, M., & Micevych, P. (2007). Release of orphanin FQ/nociceptin in the medial preoptic nucleus and ventromedial nucleus of the hypothalamus facilitates lordosis. *Hormones and Behavior, 51*(3), 406–412. http://dx.doi.org/10.1016/j.yhbeh.2006.12.008.

Sinchak, K., Dewing, P., Ponce, L., Gomez, L., Christensen, A., Berger, M., et al. (2013). Modulation of the arcuate nucleus-medial preoptic nucleus lordosis regulating circuit: A role for GABAB receptors. *Hormones and Behavior, 64*(1), 136–143. http://dx.doi.org/10.1016/j.yhbeh.2013.06.001.

Sinchak, K., Garcia, B. L., Bowlby, R., Charukulvanich, P., Garcia, M. P., & Sanathara, N. M. (2010). Mu-opioid receptor neurons and opioid receptor-like receptor neurons in the medial preoptic nucleus project to the region of the ventromedial nucleus of the hypothalamus. In *Paper presented at the Society for Neuroscience, San Diego, CA*.

Sinchak, K., Hendricks, D. G., Baroudi, R., & Micevych, P. E. (1997). Orphanin FQ/nociceptin in the ventromedial nucleus facilitates lordosis in female rats. *Neuroreport, 8*(18), 3857–3860.

Sinchak, K., & Micevych, P. E. (2001). Progesterone blockade of estrogen activation of μ-opioid receptors regulates reproductive behavior. *The Journal of Neuroscience, 21*(15), 5723–5729.

Sinchak, K., & Micevych, P. (2003). Visualizing activation of opioid circuits by internalization of G protein-coupled receptors. *Molecular Neurobiology, 27*(2), 197–222.

Sinchak, K., Romeo, H. E., & Micevych, P. E. (2006). Site-specific estrogen and progestin regulation of orphanin FQ/nociceptin and nociceptin opioid receptor mRNA expression in the female rat limbic hypothalamic system. *The Journal of Comparative Neurology, 496*(2), 252–268.

Sinchak, K., Shahedi, K., Dewing, P., & Micevych, P. (2005). Sexual receptivity is reduced in the female mu-opioid receptor knockout mouse. *Neuroreport, 16*(15), 1697–1700.

Sinchak, K., & Wagner, E. J. (2012). Estradiol signaling in the regulation of reproduction and energy balance. *Frontiers in Neuroendocrinology, 33*(4), 342–363. http://dx.doi.org/10.1016/j.yfrne.2012.08.004.

Sirinathsinghji, D. J. S. (1986). Regulation of lordosis behavior in the female rat by corticotropin-releasing factor, beta-endorphin/corticotropin and luteinizing hormone-releasing hormone neuronal systems in the medial preoptic area. *Brain Research, 375*, 149–156.

Skinner, D. C., Caraty, A., & Allingham, R. (2001). Unmasking the progesterone receptor in the preoptic area and hypothalamus of the ewe: No colocalization with gonadotropin-releasing neurons. *Endocrinology, 142*(2), 573–579.

Smith, H. O., Arias-Pulido, H., Kuo, D. Y., Howard, T., Qualls, C. R., Lee, S. J., et al. (2009). GPR30 predicts poor survival for ovarian cancer. *Gynecologic Oncology, 114*(3), 465–471. http://dx.doi.org/10.1016/j.ygyno.2009.05.015.

Smith, M. S., Freeman, M. E., & Neill, J. D. (1975). The control of progesterone secretion during the estrous cycle and early pseudopregnancy in the rat: Prolactin, gonadotropin and steroid levels associated with rescue of the corpus luteum of pseudopregnancy. *Endocrinology*, *96*(1), 219–226.

Smith, H. O., Leslie, K. K., Singh, M., Qualls, C. R., Revankar, C. M., Joste, N. E., et al. (2007). GPR30: A novel indicator of poor survival for endometrial carcinoma. *American Journal of Obstetrics and Gynecology*. *196*(4), 386 e381–386 e389. http://dx.doi.org/10.1016/j.ajog.2007.01.004, discussion: 386 e389–ȩ311.

Smith, J. T., Popa, S. M., Clifton, D. K., Hoffman, G. E., & Steiner, R. A. (2006). Kiss1 neurons in the forebrain as central processors for generating the preovulatory luteinizing hormone surge. *The Journal of Neuroscience*, *26*(25), 6687–6694. http://dx.doi.org/10.1523/JNEUROSCI.1618-06.2006.

Sodersten, P., & Eneroth, P. (1981). Serum levels of oestradiol-17 beta and progesterone in relation to receptivity in intact and ovariectomized rats. *The Journal of Endocrinology*, *89*(1), 45–54.

Sternson, S. M., Shepherd, G. M., & Friedman, J. M. (2005). Topographic mapping of VMH −> arcuate nucleus microcircuits and their reorganization by fasting. *Nature Neuroscience*, *8*(10), 1356–1363. http://dx.doi.org/10.1038/nn1550.

Tang, S. L., Tran, V., & Wagner, E. J. (2005). Sex differences in the cannabinoid modulation of an A-type K+current in neurons of the mammalian hypothalamus. *Journal of Neurophysiology*, *94*(4), 2983–2986. http://dx.doi.org/10.1152/jn.01187.2004.

Taylor, J. A., Goubillon, M. L., Broad, K. D., & Robinson, J. E. (2007). Steroid control of gonadotropin-releasing hormone secretion: Associated changes in pro-opiomelanocortin and preproenkephalin messenger RNA expression in the ovine hypothalamus. *Biology of Reproduction*, *76*(3), 524–531. http://dx.doi.org/10.1095/biolreprod.106.055533.

Tennent, B. J., Smith, E. R., & Davidson, J. M. (1980). The effects of estrogen and progesterone on female rat proceptive behavior. *Hormones and Behavior*, *14*(1), 65–75.

Terasawa, E., Noel, S. D., & Keen, K. L. (2009). Rapid action of oestrogen in luteinising hormone-releasing hormone neurones: The role of GPR30. *Journal of Neuroendocrinology*, *21*(4), 316–321. http://dx.doi.org/10.1111/j.1365-2826.2009.01839.x.

Terkel, J., & Sawyer, C. H. (1978). Male copulatory behavior triggers nightly prolactin surges resulting in successful pregnancy in rats. *Hormones and Behavior*, *11*(3), 304–309.

van den Pol, A. N., & Cassidy, J. R. (1982). The hypothalamic arcuate nucleus of rat—A quantitative Golgi analysis. *The Journal of Comparative Neurology*, *204*(1), 65–98. http://dx.doi.org/10.1002/cne.902040108.

von Zastrow, M., Keith, D., Zaki, P., & Evans, C. (1994). Intracellular trafficking of epitope-tagged opioid receptors different effects of morphine and enkephalin. *Regulatory Peptides*, *54*, 315–316.

Wagner, E. J., Ronnekleiv, O. K., Grandy, D. K., & Kelly, M. J. (1998). The peptide orphanin FQ inhibits beta-endorphin neurons and neurosecretory cells in the hypothalamic arcuate nucleus by activating an inwardly-rectifying K^+ conductance. *Neuroendocrinology*, *67*(2), 73–82.

Wang, J. B., Johnson, P. S., Imai, Y., Persico, A. M., Ozenberger, B. A., Eppler, C. M., et al. (1994). cDNA cloning of an orphan opiate receptor gene family member and its splice variant. *FEBS Letters*, *348*(1), 75–79.

Whalen, R. E. (1974). Estrogen-progesterone induction of mating in female rats. *Hormones and Behavior*, *5*(2), 157–162.

Wick, M. J., Minnerath, S. R., Lin, X., Elde, R., Law, P. Y., & Loh, H. H. (1994). Isolation of a novel cDNA encoding a putative membrane receptor with high homology to the cloned mu, delta, and kappa opioid receptors. *Brain Research. Molecular Brain Research*, *27*(1), 37–44.

Yang, L. Y., & Clemens, L. G. (1997). Function of intromissions on intromission-return latency of female rats during paced sexual behavior. *Physiology & Behavior*, *61*(6), 889–894.

Zhang, Y., Donica, C. L., & Standifer, K. M. (2012). Sex differences in the Nociceptin/Orphanin FQ system in rat spinal cord following chronic morphine treatment. *Neuropharmacology*, *63*(3), 427–433. http://dx.doi.org/10.1016/j.neuropharm.2012.04.028.

CHAPTER EIGHT

Effects of Nociceptin and Nocistatin on Uterine Contraction

Róbert Gáspár*[,1], Beáta H. Deák*, Anna Klukovits*, Eszter Ducza*, Kornélia Tekes†

*Department of Pharmacodynamics and Biopharmacy, University of Szeged, Szeged, Hungary
†Department of Pharmacodynamics, Semmelweis University, Budapest, Hungary
[1]Corresponding author: e-mail address: gaspar@pharm.u-szeged.hu

Contents

Abstract

The presence and effects of nociceptin (N/OFQ) and nocistatin (NST) in the central nervous system have been reasonably well described, but less data are available on their peripheral functions. Besides their presence in several peripheral organs (white blood cells, airway, liver, skin, vascular and intestinal smooth muscles, ovary, and testis), they have been found in the pregnant myometrium in both rat and human. The level of their precursor prepronociceptin is elevated in the preterm human myometrium as compared with full-term samples, whereas it gradually increases toward term in the pregnant rat uterus. Both N/OFQ and NST inhibit myometrial contractions, an effect which can be enhanced by naloxone and blocked by Ca^{2+}-dependent K^{+} channel (BK_{Ca}) inhibitors. Both compounds increase the myometrial cAMP level which may be responsible

Vitamins and Hormones, Volume 97
ISSN 0083-6729
http://dx.doi.org/10.1016/bs.vh.2014.10.004

for the activation of this channel and subsequent intracellular hyperpolarization. NST releases calcitonin gene-related peptide from the sensory nerve ends, which explains its cAMP-elevating effect. In contrast with the nervous system, where they behave as antagonists, N/OFQ and NST are able to potentiate the uterine-relaxing effect of each other in both rat and human tissues. Further studies are required to clarify the roles of N/OFQ and NST in the regulation of the myometrial contractions and the perception of pain during delivery.

ABBREVIATIONS

BK_{Ca} Ca^{2+}-dependent K^+ channel
CGRP calcitonin gene-related peptide
CGRP-R calcitonin gene-related peptide receptor
FSH follicle-stimulating hormone
GnRH gonadotropin-releasing hormone
hCG human chorionic gonadotropin
IBS irritable bowel syndrome
LH luteinizing hormone
N/OFQ nociceptin
NOP-R nociceptin receptor
NST nocistatin
NX naloxone
PAX paxilline
PNOC prepronociceptin
PTX pertussis toxin
RIA radioimmunoassay
SP substance P
TEA tetraethylammonium

Since the discovery of nociceptin (N/OFQ) and nocistatin (NST), their actions and physiological roles mainly have been investigated in the central nervous system. N/OFQ and NST are derived from the same precursor protein, prepronociceptin (PNOC). PNOC mRNA is expressed in the brain and spinal cord. However, similarly to endogenous opiates that are present in the peripheral nervous system, they can be found in some peripheral organs too.

1. ROLES OF PNOC, N/OFQ, AND NST IN DIFFERENT PERIPHERAL TISSUES

1.1 White blood cells

It was earlier supposed that N/OFQ receptor (NOP-R) mRNA, but not a functional receptor protein was expressed in the immune system (Pampusch et al., 2000). Evidences was later found for the existence of NOP-R in the human immune system (Arjomand, Cole, & Evans, 2002). PNOC mRNA expression has been confirmed in human blood mononuclear cells (Williams, Thompson, Rowbotham, & Lambert, 2008). It seems that activation of the N/OFQ signaling mechanism may alter the expression of inflammatory cytokines through the function of the white blood cells (Gavioli & Romão, 2011). It was recently discovered that lipopolysaccharide-induced inflammation in cultured whole blood enhances the production of different kinds of cytokines, which leads to the reduction of PNOC and NOP-R expression. Interestingly, both pro- and anti-inflammatory cytokines reduce the function of the N/OFQ system (Zhang, Stuber, & Stamer, 2013).

1.2 Airways

Animal studies suggest that N/OFQ modulates airway contraction and inflammation: upregulation of the N/OFQ-NOP-R system reduces the airway hyperresponsiveness. In contrast, inflammatory cells in asthma are both sources of N/OFQ and respond to NOP-R activation. Dysregulation of the N/OFQ-NOP-R system may have a potential role in the regulation of airway inflammation and airway tone (Singh, Sullo, D'Agostino, Brightling, & Lambert, 2013). Intravenous administration of N/OFQ was found to inhibit the cough number and the magnitude of the abdominal muscle electromyogram discharge during cough in anesthetized cat (Bolser, McLeod, Tulshian, & Hey, 2001).

1.3 Liver

Expression studies have revealed a low expression of NOP-R mRNA in zebrafish peripheral tissues, such as gills, muscle, and liver (Rivas-Boyero et al., 2011). A higher level of N/OFQ was observed in hepatic carcinoma tissue, as compared with the normal hepatic tissue from the same subject. Higher N/OFQ plasma levels were also found in patients with Wilson disease and hepatic cirrhosis (Chiou et al., 2007; Hantos et al., 2002).

1.4 Skin

PNOC and NOP-R mRNAs were found to be expressed substantially in the skin of mice; their immunoreactivities were localized in the epidermis. Administration of N/OFQ to cultures of keratinocytes from mice produced leukotriene B4, which is possibly involved in the itch-associated responses (Andoh, Yageta, Takeshima, & Kuraishi, 2004).

1.5 Vascular smooth muscle

Although the expressions of PNOC, N/OFQ, or NST have not been demonstrated in the vascular smooth muscle, some of their actions have been investigated in the vessels. N/OFQ was found to cause generalized vasodilatation. Its dilator influence was greater in the skeletal muscle than in other organs in rats (Abdelrahman & Pang, 2002). N/OFQ acts directly on blood vessels and is unlikely to cross the blood–brain barrier in view of its size. Intravenous N/OFQ produces transient hypotension and bradycardia (Doggrell, 2007). NOP-Rs are expressed in the endothelial cells of the human aortic artery, the rat aorta, and the sympathetic nerve terminals that innervate the cardiac muscle. Peripherally, N/OFQ can inhibit norepinephrine release from blood vessels (Chiou et al., 2007).

1.6 Intestinal smooth muscle

N/OFQ inhibits cholinergic transmission in the stomach and small intestine and stimulates colonic contraction by inhibiting an inhibitory neural pathway within the myenteric plexus (Yazdani, Takahashi, Bagnol, Watson, & Owyang, 1999). The contractile effect of N/OFQ on colonic smooth muscle is partially mediated by the inhibition of the tonic, neuronal release of nitrogen oxide (Menzies & Corbett, 2000). The N/OFQ system acts as a neuromodulatory system strictly relating to the modulation of excitatory or inhibitory neurotransmitters released in the guinea pig ileum enteric nervous system (Marini et al., 2012). NOP-R mRNA is detectable in the jejunum and colon of healthy subjects and is more highly expressed in the jejunum than in the colon. Its expression is lower in the jejunum and colon of patients with diarrhea-predominant irritable bowel syndrome (IBS). NOP-R and N/OFQ may be involved in the regulation of intestinal movement in healthy individuals and may have a role in the pathophysiology of IBS (Li, Dong, & Wang, 2013).

1.7 Ovary

In the human brain, the gonadotropin-releasing hormone (GnRH)-immunoreactive nerve cells are colocalized with N/OFQ in a high percentage (Foradori et al., 2007). This finding indicates that the actions of GnRH may be modulated in part by N/OFQ. This mechanism may control the reproductive functions by regulating release of the gonadotropic hormones follicle-stimulating hormone (FSH) and luteinizing hormone from the pituitary. Additionally, PNOC mRNA has been demonstrated in the rat ovary (Mollereau et al., 1996). PNOC displays high ovarian mRNA induction in response to an ovulatory human chorionic gonadotropin stimulus. This implicates N/OFQ and/or NST as paracrine factors in the periovulatory ovary (Leo, Pisarska, & Hsueh, 2001).

1.8 Testis

The expression of the PNOC gene is extremely upregulated by insulin-like 3 protein in fetal rat gubernacular bulbs. The insulin-like 3 protein has been suggested to control testis descent during development (Johnson et al., 2010). An analysis with murine Sertoli cells and testes has demonstrated that both PNOC mRNA and N/OFQ are induced by FSH. N/OFQ is induced in the testes postpartum following the FSH surge. This suggests that N/OFQ may act as a paracrine mediator of the FSH effects in the regulation of spermatogenesis (Eto, Shiotsuki, Sakai, & Abe, 2012). N/OFQ induces Rec8 phosphorylation (it mediates the association and disassociation of chromosomes occurring in synaptonemal complex formation, crossover recombination, and sister chromatid cohesion during meiosis), triggering the chromosome dynamics in the spermatocytes during meiosis in the postnatal murine testes. NOP-R is expressed in the plasma membrane of the spermatocytes. Treatment of the testes with N/OPQ results in a rapid phosphorylation of Rec8, whereas the injection of NST abolishes this action. These findings suggest that N/OPQ also plays a role in the progress of meiosis (Eto, Shiotsuki, & Abe, 2013).

2. PRESENCE OF PNOC, N/OFQ, AND NST IN UTERINE TISSUE

The expression and physiological effects of PNOC and its derivatives in the ovary and the testis suggest their putative roles in the control of reproduction. Among the reproductive organs, the uterus undergoes a dramatic change during pregnancy, which involves alterations in both the smooth muscle contractility and the adrenergic innervation, while the sensory nerves responsible for pain usually remain unchanged (Carbillon, Seince, & Uzan, 2001; Klukovits, Gaspar, Santha, Jancso, & Falkay, 2004). N/OFQ and NST are known to play crucial roles in pain mediation (Ito, Okuda-Ashitaka, & Minami, 2001) and to influence smooth muscle contractility; their presence in uterine tissue is therefore theoretically expected.

2.1 PNOC in the uterus

The expressions of PNOC mRNA have been investigated in rat and human pregnant myometrial tissues (Deak, Klukovits, Kormányos, et al., 2013; Klukovits et al., 2010). In rats, PNOC mRNA levels of the pregnant myometrium increase significantly as term approaches. A real-time PCR study has revealed that the relative quantity of PNOC mRNA is the lowest on pregnancy day 18. The expression of PNOC mRNA is changed on pregnancy day 20 and is increased significantly by the day of delivery (day 22) (Fig. 1; white columns). The PNOC mRNA levels in human myometrial tissues from Cesarean section are significantly higher in preterm uterine samples as compared with full-term pregnancy (Fig. 1; gray columns). These data suggest that the elevated expression of PNOC mRNA is a part of the normal physiological process to be matured for delivery in the rat. In contrast the elevated PNOC mRNA level in the human myometrium is linked to prematurity, while its level remains low during normal pregnancies. This contradiction is rather difficult to explain, but might be

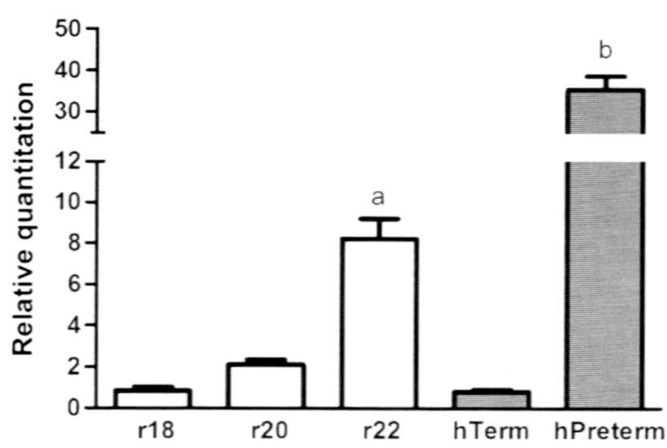

Figure 1 Levels of expression of prepronociceptin (PNOC) mRNA in pregnant rat and pregnant human uterus samples. Rat myometrial tissues were investigated on days 18, 20, and 22 of pregnancy. Human tissues were obtained from preterm birth and full-term birth by Cesarean section. The rat myometrial PNOC mRNA level (*white columns*) was the same on day 20 (r20) as on day 18 (r18), but was significantly higher on day 22 (r22) of pregnancy. In human samples (gray columns), the PNOC mRNA level was significantly lower in term pregnancy (hTerm) than at preterm (hPreterm). The endogenous control was β-actin. (A) $p < 0.001$ as compared with the rat myometrium on pregnancy day 18. (B) $p < 0.001$ as compared with human term myometrium. *Data for this figure were taken from the articles Klukovits et al. (2010) and Deak, Klukovits, Kormányos, et al. (2013).*

understood through the expressions and effects of N/OFQ and NST in both rat and human pregnant myometria. However, human data are not available concerning the myometrial expressions of N/OFQ or NST during pregnancy.

2.2 N/OFQ and NST in the uterus

Since the common precursor for N/OFQ and NST, PNOC mRNA is expressed locally in the pregnant rat and human uterus, N/OFQ and NST should be synthesized locally too.

A radioligand binding study has revealed that NOP-R is expressed in the rat myometrium (Klukovits et al., 2010). In the uteri of the nonpregnant female rat, the maximum binding capacity and the dissociation constant of the NOP-R do not differ significantly from those in the 22-day-pregnant rat myometrium. This means that the NOP-R expression and function do not change during pregnancy. A similar study cannot yet be carried out for NST, because to date no receptor for NST has been identified in any tissue. A report has indicated that the putative G-protein-coupled receptor may be the endogenous receptor for NST. Studies conducted during the past few years have revealed that NST can bind to specific molecules in the CNS as an agonist peptide. The intracellular localization of N/OFQ and NST requires a high-resolution transmission electron microscope capable of identifying these peptides and other supporting molecules that colocalize with them. A tracing technique might also elucidate a possible ability of these peptides to migrate from one cellular compartment to another (Tariq, Nurulain, Tekes, & Adeghate, 2013).

The tissue levels of N/OFQ and NST have been measured by radioimmunoassay in the uteri of nonpregnant and 22-day-pregnant rats (Deak, Klukovits, Tekes, et al., 2013; Klukovits et al., 2010). In 22-day-pregnant rats, the uterine N/OFQ concentration is higher than that in the nonpregnant rats (Fig. 2A). The myometrial NST levels are relatively low on pregnancy days 15, 18, and 20 as compared with nonpregnant samples. The myometrial NST level is elevated by day 22 as compared with previously. The NST level on day 22 does not differ from that in a nonpregnant sample (Fig. 2B). NST in the pregnant rat uterus at term is ~10 times more abundant than N/OFQ, and it therefore seems that PNOC mRNA is translated mainly to NST rather than to N/OFQ. While there is an elevation in the N/OFQ myometrial level at the end of pregnancy, the NST level is similar to the nonpregnant value, suggesting the greater importance of

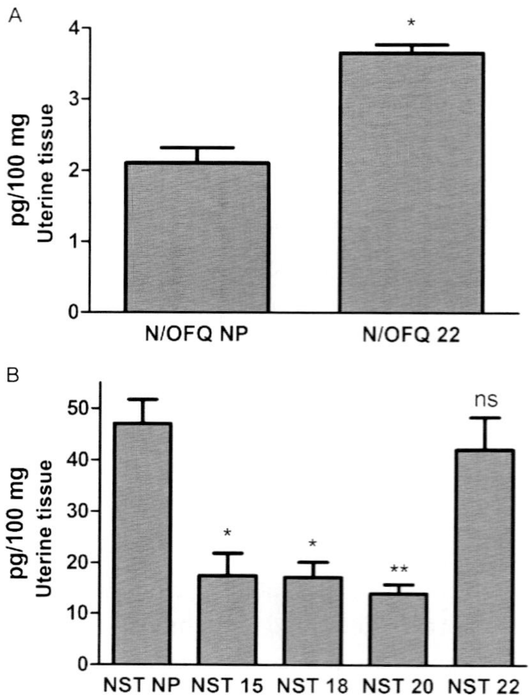

Figure 2 Myometrial nociceptin and nocistatin levels in nonpregnant and pregnant rats. Protein levels were determined by radioimmunoassay. (A) Nociceptin levels in nonpregnant and 22-day-pregnant rat uteri. N/OFQ NP, nociceptin level in nonpregnant uterus; N/OFQ 22, nociceptin level in 22-day-pregnant uterus; $*p<0.05$. (B) Nocistatin levels in nonpregnant and pregnant rat uteri. NST NP, nocistatin level in nonpregnant uterus; NST 18–22, nocistatin level in 18–22-day-pregnant uteri; $*p<0.05$; $**p<0.01$; ns, nonsignificant. Significances are expressed relative to the value for the nonpregnant myometrium. *Data for this figure were taken from the articles Klukovits et al. (2010) and Deak, Klukovits, Tekes, et al. (2013).*

myometrial N/OFQ as compared with that of NST in the process of delivery. Additionally, they play a regulatory role in pain signaling (Okuda-Ashitaka et al., 1998), which may explain their further functional importance during labor.

It is not known how the N/OFQ and NST levels change during human pregnancy, but the high PNOC mRNA expression in preterm tissues allows the assumption that their physiological function might be to act against premature contractions striving to maintain uterine quiescence until normal term. However, the cerebrospinal fluid concentration of N/OFQ does not display any association with the acute pain of labor in patients participating in elective Cesarean section or in established labor treated with epidural anesthetics (Brooks et al., 1998). The lack of an elevation in spinal N/OFQ during labor pain suggests that the uterine N/OFQ may have a more important role in both pain and contraction control.

In the nonpregnant human uterus, a correlation has been demonstrated between the high endometrial and myometrial N/OFQ levels and the severity of dysmenorrhea in patients with adenomyosis. High levels of N/OFQ are detected in both eutopic and ectopic endometrial samples (Hou, Zhang, Liu, Lu, & Zhang, 2011). These results suggest that the pain, the smooth muscle spasm, and the uterine expression of N/OFQ are connected in adenomyosis-induced dysmenorrhea.

3. THE EFFECTS AND MECHANISMS OF ACTION OF N/OFQ AND NST ON UTERINE CONTRACTILITY

The effects of N/OFQ and NST on myometrial contractions have been studied in both rat and human uterine samples in recent years (Deak, Klukovits, Kormányos, et al., 2013; Deak, Klukovits, Tekes, et al., 2013; Klukovits et al., 2010). Although the available data provide a certain picture concerning the actions of these peptides, there are several points which must be clarified in the future for a clearer understanding all of the aspects of their effects in the uterus.

3.1 The effect of N/OFQ on uterine contractility

N/OFQ elicits myometrial relaxation *in vitro* in pregnant samples from both rat and human. The extent of its maximal inhibitory effect is in all cases ~30%, independently of the species or preterm or full-term conditions. However, a one order of magnitude difference is found between the EC_{50} values of N/OFQ on the full-term rat and human myometria, with practically a one order of magnitude difference between term and preterm human samples (Table 1). These results indicate that the rat myometrium

Table 1 Maximum relaxant and EC_{50} values of nociceptin (N/OFQ) on KCl-stimulated 22-day-pregnant rat and oxytocin-stimulated full-term and preterm human myometrial contractions *in vitro*

N/OFQ	Rat, full-term	Human, full-term	Human, preterm
E_{max} (%)	27.5 ± 3.1	29.1 ± 1.6	27.5 ± 4.3
EC_{50} (M)	1.1×10^{-10}	1.0×10^{-9}	9.0×10^{-9}

Data for this table were taken from the articles Klukovits et al. (2010) and Deak, Klukovits, Kormányos, et al. (2013).

is more sensitive than the human myometrium to the relaxing effect of N/OFQ. Additionally, the preterm condition in human may worsen the sensitivity of the uterus to the action of N/OFQ.

A search for the mechanism of the relaxing effect of N/OFQ on the pregnant myometrium has revealed a multiple pathway (Klukovits et al., 2010). N/OFQ enhances the G-protein activation, which is pertussis toxin (PTX)-sensitive, the rise in the amount of G_i-protein therefore playing the predominant part in this effect. In the presence of the opiate receptor antagonist naloxone (NX), the relaxing effect of N/OFQ is not reduced, but surprisingly increases further. The G-protein-activating effect of the two compounds together is lower than that of N/OFQ alone. Moreover, the combined compounds elicit a significant cAMP elevation, pointing to some G_s-protein involvement in their effect. In the presence of PTX, their myometrial cAMP level-increasing effect is doubled, suggesting that their mutual action involves both G_i- and G_s-protein activation. A possible explanation for this phenomenon might be that NX interferes with the G_i-protein-activating potency of N/OFQ. The phenomenon that individual receptors are able to activate multiple pathways by switching between different G-proteins is well established (Knollman & Conn, 2008; Lefkowitz, Pierce, & Luttrell, 2002). The coupling of beta-adrenergic receptors to G_s-proteins leads to the activation of adenylate cyclase and the consequent phosphorylation of protein kinase A, which phosphorylates the receptor and diminishes its coupling to G_s, but increases the coupling to G_i. We presume that NX interferes with the intracellular pathways activated by N/OFQ and promotes its coupling to G_s. N/OFQ can effectively activate protein kinase C via NOP-R, through PTX-sensitive G-protein activation (G_i/0). Besides this, the phospholipase C inhibitors block this pathway, which means that the Ca^{2+} concentration is increased, and phospholipase C activation may therefore be involved in the signaling mechanism (Lou, Ma, & Pei, 1997). However, the exact explanation of this phenomenon still awaits clarification.

The potentiating effect of NX on the action of N/OFQ is somewhat unexpected, although an elevated uterine tone has been reported in humans after the intrathecal administration of opioids during labor (Abrão, Francisco, Miyadahira, Cicarelli, & Zugaib, 2009), which suggests that endogenous opioids might exert a contraction-enhancing effect on the uterus that can be attenuated by NX. However, the effect of NX alone on the uterus appears to be nonsignificant (Klukovits et al., 2010).

N/OFQ also enhances an outward K^+ conductance, which reduces neuronal excitability and transmitter release in the brain (Hawes, Graziano, & Lambert, 2000). In the uterus, BK_{Ca} channels are abundant and play an important role in limiting depolarization, thereby relaxing the uterine smooth muscle. Moreover, cAMP-dependent phosphorylation activates these BK_{Ca} channels (Chanrachakul, Pipkin, & Khan, 2004). The blockade of the K^+ channels with the nonselective blocker tetraethylammonium (TEA) or the selective BK_{Ca} channel-blocker paxilline (PAX) diminishes the uterus-relaxing effect of N/OFQ (applied either alone or in combination with NX) on the pregnant rat myometrium. This indicates that this effect of N/OFQ is mediated by the activation of K^+ channels. There is no apparent difference between the N/OFQ-inhibiting effects of TEA and PAX, which means that BK_{Ca} channels take the main role in the uterus-relaxing action of N/OFQ. In the presence of TEA or PAX, N/OFQ alone or in combination with NX still produces a limited uterus-relaxing effect, which reinforces the functions of other pathways in the relaxing effect, like the elevation in cAMP level (Klukovits et al., 2010).

3.2 The effect of NST on uterine contractility

NST elicits myometrial relaxation *in vitro* in pregnant myometria from both rat and human (Deak, Klukovits, Kormányos, et al., 2013; Deak, Klukovits, Tekes, et al., 2013). The extent of its maximal inhibitory effect is lower in rat than in human preterm and full-term conditions. Similarly as for N/OFQ, the EC_{50} values differ. A one order of magnitude difference is observed between the EC_{50} values of NST for the full-term rat and human myometria, and an order of magnitude difference between the preterm and term human samples (Table 2). Similarly to N/OFQ, the rat myometrial binding sites seem to be more sensitive to NST than those in the human

Table 2 Maximum relaxant and EC_{50} values of nocistatin (NST) on KCl-stimulated 22-day-pregnant rat and oxytocin-stimulated full-term and preterm human myometrial contractions *in vitro*

NST	Rat, full-term	Human, full-term	Human, preterm
E_{max} (%)	22.2 ± 1.7	38.0 ± 6.1	33.7 ± 2.1
EC_{50} (M)	1.2×10^{-11}	2.6×10^{-10}	1.7×10^{-9}

Data for this table were taken from the articles Deak, Klukovits, Kormányos, et al. (2013) and Deak, Klukovits, Tekes, et al. (2013).

myometrium. The preterm condition in human may worsen the sensitivity of the uterus to the action of NST as compared with full-term uteri. However, the maximal effect of NST is weaker on the rat than on the human myometrium, though this difference is not too great.

NX inhibits the contraction-inhibitory effect of NST against KCl-induced contractions. NST does not bind to any of the classical opioid receptors, or to the NOP-R receptor (Fantin, Fischetti, Trapella, & Morari, 2007; Johnson & Connor, 2007), and this inhibition is therefore not likely to be mediated by opioid receptors. On the other hand, NX induces an increase in inward Ca^{2+} currents (Kai, Wang, Hu, Shi, & Liu, 2002). These findings raise the possibility that the inhibitory effect of NX on NST-induced uterus relaxation may be mediated by the opening of inward rectifying Ca^{2+} channels. In a Ca^{2+}-poor environment, NX does not inhibit the effect of NST because it is unable to promote a Ca^{2+} influx, and hence it cannot overcome the relaxation induced by NST (Deak, Klukovits, Tekes, et al., 2013).

NST elevates the cAMP level in pregnant rat uteri, which is an NX-sensitive action. In correlation with the *in vitro* contractility studies, NX decreases the cAMP levels elevated by NST, which suggests that NX interferes with NST at the level of G-protein activation, too. PAX inhibits the relaxing effect of NST on the pregnant rat myometrium, indicating the similar role of BK_{Ca} channels in its action as for N/OFQ (Deak, Klukovits, Tekes, et al., 2013).

Although NST exhibits a contraction-inhibitory effect on KCl- and $PGF_{2\alpha}$-evoked contractions, its action against $PGF_{2\alpha}$ is weak. Additionally, NST is ineffective on oxytocin-induced contractions. Interestingly, the calcitonin gene-related peptide (CGRP) agonist adrenomedullin also fails to block oxytocin-induced contractions, whereas it acts against spontaneous and bradykinin-induced contractions (Yanagita et al., 2000). Opioid-like nociceptive peptides are known to influence the release of neurotransmitters such as CGRP or substance P (SP) from capsaicin-sensitive primary sensory neurons (Helyes, Németh, Pintér, & Szolcsányi, 1997; Peiser, Undem, & Fischer, 2000). These neuropeptides are synthesized in the dorsal root ganglia cells, stored in vesicles, and released in response to electrical (Buzzi, Carter, Shimizu, Heath, & Moskowitz, 1991) or chemical stimulation (Kilo, Harding-Rose, Hargreaves, & Flores, 1997). CGRP inhibits smooth muscle contractility, including that in the pregnant rat uterus (Pennefather, Reynoldson, & Handberg, 1990). These findings together suggest that the uterine-relaxing effect of NST may be linked to CGRP release. This hypothesis is strengthened by the fact that the CGRP-releaser capsaicin blocks the uterine contraction-inhibitory effect of NST, while the relaxing effect can be restored by incubation of the myometrial samples with CGRP (Deak, Klukovits, Tekes, et al., 2013). There is a special population of opioid receptors in dorsal root ganglia neurons, where low doses of opioids can provoke hyperalgesia, due to the activation of opioid receptors on the afferent nerve terminals and the release of neuropeptides including CGRP and SP (Crain & Shen, 1990). Similarly to opioid peptides, NST may possibly promote the release of neuropeptides from sensory nerves. In spite of the recognition of many details of the mechanism of the compound, the complete molecular mode of action of NST remains unclear (Tariq et al., 2013).

3.3 The combined effect of N/OFQ and NST on uterine contractility

In the central nervous system, NST acts as an antagonist for N/OFQ. Simultaneous intrathecal administration of NST blocks N/OFQ-induced pain, including allodynia and hyperalgesia (Minami et al., 1998). NST also reverses the N/OFQ-induced inhibition of morphine analgesia and counteracts the impairment of learning and memory induced by N/OFQ (Hiramatsu & Inoue, 1999; Okuda-Ashitaka et al., 1998). Their mutual action has been investigated peripherally only on rat and human pregnant uteri (Deak, Klukovits, Kormányos, et al., 2013; Deak, Klukovits, Tekes, et al., 2013). In the rats, N/OFQ potentiates the relaxing effect of NST and the myometrial cAMP level is also elevated. Interestingly, N/OFQ is not able to enhance the relaxing effect of NST on the human myometrium. When NST is administered first, however, the N/OFQ-induced relaxation can be increased further, especially in the preterm human uterus. NST administration must therefore precede that of N/OFQ in order to enhance the common myometrium-relaxant effect. It is assumed that the CGRP-releasing effect of NST elicits a moderate relaxation as compared with the K^+ channel opening-action of N/OFQ. NST therefore cannot exceed the relaxation caused by a single dose of N/OFQ in the human myometrium.

4. CONCLUSIONS

N/OFQ and NST seem to be synthesized locally in both the rat and the human pregnant myometrim. In contrast with the nervous system, where they mutually behave as antagonists, their action in the uterus is the same: they inhibit myometrial contractions. Although their effects in cAMP elevation and the subsequent opening of BK_{Ca} channels are basically the same, their main mechanisms of action seem to be different. N/OFQ relaxes the uterine contraction directly, while the effect of NST is mediated through the liberation of CGRP from the myometrial sensory nerve ends (Fig. 3). Each is able to potentiate the relaxing effect of the other, although

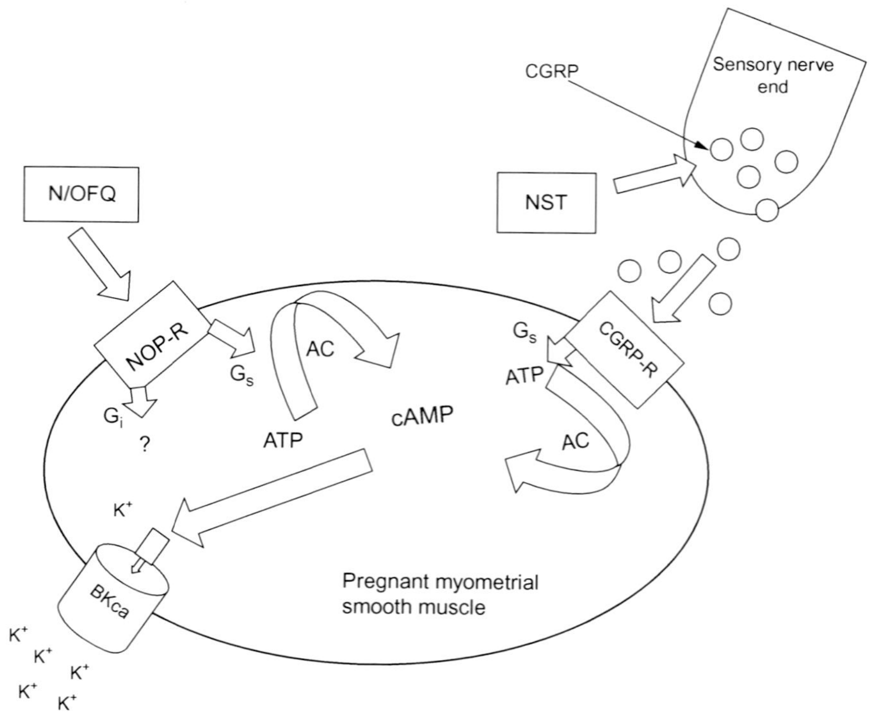

Figure 3 Putative mechanisms of relaxing action of nociceptin and nocistatin (NST) on pregnant myometrium. Nociceptin (N/OFQ) acts on the nociceptin receptor (NOP-R) and induces an elevation in the myometrial cAMP level through the activation of G_s-protein and adenylyl cyclase (AC). The level of intracellular cAMP rises in the uterine smooth muscle. The elevated cAMP level opens the Ca^{2+}-dependent K^+ channels (BK_{Ca}) and the increased outflow of K^+ ions hyperpolarizes the membrane potential, inducing relaxation in the smooth muscle. N/OFQ is able to stimulate the G_i pathway, but it is not known how this route contributes to relaxation. NST enhances the release of calcitonin gene-related peptide (CGRP) from the sensory nerve ends. CGRP binds to its receptor (CGRP-R), increases the intracellular cAMP level, and opens the BK_{Ca} channels, which is the main relaxing mechanism of NST in the pregnant myometrium. However, a receptor or myometrial binding site for NST has not yet been found and another mechanism for G_s activation and cAMP elevation cannot be excluded.

the administration of N/OFQ in the human myometrium should precede that of NST in order to achieve their enhanced action.

Further studies are required to clarify all the aspects of the physiological, pathophysiological roles, and effects of N/OFQ and NST on pregnant uterine contractility. Since both N/OFQ and NST are involved in the control of pain, one of the main questions to be answered is the roles of these compounds in the regulation of myometrial contractions and the perception of pain during delivery. It is widely accepted that pain and the uterine contractions seem to be inseparable in parturition. A deeper understanding of these natural regulatory processes and cofunctions may help reveal the secrets of the initiation of delivery and the reasons for early contractions.

REFERENCES

Abdelrahman, A. M., & Pang, C. C. (2002). Regional hemodynamic effects of nociceptin/orphanin FQ in the anesthetized rat. *European Journal of Pharmacology*, *450*, 263–266.

Abrão, K. C., Francisco, R. P., Miyadahira, S., Cicarelli, D. D., & Zugaib, M. (2009). Elevation of uterine basal tone and fetal heart rate abnormalities after labor analgesia: A randomized controlled trial. *Obstetrics & Gynecology*, *113*, 41–47.

Andoh, T., Yageta, Y., Takeshima, H., & Kuraishi, Y. (2004). Intradermal nociceptin elicits itch-associated responses through leukotriene B4 in mice. *The Journal of Investigative Dermatology*, *123*, 196–201.

Arjomand, J., Cole, S., & Evans, C. J. (2002). Novel orphanin FQ/nociceptin transcripts are expressed in human immune cells. *Journal of Neuroimmunology*, *130*, 100–108.

Bolser, D. C., McLeod, R. L., Tulshian, D. B., & Hey, J. A. (2001). Antitussive action of nociceptin in the cat. *European Journal of Pharmacology*, *430*, 107–111.

Brooks, H., Elton, C. D., Smart, D., Rowbotham, D. J., McKnight, A. T., & Lambert, D. G. (1998). Identification of nociceptin in human cerebrospinal fluid: Comparison of levels in pain and non-pain states. *Pain*, *78*, 71–73.

Buzzi, M. G., Carter, W. B., Shimizu, T., Heath, H., & Moskowitz, M. A. (1991). Dihydroergotamine and sumatriptan attenuate levels of CGRP in plasma in rat superior sagittal sinus during electrical stimulation of the trigeminal ganglion. *Neuropharmacology*, *30*, 1193–1200.

Carbillon, L., Seince, N., & Uzan, M. (2001). Myometrial maturation and labour. *Annals of Medicine*, *33*, 571–578.

Chanrachakul, B., Pipkin, F. B., & Khan, R. N. (2004). Contribution of coupling between human myometrial beta2-adrenoreceptor and the BK(Ca) channel to uterine quiescence. *American Journal of Physiology – Cell Physiology*, *287*, 1747–1752.

Chiou, L. C., Liao, Y. Y., Fan, P. C., Kuo, P. H., Wang, C. H., Riemer, C., et al. (2007). Nociceptin/orphanin FQ peptide receptors: Pharmacology and clinical implications. *Current Drug Targets*, *8*, 117–135.

Crain, S. M., & Shen, K. F. (1990). Opioids can evoke direct receptor-mediated excitatory effects on sensory neurons. *Trends in Pharmacological Sciences*, *11*, 77–81.

Deak, B. H., Klukovits, A., Kormányos, Z., Tekes, K., Ducza, E., & Gaspar, R. (2013). Uterus-relaxing effects of nociceptin and nocistatin: Studies on preterm and term-pregnant human myometrium. *Reproductive System & Sexual Disorders*, *2*, 1–5.

Deak, B. H., Klukovits, A., Tekes, K., Ducza, E., Falkay, G., & Gaspar, R. (2013). Nocistatin inhibits pregnant rat uterine contractions in vitro: Roles of calcitonin gene-related peptide and calcium-dependent potassium channel. *European Journal of Pharmacology*, *714*, 96–104.

Doggrell, S. A. (2007). Cardiovascular and renal effects of nociceptin/orphanin FQ: A new mediator to target? *Current Opinion in Investigational Drugs*, *8*, 742–749.

Eto, K., Shiotsuki, M., & Abe, S. (2013). Nociceptin induces Rec8 phosphorylation and meiosis in postnatal murine testes. *Endocrinology*, *154*, 2891–2899.

Eto, K., Shiotsuki, M., Sakai, T., & Abe, S. (2012). Nociceptin is upregulated by FSH signaling in Sertoli cells in murine testes. *Biochemical and Biophysical Research Communications*, *421*, 678–683.

Fantin, M., Fischetti, C., Trapella, C., & Morari, M. (2007). Nocistatin inhibits 5-hydroxytryptamine release in the mouse neocortex via presynaptic Gi/o protein linked pathways. *British Journal of Pharmacology*, *152*, 549–555.

Foradori, C. D., Amstalden, M., Coolen, L. M., Singh, S. R., McManus, C. J., Handa, R. J., et al. (2007). Orphanin FQ: Evidence for a role in the control of the reproductive neuroendocrine system. *Endocrinology*, *148*, 4993–5001.

Gavioli, E. C., & Romão, P. R. (2011). NOP-R receptor ligands as potential agents for inflammatory and autoimmune diseases. *Journal of Amino Acids*, *2011*, 836569.

Hantos, M. B., Szalay, F., Lakatos, P. L., Hegedus, D., Firneisz, G., Reiczigel, J., et al. (2002). Elevated plasma nociceptin level in patients with Wilson disease. *Brain Research Bulletin*, *58*, 311–313.

Hawes, B. E., Graziano, M. P., & Lambert, D. G. (2000). Cellular actions of nociceptin: Transduction mechanisms. *Peptides*, *21*, 961–967.

Helyes, Z., Németh, J., Pintér, E., & Szolcsányi, J. (1997). Inhibition by nociceptin of neurogenic inflammation and the release of SP and CGRP from sensory nerve terminals. *British Journal of Pharmacology*, *121*, 613–615.

Hiramatsu, M., & Inoue, K. (1999). Effects of nocistatin on nociceptin-induced impairment of learning and memory in mice. *European Journal of Pharmacology*, *367*, 151–155.

Hou, R., Zhang, W. Y., Liu, L. Y., Lu, J., & Zhang, W. Y. (2011). Expression of orphanin in adenomyosis and its relationship with dysmenorrhea. *Zhonghua Yi Xue Za Zhi*, *91*, 612–615.

Ito, S., Okuda-Ashitaka, E., & Minami, T. (2001). Central and peripheral roles of prostaglandins in pain and their interactions with novel neuropeptides nociceptin and nocistatin. *Neuroscience Research*, *41*, 299–332.

Johnson, E. E., & Connor, M. (2007). Towards a receptor for nocistatin? *British Journal of Pharmacology*, *152*, 415–416.

Johnson, K. J., Robbins, A. K., Wang, Y., McCahan, S. M., Chacko, J. K., & Barthold, J. S. (2010). Insulin-like 3 exposure of the fetal rat gubernaculum modulates expression of genes involved in neural pathways. *Biology of Reproduction*, *83*, 774–782.

Kai, L., Wang, Z. F., Hu, D. Y., Shi, Y. L., & Liu, L. M. (2002). Modulation of Ca^{2+} channels by opioid receptor antagonists in mesenteric arterial smooth muscle cells of rats in hemorrhagic shock. *Journal of Cardiovascular Pharmacology*, *40*, 618–624.

Kilo, S., Harding-Rose, C., Hargreaves, K. M., & Flores, C. M. (1997). Peripheral CGRP release as a marker for neurogenic inflammation: A model system for the study of neuropeptide secretion in rat paw skin. *Pain*, *73*, 201–207.

Klukovits, A., Gaspar, R., Santha, P., Jancso, G., & Falkay, G. (2004). Role of capsaicin-sensitive nerve fibers in uterine contractility in the rat. *Biology of Reproduction*, *70*, 184–190.

Klukovits, A., Tekes, K., Gündüz Cinar, O., Benyhe, S., Borsodi, A., Deák, B. H., et al. (2010). Nociceptin inhibits uterine contractions in term-pregnant rats by signaling through multiple pathways. *Biology of Reproduction*, *83*, 36–41.

Knollman, P. E., & Conn, P. M. (2008). Multiple G proteins compete for binding with the human gonadotropin releasing hormone receptor. *Archives of Biochemistry and Biophysics*, *92–97*.

Lefkowitz, R. J., Pierce, K. L., & Luttrell, L. M. (2002). Dancing with different partners: Protein kinase A phosphorylation of seven membrane-spanning receptors regulates their G protein-coupling specificity. *Molecular Pharmacology*, *62*, 971–974.

Leo, C. P., Pisarska, M. D., & Hsueh, A. J. (2001). DNA array analysis of changes in preovulatory gene expression in the rat ovary. *Biology of Reproduction*, *65*, 269–276.

Li, L., Dong, L., & Wang, S. (2013). Expression of the nociceptin/orphanin FQ receptor in the intestinal mucosa of IBS patients. *Experimental and Therapeutic Medicine*, *6*, 679–683.

Lou, L. G., Ma, L., & Pei, G. (1997). Nociceptin/orphanin FQ activates protein kinase C, and this effect is mediated through phospholipase C/Ca2+ pathway. *Biochemical and Biophysical Research Communications*, *240*, 304–308.

Marini, P., Romanelli, L., Valeri, D., Cascio, M. G., Tucci, P., Valeri, P., et al. (2012). The NOP-R receptor involvement in both withdrawal- and CCk-8-induced contracture responses of guinea pig isolated ileum after acute activation of κ-opioid receptor. *Peptides*, *38*, 418–426.

Menzies, J. R., & Corbett, A. D. (2000). Nociceptin inhibits tonic nitric oxide release in the mouse isolated proximal colon. *European Journal of Pharmacology*, *388*, 183–186.

Minami, T., Okuda-Ashitaka, E., Nishiuchi, Y., Kimura, T., Tachibana, S., Mori, H., et al. (1998). Anti-nociceptive responses produced by human putative counterpart of nocistatin. *British Journal of Pharmacology*, *124*, 1016–1018.

Mollereau, C., Simons, M. J., Soularue, P., Liners, F., Vassart, G., Meunier, J. C., et al. (1996). Structure, tissue distribution, and chromosomal localization of the prepronociceptin gene. *Proceedings of the National Academy of Sciences of the United States of America*, *93*, 8666–8670.

Okuda-Ashitaka, E., Minami, T., Tachibana, S., Yoshihara, Y., Nishiuchi, Y., Kimura, T., et al. (1998). Nocistatin, a peptide that blocks nociceptin action in pain transmission. *Nature*, *392*, 286–289.

Pampusch, M. S., Serie, J. R., Osinski, M. A., Seybold, V. S., Murtaugh, M. P., & Brown, D. R. (2000). Expression of nociceptin/OFQ receptor and prepro-nociceptin/OFQ in lymphoid tissues. *Peptides*, *21*, 1865–1870.

Peiser, C., Undem, B. J., & Fischer, A. (2000). Nociceptin effects in the airways. *Peptides*, *21*, 995–998.

Pennefather, J. N., Reynoldson, N. A., & Handberg, G. M. (1990). Inhibition of rat uterine contractions by rat and human CGRP. *Peptides*, *11*, 903–906.

Rivas-Boyero, A. A., Herrero-Turrión, M. J., Gonzalez-Nunez, V., Sánchez-Simón, F. M., Barreto-Valer, K., & Rodríguez, R. E. (2011). Pharmacological characterization of a nociceptin receptor from zebrafish (*Danio rerio*). *Journal of Molecular Endocrinology*, *46*, 111–123.

Singh, S. R., Sullo, N., D'Agostino, B., Brightling, C. E., & Lambert, D. G. (2013). The effects of nociceptin peptide (N/OFQ)-receptor (NOP) system activation in the airways. *Peptides*, *39*, 36–46.

Tariq, S., Nurulain, S. M., Tekes, K., & Adeghate, E. (2013). Deciphering intracellular localization and physiological role of nociceptin and nocistatin. *Peptides*, *43*, 174–183.

Williams, J. P., Thompson, J. P., Rowbotham, D. J., & Lambert, D. G. (2008). Human peripheral blood mononuclear cells produce pre-pro-nociceptin/orphanin FQ mRNA. *Anesthesia and Analgesia*, *106*, 865–866.

Yanagita, T., Yamamoto, R., Sugano, T., Kobayashi, H., Uezono, Y., Yokoo, H., et al. (2000). Adrenomedullin inhibits spontaneous and bradykinin-induced but not oxytocin- or prostaglandin F(2alpha)-induced periodic contraction of rat uterus. *British Journal of Pharmacology*, *130*, 1727–1730.

Yazdani, A., Takahashi, T., Bagnol, D., Watson, S. J., & Owyang, C. (1999). Functional significance of a newly discovered neuropeptide, orphanin FQ, in rat gastrointestinal motility. *Gastroenterology*, *116*, 108–117.

Zhang, L., Stuber, F., & Stamer, U. M. (2013). Inflammatory mediators influence the expression of nociceptin and its receptor in human whole blood cultures. *PLoS One*, *8*, e74138.

CHAPTER NINE

Nociceptin/Orphanin FQ-NOP Receptor System in Inflammatory and Immune-Mediated Diseases

Elaine C. Gavioli*, Iris Ucella de Medeiros*, Marta C. Monteiro†, Girolamo Calo‡, Pedro R.T. Romão§,1

*Department of Biophysic and Pharmacology, Federal University of Rio Grande do Norte, Natal, Brazil
†Laboratory of Clinical Microbiology and Immunology, Faculty of Pharmacy, Federal University of Pará, Belém, Brazil
‡Department of Medical Sciences, Section of Pharmacology and National Institute of Neuroscience, University of Ferrara, Ferrara, Italy
§Laboratory of Immunology, Department of Basic Health Sciences, Federal University of Health Sciences of Porto Alegre, Rua Sarmento Leite, Porto Alegre, Brazil
[1]Corresponding author: e-mail address: pedror@ufcspa.edu.br

Contents

Abstract

The neuropeptide nociceptin/orphanin FQ (N/OFQ) is the endogenous ligand of the G-protein-coupled receptor NOP. Cells from the immune system express the precursor preproN/OFQ and the NOP receptor, as well as secrete N/OFQ. The activation of the N/OFQ-NOP pathway can regulate inflammatory and immune responses. Several immune activities, including leukocyte migration, cytokine and chemokine production, and lymphocytes proliferation are influenced by NOP activation. It was demonstrated that cytokines and other stimuli such as Toll-like receptor agonist (e.g., lipopolysaccharide) induce N/OFQ production by cells from innate and adaptive immune response. In this context, N/OFQ could modulate the outcome of inflammatory diseases, such as sepsis and immune-mediated pathologies by mechanisms not clearly elucidated. In fact, clinical studies revealed increased levels of N/OFQ under sepsis, arthritis, and Parkinson's disease. Preclinical and clinical studies pointed to the blockade of NOP receptor

Vitamins and Hormones, Volume 97
ISSN 0083-6729
http://dx.doi.org/10.1016/bs.vh.2014.11.003

signaling as successful strategy for the treatment of inflammatory diseases. This review is focused on experimental and clinical data that suggest the participation of N/OFQ-NOP receptor activation in the modulation of the immune response, highlighting the immunomodulatory potential of NOP antagonists in the inflammatory and immunological disturbances.

1. A BRIEF OVERVIEW OF THE IMMUNE RESPONSE

The immune system has the ability to recognize antigens including invading pathogens, toxins, and allergens as well as altered self cells on injury sites and to distinguish self from nonself antigens. The initial host response is mediated by nonspecific cells (innate immunity) activated by pathogen-associated molecular patterns or damage-associated molecular patterns. Due to the release of mediators, the inflammatory response initiates and leads to the development of specific immune response mediated by T and B cells. In the early phase of the immune response, phagocytes including neutrophils, eosinophils, macrophages, and dendritic cells accumulate in the infection site or tissue injured and initiate a cascade of signaling pathways involving innate immune sensors, called pattern-recognition receptors (Beutler, 2004; Joffre, Nolte, Spörri, & Reis e Sousa, 2009). The activation of these receptors induces the production of proinflammatory cytokines and/or chemokines, which promote the recruitment and activation of inflammatory cells amplifying the inflammatory response (Kantari, Pederzoli-Ribeil, & Witko-Sarsat, 2008; Mantovani, Cassatella, Costantini, & Jaillon, 2011; Moretta, 2002; Murray & Wynn, 2011). Together with the recruited cells, resident cells are able to phagocytose and destroy microbes. On the other hand, macrophages, dendritic cells, and B lymphocyte, which express specific receptors for antigens (immunoglobulin), are professional antigen-presenting cells (APCs) responsible for capturing and processing of antigens and for T cell activation. These specific cells are defined by the cell-surface expression of the T cell receptor (TCR), which binds to processed peptide displayed by APCs associated with class I or class II Major Histocompatibility Complex (MHC) molecules. The activation of naive T lymphocytes occurs in the secondary lymphoid organs by the interaction between the MHC class II-peptide or MHC class I-peptide displayed by APCs and TCR on the surface of T $CD4^+$ (T helper) or T $CD8^+$ cells (T cytotoxic), respectively (Kapsenberg, 2003; Martin & Frevert, 2005; Murray & Wynn, 2011). The cytokines produced by cells from the innate immunity have a critical role in activating and coordinating the development of the adaptive immunity (Hoebe, Janssen, & Beutler, 2004; Medzhitov, 2007). Moreover, macrophages, dendritic cells, and B lymphocyte constitutively express MHC class I and II proteins, besides costimulatory molecules that are essential for the T-cell activation, proliferation, and differentiation into effector and memory T cells (Banchereau & Steinman, 1998; Jego et al., 2003; Steinman, 1991). Regarding the induction of protective immune response, the activation of $CD4^+$ T cells is crucial for the activation of naive $CD8^+$ T (Guidotti & Chisari, 2001) and induction of B cells to become antibody-secreting plasma cells (Bachmann & Zinkernagel, 1997). Furthermore, dendritic cells are essential for the differentiation of naive $CD4^+$ T cells into effector T cell subsets, such as Th1, Th2, Th17, and T regulatory cells. These T cell subsets secrete different cytokines that modulate the immune response and often the outcome of disease. For instance, Th1 cells, which predominantly secrete interferon-gamma (IFN-γ), tumor necrosis factor-alpha (TNF-α), and interleukin-12 (IL-12), lead to the activation of macrophages and natural killer cells, both involved in the control of intracellular pathogens, including bacteria and virus. Th1 cells are also involved in the pathogenesis of some autoimmune diseases (Jäger & Kuchroo, 2010; Jutel & Akdis, 2011; Wan, 2010). Regarding the role of Th2 cells, they are defined as source of IL-4, IL-5, IL-9, and IL-13 (Jutel & Akdis, 2011; Zhu & Paul, 2008) and play a very important role in eosinophilic inflammation and IgE-mediated hypersensitivity reactions, such as asthma, food allergies, rhinitis, and others allergic reactions. In addition, helminth antigens are strong inducers of Th2 response, which is associated with the host protection against worms. On the other hand, Th17 cells, characterized by the secretion of IL-17A, IL-17F, IL-22, IL-21, have a critical role in the host protection against extracellular bacterial and fungal pathogens. Th17 and Th1 cells mediated the deleterious effects in patients with multiple sclerosis, rheumatoid arthritis, psoriasis, and inflammatory bowel disease (IBD) (Aarvak, Chabaud, Källberg, Miossec, & Natvig, 1999; Bettelli, Korn, Oukka, & Kuchroo, 2008; Jäger & Kuchroo, 2010; Weaver et al., 2006). Finally, the regulatory T cells comprise a T cell subpopulation that can suppress or regulate the immune activation by influencing the activity of effector T cell clones or autoreactive T cells.

The regulation of the immune system is crucial to the maintenance of self-tolerance and immunocompetence. Regarding this, the immune homeostasis can be disturbed by any stimulus that affects hematopoiesis, lymphoid organ development, cell viability, lymphocyte activation, proliferation, and differentiation. Many mediators including growth factors, cytokines, chemokines, and neuropeptides released centrally or locally during the immune response can regulate the immune functions. In this context, the role of the nociceptin/orphanin FQ (N/OFQ)-N/OFQ peptide receptor (NOP) system in modulating immune responses, mainly in immune-mediated inflammatory diseases, was herein analyzed.

2. N/OFQ AND ITS RECEPTOR

It is well known that N/OFQ is an opioid-like peptide, since it shares high sequence homology with classical opioid peptides (Meunier et al., 1995; Reinscheid et al., 1995). Indeed, the NOP receptor (formerly named opioid receptor like-1) is also closely related to the opioid receptor family. Nevertheless, the N/OFQ-NOP receptor system is pharmacologically distinct from classical opioid systems since N/OFQ does not bind opioid receptors and opioid ligands, including naloxone, do not bind the NOP receptor (Cox et al., 2000). NOP, similar to the opioid receptor family, is a G-protein-coupled receptor (GPCR). The activation of NOP receptor signaling leads to the inhibition of adenylate cyclase, and Ca^{2+} channels, besides the stimulation of K^+ conductance (Lambert, 2008).

In vivo experimental studies have demonstrated that N/OFQ-NOP receptor system modulates a variety of biological functions, such as nociception, food intake, learning and memory processes, spontaneous locomotor activity, motor coordination, rewarding actions of opioids and ethanol, and responses related to stress, anxiety, and mood (for general reviews, see Calo', Guerrini, Rizzi, Salvadori, & Regoli, 2000; Lambert, 2008). Peripheral effects, such as hypotension, bradycardia, diuresis, inhibition of gastrointestinal and airway motility, and/or some reflexes such as coughing, and the micturition reflex have also been reported for N/OFQ (for a review, see Lambert, 2008). Most recently, the effects of N/OFQ-NOP receptor system in modulating immune functions have received particular attention. This interest is also based on the fact that classical opioids are expressed in immune cells, and clinical evidence suggests increased susceptibility to develop infections in opioid addicts (for a review, see Al-Hashimi, Scott, Thompson, & Lambert, 2013). Considering the similarities between N/OFQ and classical opioids, and the involvement of N/OFQ-NOP receptor system in sepsis and inflammatory and autoimmune diseases, this chapter will provide an overview of the experimental and clinical findings of the N/OFQ-NOP receptor system in the modulation of immune functions. Additionally, possible mechanisms by which N/OFQ can affect immune functions were discussed in detail. Finally, it highlighted the potential of NOP antagonists as innovative drugs for the treatment of inflammatory and autoimmune diseases.

3. N/OFQ AND NOP RECEPTOR EXPRESSION IN LEUKOCYTES

The N/OFQ peptide precursor (ppN/OFQ) and the NOP receptor are widely expressed in the nervous system as well as in peripheral organs and in immune cells. Regarding the immune system, in 1995, Halford and coworkers reported for the first time the expression of NOP mRNA in the murine helper and cytotoxic T lymphocytes (Halford, Gebhardt, & Carr, 1995). In addition, NOP transcription was observed in human monocytes and lymphocytes as well as in monocytic (U937) and T-lymphocytic lineages (Peluso et al., 1998; Wick, Minnerath, Roy, Ramakrishnan, & Loh, 1995). NOP mRNA was also detected in porcine thymus, lymph nodes, spleen, and in freshly isolated splenocytes (Pampusch, Osinski, Serie, Murtaugh, & Brown, 1998, Pampusch et al., 2000). Later, it was demonstrated that high-affinity binding sites for N/OFQ are distributed in the surface of human mononuclear and (Arjomand, Cole, & Evans, 2002) polymorphonuclear leukocytes (Fiset, Gilbert, Poubelle, & Pouliot, 2003; Serhan, Fierro, Chiang, & Pouliot, 2001) and endothelium (Brookes et al., 2013).

Considering that N/OFQ is produced in the brain, peripheral nervous system and also by cells from innate and adaptive immunity (Gavioli & Romão, 2011), the wide distribution of NOP receptor on immune system and its regulation by several immune stimuli may explain the intriguing influence of N/OFQ on the immune response.

4. EFFECTS OF NOP RECEPTOR ACTIVATION ON THE IMMUNE RESPONSE

Most microbial and environmental antigens, which could lead to immune system activation, enter the human body through the skin and respiratory and intestinal tracts, where they can be captured and engulfed by phagocytic cells to initiate the immune response. It is important to comment that N/OFQ and NOP receptor were expressed in lymphoid organs, as well as on sites of contact with antigens, such as epidermis and intestinal mucosa (Andoh, Yageta, Takeshima, & Kuraishi, 2004; Li, Dong, & Wang, 2013; Sobczak, Sałaga, Storr, & Fichna, 2013).

During immune response, cells from the immune system communicate with each other and with other nonimmune cells releasing and responding to chemical messengers called cytokines. These molecules are involved in virtually all immune activities, including leukocyte generation, differentiation, recruitment, and activation, as well as control of the effector mechanisms during innate and adaptive immune responses. Thus, as showed in Table 1, cytokines play a crucial role in N/OFQ immunomodulation, since an increasing amount of data underlines the bidirectional and dynamic interplay between immune functions and the N/OFQ-NOP receptor system.

Several studies showed that different immune stimuli can modulate ppN/OFQ and N/OFQ mRNA, NOP expression, and N/OFQ production. For example, Buzas et al. (2002) reported that inflammatory mediators, such as lipopolysaccharide (LPS), IL-1β, and TNF-α, increase ppN/OFQ mRNA and N/OFQ production in astrocytes. Miller and Fulford (2007) showed that those same cytokines enhanced the N/OFQ production in rat splenocytes. These data clearly demonstrated the influence of the immune system in the N/OFQ-NOP receptor pathway, particularly in conditions of proinflammatory response and cell activation as highlighted in the Table 1.

Concerning to the effect of NOP activation on cytokines regulation, Goldfarb and colleagues (2006) demonstrated that in mice the intraperitoneal injection of N/OFQ prior to Staphylococcal enterotoxin A challenge caused a significant increase in TNF-α and IFN-γ mRNA levels. Moreover, mice lacking the N/OFQ precursor gene showed diminished TNF-α and IFN-γ transcripts in the spleen in response to the same stimulus. Taking into account a two-way influence between the NOP receptor activation and the immune system, it was demonstrated that proinflammatory cytokines (IL-β and TNF-α), LPS, and mitogen concanavalin A enhanced the N/OFQ production by rat splenocytes *in vitro* (Miller & Fulford, 2007). In this same view, it was reported that formyl-methionyl-leucyl-phenylalanine (fMLP)-activated neutrophils rapidly secreted N/OFQ, suggesting its ability to store N/OFQ in preformed vesicles (Fiset et al., 2003).

On the other hand, N/OFQ promotes the chemotaxis of polymorphonuclear neutrophils (Serhan et al., 2001) and human monocytes (Trombella et al., 2005). In addition, N/OFQ stimulated the release of lysozyme by

Table 1 Available evidence for a proinflammatory effects of NOP activation

In vivo and *in vitro* studies	Effects of NOP activation or NOP blockage	References
C57BL/6J mice (normal and ppN/OFQ knockout)	• The administration of N/OFQ (55 nmol/kg, i.p.) 30 min prior to Staphylococcal enterotoxin A increased the expression of TNF-α and IFN-γ on the spleen • N/OFQ-deficient mice displayed attenuated TNF-α and IFN-γ mRNA levels triggered by antigen challenge	Goldfarb, Reinscheid, and Kusnecov (2006)
Rat astrocytes	• The expression of N/OFQ mRNA and protein was increased by proinflammatory mediators such as TNF-α, IL-1β, and LPS	Buzas, Rosenberger, Kim, and Cox (2002)
Rat splenocytes	• TNF-α and IL-1-β increased the N/OFQ secretion by splenocytes *in vitro*	Miller and Fulford (2007)
Anesthetized Wistar rats	• Administration of N/OFQ (0.6–60 nmol/kg i.v.) caused hypotension, vasodilatation, macromolecular leak, and leukocyte adhesion	Brookes et al. (2007)
Wistar rats and isolated mast cell	• Intradermal application of N/OFQ increased vascular permeability in rat skin by a mechanism dependent of histamine H1 receptor • *In vitro* N/OFQ stimulated the release of histamine by rat peritoneal mast cells	Kimura et al. (2000)
ICR mice and C57BL/6 NOP-deficient mice	• Intradermal inoculation of N/OFQ presented pruritogenic effect in normal but not in NOP-deficient mice. The leukotriene B_4 receptor antagonist inhibited the itch • N/OFQ stimulated the production of leukotriene B_4 by keratinocytes	Andoh et al. (2004)

Continued

Table 1 Available evidence for a proinflammatory effects of NOP activation—cont'd

***In vivo* and *in vitro* studies**	**Effects of NOP activation or NOP blockage**	**References**
Monocytes and neutrophils obtained from healthy subjects	• NOP activation stimulated the chemotaxis of human monocytes and increased the release of lysozyme by neutrophils	Trombella et al. (2005)
Neutrophils obtained from healthy volunteers	• N/OFQ exhibited a potent chemoattractant activity *in vitro*	Serhan et al. (2001)
BALB/c mice (air pouch model)	• N/OFQ at low doses (10 ng) induced significant leukocyte recruitment into the air pouch	Serhan et al. (2001)
Human neutrophils	• Neutrophils stimulated by fMLP quickly secreted N/OFQ upon exocytosis of granules	Fiset et al. (2003)
Septic rats (CLP model)	• Pharmacological blockade of NOP receptor with UFP-101 enhanced the bacterial control and decreased systemic inflammation and mortality of animals, while N/OFQ administration increased animal mortality	Carvalho et al. (2008)
Colitic mice (DSS model)	• The NOP receptor antagonist (SB612111—30 mg/kg) ameliorated the clinical signs of colitis and inhibited the production of CXCL1, IFN-γ, TNF-α, IL-1β, IL-6, and TNF-α	Alt et al. (2012)
Colitic mice (DSS model: wild-type and NOP-deficient C57BL/6 mice)	• NOP-deficient animals developed attenuated DSS-induced colitis and expressed decreased levels of mucosal addressin (MadCAM-1) and significant reduction in the number of inflammatory cells in colonic mucosa	Kato et al. (2005)

CXCL1, chemokine (C-X-C motif) ligand 1; fMLP, proinflammatory peptide *N*-formyl-L-methionyl-L-leucyl-L-phenylalanine; IFN-γ, interferon-gamma; IL-1β, interleukin 1 beta; LPSs, lipopolysaccharides; MadCAM-1, mucosal addressin cell adhesion molecule-1; N/OFQ, nociceptin/orphanin FQ peptide; NOP, N/OFQ receptor; UFP-101, University of Ferrara Peptide-101; TNF-α, tumor necrosis factor-alpha.

neutrophils in a concentration-dependent manner (Trombella et al., 2005). Corroborating the *in vitro* studies, reporting the chemotactic and proinflammatory role of N/OFQ, it was demonstrated this neuropeptide induces leukocyte migration in the air pouch model in mice (Serhan et al., 2001), elicits itch by a mechanism mediated by leukotriene B_4 (Andoh et al., 2004), increases vascular permeability in rats (Kimura et al., 2000), and causes hypotension, vasodilation of arterioles and venules, and macromolecular leak and leukocyte rolling in normal rats (Brookes et al., 2013, 2007). In rats, the vasodilatation and macromolecular leak induced by N/OFQ-NOP activation occur by a mechanism dependent of mast cell and histamine secretion (Brookes et al., 2007). In this context, Kimura and coworkers (2000) showed that N/OFQ stimulates the histamine secretion by rat mast cell *in vitro*. Furthermore, in acutely inflamed rat knee joints (induced by the intra-articular injection of kaolin 2% and carrageenan 2% combined with manual extension/flexion), the local application of N/OFQ caused a dose-dependent increase in synovial blood flow by a mechanism dependent of mast cell and leukocytes activation (Zhang & McDougall, 2006).

Despite the proinflammatory and immune stimulant activities of N/OFQ, other studies have reported negative/suppressive role for the immune system. Kaminsky and Rogers (2008) showed that N/OFQ suppressed the production of chemokine (C-C motif) ligand 2 (CCL2/MCP-1) and CCL5/RANTES by human $CD14^+$ monocytes. Interestingly, N/OFQ enhanced or decreased the proliferative response of T cells stimulated by Staphylococcal enterotoxin B (Waits, Purcell, Fulford, & McLeod, 2004). The authors showed that N/OFQ increased the expression of markers of T cell activation (CD28, CD25, and CD69) in SEB-activated human T lymphocytes. However, when day 4 SEB-activated T cells were restimulated with SEB in the presence of CHO cells ($CD80/CR4^+$ cells), N/OFQ caused a significant decrease in proliferation and induced the expression of CTLA-4, a negative regulator of T cell activation. These results indicate that N/OFQ may influence the T cell activation by a balance between CD28 and CTLA-4 (Waits et al., 2004). Thus, the role of NOFQ in inflammation, cell activation, or other immune functions needs to be better investigated.

Recently, using a mouse model of allergic asthma, it was demonstrated that the potent and selective NOP agonist UFP-112 (Rizzi et al., 2007) administrated during ovalbumin (OVA) sensitization significantly inhibited bronchoconstriction and bronchial reactivity to acetylcholine and eosinophil migration to lung in response to OVA challenge (Sullo et al., 2013).

Additionally, the *in vitro* proliferative response of lymphocytes to OVA was significantly reduced when animals were pretreated with UFP-112 during the sensitization with OVA. Moreover, while the production of Th2 cytokines (IL-4 and IL-5) was significantly reduced in both pulmonary tissues and in the supernatant of OVA-restimulated lymphocytes, the IFN-γ secretion was enhanced by the treatment with NOP agonist (Sullo et al., 2013). Based on these data, we hypothesized that in a classic model of Th2 response, NOP activation seems to modulate negatively the establishment or maintenance of the Th2 response, since Th2 cytokines are involved in both eosinophilic inflammation and the bronchial hyperresponsiveness.

Taken together, a growing body of evidence suggests a complex role for N/OFQ-NOP receptor system in modulating immune functions, acting either as a stimulator or an inhibitor depending on the nature of the stimulus, N/OFQ concentrations, and timing of NOP activation, besides the sites of immune activation.

5. NOP RECEPTOR ACTIVATION AND INFLAMMATORY AND AUTOIMMUNE DISEASES

Considering the bidirectional relationship between NOP signaling and the immune system, it is plausible that NOP receptor activation might modulate the severity and/or the outcome of systemic inflammatory diseases such as sepsis and other immune-mediated diseases as summarized in Table 2.

There are strong correlation between the plasmatic NOFQ levels and severity of sepsis (Stamer et al., 2011; Thompson et al., 2013; Williams et al., 2008), Parkinson's disease (PD) (Mabrouk, Marti, & Morari, 2010; Marti et al., 2005, 2010; Volta, Mabrouk, Bido, Marti, & Morari, 2010), arthritis (Fiset et al., 2003), and IBD (Alt et al., 2012; Kato et al., 2005; Petrella et al., 2013) (Table 2).

Sepsis often is a consequence of uncontrolled bacterial infection, although it may occur in response to disseminated viruses, fungi, and protozoa infections. Bacterial sepsis is one of the most common causes of morbidity and mortality in intensive care units. Its hallmark is the neutrophil paralysis, which is directly related with bacterial dissemination, systemic inflammation, and multiple organ failure (Alves-Filho, de Freitas, Russo, & Cunha, 2006; Hotchkiss & Karl, 2003; Tavares-Murta et al., 2002). The systemic inflammatory response during sepsis is mediated by proinflammatory cytokines such as TNF-α, IL-1β, chemokines, nitric oxide

Table 2 Highlights of the interplay among N/OFQ-NOP receptor pathway and inflammatory and autoimmune diseases

Sepsis
• Sepsis is characterized by systemic inflammation, neutrophil paralysis, cardiovascular dysfunction, and multiple organ failure • The mortality of septic rats was significant reduced by the treatment with a single dose of NOP antagonist UFP-101 • N/OFQ was higher in critically ill patients and who died of sepsis • Patients critically ill with severe sepsis presented high systemic levels of N/OFQ and IL-8 compared to levels detected in volunteers and after recovery • High expression of NOP receptor mRNA was detected in peripheral blood cells of nonsurvivor septic patients compared to survivors
Arthritis
• High levels of N/OFQ were found in the synovial fluid of patients with rheumatoid arthritis and osteoarthritis
Inflammatory bowel disease
• Inflammatory bowel disease (IBD) is an inflammatory and immunological disturbance affecting the gastrointestinal tract • In a dextran sulfate sodium model of bowel disease, the severity of disease is associated with the upregulation of N/OFQ expression. NOP knockout mice developed a less intense colitis; NOP antagonist ameliorated the signs of colitis • In 2.4.6-trinitrobenzenesulfonic acid (TNBS)-induced colitis rats, N/OFQ administration had protective or prejudicial effects depending on the dose. Low doses significantly decreased the colonic inflammation, while high doses aggravated signs of colitis
Parkinson's disease
• Parkinson's disease (PD) is a progressive neurodegenerative disease characterized by microglial activation, astrogliosis, production of proinflammatory cytokines, nitric oxide, and infiltration of $CD4^+$ T cells • The N/OFQ production was detected in the lesioned *substantia nigra* of 6-hydroxydopamine-hemilesioned rats. The blockade of NOP receptor signaling attenuates parkinsonian-like behavior in 6-hydroxydopamine-hemilesioned, haloperidol, and reserpine-treated rodents • Elevated levels of N/OFQ were found in the cerebrospinal fluid of parkinsonian patients

(NO), and leukotrienes (Alves-Filho, Benjamim, Tavares-Murta, & Cunha, 2005; de Jong, van der Poll, & Wiersinga, 2010). The failure of neutrophil migration to infectious focus seems to be mediated at least in part by down-modulation of IL-8 receptors, chemokine (C-X-C motif) receptor 1 (CXCR1), and chemokine (C-X-C motif) receptor 2 (CXCR2) in a NO-dependent manner (Paula-Neto et al., 2011).

The cardiovascular dysfunction in septic patients is characterized by severe hypotension, which may be mediated at least in part by NO, a potent vasodilator (Bateman, Sharpe, & Ellis, 2003). It was demonstrated that N/OFQ induces hypotension and bradycardia (Kapusta, 2000; Malinowska, Godlewski, & Schlicker, 2002), decreasing heart rate and mean arterial pressure in mice (Burmeister, Ansonoff, Pintar, & Kapusta, 2008).

Using the cecal ligation and puncture (CLP) model of sepsis in rats, our group has demonstrated that the mortality rate of septic animals was significantly reduced by the treatment with a single dose of the NOP antagonist UFP-101 (Carvalho et al., 2008). We verified that the protective effect of UFP-101 was correlated with the inhibition of cell migration to lung, decreased bacterial dissemination, and inhibition of TNF-α, IL-1β, and CCL2/MCP-1 production. As the beneficial effects of UFP-101 were obtained with a single dose of NOP antagonist immediately after the induction of CLP, we believed that the N/OFQ acts in the early phase of sepsis.

Williams and colleagues (2008) first demonstrated that patients critically ill with severe sepsis presented high systemic levels of N/OFQ. In agreement, Stamer and colleagues (2011) found a higher expression of mRNA for NOP receptor in peripheral blood cells of nonsurvivor septic patients compared to survivors and healthy controls. Recently, reinforcing the participation of N/OFQ on systemic inflammatory response, in a study conducted with 82 critically ill septic patients (based on the presence of infection and systemic inflammatory response), Thompson and colleagues (2013) demonstrated that plasmatic levels of N/OFQ and IL-8 were higher in septic patients at days 1 and 2 of the admission at the intensive care unit, compared to levels detected in volunteers and after recovery. However, in contrast to data found by Stamer and colleagues, NOP and ppN/OFQ mRNA levels in neutrophils from septic patients were lower compared to healthy controls. Moreover, there were no significant differences in plasma N/OFQ or NOP expression in neutrophils between 30-day survivors and nonsurvivors. Thereby, the data suggest that N/OFQ levels are altered during sepsis; however, the role of N/OFQ in sepsis needs to be better investigated. A schematic view of the role played by N/OFQ in sepsis is summarized in Fig. 1.

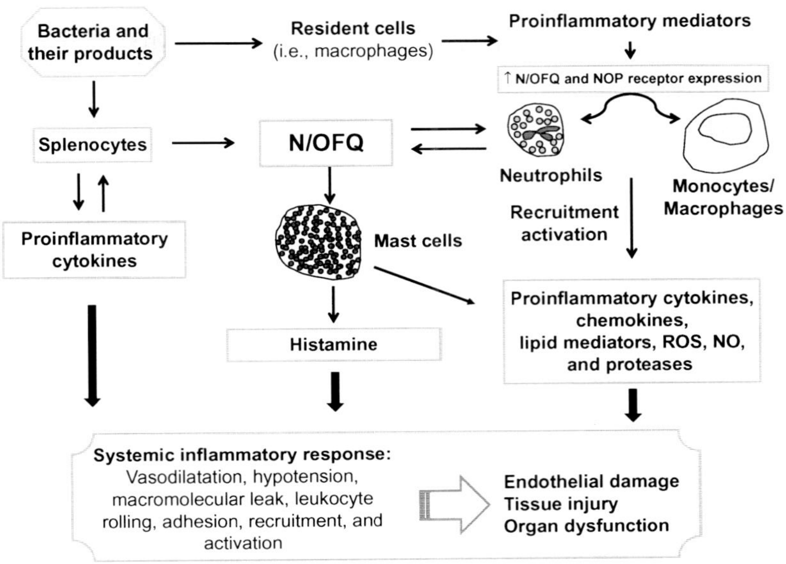

Figure 1 Relationship between N/OFQ-NOP receptor system, systemic inflammation, and sepsis. In response to microbial invasion, macrophages release inflammatory mediators, which induce NOP receptor expression and N/OFQ release from mononuclear and polymorphonuclear cells. Splenocytes are also stimulated by bacterial products and then releasing N/OFQ, besides proinflammatory cytokines. N/OFQ may activate neutrophils, monocytes, macrophages, and mast cells, then contributing to the release of histamine, proinflammatory cytokines, chemokines, lipid mediators, reactive oxygen species (ROS), nitric oxide (NO), and proteases. All these mediators acting together evoke a systemic inflammation response, thus contributing to endothelial damage and organ dysfunction in sepsis. (See the color plate.)

Regarding the involvement of NOFQ-NOP in immunopathogenesis of other immune-mediated diseases, Fiset and colleagues (2003) found high levels of N/OFQ in the synovial fluid of patients with arthritis. Considering the inflammatory and immunomodulatory effects of N/OFQ and the knowledge about the immunopathogenesis of arthritis, further studies on basic and clinical research are imperative to elucidate the involvement of N/OFQ-NOP receptor in this autoimmune disease.

IBD is an inflammatory condition of gastrointestinal tract that is chronic, remitting and relapsing, and also progressive in its course. IBD includes two major clinical entities: Crohn's disease (CD) and ulcerative colitis (UC) that may affect the entire gastrointestinal tract and the colonic mucosa, respectively. Although the etiology of IBD is unknown, its pathogenesis is affected by genetic susceptibility, intestinal flora, and the immune system. CD is characterized by a dense infiltration of lymphocytes and macrophages, presence of granulomas in up to 60% of patients, fissuring ulceration, and submucosal fibrosis (for review, see Bouma & Strober, 2003). On the other hand, UC is marked by the presence of lymphocytes infiltration and granulocytes. Using a mouse model of bowel disease induced by the injection of dextran sulfate sodium (DSS), Kato and colleagues (2005) demonstrated that the inflammatory response and the severity of disease were associated with the upregulation of N/OFQ expression.

In addition, NOP knockout mice developed a less intense colitis, indicating the participation of N/OFQergic signaling in the pathogenesis (Kato et al., 2005). Moreover, it was demonstrated that the NOP antagonist SB-612111 (Zaratin et al., 2004) significantly ameliorated the signs of colitis induced by DSS, which was correlated with decreased production of chemokine (C-X-C motif) ligand 1 (CXCL1/KC/GRO-alpha), IFN-γ, IL-1β, IL-6, and TNF-α in the colon tissue (Alt et al., 2012). Interestingly, in 2.4.6-trinitrobenzenesulfonic acid-induced colitis rats, that resemble CD, which is mediated by the activation of both Th1 and Th17 response and proinflammatory cytokines, N/OFQ administration had protective or prejudicial effects depending on the dose used. At low doses (maximal 0.2 nmol/kg), it induced a significant decrease in the colonic inflammatory profile (colon damage score, myeloperoxidase activity, and IL-1β levels), while at high dose (20 nmol/kg) it showed an aggravating effect (Petrella et al., 2013). To date, no studies correlate the N/OFQ plasmatic or colonic levels and IBD in humans. However, the animal studies clearly have demonstrated the participation of N/OFQ on the modulation of the inflammatory process during colitis.

PD is a progressive neurodegenerative disease characterized by motor disturbance such as akinesia, bradykinesia, and tremor, often accompanied by cognitive impairment and depression. In the last years, accumulating evidence shows the involvement of autoimmune response in PD pathogenesis (Benkler, Agmon-Levin, & Shoenfeld, 2009). The immunological markers of PD include microglial activation, astrogliosis, and production of TNF-α, IL-1β, IL-2, IL-6, CCL5/RANTES, NO, and infiltration of $CD4^+$ T cells (Blum-Degen et al., 1995; Dobbs et al., 1999; Hirsch & Hunot, 2009; Hisanaga, Asagi, Itoyama, & Iwasaki, 2001; Lee, Tran, & Tansey, 2009; McGeer & McGeer, 2004; Qian, Flood, & Hong, 2010; Rentzos et al., 2007; Tansey, McCoy, & Frank-Cannon, 2007). In these patients, the inflammation, oxidative stress, and microglia-mediated neurotoxicity of dopaminergic neurons in the *substantia nigra* are considered the hallmark of disease (Lee et al., 2009; McGeer & McGeer, 2004; Ransohoff & Perry, 2009; Tansey et al., 2007).

N/OFQ and its receptor are widely expressed in cortical and subcortical motor areas (Norton, Neal, Kumar, Akil, & Watson, 2002), particularly in the *substantia nigra*, in which occurs progressive degeneration of dopaminergic neurons. Several preclinical studies have suggested the participation of N/OFQ in the physiopathology of PD (Marti et al., 2005, 2004; Mabrouk et al., 2010; Viaro et al., 2008; Visanji et al., 2008). The enhanced N/OFQ production in 6-hydroxydopamine-hemilesioned rats was detected in the lesioned *substantia nigra*, indicating that parkinsonism may be associated with overactivation of NOP receptor pathway (Marti et al., 2005). In addition, the pharmacological blockade of NOP receptor signaling attenuates parkinsonian-like behavior in 6-hydroxydopamine-hemilesioned, haloperidol, and reserpine-treated rodents, whereas deletion of the NOP receptor gene conferred mice protection from these symptoms (Mabrouk et al., 2010; Marti et al., 2005; Volta et al., 2010). Corroborating the view that N/OFQ-NOP receptor system plays a pathogenic role in PD, a clinical study found elevated levels of N/OFQ in the cerebrospinal fluid of parkinsonian patients (Marti et al., 2010). In addition, N/OFQ and its receptor are expressed in the brain area containing dopaminergic neurons, which is degenerated in Parkinson disease (Marti, Guerrini, Beani, Bianchi, & Morari, 2002). In addition, beneficial effects of NOP receptor antagonist were reported in nonhuman primate models of Parkinson's disease (Viaro et al., 2008). Together, these findings suggest a pathogenic role for NOP activation during disease and that NOP receptor antagonist could represent innovative drugs for patients with Parkinson.

6. MOLECULAR MECHANISMS UNDERLYING N/OFQ ACTIONS ON IMMUNE FUNCTIONS

The N/OFQ system modulates many functions in a variety of immune cells including monocytes, macrophages, neutrophils, mast cells, and T lymphocytes by a mechanism not clearly elucidated yet. Table 1 summarized some of these functions as well as the mechanisms underlying some immunoregulatory activities triggered by NOP receptor activation.

NOP receptor together with the classical opioid receptors (MOP, KOP, and DOP) belongs to the GPCR family, which plays a vital role in the transduction of signals regulating several effectors. Regarding N/OFQ, some authors have described that the NOP activation by its natural ligand or synthetic agonists induces the activation of K^+ conductance, inhibition of voltage-gated Ca^{2+} channels, and decrease cAMP formation in a variety of cells including immune cells (Connor, Yeo, & Henderson, 1996; Lambert, 2008; Matthes, Seward, Kieffer, & North, 1996; Meunier, 1997;

Meunier et al., 1995; Reinscheid et al., 1995). N/OFQ also induces stimulation of phospholipase C (via α-subunit of G_q), which leads to 1,4,5-triphosphate (IP_3) and diacylglycerol production, and also to Ca^{2+}-dependent protein kinase C (PKC) activation (for a review, see Chan et al., 1998; Hawes, Graziano, & Lambert, 2000; New & Wong, 2002). It was also demonstrated that in CHO cells, the type of G-protein involved in PKC activation by N/OFQ was a $G_{i/o}$ protein (Lou, Ma, & Pei, 1997). Moreover, N/OFQ modulates extracellular signal-regulated kinase (ERK), p38, and c-Jun N-terminal kinase (JNK) isoforms of mitogen-activated protein kinase (Armstead, 2006; Fukuda, Shoda, Morikawa, Kato, & Mori, 1997; Hawes, Fried, Yao, Weig, & Graziano, 1998; New & Wong, 2002), as well as the transcription of a variety of genes involved in immune and inflammatory responses (Harrison & Grandy, 2000; Hawes et al., 2000). It had been suggested that the signal transducer and activator of transcription (STAT3) may be involved in the transduction of NOP signaling (New & Wong, 2002; Wu, Lo, & Wong, 2003).

Considering the glia–immune cell communication, it was demonstrated that LPS, IL-1β, and TNF-α increase the levels of N/OFQ mRNA and immunoreactivity in rat astrocytes in culture by a mechanism dependent of the activation of ERK 1/2, p38 MAP kinases, and the transcription factor CREB. It was demonstrated that NFκB pathway appears to be involved in the induction of N/OFQ transcription by LPS (Buzas et al., 2002). N/OFQ has been shown to cause IκB kinase phosphorylation and IκB degradation in SH-SY5Y human neuroblastoma cells (Liu & Wong, 2005). Recently, Donica and colleagues (2011) showed that N/OFQ increases the nuclear translocation, binding to DNA, and activation of transcription. Hence, the activation of NFκB by N/OFQ may be critical for many immune functions.

7. RELATIONSHIP BETWEEN N/OFQ, STRESS, AND HPA AXIS

A body of evidence strongly suggests the N/OFQ-NOP receptor system in regulation of the stress response. In particular, some studies have shown a possible contribution of the N/OFQ system in feed-forward regulation of the hypothalamic–pituitary–adrenal axis (HPA) specifically, which could affect immune functions. It should be mentioned that in the central nervous system, the NOP receptor is expressed in the forebrain, including cortical areas, olfactory regions, thalamus, and a variety of limbic structures, such as the hippocampus, the amygdaloid complex, and in several nuclei of the hypothalamus, that are involved in the processing of endocrine and emotional stimuli (for a review, see Mollereau & Mouledous, 2000).

Regarding the expression of N/OFQ, Neal et al. (1999) using *in situ* hybridization and immunohistochemistry showed the distribution of N/OFQ peptide and mRNA in the central nervous system of the adult rat. N/OFQ immunoreactivity and preproN/OFQ mRNA expression correlated virtually in all brain areas studied. N/OFQ was found expressed in several limbic structures, such as lateral septum, hypothalamus, mammillary bodies, central and medial nuclei of the amygdala, hippocampal formation, reticular nuclei of the thalamus, medial habenula, and zona incerta. Additionally, in the brainstem, N/OFQ was prominent in the ventral tegmental area, *substantia nigra*, *locus coeruleus*, and raphe complex, besides many other brainstem nuclei. The wide distribution of this peptide provides support for its role in a multitude of biological functions (Neal et al., 1999).

Stress events can affect the functionality of nervous, endocrine, and immune systems. The major branch of neuroendocrine system and the main endocrine component of the stress response is the HPA. Exposure to stress activates the HPA axis that culminates in the release of glucocorticoids. Stressor exposure activates neurosecretory cells of the paraventricular nucleus (PVN) of the hypothalamus, which secrete corticotropin-releasing hormone (CRH) into the hypophyseal portal circulation. Once in the anterior pituitary, CRH stimulates CRH1 receptor and induces the release of adrenocorticotropic hormone (ACTH), which is released into the system circulation and stimulates the secretion of glucocorticoids from the adrenal glands. A variety of stimuli can serve as stressors to the activation of the HPA axis, including psychogenic, physical, and immunologic stressors. In most cases, such stressors activate encephalic areas mediated by limbic structures, such as amygdala, bed nucleus of the stria terminalis, and medial prefrontal cortex, all of which either directly or indirectly form connections with the PVN of hypothalamus (Ulrich-Lai & Herman, 2009). It should be mentioned that the activation of the HPA axis has the potential to depress the immune response by the resulting increased plasma concentrations of glucocorticoids (Franchimont, 2004). However, a crucial point to consider is whether plasma cortisol concentrations are a relevant marker of immune suppression, since it is not clear whether plasma cortisol concentrations are increased in patients with depressed immune function (Al-Hashimi et al., 2013).

In rodents, the acute administration of N/OFQ generally increases ACTH and glucocorticoids; this may have immunosuppressive actions (Devine, Watson, & Akil, 2001; Fernandez, Misilmeri, Felger, & Devine, 2004; Green, Barbieri, Brown, Chen, & Devine, 2007; Leggett, Harbuz, Jessop, & Fulford, 2006; Nicholson, Akil, & Watson, 2002; Vitale, Arletti, Ruggieri, Cifani, & Massi, 2006). However, opposite effects have also been reported (Le Cudennec, Naudin, Do Rego, & Costentin, 2002). It is difficult to separate stress-induced increases in corticosterone (because of animal handling/injection) from drug-dependent changes. In contrast to pharmacological findings, mice lacking the preproN/OFQ gene exhibited elevated basal and poststress levels of plasma corticosterone compared to wild-type mice (Köster et al., 1999), while the genetic blockade of NOP receptor did not affect corticosterone plasma levels in mice (Uezu et al., 2004).

Few studies were developed to investigate the role of endogenous N/OFQ-NOP receptor system in mediating biological actions under psychological and immune challenges. Under unpredictable stressful conditions, the chronic blockade of NOP receptor, with the central administration of the antagonist UFP-101, reduced corticosterone to normal levels (Vitale et al., 2009). However, during acute restraint stress, i.c.v. administration of UFP-101 enhanced and prolonged ACTH and corticosterone levels (Leggett, Jessop, & Fulford, 2007). More recently, Delaney et al. (2012) showed stimulant effects of the NOP antagonist JTC-801 (intravenous injected) on the HPA axis during basal, but not under stressful conditions. These effects of NOP antagonists were further explained based on the possible partial agonistic activity of UFP-101 and low selectivity of action of JTC-801 (for more information, see Mallimo & Kusnecov, 2013).

Still regarding the central effects of N/OFQ on immune functions, a study showed that the i.c.v. administration of N/OFQ in laparotomized rats significantly reduced IL-1β and TNF-α produced by peritoneal macrophages (Zhao, Wu, & Cao, 2002). Considering the administration of N/OFQ was done centrally, these effects could be explained in part due to the activation effects of N/OFQ on HPA axis.

Just one study is available about the effects of LPS-induced immunological stress on N/OFQ-NOP receptor system. Leggett, Dawe, Jessop, and Fulford (2009) found that LPS significantly increased preproN/OFQ transcript expression in the hypothalamus 4 h after injection compared to saline. Regarding hormone plasma levels in LPS-treated rats, i.c.v. N/OFQ had no significant effect on LPS-induced plasma corticosterone release at 30 or

60 min postinjection, while i.c.v. UFP-101/LPS significantly attenuated plasma ACTH hormone and corticosterone at the 30-min time point compared to i.c.v. saline/LPS. However, in the same study, authors showed that the i.c.v. administration of UFP-101 in LPS-treated rats increased POMC mRNA expression 4 h after injection, and a clear trend toward increased parvocellular CRH mRNA (Leggett et al., 2009).

Collectively, the presented findings suggest that when injected into the central nervous system, N/OFQ would stimulate HPA responses, thus increasing plasma corticosterone levels and consequently suppressing immune functions. Little literature information supports a role for central injected NOP antagonists in reverting stress-induced corticosterone levels. In conclusion, while it would seem that the HPA is an indirect site for immune modulation by N/OFQ, the direct evidence for this hypothesis is still lacking.

8. CONCLUSIONS

Preclinical and clinical data support the involvement of NOFQ-NOP receptor pathway in sepsis, IBD, Parkinson's disease, and arthritis. However, the effects of proinflammatory mediators on the NOP receptor expression and NOFQ secretion as well as the NOP influence in immune responses need to be better investigated. Finally, it should be outlined that the proinflammatory profile of action of N/OFQ candidates NOP antagonists as innovative drugs for the treatment of inflammatory and immune-mediated diseases.

ACKNOWLEDGMENTS

The authors would like to thank National Council for Scientific and Technological Development (CNPq), FAPERGS and CAPES for the financial support. P. R. T. R., M. C. M., and E. C. G. are recipients from fellowship of CNPq-Brazil.

REFERENCES

Aarvak, T., Chabaud, M., Källberg, E., Miossec, P., & Natvig, J. B. (1999). Change in the Th1/Th2 phenotype of memory T-cell clones from rheumatoid arthritis synovium. *Scandinavian Journal of Immunology*, *50*, 1–9.

Al-Hashimi, M., Scott, S. W., Thompson, J. P., & Lambert, D. G. (2013). Opioids and immune modulation: More questions than answers. *British Journal of Anaesthesia*, *111*, 80–88.

Alt, C., Lam, J. S., Harrison, M. T., Kershaw, K. M., Samuelsson, S., Toll, L., et al. (2012). Nociceptin/orphanin FQ inhibition with SB612111 ameliorates dextran sodium sulfate-induced colitis. *European Journal of Pharmacology*, *683*, 285–293.

Alves-Filho, J. C., Benjamim, C., Tavares-Murta, B. M., & Cunha, F. Q. (2005). Failure of neutrophil migration toward infectious focus in severe sepsis: A critical event for the outcome of this syndrome. *Memórias do Instituto Oswaldo Cruz, 100*, 223–226.

Alves-Filho, J. C., de Freitas, A., Russo, M., & Cunha, F. Q. (2006). Toll-like receptor 4 signaling leads to neutrophil migration impairment in polymicrobial sepsis. *Critical Care Medicine, 34*, 461–470.

Andoh, T., Yageta, Y., Takeshima, H., & Kuraishi, Y. (2004). Intradermal nociceptin elicits itch-associated responses through leukotriene B(4) in mice. *The Journal of Investigative Dermatology, 123*, 196–201.

Arjomand, J., Cole, S., & Evans, C. J. (2002). Novel orphanin FQ/nociceptin transcripts are expressed in human immune cells. *Journal of Neuroimmunology, 130*, 100–108.

Armstead, W. M. (2006). Differential activation of ERK, p38, and JNK MAPK by nociceptin/orphanin FQ in the potentiation of prostaglandin cerebrovasoconstriction after brain injury. *European Journal of Pharmacology, 529*, 129–135.

Bachmann, M. F., & Zinkernagel, R. M. (1997). Neutralizing antiviral B cell responses. *Annual Review of Immunology, 15*, 235–270.

Banchereau, J., & Steinman, R. M. (1998). Dendritic cells and the control of immunity. *Nature, 19*, 245–252.

Bateman, R. M., Sharpe, M. D., & Ellis, C. G. (2003). Bench-to-bedside review: Microvascular dysfunction in sepsis—Hemodynamics, oxygen transport, and nitric oxide. *Critical Care, 7*, 359–373.

Benkler, M., Agmon-Levin, N., & Shoenfeld, Y. (2009). Parkinson's disease, autoimmunity, and olfaction. *International Journal of Neuroscience, 119*, 2133–2143.

Bettelli, E., Korn, T., Oukka, M., & Kuchroo, V. K. (2008). Induction and effector functions of T(H)17 cells. *Nature, 453*, 1051–1057.

Beutler, B. (2004). Innate immunity: An overview. *Molecular Immunology, 40*, 845–859.

Blum-Degen, D., Müller, T., Kuhn, W., Gerlach, M., Przuntek, H., & Riederer, P. (1995). Interleukin-1 beta and interleukin-6 are elevated in the cerebrospinal fluid of Alzheimer's and de novo Parkinson's disease patients. *Neuroscience Letters, 202*, 17–20.

Bouma, G., & Strober, W. (2003). The immunological and genetic basis of inflammatory bowel disease. *Nature Reviews. Immunology, 3*, 521–533.

Brookes, Z. L., Stedman, E. N., Brown, N. J., Hebbes, C. P., Guerrini, R., Calo, G., et al. (2013). The nociceptin/orphanin FQ receptor antagonist UFP-101 reduces microvascular inflammation to lipopolysaccharide *in vivo*. *PLoS One, 8*, e74943.

Brookes, Z. L., Stedman, E. N., Guerrini, R., Lawton, B. K., Calo, G., & Lambert, D. G. (2007). Proinflammatory and vasodilator effects of nociceptin/orphanin FQ in the rat mesenteric microcirculation are mediated by histamine. *American Journal of Physiology. Heart and Circulatory Physiology, 293*, H2977–H2985.

Burmeister, M. A., Ansonoff, M. A., Pintar, J. E., & Kapusta, D. R. (2008). Nociceptin/orphanin FQ (N/OFQ)-evoked bradycardia, hypotension, and diuresis are absent in N/OFQ peptide (NOP) receptor knockout mice. *The Journal of Pharmacology and Experimental Therapeutics, 326*, 897–904.

Buzas, B., Rosenberger, J., Kim, K. W., & Cox, B. M. (2002). Inflammatory mediators increase the expression of nociceptin/orphanin FQ in rat astrocytes in culture. *Glia, 39*, 237–246.

Calo', G., Guerrini, R., Rizzi, A., Salvadori, S., & Regoli, D. (2000). Pharmacology of nociceptin and its receptor: A novel therapeutic target. *British Journal of Pharmacology, 129*, 1261–1283.

Carvalho, D., Petronilho, F., Vuolo, F., Machado, R. A., Constantino, L., Guerrini, R., et al. (2008). The nociceptin/orphanin FQ-NOP receptor antagonist effects on an animal model of sepsis. *Intensive Care Medicine, 34*, 2284–2290.

Chan, J. S., Yung, L. Y., Lee, J. W., Wu, Y. L., Pei, G., & Wong, Y. H. (1998). Pertussis toxin-insensitive signaling of the ORL1 receptor: Coupling to Gz and G16 proteins. *Journal of Neurochemistry*, *71*, 2203–2210.

Connor, M., Yeo, A., & Henderson, G. (1996). The effect of nociceptin on Ca2+ channel current and intracellular Ca2+ in the SH-SY5Y human neuroblastoma cell line. *British Journal of Pharmacology*, *118*, 205–207.

Cox, B. M., Chavkin, C., Christie, M. J., Civelli, O., Evans, C., Hamon, M. D., et al. (2000). Opioid receptors. In D. Girdlestone (Ed.), *The IUPHAR compendium of receptor characterization and classification* (pp. 321–333). London: IUPHAR Media.

de Jong, H. K., van der Poll, T., & Wiersinga, W. J. (2010). The systemic pro-inflammatory response in sepsis. *Journal of Innate Immunity*, *2*, 422–430.

Delaney, G., Dawe, K. L., Hogan, R., Hunjan, T., Roper, J., Hazell, G., et al. (2012). Role of nociceptin/orphanin FQ and NOP receptors in the response to acute and repeated restraint stress in rats. *Journal of Neuroendocrinology*, *24*, 1527–1541.

Devine, D. P., Watson, S. J., & Akil, H. (2001). Nociceptin/orphanin FQ regulates neuroendocrine function of the limbic-hypothalamic-pituitary-adrenal axis. *Neuroscience*, *102*, 541–553.

Dobbs, R. J., Charlett, A., Purkiss, A. G., Dobbs, S. M., Weller, C., & Peterson, D. W. (1999). Association of circulating TNF-alpha and IL-6 with ageing and parkinsonism. *Acta Neurologica Scandinavica*, *100*, 34–41.

Donica, C. L., Ramirez, V. I., Awwad, H. O., Zaveri, N. T., Toll, L., & Standifer, K. M. (2011). Orphanin FQ/nociceptin activates nuclear factor kappa B. *Journal of Neuroimmune Pharmacology*, *6*, 617–625.

Fernandez, F., Misilmeri, M. A., Felger, J. C., & Devine, D. P. (2004). Nociceptin/orphanin FQ increases anxiety-related behavior and circulating levels of corticosterone during neophobic tests of anxiety. *Neuropsychopharmacology*, *29*, 59–71.

Fiset, M. E., Gilbert, C., Poubelle, P. E., & Pouliot, M. (2003). Human neutrophils as a source of nociceptin: A novel link between pain and inflammation. *Biochemistry*, *42*, 10498–10505.

Franchimont, D. (2004). Overview of the actions of glucocorticoids on the immune response: A good model to characterize new pathways of immunosuppression for new treatment strategies. *Annals of the New York Academy of Sciences*, *1024*, 124–137.

Fukuda, K., Shoda, T., Morikawa, H., Kato, S., & Mori, K. (1997). Activation of mitogen-activated protein kinase by the nociceptin receptor expressed in Chinese hamster ovary cells. *FEBS Letters*, *412*, 290–294.

Gavioli, E. C., & Romão, P. R. (2011). NOP receptor ligands as potential agents for inflammatory and autoimmune diseases. *Journal of Amino Acids*, *2011*, 836569.

Goldfarb, Y., Reinscheid, R. K., & Kusnecov, A. W. (2006). Orphanin FQ/nociceptin interactions with the immune system in vivo: Gene expression changes in lymphoid organs and regulation of the cytokine response to staphylococcal enterotoxin A. *Journal of Neuroimmunology*, *176*, 76–85.

Green, M. K., Barbieri, E. V., Brown, B. D., Chen, K. W., & Devine, D. P. (2007). Roles of the bed nucleus of stria terminalis and of the amygdala in N/OFQ-mediated anxiety and HPA axis activation. *Neuropeptides*, *41*, 399–410.

Guidotti, L. G., & Chisari, F. V. (2001). Noncytolytic control of viral infections by the innate and adaptive immune response. *Annual Review of Immunology*, *19*, 65–91.

Halford, W. P., Gebhardt, B. M., & Carr, D. J. (1995). Functional role and sequence analysis of a lymphocyte orphan opioid receptor. *Journal of Neuroimmunology*, *59*, 91–101.

Harrison, L. M., & Grandy, D. K. (2000). Opiate modulating properties of nociceptin/orphanin FQ. *Peptides*, *21*, 151–172.

Hawes, B. E., Fried, S., Yao, X., Weig, B., & Graziano, M. P. (1998). Nociceptin (ORL-1) and mu-opioid receptors mediate mitogen-activated protein kinase activation in CHO cells through a Gi-coupled signaling pathway: Evidence for distinct mechanisms of agonist-mediated desensitization. *Journal of Neurochemistry, 71*, 1024–1033.

Hawes, B. E., Graziano, M. P., & Lambert, D. G. (2000). Cellular actions of nociceptin: Transduction mechanisms. *Peptides, 21*, 961–967.

Hirsch, E. C., & Hunot, S. (2009). Neuroinflammation in Parkinson's disease: A target for neuroprotection? *Lancet Neurology, 8*, 382–397.

Hisanaga, K., Asagi, M., Itoyama, Y., & Iwasaki, Y. (2001). Increase in peripheral CD4 bright + CD8 dull + T cells in Parkinson disease. *Archives of Neurology, 58*, 1580–1583.

Hoebe, K., Janssen, E., & Beutler, B. (2004). The interface between innate and adaptive immunity. *Nature Immunology, 5*, 971–974.

Hotchkiss, R. S., & Karl, I. E. (2003). The pathophysiology and treatment of sepsis. *New England Journal of Medicine, 348*, 138–150.

Jäger, A., & Kuchroo, V. K. (2010). Effector and regulatory T-cell subsets in autoimmunity and tissue inflammation. *Scandinavian Journal of Immunology, 72*, 173–184.

Jego, G., Palucka, A. K., Blanck, J. P., Chalouni, C., Pascual, V., & Banchereau, J. (2003). Plasmacytoid dendritic cells induce plasma cell differentiation through type I interferon and interleukin 6. *Immunity, 19*, 225–234.

Joffre, O., Nolte, M. A., Spörri, R., & Reis e Sousa, C. (2009). Inflammatory signals in dendritic cell activation and the induction of adaptive immunity. *Immunological Reviews, 227*, 234–247.

Jutel, M., & Akdis, C. A. (2011). T-cell subset regulation in atopy. *Current Allergy and Asthma Reports, 11*, 139–145.

Kaminsky, D. E., & Rogers, T. J. (2008). Suppression of CCL2/MCP-1 and CCL5/RANTES expression by nociceptin in human monocytes. *Journal of Neuroimmune Pharmacology, 3*, 75–82.

Kantari, C., Pederzoli-Ribeil, M., & Witko-Sarsat, V. (2008). The role of neutrophils and monocytes in innate immunity. *Contributions to Microbiology, 15*, 118–146.

Kapsenberg, M. L. (2003). Dendritic-cell control of pathogen-driven T-cell polarization. *Nature Reviews. Immunology, 3*, 984–993.

Kapusta, D. R. (2000). Neurohumoral effects of orphanin FQ/nociceptin: Relevance to cardiovascular and renal function. *Peptides, 21*, 1081–1099.

Kato, S., Tsuzuki, Y., Hokari, R., Okada, Y., Miyazaki, J., Matsuzaki, K., et al. (2005). Role of nociceptin/orphanin FQ (Noc/oFQ) in murine experimental colitis. *Journal of Neuroimmunology, 161*, 21–28.

Kimura, T., Kitaichi, K., Hiramatsu, K., Yoshida, M., Ito, Y., & Kume, H. (2000). Intradermal application of nociceptin increases vascular permeability in rats: The possible involvement of histamine release from mast cells. *European Journal of Pharmacology, 407*, 327–332.

Köster, A., Montkowski, A., Schulz, S., Stübe, E. M., Knaudt, K., Jenck, F., et al. (1999). Targeted disruption of the orphanin FQ/nociceptin gene increases stress susceptibility and impairs stress adaptation in mice. *Proceedings of the National Academy of Sciences of the United States of America, 96*, 10444–10449.

Lambert, D. G. (2008). The nociceptin/orphanin FQ receptor: A target with broad therapeutic potential. *Nature Reviews. Drug Discovery, 7*, 694–710.

Le Cudennec, C., Naudin, B., Do Rego, J. C., & Costentin, J. (2002). Nociceptin/orphanin FQ and related peptides reduce the increase in plasma corticosterone elicited in mice by an intracerebroventricular injection. *Life Sciences, 72*, 163–171.

Lee, J. K., Tran, T., & Tansey, M. G. (2009). Neuroinflammation in Parkinson's disease. *Journal of NeuroImmune Pharmacology, 4*, 419–429.

Leggett, J. D., Dawe, K. L., Jessop, D. S., & Fulford, A. J. (2009). Endogenous nociceptin/orphanin FQ system involvement in hypothalamic-pituitary-adrenal axis responses: Relevance to models of inflammation. *Journal of Neuroendocrinology, 21*, 888–897.

Leggett, J. D., Harbuz, M. S., Jessop, D. S., & Fulford, A. J. (2006). The nociceptin receptor antagonist [Nphe1, Arg14, Lys15]nociceptin/orphanin FQ-NH2 blocks the stimulatory effects of nociceptin/orphanin FQ on the HPA axis in rats. *Neuroscience*, *141*, 2051–2057.

Leggett, J. D., Jessop, D. S., & Fulford, A. J. (2007). The nociceptin/orphanin FQ antagonist UFP-101 differentially modulates the glucocorticoid response to restraint stress in rats during the peak and nadir phases of the hypothalamo-pituitary-adrenal axis circadian rhythm. *Neuroscience*, *147*, 757–764.

Li, L., Dong, L., & Wang, S. (2013). Expression of the nociceptin/orphanin FQ receptor in the intestinal mucosa of IBS patients. *Experimental and Therapeutic Medicine*, *6*, 679–683.

Liu, A. M., & Wong, Y. H. (2005). Mu-opioid receptor-mediated phosphorylation of IkappaB kinase in human neuroblastoma SH-SY5Y cells. *Neurosignals*, *14*, 136–142.

Lou, L. G., Ma, L., & Pei, G. (1997). Nociceptin/orphanin FQ activates protein kinase C, and this effect is mediated through phospholipase C/Ca2+ pathway. *Biochemical and Biophysical Research Communications*, *240*, 304–308.

Mabrouk, O. S., Marti, M., & Morari, M. (2010). Endogenous nociceptin/orphanin FQ (N/OFQ) contributes to haloperidol-induced changes of nigral amino acid transmission and parkinsonism: A combined microdialysis and behavioral study in naïve and nociceptin/orphanin FQ receptor knockout mice. *Neuroscience*, *166*, 40–48.

Malinowska, B., Godlewski, G., & Schlicker, E. (2002). Function of nociceptin and opioid OP4 receptors in the regulation of the cardiovascular system. *Journal of Physiology and Pharmacology*, *53*, 301–324.

Mallimo, E. M., & Kusnecov, A. W. (2013). The role of orphanin FQ/nociceptin in neuroplasticity: Relationship to stress, anxiety and neuroinflammation. *Frontiers in Cellular Neuroscience*, 7, 173.

Mantovani, A., Cassatella, M. A., Costantini, C., & Jaillon, S. (2011). Neutrophils in the activation and regulation of innate and adaptive immunity. *Nature Reviews Immunology*, *11*, 519–531.

Marti, M., Guerrini, R., Beani, L., Bianchi, C., & Morari, M. (2002). Nociceptin/orphanin FQ receptors modulate glutamate extracellular levels in the substantia nigra pars reticulata. A microdialysis study in the awake freely moving rat. *Neuroscience*, *12*, 153–160.

Marti, M., Mela, F., Fantin, M., Zucchini, S., Brown, J. M., & Witta, J. (2005). Blockade of nociceptin/orphanin FQ transmission attenuates symptoms and neurodegeneration associated with Parkinson's disease. *Journal of Neuroscience*, *25*, 9591–9601.

Marti, M., Mela, F., Veronesi, C., Guerrini, R., Salvadori, S., & Federici, M. (2004). Blockade of nociceptin/orphanin FQ receptor signaling in rat substantia nigra pars reticulata stimulates nigrostriatal dopaminergic transmission and motor behavior. *Journal of Neuroscience*, *24*, 6659–6666.

Marti, M., Sarubbo, S., Latini, F., Cavallo, M., Eleopra, R., & Biguzzi, S. (2010). Brain interstitial nociceptin/orphanin FQ levels are elevated in Parkinson's disease. *Movement Disorders*, *25*, 1723–1732.

Martin, T. R., & Frevert, C. W. (2005). Innate immunity in the lungs. *Proceedings of the American Thoracic Society*, *2*, 403–411.

Matthes, H., Seward, E. P., Kieffer, B., & North, R. A. (1996). Functional selectivity of orphanin FQ for its receptor coexpressed with potassium channel subunits in Xenopus laevis oocytes. *Molecular Pharmacology*, *50*, 447–450.

McGeer, P. L., & McGeer, E. G. (2004). Inflammation and neurodegeneration in Parkinson's disease. *Parkinsonism & Related Disorders*, *10*(Suppl. 1), S3–S7.

Medzhitov, R. (2007). Recognition of microorganisms and activation of the immune response. *Nature*, *449*, 819–826.

Meunier, J. C. (1997). Nociceptin/orphanin FQ and the opioid receptor-like ORL1 receptor. *European Journal of Pharmacology*, *340*, 1–15.

Meunier, J. C., Mollereau, C., Toll, L., Suaudeau, C., Moisand, C., Alvinerie, P., et al. (1995). Isolation and structure of the endogenous agonist of opioid receptor-like ORL1 receptor. *Nature*, *377*, 532–535.
Miller, T. R., & Fulford, A. J. (2007). Regulation of nociceptin/orphaninFQ secretion by immune cells and functional modulation of interleukin-2. *Peptides*, *28*, 2243–2252.
Mollereau, C., & Mouledous, L. (2000). Tissue distribution of the opioid receptor-like (ORL1) receptor. *Peptides*, *21*, 907–917.
Moretta, A. (2002). Natural killer cells and dendritic cells: Rendezvous in abused tissues. *Nature Reviews. Immunology*, *2*, 957–965.
Murray, P. J., & Wynn, T. A. (2011). Protective and pathogenic functions of macrophage subsets. *Nature Reviews. Immunology*, *11*, 723–737.
Neal, C. R., Jr., Mansour, A., Reinscheid, R., Nothacker, H. P., Civelli, O., & Watson, S. J., Jr. (1999). Localization of orphanin FQ (nociceptin) peptide and messenger RNA in the central nervous system of the rat. *Journal of Comparative Neurology*, *406*, 503–547.
New, D. C., & Wong, Y. H. (2002). The ORL1 receptor: Molecular pharmacology and signalling mechanisms. *Neurosignals*, *11*, 197–212.
Nicholson, J. R., Akil, H., & Watson, S. J. (2002). Orphanin FQ-induced hyperphagia is mediated by corticosterone and central glucocorticoid receptors. *Neuroscience*, *115*, 637–643.
Norton, C. S., Neal, C. R., Kumar, S., Akil, H., & Watson, S. J. (2002). Nociceptin/orphanin FQ and opioid receptor-like receptor mRNA expression in dopamine systems. *Journal of Comparative Neurology*, *444*, 358–368.
Pampusch, M. S., Osinski, M. A., Serie, J. R., Murtaugh, M. P., & Brown, D. R. (1998). Opioid receptor gene expression in the porcine immune system. *Advances in Experimental Medicine and Biology*, *437*, 59–65.
Pampusch, M. S., Serie, J. R., Osinski, M. A., Seybold, V. S., Murtaugh, M. P., & Brown, D. R. (2000). Expression of nociceptin/OFQ receptor and prepro-nociceptin/-OFQ in lymphoid tissues. *Peptides*, *21*, 1865–1870.
Paula-Neto, H. A., Alves-Filho, J. C., Souto, F. O., Spiller, F., Amêndola, R. S., Freitas, A., et al. (2011). Inhibition of guanylyl cyclase restores neutrophil migration and maintains bactericidal activity increasing survival in sepsis. *Shock*, *35*, 17–27.
Peluso, J., LaForge, K. S., Matthes, H. W., Kreek, M. J., Kieffer, B. L., & Gavériaux-Ruff, C. (1998). Distribution of nociceptin/orphanin FQ receptor transcript in human central nervous system and immune cells. *Journal of Neuroimmunology*, *81*, 184–192.
Petrella, C., Giuli, C., Broccardo, M., Eutamene, H., Cartier, C., Leveque, M., et al. (2013). Protective and worsening peripheral nociceptin/orphanin FQ receptor-mediated effect in a rat model of experimental colitis. *Pharmacological Research*, *70*, 72–79.
Qian, L., Flood, P. M., & Hong, J. S. (2010). Neuroinflammation is a key player in Parkinson's disease and a prime target for therapy. *Journal of Neural Transmission*, *117*, 971–979.
Ransohoff, R. M., & Perry, V. H. (2009). Microglial physiology: Unique stimuli, specialized responses. *Annual Review of Immunology*, *27*, 119–145.
Reinscheid, R. K., Nothacker, H. P., Bourson, A., Ardati, A., Henningsen, R. A., Bunzow, J. R., et al. (1995). Orphanin FQ: A neuropeptide that activates an opioid like G protein-coupled receptor. *Science*, *270*, 792–794.
Rentzos, M., Nikolaou, C., Andreadou, E., Paraskevas, G. P., Rombos, A., & Zoga, M. (2007). Circulating interleukin-15 and RANTES chemokine in Parkinson's disease. *Acta Neurologica Scandinavica*, *116*, 374–379.
Rizzi, A., Spagnolo, B., Wainford, R. D., Fischetti, C., Guerrini, R., Marzola, G., et al. (2007). *In vitro* and *in vivo* studies on UFP-112, a novel potent and long lasting agonist selective for the nociceptin/orphanin FQ receptor. *Peptides*, *28*, 1240–1251.

Serhan, C. N., Fierro, I. M., Chiang, N., & Pouliot, M. (2001). Cutting edge: Nociceptin stimulates neutrophil chemotaxis and recruitment: Inhibition by aspirin-triggered-15-epi-lipoxin A4. *Journal of Immunology*, *166*, 3650–3654.

Sobczak, M., Sałaga, M., Storr, M., & Fichna, J. (2013). Nociceptin/orphanin FQ (NOP) receptors as novel potential target in the treatment of gastrointestinal diseases. *Current Drug Targets. Inflammation & Allergy*, *14*, 1203–1209.

Stamer, U. M., Book, M., Comos, C., Zhang, L., Nauck, F., & Stüber, F. (2011). Expression of the nociceptin precursor and nociceptin receptor is modulated in cancer and septic patients. *British Journal of Anaesthesia*, *106*, 566–572.

Steinman, R. M. (1991). The dendritic cell system and its role in immunogenicity. *Annual Review of Immunology*, *9*, 271–296.

Sullo, N., Roviezzo, F., Matteis, M., Ianaro, A., Calo, G., Guerrini, R., et al. (2013). Nociceptin/orphanin FQ receptor activation decreases the airway hyperresponsiveness induced by allergen in sensitized mice. *American Journal of Physiology—Lung Cellular and Molecular Physiology*, *304*, L657–L664.

Tansey, M. G., McCoy, M. K., & Frank-Cannon, T. C. (2007). Neuroinflammatory mechanisms in Parkinson's disease: Potential environmental triggers, pathways, and targets for early therapeutic intervention. *Experimental Neurology*, *208*, 1–25.

Tavares-Murta, B. M., Zaparoli, M., Ferreira, R. B., Silva-Vergara, M. L., Oliveira, C. H., Murta, E. F., et al. (2002). Failure of neutrophil chemotactic function in septic patients. *Critical Care Medicine*, *30*, 1056–1061.

Thompson, J. P., Serrano-Gomez, A., Mcdonald, J., Ladak, N., Bowrey, S., & Lambert, D. G. (2013). The Nociceptin/Orphanin FQ system is modulated in patients admitted to ICU with sepsis and after cardiopulmonary bypass. *PLoS One*, *8*, e76682.

Trombella, S., Vergura, R., Falzarano, S., Guerrini, R., Calo, G., & Spisani, S. (2005). Nociceptin/orphanin FQ stimulates human monocyte chemotaxis via NOP receptor activation. *Peptides*, *26*, 1497–1502.

Uezu, K., Sei, H., Sano, A., Toida, K., Suzuki-Yamamoto, T., Houtani, T., et al. (2004). Lack of nociceptin receptor alters body temperature during resting period in mice. *Neuroreport*, *15*, 751–755.

Ulrich-Lai, Y. M., & Herman, J. P. (2009). Neural regulation of endocrine and autonomic stress responses. *Nature Reviews. Neuroscience*, *10*, 397–409.

Viaro, R., Sanchez-Pernaute, R., Marti, M., Trapella, C., Isacson, O., & Morari, M. (2008). Nociceptin/orphanin FQ receptor blockade attenuates MPTP-induced parkinsonism. *Neurobiology of Disease*, *30*, 430–438.

Visanji, N. P., de Bie, R. M., Johnston, T. H., McCreary, A. C., Brotchie, J. M., & Fox, S. H. (2008). The nociceptin/orphanin FQ (NOP) receptor antagonist J-113397 enhances the effects of levodopa in the MPTP-lesioned nonhuman primate model of Parkinson's disease. *Movement Disorders*, *23*, 1922–1925.

Vitale, G., Arletti, R., Ruggieri, V., Cifani, C., & Massi, M. (2006). Anxiolytic-like effects of nociceptin/orphanin FQ in the elevated plus maze and in the conditioned defensive burying test in rats. *Peptides*, *27*, 2193–2200.

Vitale, G., Ruggieri, V., Filaferro, M., Frigeri, C., Alboni, S., Tascedda, F., et al. (2009). Chronic treatment with the selective NOP receptor antagonist [Nphe 1, Arg 14, Lys 15]N/OFQ-NH 2 (UFP-101) reverses the behavioural and biochemical effects of unpredictable chronic mild stress in rats. *Psychopharmacology*, *207*, 173–189.

Volta, M., Mabrouk, O. S., Bido, S., Marti, M., & Morari, M. (2010). Further evidence for an involvement of nociceptin/orphanin FQ in the pathophysiology of Parkinson's disease: A behavioral and neurochemical study in reserpinized mice. *Journal of Neurochemistry*, *115*, 1543–1555.

Waits, P. S., Purcell, W. M., Fulford, A. J., & McLeod, J. D. (2004). Nociceptin/orphanin FQ modulates human T cell function *in vitro*. *Journal of Neuroimmunology*, *149*, 110–120.

Wan, Y. Y. (2010). Multi-tasking of helper T cells. *Immunology, 130*, 166–171.
Weaver, A. L., Lautzenheiser, R. L., Schiff, M. H., Gibofsky, A., Perruquet, J. L., Luetkemeyer, J., et al. (2006). Real-world effectiveness of select biologic and DMARD monotherapy and combination therapy in the treatment of rheumatoid arthritis: Results from the RADIUS observational registry. *Current Medical Research and Opinion, 22*, 185–198.
Wick, M. J., Minnerath, S. R., Roy, S., Ramakrishnan, S., & Loh, H. H. (1995). Expression of alternate forms of brain opioid 'orphan' receptor mRNA in activated human peripheral blood lymphocytes and lymphocytic cell lines. *Brain Research. Molecular Brain Research, 32*, 342–347.
Williams, J. P., Thompson, J. P., Young, S. P., Gold, S. J., McDonald, J., Rowbotham, D. J., et al. (2008). Nociceptin and urotensin-II concentrations in critically ill patients with sepsis. *British Journal of Anaesthesia, 100*, 810–814.
Wu, E. H., Lo, R. K., & Wong, Y. H. (2003). Regulation of STAT3 activity by G16-coupled receptors. *Biochemical and Biophysical Research Communications, 303*, 920–925.
Zaratin, P. F., Petrone, G., Sbacchi, M., Garnier, M., Fossati, C., Petrillo, P., et al. (2004). Modification of nociception and morphine tolerance by the selective opiate receptor-like orphan receptor antagonist (−)-cis-1-methyl-7-[[4-(2,6-dichlorophenyl) piperidin-1-yl]methyl]-6,7,8,9-tetrahydro-5H-benzocyclohepten-5-ol (SB-612111). *Journal of Pharmacology and Experimental Therapeutics, 308*, 454–461.
Zhang, C., & McDougall, J. J. (2006). Stimulation of sensory neuropeptide release by nociceptin/orphanin FQ leads to hyperaemia in acutely inflamed rat knees. *British Journal of Pharmacology, 148*, 938–946.
Zhao, H., Wu, G. C., & Cao, X. D. (2002). Immunomodulatory activity of orphanin FQ/nociceptin on traumatic rats. *Acta Pharmacologica Sinica, 23*, 343–348.
Zhu, J., & Paul, W. E. (2008). CD4 T cells: Fates, functions, and faults. *Blood, 112*, 1557–1569.

CHAPTER TEN

Endogenous Nociceptin System Involvement in Stress Responses and Anxiety Behavior

Allison Jane Fulford[1]
Centre for Comparative and Clinical Anatomy, University of Bristol, Bristol, BS2 8EJ, United Kingdom
[1]Corresponding author: e-mail address: a.j.fulford@bristol.ac.uk

Contents

Abstract

The mechanisms underpinning stress-related behavior and dysfunctional events leading to the expression of neuropsychiatric disorders remain incompletely understood. Novel candidates involved in the neuromodulation of stress, mediated both peripherally and centrally, provide opportunities for improved understanding of the neurobiological basis of stress disorders and may represent targets for novel therapeutic development. This chapter provides an overview of the mechanisms by which the opioid-related peptide, nociceptin, regulates the neuroendocrine stress response and stress-related behavior. In our research, we have employed nociceptin receptor antagonists to investigate endogenous nociceptin function in tonic control over stress-induced activity of the hypothalamo-pituitary–adrenal axis. Nociceptin demonstrates a wide range of functions, including modulation of psychological and inflammatory stress responses, modulation of neurotransmitter release, immune homeostasis, in addition to anxiety and cognitive behaviors. Greater appreciation of the complexity of limbic–hypothalamic neuronal networks, together with attention toward gender

Vitamins and Hormones, Volume 97
ISSN 0083-6729
http://dx.doi.org/10.1016/bs.vh.2014.12.012

differences and the roles of steroid hormones, provides an opportunity for deeper understanding of the importance of the nociceptin system in the context of the neurobiology of stress and behavior.

ABBREVIATIONS

ACTH adrenocorticotrophic hormone
BLA basolateral amygdala
BNST bed nucleus of stria terminalis
CeA central amygdala
CRH corticotrophin-releasing hormone
HPA hypothalamo-pituitary–adrenal
LA lateral amygdala
N/OFQ nociceptin/orphanin FQ
NOP nociceptin/orphanin FQ peptide receptor
OT oxytocin
POMC pro-opiomelanocortin
PRL prolactin

1. INTRODUCTION

1.1 Nociceptin peptide and receptor system

Despite bearing striking sequence homology with the opioid peptides, the heptadecapeptide, nociceptin/orphanin FQ (N/OFQ) appears to act through pharmacologically distinct mechanisms by selectively binding to the G-protein-coupled receptor (NOP—nociceptin/orphanin FQ peptide receptor) (Meunier et al., 1995; Reinscheid et al., 1995). The NOP receptor-coupling is analogous to that seen with opioid peptides and involves inhibition of adenylyl cyclase activity (Reinscheid et al., 1995), activation of K^+ conductances (Matthes, Seward, Kieffer, & North, 1996), and modulation of voltage-dependent calcium currents (Knoflach, Reinscheid, Civelli, & Kemp, 1996) that have been linked to N/OFQ-mediated inhibition of neurotransmitter release (Schlicker & Morari, 2000). Intracerebroventricular (i.c.v.) injection of N/OFQ causes a variety of behavioral effects in rodents (Calò et al., 1998; Gavioli & Calò, 2006; Meunier et al., 1995; Reinscheid et al., 1995) consistent with a broad central nervous system distribution of the peptide and its receptor (Boom et al., 1999; Mollereau et al., 1996; Neal et al., 1999a, 1999b; Nothacker et al., 1996). Diverse central actions of N/OFQ have been reported including regulation of locomotor activity (Rizzi et al., 2001), feeding (Polidori, de

Caro, & Massi, 2000), cognition (Higgins et al., 2002; Nabeshima, Noda, & Mamiya, 1999), the response to psychoactive drugs (Ciccocioppo, Angeletti, Panocka, & Massi, 2000; Martin-Fardon, Ciccocioppo, Massi, & Weiss, 2000), and regulation of cardiovascular and renal function (Kapusta, 2000; Kapusta, Dayan, & Kenigs, 2002). As its name suggests, N/OFQ also regulates pain but in a complex way reflecting the intriguing pharmacology of this peptide system. i.c.v. N/OFQ induces hyperalgesia (Meunier et al., 1995; Reinscheid et al., 1995) or has antinociceptive effects (Grisel, Mogil, Belknap, & Grandy, 1996; Mogil et al., 1996). In rats experiencing inflammation induced by Freund's adjuvant i.c.v. N/OFQ or NOP analogs potently reverse morphine-induced analgesia (Bertorelli et al., 1999) demonstrating antiopioid peptide effects. N/OFQ and its receptor are present in dorsal spinal cord (Xu, Grass, Hao, Xu, & Wiesenfeld-Hallin, 2000) and when given intraspinally N/OFQ exerts bidirectional effects. At nanomolar doses, N/OFQ induces analgesia (Erb et al., 1997; Tian et al., 1997; Yamamoto, Nozaki-Taguchi, & Kimura, 1997) and promotes analgesic effects of morphine (Tian et al., 1997), whereas in the femtomolar range N/OFQ causes hyperalgesia or allodynia (Hara et al., 1997; Sakurada et al., 1999). Effects are NOP receptor-mediated as they are naloxone-insensitive and absent in mice lacking the *NOP* gene (Ahmadi et al., 2001). Despite these findings, N/OFQ is not essential for nociceptive function since mutant mice deficient for NOP display normal nociceptive thresholds (Nishi et al., 1997).

1.2 Nociceptin and NOP receptor: Relevance to inflammation

Compelling evidence suggests that N/OFQ has significant potential in the regulation of inflammation, N/OFQ reduces inflammation-induced thermal hyperalgesia (Yamamoto et al., 1997; Hao, Xu, Wiesenfeld-Hallin, & Xu, 1998), exerts depression on spinal nociceptive input during peripheral inflammation (Xu, Grass, Wiesenfeld-Hallin, & Xu, 1999), inhibits antidromic vasodilatation (Häbler et al., 1999), and attenuates pronociceptive and proinflammatory tachykinins from peripheral sensory nerve endings (Giuliani & Maggi, 1996; Helyes, Németh, Pintér, & Szolcsányi, 1997; Németh et al., 1998) that could account for N/OFQ-induced antinociceptive and anti-inflammatory effects. Carrageenan-induced peripheral inflammation upregulates *preproN/OFQ* mRNA, N/OFQ peptide, and NOP binding in primary sensory neurones of dorsal root ganglia (Andoh, Itoh, & Kuraishi, 1997; Itoh et al., 2001), and dorsal horn (Fu, Wang, Wang, Yu, & Wu, 2007; Fu, Zhu, Wang, & Wu, 2007) in addition to the hypothalamus (Rosén, Lundeberg, Bytner, & Nylander, 2000) revealing plasticity in the N/OFQ system. Pain patients have elevated serum N/OFQ that is augmented in patients with long-lasting chronic pain (Ko, Kim, Woo, & Kim, 2002), although females with fibromyalgia syndrome have lower plasma N/OFQ (Anderberg, Liu, Berglund, & Nyberg, 1998) suggesting that a correlation between pain and serum N/OFQ may depend on factors such as the presence of inflammation, gonadal steroids, and stress. In humans, N/OFQ and NOP are present in the brainstem (Mollereau et al., 1994) and trigeminal ganglion (Mørk et al., 2002) and therefore likely to regulate nociception. A study by Kumar et al. (1999) failed to detect [3H]-N/OFQ binding to human synovial tissue in osteoarthritis patients or N/OFQ in synovial fluid of osteo- or rheumatoid arthritis as measured by radioimmunoassay. It was suggested that N/OFQ may not represent a useful target for peripheral pain, although techniques with improved sensitivity may have revealed significant differences.

PreproN/OFQ and *NOP* receptor mRNAs are also expressed in mammalian immune cells and mitogens or releasing agents (e.g., cytokines) upregulate their expression (Halford, Gebhardt, & Carr, 1995). The immunological NOP receptor appears to be functionally active since N/OFQ and NOP ligands modulate proliferation of activated human and rat lymphocytes *in vitro* (Miller & Fulford, 2007; Peluso, Gaveriaux-Ruff, Matthes, Filliol, & Kieffer, 2001; Waits, Purcell, Fulford, & McLeod, 2004). NOP expression is implicated in the regulation of antibody production (Halford et al., 1995) and neutrophil chemotaxis (Serhan, Fierro, Chiang, & Pouliot, 2001). N/OFQ may also stimulate proinflammatory responses by triggering mast cell histamine release and increased vascular permeability *in vivo* (Kimura et al., 2000), although reports of inhibitory effects of N/OFQ on mast cell-mediated plasma extravasation exist (Németh et al., 1998) and we have shown that N/OFQ modulates rat splenocyte proliferation and proinflammatory cytokine production *in vitro* (Miller & Fulford, 2007). Powerful inflammatory stimuli, such as bacterial endotoxin, lipolysaccharide, appear to stimulate splenocyte release of N/OFQ (Miller & Fulford, 2007), thus immune-derived N/OFQ may act as a paracrine/autocrine regulator of local inflammation or could interact with NOP receptors on peripheral nerves to modulate nociceptive neuronal transmission *in vivo*. Endogenous N/OFQ function is also implicated in the response to systemic inflammation following lipopolysaccharide administration. Peptidic NOP receptor antagonist, UFP-101, has been shown to inhibit microvascular inflammation *in vivo* (Brookes et al., 2013), and we have shown that UFP-101 suppresses the hypothalamo-pituitary–adrenal (HPA) axis response to peripheral LPS challenge (Leggett, Dawe, Jessop, & Fulford, 2009). These findings are potentially of significant clinical relevance in the context that patients with more severe cases of sepsis have significantly higher concentrations of plasma N/OFQ and worse outcomes (Williams et al., 2008). Clearly, the net effect of N/OFQ activity in inflammatory responses is complex, possibly involving bidirectional effects dependent on the activation state of the immune system. Emerging data are highly interesting and indicate that N/OFQ system warrants detailed scrutiny since this may represent a novel target for systemic inflammation.

1.3 Nociceptin and NOP receptor: Relevance to anxiety and stress

The N/OFQ system may exert effects on animal behavior via interaction with the HPA axis in addition to the autonomic nervous system. These two physiological systems can be considered as major effector pathways for maintenance of homeostasis during stress. On the basis of the preproN/OFQ knockout phenotype, N/OFQ was proposed as an endogenous antistress peptide opposing the stress-promoting effects of the HPA axis (Koster et al., 1999), however, we and others have provided evidence indicating that central N/OFQ in fact activates the HPA axis in rats (Devine, Watson, & Akil, 2001; Leggett, Harbuz, Jessop, & Fulford, 2006; Leggett, Jessop, & Fulford, 2007). i.c.v. N/OFQ stimulates basal adrenocorticotrophic hormone (ACTH) and corticosterone release in conscious rats and augments the corticosterone response to exposure to a novel environment (Devine et al., 2001). We confirmed these findings in our own laboratory and also examined the mechanisms whereby N/OFQ activates the stress axis. We have shown that central N/OFQ injection stimulates corticotrophin-releasing hormone (CRH) neurones of the parvocellular paraventricular nucleus (PVN) and synthesis of the ACTH precursor, pro-opiomelanocortin (POMC), in pituitary corticotrophs (Leggett et al., 2006). The increased hypothalamic *CRH* and pituitary *POMC* mRNAs are associated with elevated plasma corticosterone levels at 30 min post-injection. Although *CRH* mRNA expression was stimulated by i.c.v. N/OFQ injection, no changes in arginine vasopressin mRNA were seen (Leggett et al., 2006). The evidence for N/OFQ and CRH interactions is significant given the role of CRH in the integration of neurobehavioral responses to stress and anxiety. The function of N/OFQ in stress and anxiety is therefore complicated by the fact that in normal resting animals, N/OFQ activates the HPA axis, which at first sight is incompatible with an antistress role. However, these data are consistent with reports of the effects of acute opiate or opioid peptide injection and emphasize the fact that corticosterone levels do not necessarily correlate with the presence of fear or anxiety. As integrity of the HPA axis is essential for internal homeostasis, the fact that central N/OFQ-induces activation of the HPA axis identifies N/OFQ as a neuropeptide with major significance for stress regulation. As prolonged stress can predispose to physical or mental disorders including inflammatory diseases, anxiety, and depression (Sternberg, Chrousos, Wilder, & Gold, 1992), N/OFQergic mechanisms that contribute significantly to stress adaptation have attracted attention. Central N/OFQ peptide infusion activates the HPA axis acutely via binding to NOP receptors, as we have shown that coadministration with the selective peptidic antagonist, UFP-101, can fully attenuate the N/OFQ-induced stimulatory effect on the HPA axis in rats (Leggett et al., 2006). Our studies employing the UFP-101 antagonist revealed an interesting role of endogenous N/OFQ function in the regulation of HPA axis activity depending on the type of stressor and time of day of experiment. We reported a differential effect of UFP-101 on restraint stress-induced HPA axis activity characterized by enhancement of stress-induced activity in the morning but no significant effect on the response to restraint in the evening. This finding is consistent with restraint stress being superimposed upon existing diurnal rhythms in basal HPA axis activity and glucocorticoid release (see Dallman et al., 1992). In contrast to the impact of UFP-101 on restraint stress-induced HPA axis regulation, central UFP-101 administration was associated with suppression of the HPA axis response to the inflammatory stressor, LPS administration (Leggett et al., 2009), demonstrating stressor-specificity in HPA axis modulation by endogenous NOP receptors.

Central N/OFQ peptide infusion also modulates stress behavior, cognition, nociception, cardiovascular, and endocrine systems. Effects are absent in NOP receptor deficient mice. The highest central expression of N/OFQ precursor (preproN/OFQ) and its GPCR, *NOP* mRNA, N/OFQ immunoreactivity, and NOP binding density are found in anxiety-associated regions including hypothalamic nuclei, hippocampus, amygdaloid complex, and bed nucleus of stria terminalis (BNST) (see Mollereau et al., 1996; Neal et al., 1999a, 1999b). We have also reported significant changes in expression of *preproN/OFQ* mRNA transcript following acute and repeated restraint stress in rats. Acute restraint significantly reduces *preproN/OFQ* mRNA expression in the central amygdala (CeA), whereas repeated restraint significantly increases precursor transcript levels in the BNST and dorsal reticular thalamic nucleus (Delaney et al., 2012). In contrast, *NOP* receptor mRNA expression appears to be quite resistant to stress-induced changes, at least when monitored using *in situ* hybridization histochemistry (Delaney et al., 2012). The impact of chronic stressors on mRNA expression implies that robust changes in N/OFQergic neuronal regulation contribute to habituation to stress.

Endogenous N/OFQ function is relevant to mood, with NOP antagonists eliciting antidepressant-like effects in the forced swim test, attenuation of chronic mild stress-induced depressive symptoms (Gavioli et al., 2003; Vitale et al., 2009), and plasma N/OFQ level being implicated in postpartum depression. N/OFQ infusions evoke robust, dose-related effects in anxiety tests with the great majority reporting anxiolysis in male rodents. However, Fernandez, Misilmeri, Felger, and Devine (2004) have also documented N/OFQ-induced anxiogenesis. Acute N/OFQ action in behavioral tests may be influenced by an N/OFQ-induced hypolocomotor response that is subject to tolerance, thus recent behavioral studies have examined the impact of paired doses of N/OFQ to counteract the acute hypolocomotor effect. A further consideration is that just a handful of reports have considered sex effects on N/OFQ behavioral responses and there is a major bias toward the use of male animal models (Chesterfield et al., 2006). In female rats, however, neonatal handling does increase N/OFQ levels in periaqueductal gray (a midbrain defense region) (Ploj, Roman, Bergstrom, & Nylander, 2001), and N/OFQ induces lateral and central amygdala neuronal hyperpolarization (Meis & Pape, 1998) providing potential too for N/OFQ neuromodulation of aversion in females.

Long-term stress may induce pathophysiological anxiety, for example, repeated restraint induces high social anxiety (Doremus-Fitzwater, Varlinskaya, & Spear, 2009). Many brain regions associated with stressor responding also mediate anxiety-like behavior and are most notably high in N/OFQ expression. Our *in vivo* studies using peptide infusions have shown that N/OFQ regulates hypothalamic CRH expression, and NOP antagonists prolong HPA axis responses to acute restraint (Leggett et al., 2007). Recently, we reported that stress modulates endogenous N/OFQ and decreases *NOP* mRNA expression (Delaney et al., 2012; Leggett et al., 2009), whereas acute glucocorticoid treatment completely suppresses N/OFQ action on BNST neuronal activity (Dawe, Wakerley, & Fulford, 2010). Chronic restraint stress is also associated with marked increases in limbic BNST expression of preproN/OFQ, presumably in response to stress-induced release of N/OFQ peptide (Delaney et al., 2012). *NOP* receptor mRNA in BNST is correspondingly reduced by chronic restraint, providing provocative evidence that chronic stress (and associated anxiety) elicits adaptations in the N/OFQ system. The precise mechanisms underlying N/OFQ's role in anxiety remains unclear, however, site-specific deficits in limbic endogenous N/OFQ function could precipitate high anxiety. N/OFQ–CRH interactions may be relevant to the modulation of stress responses and behavioral states. Functional interactions between these two peptide systems have been localized to the limbic BNST (Rodi et al., 2008), and more recently to the dorsal raphe nucleus (DRN), a major site for 5-HT neurones in the brainstem (Nazzaro et al., 2010). N/OFQ-mediated suppression of firing of DRN neurones has been robustly shown and is correlated with a reduction in 5-HT efflux *in vitro* (Nazzaro, Marino, Barbieri, & Siniscalchi, 2009). This data is particularly interesting given that aberrant 5-HT neurotransmission has been strongly implicated in mood and aversion, and stress-induced adaptations in the N/OFQ peptidergic system may impact 5-HT neuronal function.

A widely employed test for neurobiological studies of fear, memory, and learning is the Pavlovian (or classical) conditioning task. Fear, albeit an adaptive response to a threatening stimulus, is associated with common neuropsychiatric conditions when inappropriately expressed. A huge amount of effort has gone into attempts to identify the neurocircuitry of fear memory processing, although the factors controlling aversive conditioning remain to be elucidated. Identification of novel candidates involved in fear regulation will benefit understanding of the basis of mental disorders associated with intense fear such as post-traumatic stress and anxiety disorders, in addition to other disorders associated with aberrant behavior like schizophrenia and drug addiction (Sato, 1992; Sinha, Catapano, & O'Malley, 1999). Recent research into the role of N/OFQ in the amygdala (Roozendaal, Lengvilas, McGaugh, Civelli, & Reinscheid, 2007; Uchiyama, Toda, Hiranita, Watanabe, & Eyanagi, 2008), in memory (Higgins et al., 2002), anxiety (Fernandez et al., 2004; Green, Barbieri, Brown, Chen, & Devine, 2008; Jenck et al., 2000; Uchiyama et al., 2008; Varty et al., 2005; Vitale, Arletti, Ruggieri, Cifani, & Massi, 2006), and stress (Dawe et al., 2010; Delaney et al., 2012; Devine et al., 2001; Leggett et al., 2006, 2007, 2009; Rodi et al., 2008) demonstrates the clear potential of this peptidergic system with regard to emotional functions. The emerging data is compelling, demanding further investigation of N/OFQ's contribution to the basic fear network and fear learning.

2. THE NEUROANATOMICAL BASIS OF FEAR CONDITIONING

Telencephalic regions involved in fear conditioning include the hippocampus, amygdala, medial prefrontal cortex (MPFC), and nucleus accumbens (NAcc) (Antoniadis & McDonald, 2000, 2006) and these are closely anatomically and functionally connected. The role of the amygdala in fear conditioning is well established and is known to mediate the acquisition of fear associations and expression of behavioral, autonomic, and endocrine fear responses (Huff, Wright-Hardesty, Higgins, Matus-Amat, & Rudy, 2005; Le Doux, 2000; Lee, Dickinson, & Everitt, 2005). The MPFC is important in fear extinction and bilateral neurotoxic lesions of the MPFC produce robust deficits in context-specific fear responses, including freezing (Antoniadis & McDonald, 2006). The ventral, limbically-innervated striatum, or NAcc, is a common output from areas such as the basolateral amygdala (BLA), hippocampus and MPFC, implicated in gateway control of emotional state and motivated behavior (Alexander, DeLong, & Strick, 1986). Many ascribe to the view that the NAcc is primarily involved in response to rewards, although present evidence points toward an, as yet, unspecified role for the NAcc in aversive conditioning. With respect to conditioned aversion, inconsistencies in the literature with respect to experimental approach, in addition to functional segregation of subcircuits in the NAcc (McCittrick & Abercrombie, 2007) provides a basis for further investigation of the importance of the NAcc in associative learning. Electrolytic lesions of the NAcc core and shell have dissociable effects on cued and contextual conditioning in appetitive or aversive tasks (Cassaday, Horsley, & Norman, 2005). More recent evidence suggests that NAcc neurones may be uniquely tuned to selective rewarding or aversive stimuli (Reynolds & Berridge, 2008; Roitman, Wheeler, & Carelli, 2005) and the NAcc may be important for predictive learning involving either aversive or appetitive cues (Schoenbaum & Setlow, 2003; Schultz, 2007). In the context of limbic loop circuits, the abundant expression of N/OFQ precursor (preproN/OFQ) and NOP receptor mRNA transcripts, N/OFQ peptide-expressing cells and NOP receptor binding density (Neal et al., 1999a, 1999b; Sim-Selley, Vogt, Childers, & Vogt, 2003; Sinchak, Romeo, & Micevych, 2006) in the MPFC (infralimbic and cingulate), amygdala subregions and NAcc, in addition to our recent findings showing acute stress-induced modulation of limbic N/OFQ and NOP receptor expression (Delaney et al., 2012) are entirely consistent with substantive roles for this neuropeptide in fear.

3. EVIDENCE FOR A ROLE OF NOCICEPTIN IN FEAR LEARNING AND MEMORY

Several studies have identified that central or peripheral administration of N/OFQ or N/OFQ agonists prior to training impairs memory performance in a broad range of spatial learning tasks in rodents (Higgins et al., 2002; Hiramatsu & Inoue, 1999). NOP receptor knockout mice display enhanced learning and memory (Mamiya et al., 2003; Nagal, Kurokawa, Takeshima, Kieffer, & Ueda, 2007) and greater hippocampal CA1 long-term potentiation (Manabe et al., 1998), whereas NOP agonist, Ro64-6198, administration elicits anxiolysis in the Geller-Seifter and conditioned lick-suppression paradigms (Higgins et al., 2002). Together with findings from N/OFQ precursor knockout mice that show enhanced freezing compared to wildtype controls in passive avoidance and cued fear conditioning tasks, interest in the precise role of endogenous N/OFQ and NOP receptors in aversive learning and memory has grown. Classical fear conditioning involves presentation of a conditioned stimulus, such as tone, paired with an unconditioned stimulus, such as footshock, in a determined context. Subsequent exposure to the tone or same context elicits conditioned fear characterized by freezing. Contextual and cued fear conditioning represent different types of memory, the former involving hippocampal processing with the latter being hippocampal-independent, however importantly, both types are dependent on the amygdala (Phillips & Le Doux, 1992). Most recently intracerebroventricular (i.c.v.) injection of N/OFQ has been shown to dose-relatedly impair the acquisition of contextual and tone fear conditioning in rats (Fornari, Soares, Ferreira, Moreira, & Oliveira, 2008). Low-moderate (0.1–2.5 nmol i.c.v.) doses of peptide enhanced suppression of freezing in a contextual fear paradigm, with a higher dose (5.0 nmol i.c.v.) impairing both tone and contextual fear responses. Importantly, only one study has attempted site-specific injections to investigate the role of N/OFQ in an inhibitory avoidance test. Post-training bilateral N/OFQ infusion into the BLA was shown to dose-relatedly impair inhibitory avoidance retention performance (Roozendaal et al., 2007). However, a single dose (1 pmol/0.2 μl) of N/OFQ bilaterally infused into the CeA immediately post-training was without effect. These findings are important given that Pavlovian fear conditioning and inhibitory avoidance have been proposed as different forms of aversive learning and theoretically could involve distinct neurobiological processes (Le Doux, 2000; Lee et al., 2005;

Wilensky, Schafe, Kristensen, & LeDoux, 2006). Thus, there will be a need to examine whether the roles and neuroanatomical substrates of the effects of N/OFQ in the acquisition and expression of conditioned fear are task-dependent.

4. NOCICEPTIN AND NEUROCHEMICAL SUBSTRATES OF FEAR CONDITIONING: FOCUS ON BIOGENIC AMINES

Although the roles of the BLA and NAcc in conditioned behavior have been investigated, there remain gaps in knowledge related to the precise neurotransmitter mechanisms subserving expression of Pavlovian conditioning. 5-HT has been widely studied in the context of fear and anxiety. Selective serotonin reuptake inhibitors (SSRIs) increase amygdala extracellular 5-HT levels (Burghardt, Bush, McEwen, & LeDoux, 2007), increase acquisition of conditioned fear (Burghardt et al., 2007), whereas 5-HT2 antagonists also modulate conditioned behavior (Macedo, Martinez, Albrechet-Souza, Molina, & Brandao, 2007). Conditioned fear is associated with increased extracellular 5-HT in the NAcc shell (Fulford & Marsden, 1998a) and increased 5-HT levels in the BLA (Martinez, Ribeiro de Oliveira, & Brandão, 2007). Conditioned 5-HT efflux is dependent on NAcc catecholamines (Fulford & Marsden, 2007) and N/OFQ modulates presynaptic 5-HT release in the NAcc (Tao, Ma, Thakkar, McCarley, & Auerbach, 2007). This raises the question that N/OFQ-mediated impairments in memory may involve effects on 5-HT neurotransmission in the BLA or NAcc, both sites of major dorsal raphe inputs. NOP activation has also been shown to tonically inhibit noradrenaline release in the BLA (Kawahara, Hesselink, van Scharrenburg, & Westerink, 2004), glutamate and GABA release in rat lateral amygdala (LA) (Meis & Pape, 2001), and neuronal excitability in the BNST (Dawe et al., 2010). Despite this, very little is known about neurotransmitter substrates underpinning N/OFQ effects specifically on conditioned fear. With respect to major catecholamines, post-training intraBLA N/OFQ suppression of inhibitory avoidance retention is enhanced by β1-adrenoceptor antagonism (Roozendaal et al., 2007), consistent with the proposed role of noradrenaline in memory consolidation (Haycock, Van Buskirk, Ryan, & McGaugh, 1977; LaLumiere, Buen, & McGaugh, 2003; Murchison et al., 2004). Importantly, the BLA and NAcc also receive rich, mesolimbic dopaminergic inputs originating in the ventrotegmental area, VTA (Le Moal & Simon, 1991) and dopamine (DA) has most recently been implicated in fear memory consolidation and fear predictive learning (Iordanova, Westbrook, & Kilcross, 2006).

Furthermore, the BLA regulates NAcc function by glutamate and DA mechanisms (Floresco, Blaha, Yang, & Phillips, 2001; Howland, Taepavarapunk, & Phillips, 2002) and BLA neurones regulate NAcc shell DA release via glutamatergic inputs independently of DA neuronal firing (Floresco, Yang, Phillips, & Blaha, 1998). The particular role of DA systems in the NAcc in aversive conditioning remains undetermined (Levita, Dalley, & Robbins, 2002; Young, 2004), yet existing findings for conditioned reinforcement suggest that dopaminergic mechanisms are crucial in the modulation of aversively motivated behavior (Wilkinson, 1997). Low doses of the indirect DA agonist amphetamine enhance conditioned punishment responding, an effect that can be blocked using the D1/D2 antagonist, alpha-flupenthixol (Killcross, Everitt, & Robbins, 1997). IntraBLA infusions of DA enhance, whereas DA receptor antagonists impair, inhibitory avoidance retention (LaLumiere, Nguyen, & McGaugh, 2004), and intraNAcc infusions of amphetamine impair acquisition of conditioned suppression. Despite such strong behavioral evidence, controversy surrounds the precise stimuli that excite midbrain DA neurones, with arguments favoring rapid activation of neuronal firing by unexpected rewards (Mirenowicz & Schultz, 1996; Salamone, 1994; Schoenbaum & Setlow, 2003; Schultz, 2007; Ungless, 2004; Ungless, Magill, & Bolam, 2004) and suppression by aversive stimuli (Ungless et al., 2004). A wider, attentional/motivational role for mesolimbic DA is, however, robustly supported by voltammetric (Louilot & Besson, 2000) and microdialysis studies (Datla, Ahier, Young, Gray, & Joseph, 2002; Fulford & Marsden, 1998b, 2007; Saulskaya & Marsden, 1995; Young, Joseph, & Gray, 1993) that show increased NAcc DA efflux associated with appetitive or aversive conditioning. Rapid 1 min microdialysis sampling also found that NAcc shell DA efflux increased in parallel with the onset of tone-conditioned fear (Young, Moran, & Joseph, 2005). Discrimination between tonic DA activity and phasic DA neuronal firing may therefore be critical to understanding conditioned behavior, and evidence that NAcc μ and κ-opiate receptors differentially modulate predictive learning (Iordanova, McNally, & Westbrook, 2006) emphasizes the relevance of opioids in this structure. I.c.v. N/OFQ modulation of NAcc DA efflux (Koizumi, Midorikawa, Takeshima, & Murphy, 2004) is of interest requiring further site-specific studies of NOP-dependent effects on monoamine transmission. This will provide valuable insight into the peptidergic mechanisms governing integration of conditioned responses. NOP receptor function is also reported to be decreased in the NAcc of anxious mice (Le Maitre, Daubeuf, Duterte-Boucher, Costentin, & Leroux-Nicollet, 2006).

Although the roles of the N/OFQ and NOP receptors in fear and anxiety have been studied, little endeavor has been given to explain the neurobiology of the N/OFQ–NOP system in specific forms of aversive and appetitive conditioning. That NOP-mediated effects on conditioned behaviors are dependent on changes in presynaptic monoamine transmission is thus an area of research interest. Deeper understanding of the neural mechanisms underlying emotional memory and critically, the neuromodulatory role of N/OFQ–NOP system in fear learning, will be valuable as NOP receptor ligands have theoretical potential for the treatment of stress-related disorders.

5. MATERNAL ADAPTATIONS OF THE NOCICEPTIN SYSTEM

The large majority of research concerning the N/OFQergic system has focussed on adult male animals. Nevertheless there is emerging evidence for a role of N/OFQ, like for other opioids and peptides, in females. This may have relevance for human conditions, like pregnancy and lactation. High anxiety is associated with perinatal depression (Austin, Tully, & Parker, 2007), susceptibility to stress, and other pregnancy and birth complications in women, therefore it is of interest to understand the mechanisms involved in controlling anxiety in the perinatal period. One aspect of interest is the central theme of N/OFQ in the control of anxiety during pregnancy, which has implications for understanding central N/OFQ actions relevant to maternal behavioral responses.

HPA axis responses to stress in rodents are attenuated in late pregnancy, providing some protection for mother and offspring (da Costa, Ma, Ingram, Lightman, & Aguilera, 2001; Douglas, Brunton, Bosch, Russell, & Neumann, 2003; Douglas, Meddle, Toschi, Bosch, & Neumann, 2005). Brain mechanisms underlying this adaptation are fairly well understood (Brunton, Russell, & Hirst, 2014; Douglas et al., 1998, 2005; Neumann et al., 1998). Anxiety in pregnancy, however has received comparatively little study despite maternal anxiety being commonly reported during pregnancy and postpartum, with up to half of women presenting with symptoms (Goedhart et al., 2010). Women with higher prenatal anxiety deliver babies earlier than women with lower anxiety (Mancuso, Schetter, Rini, Roesch, & Hobel, 2004; Rini, Dunkel-Schetter, Wadhwa, & Sandman, 1999). Pregnancy anxiety is a good predictor of offspring cognition, behavior and developmental outcomes, and may be a distinctive behavioral anxiety syndrome (Huizink, Mulder, Robles de Medina, Visser, & Buitelaar, 2004). Numerous neuropeptides regulate anxiety and N/OFQ has attained prominence, being consistently and robustly anxiolytic in diverse tests in male rodents (Jenck et al., 1997, 2000; Ouagazzal, Moreau, Pauly-Evers, & Jenck, 2008; Uchiyama et al., 2008; Varty, Lu, Morgan, Cohen-Williams, & Hodgson, 2008; Varty et al., 2005; Vitale et al., 2006), and influencing other mediators of anxiety, such as oxytocin, OT (Doi, Dutia, & Russell, 1998), prolactin, PRL (Chesterfield et al., 2006), and CRH (Rodi et al., 2008). In animals, anxiety decreases in pregnancy (de Brito Faturi, Teixeira-Silva, & Leite, 2006; Macbeth, Gautreaux, & Luine, 2008), although the neurobiological mechanisms are poorly understood. As N/OFQ expression is positively influenced by sex steroids which rise dramatically through gestation (Quesada & Micevych, 2008), it is conceivable that N/OFQ will exert a major role in modulating anxiety through reproduction in females.

5.1 Maternal adaptations in neuroendocrine behavioral and stress responses

Late pregnancy is accompanied by remarkable plasticity in behavioral and neuroendocrine systems that serve to optimize survival of the offspring. These have considerable importance for ensuring stability during sensitive periods of fetal organ development. Peptide hormones are involved in the characteristic changes associated with these physiological states and many of the maternal adaptations are initiated in pregnancy. The mechanisms underlying maternal behavioral adaptations are not understood, and these may involve enhanced N/OFQ action during pregnancy. Anxiety evidently decreases in advanced pregnancy (de Brito Faturi et al., 2006; Macbeth et al., 2008; Wartella et al., 2003) and this adaptation may be necessary to avoid extreme perinatal emotional disturbances such as postpartum depression. Conversely, chronic stress may induce maternal anxiety and predispose to postnatal depression.

Neuroendocrine adaptations in pregnancy have focussed on the role of classical peptides such as OT, CRH, and enkephalins that are abundantly expressed in hypothalamic supraoptic (SON) and paraventricular (PVN) nuclei. These neuropeptides modulate anxiety in rodents, and human studies also indicate roles in emotionality, particularly for OT (Heinrichs, Baumgartner, Kirschbaum, & Ehlert, 2003). OT anxiolysis is well-known and in pregnancy OT action has been implicated in restraining maternal anxiety (Bosch, Meddle, Beiderbeck, & Douglas, 2005; Neumann,

Torner, & Wigger, 2000). Little is known about the role of CRH or enkephalins in the control of anxiety in pregnancy, but CRH may not play a critical role (Bosch, Krömer, & Neumann, 2006). Plasma PRL increases at the end of pregnancy (Grattan, 2001) along with hypothalamic PRL and PRL receptor expression (Pi & Grattan, 1999) and these may subserve PRL's role in perinatal stress hyporesponsiveness. In parallel with anxiety, basal and stress-induced HPA axis activity is suppressed during late pregnancy, involving blunted responses to psychological (Douglas et al., 1998, 2005) and physical/immune stressors (Brunton & Russell, 2008). CRH expression and PVN neurone responses are profoundly inhibited in late pregnancy (da Costa et al., 2001; Douglas et al., 2003), further indicating that CRH action does not underlie gestational anxiolysis. Evidence implicates reduced noradrenergic afferent excitation of parvocellular PVN CRH neurones, commensurate with an increased production of inhibitory opioid peptides (Douglas et al., 1995, 1998). Peripheral OT secretory responses to psychological stressors are also slightly attenuated in late pregnant rats (Douglas et al., 1995; Neumann et al., 2000), although responses to physical stress such as immune challenge are more restrained. Despite that, OT neurones increase their OT content, dendritic OT efflux of the nuclei slightly increases and OT neurones exhibit increased responsiveness to some stimuli toward the end of pregnancy (Douglas et al., 1995; Leng, Meddle, & Douglas, 2008; Lipschitz, Crowley, & Bealer, 2004). Together with the evidence that OT receptor expression and binding increase in late pregnancy (Bealer, Lipschitz, Ramoz, & Crowley, 2006), including in BNST and amygdala, these data further support a role for OT in gestational anxiolysis. Since N/OFQ influences OT neurone firing rate, increasing burst-like firing (Doi et al., 1998), there is some indication that N/OFQ may act via OT. Therefore, N/OFQ may inhibit anxiety perinatally by acting via central OT mechanisms.

5.2 Prepartum adaptations and changes in N/OFQ expression and function

Several important indicators support a role for N/OFQ–NOP system in females and pregnancy. Sex steroids (i.e., estrogens and progestagens such as those secreted during pregnancy) are well known to inhibit anxiety (Frye & Walf, 2002, 2004). Relevant interactions have already been identified between sex steroids and the N/OFQ system. Intraspinal N/OFQ blocks pregnancy and ovarian sex steroid-induced antinociception and induces hyperalgesia in rats and N/OFQ effects on basal pain thresholds are absent in nonpregnancy (Dawson-Basoa & Gintzler, 1997) implicating a changing ovarian steroid milieu for elevated N/OFQ responsiveness in advanced gestation. In addition, N/OFQ peptide levels are significantly higher in selected brain regions from proestrus rats (Roman, Ploj, Gustafsson, Meyerson, & Nylander, 2006), the phase with highest circulating oestradiol. *NOP* mRNA is increased in the anteroventral PVN, ventromedial hypothalamus (VMH) and medial preoptic nucleus by 17β-oestradiol treatment alone or combined with progesterone. Ovarian steroids correspondingly increase NOP density, NOP binding affinity and stimulate NOP [^{35}S]GTPγS binding in mediobasal hypothalamus (Quesada & Micevych, 2008). *PreproN/OFQ* mRNA expression is also positively regulated by ovarian steroids in limbic and hypothalamic sites including medial amygdala (Sinchak et al., 2006). Furthermore, N/OFQ infusion into the medial preoptic area or VMH facilitates sexual receptivity in oestradiol-primed rats (Sinchak, Hendricks, Baroudi, & Micevych, 1997). Substantial colocalization between estrogen receptor *(ER)β*-mRNA and *N/OFQ* mRNA exists in ventromedial parvocellular and magnocellular PVN, and to a lesser extent SON (Isgor, Shieh, Akil, & Watson, 2003). ERβ and NOP are also coexpressed in limbic regions. Such associations are highly relevant for neuroendocrine and behavioral integration (Walf, Ciriza, Garcia-Segura, & Frye, 2008) and oestradiol control of N/OFQ and/or OT expression, release and action given that *ERα* mRNA is largely absent in PVN and SON (Laflamme, Nappi, Drolet, Labrie, & Rivest, 1998). Sex steroids enhance OT and OT receptor expression and action in anxiety (Choleris, Devidze, Kavaliers, & Pfaff, 2008). Since OT neurones express few estrogen or progesterone receptors it is thought that the steroids act indirectly via other inputs or via neuroactive steroids such as allopregnanolone. From the above evidence, such networks could involve N/OFQ and/or NOP expressing inputs. In this regard, high expression of *preproN/OFQ* mRNA and allopregnanolone synthetic enzymes overlaps in amygdala, BNST and reticular thalamus (Agís-Balboa et al., 2006; Pinna et al., 2008; Toufexis, Davis, Hammond, & Davis, 2004), providing a basis for interaction between neurosteroid and N/OFQ neurones during pregnancy. Pregnancy steroids have the direct potential to decrease anxiety via N/OFQ and OT and it is possible that they are responsible for enhanced N/OFQ action in late gestation. Furthermore, in chronic stress, relatively lower levels of allopregnanolone in pregnancy could attenuate N/OFQ function to facilitate high anxiety. In summary, it is interesting to consider if hormonal steroids are required for enhanced N/OFQ action, and indeed the transmitter substrates required for pregnancy anxiolysis.

6. CONCLUSIONS

Although the role of opioids in the context of HPA axis control, maternal adaptations, and stress-related behaviors has been well studied, attention has shifted toward an explanation of the neurobiological roles of the N/OFQ–NOP system in stress and anxiety states. The combination of well-characterized behavioral models in rodents, *in vivo* approaches and cutting-edge molecular biochemical techniques has enabled a robust investigation into this intriguing opioid-like peptide and its G-protein-coupled receptor. Research has established that nociceptin is a potent regulatory peptide, critical to the normal integration of stress and anxiety responses in a plethora of physiological contexts. Research into the neuroendocrine functions, immunological roles and behavioral integrations of N/OFQ peptide and the NOP receptor has enabled a deeper understanding of the limbic and hypothalamic circuitry underpinning responses to psychological and inflammatory challenges. N/OFQ and the NOP receptor are implicated in adaptation to chronic stress, in addition to the encoding and processing of emotional memories. Recent studies highlighting the interactions between N/OFQ–NOP system and gonadal steroid hormones provides an opportunity to explore sex-specific functions of this modulatory peptide. Further research is timely given that NOP receptor ligands have theoretical potential in the treatment of stress-related disorders.

REFERENCES

Agís-Balboa, R. C., Pinna, G., Zhubi, A., Maloku, E., Veldic, M., Costa, E., et al. (2006). Characterization of brain neurons that express enzymes mediating neurosteroid biosynthesis. *Proceedings of the National Academy of Sciences of the United States of America*, *103*, 14602–14607.

Ahmadi, S., Kotalla, C., Gühring, H., Takeshima, H., Pahl, A., & Zeilhofer, H. U. (2001). Modulation of synaptic transmission by nociceptin/orphanin FQ and nocistatin in the spinal cord dorsal horn of mutant mice lacking the nociceptin/orphanin FQ receptor. *Molecular Pharmacology*, *59*(3), 612–618.

Alexander, G. E., DeLong, M., & Strick, B. E. (1986). Parallel organisation of functionally segregated circuits linking basal ganglia and cortex. *Annual Review of Neuroscience*, *9*, 357–381.

Anderberg, U. M., Liu, Z., Berglund, L., & Nyberg, F. (1998). Plasma levels on nociceptin in female fibromyalgia syndrome patients. *Zeitschrift für Rheumatologie*, *57*(Suppl. 2), 77–80.

Andoh, T., Itoh, M., & Kuraishi, Y. (1997). Nociceptin gene expression in rat dorsal root ganglia induced by peripheral inflammation. *Neuroreport*, *8*(12), 2793–2796.

Antoniadis, E. A., & McDonald, R. J. (2000). Amygdala, hippocampus and discriminative fear conditioning to context. *Behavioural Brain Research*, *108*, 25–37.

Antoniadis, E. A., & McDonald, R. J. (2006). Fornix, medial prefrontal cortex, nucleus accumbens and mediodorsal thalamic nucleus: Roles in a fear-based context discrimination task. *Neurobiology of Learning and Memory, 85*, 71–85.

Austin, M. P., Tully, L., & Parker, G. (2007). Examining the relationship between antenatal anxiety and postnatal depression. *Journal of Affective Disorders, 101*, 169–174.

Bealer, S. L., Lipschitz, D. L., Ramoz, G., & Crowley, W. R. (2006). Oxytocin receptor binding in the hypothalamus during gestation in rats. *American Journal of Physiology Regulatory, Integrative and Comparative Physiology, 291*(1), R53–R58.

Bertorelli, R., Corradini, L., Rafiq, K., Tupper, J., Calò, G., & Ongini, E. (1999). Nociceptin and the ORL-1 ligand [Phe1psi (CH2-NH)Gly2]nociceptin(1-13)NH2 exert anti-opioid effects in the Freund's adjuvant-induced arthritic rat model of chronic pain. *British Journal of Pharmacology, 128*(6), 1252–1258.

Boom, A., Mollereau, C., Meunier, J. C., Vassart, G., Parmentier, M., Vanderhaeghen, J. J., et al. (1999). Distribution of the nociceptin and nocistatin precursor transcript in the mouse central nervous system. *Neuroscience, 91*(3), 991–1007.

Bosch, O. J., Krömer, S. A., & Neumann, I. D. (2006). Prenatal stress: Opposite effects on anxiety and hypothalamic expression of vasopressin and corticotropin-releasing hormone in rats selectively bred for high and low anxiety. *The European Journal of Neuroscience, 23*(2), 541–551.

Bosch, O. J., Meddle, S. L., Beiderbeck, D. I., & Douglas, A. J. (2005). Brain oxytocin correlates with maternal aggression: Link to anxiety. *The Journal of Neuroscience, 25*, 6807–6815.

Brookes, Z. L., Stedman, E. N., Brown, N. J., Hebbes, C. P., Guerrini, R., Calo, G., et al. (2013). The nociceptin/orphanin FQ receptor antagonist UFP-101 reduces microvascular inflammation to lipopolysaccharide in vivo. *PLoS One, 8*(9), e74943.

Brunton, P. J., & Russell, J. A. (2008). Attenuated hypothalamo-pituitary-adrenal axis responses to immune challenge during pregnancy: The neurosteroid opioid connection. *The Journal of Physiology, 586*(2), 369–375.

Brunton, P. J., Russell, J. A., & Hirst, J. J. (2014). Allopregnanolone in the brain: Protecting pregnancy and birth outcomes. *Progress in Neurobiology, 113*, 106–136.

Burghardt, N. S., Bush, D. E. A., McEwen, B. S., & LeDoux, J. E. (2007). Acute selective serotonin reuptake inhibitors increase conditioned fear expression: Blockade with a 5-HT2C receptor antagonist. *Biological Psychiatry, 62*, 1111–1118.

Calò, G., Rizzi, A., Marzola, G., Guerrini, R., Salvadori, S., Beani, L., et al. (1998). Pharmacological characterization of the nociceptin receptor mediating hyperalgesia in the mouse tail withdrawal assay. *British Journal of Pharmacology, 125*(2), 373–378.

Cassaday, H. J., Horsley, R. R., & Norman, C. (2005). Electrolytic lesions to nucleus accumbens core and shell have dissociable effects on conditioning to discrete and contextual cues in aversive and appetitive procedures respectively. *Behavioural Brain Research, 160*, 222–235.

Chesterfield, M., Janik, J., Murphree, E., Lynn, C., Schmidt, E., & Callahan, P. (2006). Orphanin FQ/nociceptin is a physiological regulator of prolactin secretion in female rats. *Endocrinology, 147*(11), 5087–5093.

Choleris, E., Devidze, N., Kavaliers, M., & Pfaff, D. W. (2008). Steroidal/neuropeptide interactions in hypothalamus and amygdala related to social anxiety. *Progress in Brain Research, 170*, 291–303.

Ciccocioppo, R., Angeletti, S., Panocka, I., & Massi, M. (2000). Nociceptin/orphanin FQ and drugs of abuse. *Peptides, 21*(7), 1071–1080.

da Costa, A. P., Ma, X., Ingram, C. D., Lightman, S. L., & Aguilera, G. (2001). Hypothalamic and amygdaloid corticotropin-releasing hormone (CRH) and CRH receptor-1 mRNA expression in the stress-hyporesponsive late pregnant and early lactating rat. *Brain Research Molecular Brain Research, 91*, 119–130.

Dallman, M. F., Akana, S. F., Scribner, K. A., Bradbury, M. J., Walker, C. D., Strack, A. M., et al. (1992). Stress, feedback and facilitation in the hypothalamo-pituitary-adrenal axis. *Journal of Neuroendocrinology*, *4*, 517–526.

Datla, K. P., Ahier, R. G., Young, A. M. J., Gray, J. A., & Joseph, M. (2002). A conditioned appetitive stimulus increases dialysate dopamine in the nucleus accumbens of the rat. *The European Journal of Neuroscience*, *16*, 1987–1993.

Dawe, K., Wakerley, J. B., & Fulford, A. J. (2010). Nociceptin/orphanin FQ and the regulation of neuronal excitability in the bed nucleus of the stria terminalis: Interaction with glucocorticoids. *Stress*, *13*(6), 516–527.

Dawson-Basoa, M., & Gintzler, A. R. (1997). Nociceptin (Orphanin FQ) abolishes gestational and ovarian sex steroid-induced antinociception and induces hyperalgesia. *Brain Research*, *750*(1–2), 48–52.

de Brito Faturi, C., Teixeira-Silva, F., & Leite, J. R. (2006). The anxiolytic effect of pregnancy in rats is reversed by finasteride. *Pharmacology, Biochemistry, and Behavior*, *85*, 569–574.

Delaney, G., Dawe, K. L., Hogan, R., Hunjan, T., Roper, J., Hazell, G., et al. (2012). Role of nociceptin/orphanin FQ and NOP receptors in the response to acute and repeated restraint stress in rats. *Journal of Neuroendocrinology*, *24*(12), 1527–1541.

Devine, D. P., Watson, S. J., & Akil, H. (2001). Nociceptin/orphaninFQ regulates neuroendocrine function of the limbic-hypothalamic-pituitary-adrenal axis. *Neuroscience*, *102*, 541–553.

Doi, N., Dutia, M. B., & Russell, J. A. (1998). Inhibition of rat oxytocin and vasopressin supraoptic nucleus neurons by nociceptin in vitro. *Neuroscience*, *84*, 913–921.

Doremus-Fitzwater, T. L., Varlinskaya, E. I., & Spear, L. P. (2009). Social and non-social anxiety in adolescent and adult rats after repeated restraint. *Physiology & Behavior*, *97*(3–4), 484–494.

Douglas, A. J., Brunton, P. J., Bosch, O. J., Russell, J. A., & Neumann, I. D. (2003). Neuroendocrine responses to stress in mice: Hyporesponsiveness in pregnancy and parturition. *Endocrinology*, *144*, 5268–5276.

Douglas, A. J., Johnstone, H. A., Wigger, A., Landgraf, R., Russell, J. A., & Neumann, I. D. (1998). The role of endogenous opioids in neurohypophysial and hypothalamo-pituitary-adrenal axis hormone secretory responses to stress in pregnant rats. *The Journal of Endocrinology*, *158*, 285–293.

Douglas, A. J., Meddle, S. L., Toschi, N., Bosch, O. J., & Neumann, I. D. (2005). Reduced activity of the noradrenergic system in the paraventricular nucleus at the end of pregnancy: Implications for stress hyporesponsiveness. *Journal of Neuroendocrinology*, *17*, 40–48.

Douglas, A. J., Neumann, I., Meeren, H. K., Leng, G., Johnstone, L. E., Munro, G., et al. (1995). Central endogenous opioid inhibition of supraoptic oxytocin neurons in pregnant rats. *The Journal of Neuroscience*, *15*, 5049–5057.

Erb, K., Liebel, J. T., Tegeder, I., Zeilhofer, H. U., Brune, K., & Geisslinger, G. (1997). Spinally delivered nociceptin/orphanin FQ reduces flinching behaviour in the rat formalin test. *Neuroreport*, *8*(8), 1967–1970.

Fernandez, F., Misilmeri, M. A., Felger, J. C., & Devine, D. P. (2004). Nociceptin/orphanin FQ increases anxiety-related behavior and circulating levels of corticosterone during neophobic tests of anxiety. *Neuropsychopharmacology*, *29*(1), 59–71.

Floresco, S. B., Blaha, C. D., Yang, C. R., & Phillips, A. G. (2001). Dopamine D1 and NMDA receptors mediate potentiation of basolateral amygdala-evoked firing of nucleus accumbens neurons. *The Journal of Neuroscience*, *21*, 6370–6376.

Floresco, S. B., Yang, C. R., Phillips, A. G., & Blaha, C. D. (1998). Basolateral amygdala stimulation evokes glutamate receptor-dependent dopamine efflux in the nucleus accumbens of the anaesthetised rat. *The European Journal of Neuroscience*, *10*, 1241–1251.

Fornari, R. V., Soares, J. C., Ferreira, T. L., Moreira, K. M., & Oliveira, M. G. (2008). Effects of nociceptin/orphanin FQ in the acquisition of contextual and tone fear conditioning in rats. *Behavioral Neuroscience*, *122*, 98–106.

Frye, C. A., & Walf, A. A. (2002). Changes in progesterone metabolites in the hippocampus can modulate open field and forced swim test behavior of proestrous rats. *Hormones and Behavior*, *41*, 306–315.

Frye, C. A., & Walf, A. A. (2004). Estrogen and/or progesterone administered systemically or to the amygdala can have anxiety-, fear-, and pain-reducing effects in ovariectomized rats. *Behavioral Neuroscience*, *118*(2), 306–313.

Fu, X., Wang, Y. Q., Wang, J., Yu, J., & Wu, G. C. (2007). Changes in expression of nociceptin/orphanin FQ and its receptor in spinal dorsal horn during electroacupuncture treatment for peripheral inflammatory pain in rats. *Peptides*, *28*(6), 1220–1228.

Fu, X., Zhu, Z. H., Wang, Y. Q., & Wu, G. C. (2007). Regulation of proinflammatory cytokines gene expression by nociceptin/orphanin FQ in the spinal cord and the cultured astrocytes. *Neuroscience*, *144*, 275–285.

Fulford, A. J., & Marsden, C. A. (1998a). Conditioned release of 5-hydroxytryptamine in vivo in the nucleus accumbens following isolation-rearing in the rat. *Neuroscience*, *83*(2), 481–487.

Fulford, A. J., & Marsden, C. A. (1998b). Effect of isolation-rearing on conditioned dopamine release in vivo in the nucleus accumbens of the rat. *Journal of Neurochemistry*, *70*(1), 384–390.

Fulford, A. J., & Marsden, C. A. (2007). An intact dopaminergic system is required for context-conditioned release of 5-HT in the nucleus accumbens of postweaning isolation-reared rats. *Neuroscience*, *149*(2), 392–400.

Gavioli, E. C., & Calò, G. (2006). Antidepressant- and anxiolytic-like effects of nociceptin/orphanin FQ receptor ligands. *Naunyn-Schmiedeberg's Archives of Pharmacology*, *372*, 319–330.

Gavioli, E. C., Marzola, G., Guerrini, R., Bertorelli, R., Zucchini, S., De Lima, T. C., et al. (2003). Blockade of nociceptin/orphanin FQ-NOP receptor signalling produces antidepressant-like effects: Pharmacological and genetic evidences from the mouse forced swimming test. *The European Journal of Neuroscience*, *17*(9), 1987–1990.

Giuliani, S., & Maggi, C. A. (1996). Inhibition of tachykinin release from peripheral endings of sensory nerves by nociceptin, a novel opioid peptide. *British Journal of Pharmacology*, *118*(7), 1567–1569.

Goedhart, G., Vrijkotte, T. G., Roseboom, T. J., van der Wal, M. F., Cuijpers, P., & Bonsel, G. J. (2010). Maternal cortisol and offspring birthweight: Results from a large prospective cohort study. *Psychoneuroendocrinology*, *35*(5), 644–652.

Grattan, D. (2001). The actions of prolactin in the brain during pregnancy and lactation. *Progress in Brain Research*, *133*, 153–157.

Green, M. K., Barbieri, E. V., Brown, B. D., Chen, K.-W., & Devine, D. P. (2008). Roles of the bed nucleus of stria terminalis and of the amygdala in N/OFQ-mediated anxiety and HPA axis activation. *Neuropeptides*, *41*, 399–410.

Grisel, J. E., Mogil, J. S., Belknap, J. K., & Grandy, D. K. (1996). Orphanin FQ acts as a supraspinal, but not a spinal, anti-opioid peptide. *Neuroreport*, 7(13), 2125–2129.

Häbler, H., Timmermann, L., Stegmann, J., & Jänig, W. (1999). Effects of nociceptin and nocistatin on antidromic vasodilatation in hairless skin of the rat hindlimb in vivo. *British Journal of Pharmacology*, *27*(7), 1719–1727.

Halford, W. P., Gebhardt, B. M., & Carr, D. J. (1995). Functional role and sequence analysis of a lymphocyte orphan opioid receptor. *Journal of Neuroimmunology*, *59*, 91–101.

Hao, J. X., Xu, I. S., Wiesenfeld-Hallin, Z., & Xu, X. J. (1998). Anti-hyperalgesic and anti-allodynic effects of intrathecal nociceptin/orphanin FQ in rats after spinal cord injury, peripheral nerve injury and inflammation. *Pain*, *76*(3), 385–393.

Hara, N., Minami, T., Okuda-Ashitaka, E., Sugimoto, T., Sakai, M., Onaka, M., et al. (1997). Characterization of nociceptin hyperalgesia and allodynia in conscious mice. *British Journal of Pharmacology, 21*(3), 401–408.

Haycock, J. W., Van Buskirk, R., Ryan, J. R., & McGaugh, J. L. (1977). Enhancement of retention with centrally administered catecholamines. *Experimental Neurology, 54*, 199–208.

Heinrichs, M., Baumgartner, T., Kirschbaum, C., & Ehlert, U. (2003). Social support and oxytocin interact to suppress cortisol and subjective responses to psychosocial stress. *Biological Psychiatry, 54*, 1389–1398.

Helyes, Z., Németh, J., Pintér, E., & Szolcsányi, J. (1997). Inhibition by nociceptin of neurogenic inflammation and the release of SP and CGRP from sensory nerve terminals. *British Journal of Pharmacology, 121*(4), 613–615.

Higgins, G. A., Kew, J. N., Richards, J. G., Takeshima, H., Jenck, F., Adams, G., et al. (2002). A combined pharmacological and genetic approach to investigate the role of orphanin FQ in learning and memory. *The European Journal of Neuroscience, 15*, 911–922.

Hiramatsu, M., & Inoue, K. (1999). Nociceptin/orphanin FQ and nocistatin on learning and memory impairment induced by scopolamine in mice. *British Journal of Pharmacology, 127*(3), 655–660.

Howland, J. G., Taepavarapunk, P., & Phillips, A. G. (2002). Glutamate receptor-dependent modulation of dopamine efflux in the nucleus accumbens by basolateral, but not central, nucleus of the amygdala in rats. *The Journal of Neuroscience, 22*, 1137–1145.

Huff, N. C., Wright-Hardesty, K., Higgins, E. A., Matus-Arnat, P., & Rudy, J. W. (2005). Context pre-exposure obscures amygdala modulation of contextual fear conditioning. *Learning & Memory, 12*, 456–460.

Huizink, A. C., Mulder, E. J., Robles de Medina, P. G., Visser, G. H., & Buitelaar, J. K. (2004). Is pregnancy anxiety a distinctive syndrome? *Early Human Development, 79*(2), 81–91.

Iordanova, M. D., McNally, G. P., & Westbrook, R. F. (2006). Opioid receptors in the nucleus accumbens regulate attentional learning in the blocking paradigm. *The Journal of Neuroscience, 26*, 4036–4045.

Iordanova, M. D., Westbrook, R. F., & Kilcross, A. S. (2006). Dopamine activity in the nucleus accumbens modulates blocking in fear conditioning. *The European Journal of Neuroscience, 24*, 3265–3270.

Isgor, C., Shieh, K. R., Akil, H., & Watson, S. J. (2003). Colocalization of estrogen beta-receptor messenger RNA with orphanin FQ, vasopressin and oxytocin in the rat hypothalamic paraventricular and supraoptic nuclei. *Anatomy and Embryology, 206*, 461–469.

Itoh, M., Takasaki, I., Andoh, T., Nojima, H., Tominaga, M., & Kuraishi, Y. (2001). Induction by carrageenan inflammation of prepronociceptin mRNA in VR1-immunoreactive neurons in rat dorsal root ganglia. *Neuroscience Research, 40*(3), 227–233.

Jenck, F., Moreau, J. L., Martin, J. R., Kilpatrick, G. J., Reinscheid, R. K., Monsma, F. J., Jr., et al. (1997). Orphanin FQ acts as an anxiolytic to attenuate behavioural responses to stress. *Proceedings of the National Academy of Sciences of the United States of America, 94*, 14854–14858.

Jenck, F., Wichmann, J., Dautzenberg, F. M., Moreau, J. L., Ouagazzal, A. M., Martin, J. R., et al. (2000). A synthetic agonist at the orphanin FQ/nociceptin receptor ORL1: Anxiolytic profile in the rat. *Proceedings of the National Academy of Sciences of the United States of America, 97*, 4938–4943.

Kapusta, D. R. (2000). Neurohumoral effects of orphanin FQ/nociceptin: Relevance to cardiovascular and renal function. *Peptides, 21*(7), 1081–1099.

Kapusta, D. R., Dayan, L. A., & Kenigs, V. A. (2002). Nociceptin/orphanin FQ modulates the cardiovascular, but not renal, responses to stress in spontaneously hypertensive rats. *Clinical and Experimental Pharmacology & Physiology, 29*(3), 254–259.

Kawahara, Y., Hesselink, M. B., van Scharrenburg, G., & Westerink, B. H. (2004). Tonic inhibition by orphanin FQ/nociceptin of noradrenaline neurotransmission in the amygdala. *European Journal of Pharmacology, 485*, 197–200.

Killcross, A. S., Everitt, B. J., & Robbins, T. W. (1997). Symmetrical effects of amphetamine and alphaflupenthixol on conditioned punishment and reinforcement: Contrasts with midazolam. *Psychopharmacology, 129*, 141–152.

Kimura, T., Kitaichi, K., Hiramatsu, K., Yoshida, M., Ito, Y., Kume, H., et al. (2000). Intradermal application of nociceptin increases vascular permeability in rats: The possible involvement of histamine release from mast cells. *European Journal of Pharmacology, 407*(3), 327–332.

Knoflach, F., Reinscheid, R. K., Civelli, O., & Kemp, J. A. (1996). Modulation of voltage-gated calcium channels by orphanin FQ in freshly dissociated hippocampal neurons. *The Journal of Neuroscience, 16*(21), 6657–6664.

Ko, M. H., Kim, Y. H., Woo, R. S., & Kim, K. W. (2002). Quantitative analysis of nociceptin in blood of patients with acute and chronic pain. *Neuroreport, 13*(13), 1631–1633.

Koizumi, M., Midorikawa, N., Takeshima, H., & Murphy, N. P. (2004). Exogenous, but not endogenous nociceptin modulates mesolimbic dopamine release in mice. *Journal of Neurochemistry, 89*, 257–263.

Koster, A., Montkowski, A., Schulz, S., Stube, E.-M., Knaudt, K., Jenck, F., et al. (1999). Targeted disruption of the orphaninFQ/nociceptin gene increases stress susceptibility and impairs stress adaptation in mice. *Proceedings of the National Academy of Sciences of the United States of America, 96*, 10444–10449.

Kumar, N., Smart, D., Mason, S., McKnight, A. T., Rowbotham, D. J., & Lambert, D. G. (1999). Neither nociceptin nor its receptor are present in human synovial fluid or tissue. *British Journal of Anaesthesia, 83*(3), 470–471.

Laflamme, N., Nappi, R. E., Drolet, G., Labrie, C., & Rivest, S. (1998). Expression and neuropeptidergic characterization of estrogen receptors (ERalpha and ERbeta) throughout the rat brain: Anatomical evidence of distinct roles of each subtype. *Journal of Neurobiology, 36*(3), 357–378.

LaLumiere, R. T., Buen, T. V., & McGaugh, J. L. (2003). Post-training intra-basolateral amygdala infusions of norepinephrine enhance consolidation on memory for contextual fear conditioning. *The Journal of Neuroscience, 23*, 6754–6758.

LaLumiere, R. T., Nguyen, L. T., & McGaugh, J. L. (2004). Post-training intrabasolateral amygdala infusions of dopamine modulate consolidation of inhibitory avoidance memory: Involvement of noradrenergic and cholinergic systems. *The European Journal of Neuroscience, 20*, 2804–2810.

Le Doux, J. E. (2000). Emotion circuits in the brain. *Annual Review of Neuroscience, 23*, 155–184.

Lee, J. L. C., Dickinson, A., & Everitt, B. J. (2005). Conditioned suppression and freezing as measures of aversive Pavlovian conditioning: Effects of discrete amygdala lesions and overtraining. *Behavioural Brain Research, 159*, 221–233.

Leggett, J. D., Dawe, K. L., Jessop, D. S., & Fulford, A. J. (2009). Endogenous nociceptin/orphanin FQ system involvement in hypothalamic-pituitary-adrenal axis responses: Relevance to models of inflammation. *Journal of Neuroendocrinology, 21*(11), 888–897.

Leggett, J. D., Harbuz, M. S., Jessop, D. S., & Fulford, A. J. (2006). The nociceptin receptor antagonist UFP-101 blocks the stimulatory effects of nociceptin/orphaninFQ on the HPA axis in rats. *Neuroscience, 141*, 2051–2057.

Leggett, J. D., Jessop, D. S., & Fulford, A. J. (2007). The nociceptin/orphanin FQ antagonist UFP-101 differentially modulates the glucocorticoid response to restraint stress in rats during the peak and nadir phases of the HPA axis circadian rhythm. *Neuroscience, 147*, 757–776.

Le Maitre, E., Daubeuf, F., Duterte-Boucher, D., Costentin, J., & Leroux-Nicollet, I. (2006). Coupling of ORL1 (NOP) receptor to G proteins is decreased in the nucleus accumbens of anxious relative to non-anxious mice. *Brain Research*, *1110*, 144–149.

Le Moal, M., & Simon, H. (1991). Mesocorticolimbic dopaminergic network: Functional and regulatory roles. *Physiological Reviews*, *71*, 155–234.

Leng, G., Meddle, S. L., & Douglas, A. J. (2008). Oxytocin and the maternal brain. *Current Opinion in Pharmacology*, *8*, 731–734.

Levita, L., Dalley, J., & Robbins, T. W. (2002). Nucleus accumbens dopamine and learned fear revisited: A review and some new findings. *Behavioural Brain Research*, *137*, 115–127.

Lipschitz, D. L., Crowley, W. R., & Bealer, S. L. (2004). Differential sensitivity of intranuclear and systemic oxytocin release to central noradrenergic receptor stimulation during mid- and late gestation in rats. *American Journal of Physiology Endocrinology and Metabolism*, *287*(3), E523–E528.

Louilot, A., & Besson, C. (2000). Specificity of amygdalostriatal interactions in the involvement of mesencephalic dopaminergic neurons in affective perception. *Neuroscience*, *96*, 72–82.

Macbeth, A. H., Gautreaux, C., & Luine, V. N. (2008). Pregnant rats show enhanced spatial memory, decreased anxiety, and altered levels of monoaminergic neurotransmitters. *Brain Research*, *1241*, 136–147.

Macedo, C. E., Martinez, R. C. R., Albrechet-Souza, L., Molina, V. A., & Brandao, M. L. (2007). 5-HT2 and D1-mechanisms of the basolateral nucleus of the amygdala enhance conditioned fear and impair unconditioned fear. *Behavioural Brain Research*, *177*, 100–108.

Mamiya, T., Yamada, K., Miyamoto, Y., König, N., Watanabe, Y., Noda, Y., et al. (2003). Neuronal mechanism of nociceptin-induced modulation of learning and memory: Involvement of N-methyl-D-aspartate receptors. *Molecular Psychiatry*, *8*, 752–765.

Manabe, T., Noda, Y., Mamiya, T., Katagiri, H., Houtani, T., Nishi, M., et al. (1998). Facilitation of long-term potentiation and memory in mice lacking nociceptin receptors. *Nature*, *394*, 577–581.

Mancuso, R. A., Schetter, C. D., Rini, C. M., Roesch, S. C., & Hobel, C. J. (2004). Maternal prenatal anxiety and corticotropin-releasing hormone associated with timing of delivery. *Psychosomatic Medicine*, *66*(5), 762–769.

Martinez, R. C., Ribeiro de Oliveira, A., & Brandão, M. L. (2007). Serotonergic mechanisms in the basolateral amygdala differentially regulate the conditioned and unconditioned fear organized in the periaqueductal gray. *European Neuropsychopharmacology*, *17*, 717–724.

Martin-Fardon, R., Ciccocioppo, R., Massi, M., & Weiss, F. (2000). Nociceptin prevents stress-induced ethanol- but not cocaine-seeking behavior in rats. *Neuroreport*, *11*(9), 1939–1943.

Matthes, H., Seward, E. P., Kieffer, B., & North, R. A. (1996). Functional selectivity of orphanin FQ for its receptor coexpressed with potassium channel subunits in *Xenopus laevis* oocytes. *Molecular Pharmacology*, *50*(3), 447–450.

McCittrick, C. R., & Abercrombie, E. D. (2007). Catecholamine mapping within nucleus accumbens: Differences in basal and amphetamine-stimulated efflux of norepinephrine and dopamine in shell and core. *Journal of Neurochemistry*, *100*, 1247–1256.

Meis, S., & Pape, H. C. (1998). Postsynaptic mechanisms underlying responsiveness of amygdaloid neurons to nociceptin/orphanin FQ. *The Journal of Neuroscience*, *18*, 8133–8144.

Meis, S., & Pape, H. C. (2001). Control of glutamate and GABA release by nociceptin/orphanin FQ in the rat lateral amygdala. *The Journal of Physiology*, *532*(Pt 3), 701–712.

Meunier, J. C., Mollereau, C., Toll, L., Suandeau, C., Moisand, C., Alvinerie, P., et al. (1995). Isolation and structure of the endogenous agonist of opioid receptor-like ORL1 receptor. *Nature*, *377*, 532–535.

Miller, T. R., & Fulford, A. J. (2007). Regulation of nociceptin/orphanin FQ secretion by immune cells and functional modulation of interleukin-2. *Peptides, 28*, 2243–2252.
Mirenowicz, J., & Schultz, W. (1996). Preferential activation of midbrain dopamine neurons by appetitive rather than aversive stimuli. *Nature, 379*, 449–451.
Mogil, J. S., Grisel, J. E., Reinscheid, R. K., Civelli, O., Belknap, J. K., & Grandy, D. K. (1996). Orphanin FQ is a functional anti-opioid peptide. *Neuroscience, 75*(2), 333–337.
Mollereau, C., Parmentier, M., Mailleux, P., Butour, J. L., Moisand, C., Chalon, P., et al. (1994). ORL1, a novel member of the opioid receptor family. Cloning, functional expression and localization. *FEBS Letters, 341*(1), 33–38.
Mollereau, C., Simons, M. J., Soularue, P., Liners, F., Vassart, G., Meunier, J. C., et al. (1996). Structure, tissue distribution, and chromosomal localization of the prepronociceptin gene. *Proceedings of the National Academy of Sciences of the United States of America, 93*(16), 8666–8670.
Mørk, H., Hommel, K., Uddman, R., Edvinsson, L., & Jensen, R. (2002). Does nociceptin play a role in pain disorders in man? *Peptides, 23*(9), 1581–1587.
Murchison, C. F., Zhang, X.-Y., Zhang, W.-P., Ouyang, M., Lee, A., & Thomas, S. A. (2004). A distinct role for norepinephrine in memory retrieval. *Cell, 117*, 131–143.
Nabeshima, T., Noda, Y., & Mamiya, T. (1999). The role of nociceptin in cognition. *Brain Research, 48*(1–2), 167–173.
Nagal, J., Kurokawa, M., Takeshima, H., Kieffer, B. L., & Ueda, H. (2007). Circadian-dependent learning and memory enhancement in nociceptin receptor deficient mice with a novel Kurobox apparatus using stress-free positive cue task. *The Journal of Pharmacology and Experimental Therapeutics, 321*, 195–201.
Nazzaro, C., Barbieri, M., Varani, K., Beani, L., Valentino, V., & Siniscalchi, A. (2010). Swim stress enhances nociceptin/orphanin FQ-induced inhibition of rat dorsal raphe nucleus activity in vivo and in vitro: Role of corticotropin releasing factor. *Neuropharmacology, 58*, 457–464.
Nazzaro, C., Marino, S., Barbieri, M., & Siniscalchi, A. (2009). Inhibition of serotonin outflow by nociceptin/orphaninFQ in dorsal raphe nucleus slices from normal and stressed rats: Role of corticotropin releasing factor. *Neurochemistry International, 54*(5–6), 378–384.
Neal, C. R., Jr., Mansour, A., Reinscheid, R., Nothacker, H. P., Civelli, O., Akil, H., et al. (1999a). Opioid receptor-like (ORL1) receptor distribution in the rat central nervous system: Comparison of ORL1 receptor mRNA expression with (125)I-[(14)Tyr]-orphanin FQ binding. *The Journal of Comparative Neurology, 412*, 563–605.
Neal, C. R., Jr., Mansour, A., Reinscheid, R., Nothacker, H. P., Civelli, O., & Watson, S. J. (1999b). Localization of orphanin FQ (nociceptin) peptide and messenger RNA in the central nervous system of the rat. *The Journal of Comparative Neurology, 404*, 503–547.
Németh, J., Helyes, Z., Oroszi, G., Thán, M., Pintér, E., & Szolcsányi, J. (1998). Inhibition of nociceptin on sensory neuropeptide release and mast cell-mediated plasma extravasation in rats. *European Journal of Pharmacology, 347*(1), 101–104.
Neumann, I. D., Johnstone, H. A., Hatzinger, M., Liebsch, G., Shipston, M., Russell, J. A., et al. (1998). Attenuated neuroendocrine responses to emotional and physical stressors in pregnant rats involve adenohypophysial changes. *The Journal of Physiology, 508*(Pt 1), 289–300.
Neumann, I. D., Torner, L., & Wigger, A. (2000). Brain oxytocin: Differential inhibition of neuroendocrine stress responses and anxiety-related behaviour in virgin, pregnant and lactating rats. *Neuroscience, 95*(2), 567–575.
Nishi, M., Houtani, T., Noda, Y., Mamiya, T., Sato, K., Doi, T., et al. (1997). Unrestrained nociceptive response and disregulation of hearing ability in mice lacking the nociceptin/orphaninFQ receptor. *The EMBO Journal, 16*(8), 1858–1864.

Nothacker, H. P., Reinscheid, R. K., Mansour, A., Henningsen, R. A., Ardati, A., Monsma, F. J., Jr., et al. (1996). Primary structure and tissue distribution of the orphanin FQ precursor. *Proceedings of the National Academy of Sciences of the United States of America, 93*(16), 8677–8682.

Ouagazzal, A.-M., Moreau, J.-L., Pauly-Evers, M., & Jenck, F. (2008). Impact of environmental housing conditions on the emotional responses of mice deficient for nociceptin/orphaninFQ peptide precursor gene. *Behavioural Brain Research, 144*, 111–117.

Peluso, J., Gaveriaux-Ruff, C., Matthes, H. W., Filliol, D., & Kieffer, B. L. (2001). Orphanin FQ/nociceptin binds to functionally coupled ORL1 receptors on human immune cell lines and alters peripheral blood mononuclear cell proliferation. *Brain Research Bulletin, 54*, 655–660.

Phillips, R. G., & Le Doux, J. E. (1992). Differential contribution of the amygdala and hippocampus to cued and contextual fear conditioning. *Behavioral Neuroscience, 106*, 274–285.

Pi, X. J., & Grattan, D. R. (1999). Increased expression of both short and long forms of prolactin receptor mRNA in hypothalamic nuclei of lactating rats. *Journal of Molecular Endocrinology, 23*(1), 13–22.

Pinna, G., Agis-Balboa, R. C., Pibiri, F., Nelson, M., Guidotti, A., & Costa, E. (2008). Neurosteroid biosynthesis regulates sexually dimorphic fear and aggressive behavior in mice. *Neurochemical Research, 33*, 1990–2007.

Ploj, K., Roman, E., Bergstrom, L., & Nylander, I. (2001). Effects of neonatal handling on nociceptin/orphanin FQ and opioid peptide levels in female rats. *Pharmacology, Biochemistry, and Behavior, 69*, 173–179.

Polidori, C., de Caro, G., & Massi, M. (2000). The hyperphagic effect of nociceptin/orphanin FQ in rats. *Peptides, 21*(7), 1051–1062.

Quesada, A., & Micevych, P. (2008). Estrogen and progesterone modulate [^{35}S] GTP gammaS binding to nociceptin receptors. *Neuroendocrinology, 88*, 35–42.

Reinscheid, R. K., Nothacker, H. P., Bourson, A., Ardati, A., Henningsen, R. A., Bunzow, J. R., et al. (1995). Orphanin FQ: A neuropeptide that activates an opioidlike G protein-coupled receptor. *Science, 270*(5237), 792–794.

Reynolds, S. M., & Berridge, K. C. (2008). Emotional environments retune the valence of appetitive versus fearful functions in nucleus accumbens. *Nature Neuroscience, 11*, 423–425.

Rini, C. K., Dunkel-Schetter, C., Wadhwa, P. D., & Sandman, C. A. (1999). Psychological adaptation and birth outcomes: The role of personal resources, stress, and sociocultural context in pregnancy. *Health Psychology, 18*(4), 333–345.

Rizzi, A., Bigoni, R., Marzola, G., Guerrini, R., Salvadori, S., Regoli, D., et al. (2001). Characterization of the locomotor activity-inhibiting effect of nociceptin/orphanin FQ in mice. *Naunyn-Schmiedeberg's Archives of Pharmacology, 363*(2), 161–165.

Rodi, D., Zucchini, S., Simonato, M., Cifani, C., Massi, M., & Polidori, C. (2008). Functional antagonism between nociceptin/orphanin FQ (N/OFQ) and corticotropin-releasing factor (CRF) in the rat brain: Evidence for involvement of the bed nucleus of the stria terminalis. *Psychopharmacology, 196*(4), 523–531.

Roitman, M. F., Wheeler, R. A., & Carelli, R. M. (2005). Nucleus accumbens neurons are innately tuned for rewarding and aversive taste stimuli, encode their predictors, and are linked to motor output. *Neuron, 45*, 587–597.

Roman, E., Ploj, K., Gustafsson, L., Meyerson, B. J., & Nylander, I. (2006). Variations in the opioid peptide levels during the estrous cycle in Sprague-Dawley rats. *Neuropeptides, 40*, 195–206.

Roozendaal, B., Lengvilas, R., McGaugh, J. L., Civelli, O., & Reinscheid, R. K. (2007). Orphanin FQ/nociceptin interacts with the basolateral amygdala noradrenergic system in memory consolidation. *Learning & Memory, 14*, 29–35.

Rosén, A., Lundeberg, T., Bytner, B., & Nylander, I. (2000). Central changes in nociceptin dynorphin B and Met-enkephalin-Arg-Phe in different models of nociception. *Brain Research*, *857*(1–2), 212–218.

Sakurada, T., Sakurada, S., Katsuyama, S., Sakurada, C., Tan-No, K., & Terenius, L. (1999). Nociceptin (1–7) antagonizes nociceptin-induced hyperalgesia in mice. *British Journal of Pharmacology*, *28*(5), 941–944.

Salamone, J. D. (1994). The involvement of nucleus accumbens dopamine in appetitive and aversive motivation. *Behavioural Brain Research*, *61*, 117–133.

Sato, M. (1992). A lasting vulnerability to psychosis in patients with previous methamphetamine psychosis. *Annals of the New York Academy of Sciences*, *654*, 160–170.

Saulskaya, N., & Marsden, C. A. (1995). Conditioned dopamine release- dependence on N-methyl-D-aspartate receptors. *Neuroscience*, *67*, 57–63.

Schlicker, E., & Morari, M. (2000). Nociceptin/orphanin FQ and neurotransmitter release in the central nervous system. *Peptides*, *21*(7), 1023–1029.

Schoenbaum, G., & Setlow, B. (2003). Lesions of the nucleus accumbens disrupt learning about aversive outcomes. *The Journal of Neuroscience*, *23*, 9833–9841.

Schultz, W. (2007). Mutliple dopamine functions at different time-courses. *Annual Review of Neuroscience*, *30*, 259–288.

Serhan, C. N., Fierro, I. M., Chiang, N., & Pouliot, M. (2001). Cutting edge: Nociceptin stimulates neutrophil chemotaxis and recruitment: Inhibition by aspirin-triggered-15-epi-lipoxin A4. *Journal of Immunology*, *166*(6), 3650–3654.

Sim-Selley, L. J., Vogt, L. J., Childers, S. R., & Vogt, B. A. (2003). Distribution of ORL-1 receptor binding and receptor-activated G-proteins in rat forebrain and their experimental localisation in anterior cingulate cortex. *Neuropharmacology*, *45*, 220–230.

Sinchak, K., Hendricks, D. G., Baroudi, R., & Micevych, P. E. (1997). Orphanin FQ/nociceptin in the ventromedial nucleus facilitates lordosis in female rats. *Neuroreport*, *8*, 3857–3860.

Sinchak, K., Romeo, H. E., & Micevych, P. E. (2006). Site-specific estrogen and progestin regulation of orphanin FQ/nociceptin and nociceptin opioid receptor mRNA expression in the female rat limbic hypothalamic system. *The Journal of Comparative Neurology*, *496*, 252–268.

Sinha, R., Catapano, D., & O'Malley, S. (1999). Stress-induced craving and stress response in cocaine-dependent individuals. *Psychopharmacology*, *142*, 343–351.

Sternberg, E. M., Chrousos, G. P., Wilder, R. L., & Gold, P. W. (1992). The stress response and the regulation of inflammatory disease. *Annals of Internal Medicine*, *117*(10), 854–866.

Tao, R., Ma, Z., Thakkar, M. M., McCarley, R. W., & Auerbach, S. B. (2007). Nociceptin/orphanin FQ decreases serotonin efflux in the rat brain but in contrast to a kappa-opioid has no antagonistic effect on mu-opioid-induced increases in serotonin efflux. *Neuroscience*, *147*, 106–116.

Tian, J. H., Xu, W., Fang, Y., Mogil, J. S., Grisel, J. E., Grandy, D. K., et al. (1997). Bidirectional modulatory effect of orphanin FQ on morphine-induced analgesia: Antagonism in brain and potentiation in spinal cord of the rat. *British Journal of Pharmacology*, *120*(4), 676–680.

Toufexis, D. J., Davis, C., Hammond, A., & Davis, M. (2004). Progesterone attenuates corticotropin-releasing factor-enhanced but not fear-potentiated startle via the activity of its neuroactive metabolite, allopregnanolone. *The Journal of Neuroscience*, *24*, 10280–10287.

Uchiyama, H., Toda, A., Hiranita, T., Watanabe, S., & Eyanagi, R. (2008). Role of amygdaloid nuclei in anxiolytic-like effect of nociceptin/orphaninFQ in rats. *Neuroscience Letters*, *431*, 66–70.

Ungless, M. A. (2004). Dopamine: The salient issue. *Trends in Neurosciences*, *27*, 702–706.

Ungless, M. A., Magill, P. J., & Bolam, J. P. (2004). Uniform inhibition of dopamine neurons in the ventral tegmental area by aversive stimuli. *Science*, *303*, 2040–2042.

Varty, G. B., Hyde, L. A., Hodgson, R. A., Lu, S. X., McCool, M. F., Kazdoba, T. M., et al. (2005). Characterization of the nociceptin receptor (ORL-1) agonist, Ro64-6198, in tests of anxiety across multiple species. *Psychopharmacology, 182*, 132–143.

Varty, G. B., Lu, S. X., Morgan, C. A., Cohen-Williams, M. E., & Hodgson, R. A. (2008). The anxiolytic-like effects of the novel, orally active nociceptin opioid receptor agonist 8-[bis(2-methylphenyl)methyl]-3-phenyl-8-azabicyclo[3.2.1]octan-3-ol (SCH 221510). *The Journal of Pharmacology and Experimental Therapeutics, 326*, 672–682.

Vitale, G., Arletti, R., Ruggieri, V., Cifani, C., & Massi, M. (2006). Anxiolytic-like effects of nociceptin/orphanin FQ in the elevated plus maze and conditioned defensive burying test in rats. *Peptides, 27*, 2193–2200.

Vitale, G., Ruggieri, V., Filaferro, M., Frigeri, C., Alboni, S., Tascedda, F., et al. (2009). Chronic treatment with the selective NOP receptor antagonist [Nphe 1, Arg 14, Lys 15]N/OFQ-NH 2 (UFP-101) reverses the behavioural and biochemical effects of unpredictable chronic mild stress in rats. *Psychopharmacology, 207*(2), 173–189.

Waits, P. S., Purcell, W. M., Fulford, A. J., & McLeod, J. D. (2004). Nociceptin/orphanin FQ modulates human T cell function in vitro. *Journal of Neuroimmunology, 149*, 110–120.

Walf, A. A., Ciriza, I., Garcia-Segura, L. M., & Frye, C. A. (2008). Antisense oligodeoxynucleotides for estrogen receptor-α and β attenuate estradiol's modulation of affective and sexual behaviour, respectively. *Neuropsychopharmacology, 33*, 431–440.

Wartella, J., Amory, E., Lomas, L. M., Macbeth, A., McNamara, I., Stevens, L., et al. (2003). Single or multiple reproductive experiences attenuate neurobehavioral stress and fear responses in the female rat. *Physiology & Behavior, 79*, 373–381.

Wilensky, A. E., Schafe, G. E., Kristensen, M. P., & LeDoux, J. E. (2006). Rethinking the fear circuit: The central nucleus of the amygdala is required for the acquisition, consolidation, and expression of Pavlovian fear conditioning. *The Journal of Neuroscience, 26*, 12387–12396.

Wilkinson, L. S. (1997). The nature of interactions involving prefrontal and striatal dopamine systems. *Journal of Psychopharmacology, 11*, 143–150.

Williams, J. P., Thompson, J. P., Young, S. P., Gold, S. J., McDonald, J., Rowbotham, D. J., et al. (2008). Nociceptin and urotensin-II concentrations in critically ill patients with sepsis. *British Journal of Anaesthesia, 100*(6), 810–814.

Xu, X., Grass, S., Hao, J., Xu, I. S., & Wiesenfeld-Hallin, Z. (2000). Nociceptin/orphanin FQ in spinal nociceptive mechanisms under normal and pathological conditions. *Peptides, 21*(7), 1031–1036.

Xu, I. S., Grass, S., Wiesenfeld-Hallin, Z., & Xu, X. J. (1999). Effects of intrathecal orphanin FQ on a flexor reflex in the rat after inflammation or peripheral nerve section. *European Journal of Pharmacology, 370*(1), 17–22.

Yamamoto, T., Nozaki-Taguchi, N., & Kimura, S. (1997). Analgesic effect of intrathecally administered nociceptin, an opioid receptor-like1 receptor agonist, in the rat formalin test. *Neuroscience, 81*(1), 249–254.

Young, A. M. J. (2004). Increased extracellular dopamine in nucleus accumbens in response to unconditioned and conditioned aversive stimuli: Studies using 1 min microdialysis in rats. *Journal of Neuroscience Methods, 138*, 57–63.

Young, A. M. J., Joseph, M. H., & Gray, J. A. (1993). Latent inhibition of conditioned dopamine release in rat nucleus accumbens. *Neuroscience, 54*, 5–9.

Young, A. M. J., Moran, P. M., & Joseph, M. H. (2005). The role of dopamine in conditioning and latent inhibition: What, when, where and how? *Neuroscience and Biobehavioral Reviews, 29*, 963–976.

CHAPTER ELEVEN

The Neuronal Circuit Between Nociceptin/Orphanin FQ and Hypocretins/Orexins Coordinately Modulates Stress-Induced Analgesia and Anxiety-Related Behavior

Xinmin (Simon) Xie[*,†,1]

*AfaSci Research Laboratories, Redwood City, California, USA
†Department of Anesthesia, Stanford University School of Medicine, Stanford, California, USA
[1]Corresponding author: e-mail address: simonxie@afasci.com

Contents

Abstract

The neuropeptide nociceptin/orphanin FQ (N/OFQ), acting on its receptors (NOP), modulates a variety of biological functions and neurobehavior including nociception, stress

Vitamins and Hormones, Volume 97
ISSN 0083-6729
http://dx.doi.org/10.1016/bs.vh.2014.11.004

responses, water and food-intake, locomotor activity, and spatial attention. N/OFQ is conventionally regarded as an "antiopiate" peptide in the brain because central administration of N/OFQ attenuates stress-induced analgesia (SIA) and produces anxiolytic effects. However, naloxone-irreversible SIA and anxiolytic action are unlikely to be mediated by the opiate system.

Both N/OFQ and NOP receptors are expressed most abundantly in the hypothalamus, where two other neuropeptides, the hypocretins/orexins (Hcrts), are exclusively synthesized in the lateral hypothalamic area. N/OFQ and Hcrt regulate most cellular physiological responses in opposite directions (e.g., ion channel modulation and second messenger coupling), and produce differential modulations for almost all neurobehavior assessed, including sleep/wake, locomotion, and rewarding behaviors. This chapter focuses on recent studies that provide evidence at a neuroanatomical level showing that a local neuronal circuit linking N/OFQ to Hcrt neurons exists. Functionally, N/OFQ depresses Hcrt neuronal activity at the cellular level, and modulates stress responses, especially SIA and anxiety-related behavior in the whole organism. N/OFQ exerts its attenuation of SIA and anxiolytic action on fear-induced anxiety through direct modulation of Hcrt neuronal activity. The information obtained from these studies has provided insights into how interaction between the Hcrt and N/OFQ systems positively and negatively modulates the complex and integrated stress responses.

1. INTRODUCTION

The endogenous 17-amino acid neuropeptide nociceptin/orphanin FQ (N/OFQ), via acting on its receptor NOP, modulates a variety of biological functions and neurobehavior including nociception, stress responses, drinking and feeding behavior, locomotor activity, and spatial attention (Leon-Olea et al., 2013; Meunier et al., 1995; Rizzi, Molinari, Marti, Marzola, & Calo, 2011; Witkin et al., 2013; Zhang, Gandhi, & Standifer, 2012). N/OFQ is conventionally regarded as an "antiopiate" peptide in the brain because central N/OFQ causes pro-nociception and reversal of stress-induced analgesia (SIA). However, naloxone- irreversible SIA (Meunier, 1997; Watanabe, Kuwaki, Yanagisawa, Fukuda, & Shimoyama, 2005) and anxiolytic effects are unlikely to be mediated by the opiate system. The neuronal pathways that mediate the N/OFQ effects on these nonopioid-mediated actions are poorly understood.

Both N/OFQ and NOP receptors are expressed throughout the central nervous system (CNS) and distributed most abundantly in the hypothalamus (Anton et al., 1996; Meunier, 1997; Mollereau & Mouledous, 2000; Neal et al., 1999), where two more recently discovered neuropeptides, the hypocretins/orexins (Hcrts), are exclusively synthesized (de Lecea et al., 1998;

Sakurai et al., 1998). The Hcrts play an important role in regulation of wakefulness and sleep, the integration of adaptive behavior: arousal and vigilance, threat defense responses, stress adaption, nociceptive processing and respiratory regulation and metabolic control (Burdakov, Karnani, & Gonzalez, 2013; Kilduff, 2005; Li, Hu, & de Lecea, 2014; Peyron et al., 1998; Sakurai & Mieda, 2011). A study using prepro-orexin knockout (KO) mice has demonstrated that the Hcrt system contributes to SIA (Watanabe et al., 2005). Since N/OFQ and Hcrt regulate most physiological responses in opposite directions and produce differential modulations for almost all cellular actions assessed (e.g., ion channel modulation and second messenger coupling), we hypothesized that N/OFQ exerts its effects on neurobehavior, primarily SIA and fear stress-induced anxiety-related behavior, through direct modulation of Hcrt neuronal activity in the LHA (Gerashchenko, Horvath, & Xie, 2011; Xie et al., 2008). This chapter reviews recent studies and presents evidence that supports the hypotheses that a local neuronal circuit directly linking the N/OFQ to Hcrt systems in fine modulation of neuronal activity and neurobehavior.

2. THE N/OFQ SYSTEM

2.1 The discovery of N/OFQ

The cloning of the μ-, δ-, and κ-opioid receptors led to identification of a closely related orphan receptor, initially termed opioid receptor-like 1 (ORL1), XOR1, or LC132 (Bunzow et al., 1994; Fernandez, Misilmeri, Felger, & Devine, 2004; Mollereau & Mouledous, 2000; Shoblock, 2007). This receptor, now officially named NOP receptor, has nucleotide and amino acid homology to the three opioid receptors (Chiou et al., 2007). NOP and opiate receptors couple via the same $G_{i/o}$ type of G-proteins to similar effectors such as K^+ channels, Ca^{2+} channels, adenylyl cyclase, and several protein kinases. Meunier and his colleagues purified and sequenced an endogenous ligand for NOP from rat brain and named this 17-amino acid peptide *nociceptin* because of its pronociceptive activity (Meunier et al., 1995). At the same period, Civelli and his colleagues independently identified the identical peptide from bovine brain and termed the peptide as *orphanin FQ* because it binds to an orphan receptor with its first and last amino acids being Phe (F) and Gln (Q) (Reinscheid et al., 1995). In addition to inhibition of cAMP production, N/OFQ via its receptor activation produces inhibitory cellular actions in widespread areas of the brain through activation of inwardly rectifying K^+ channels and inhibition of

L-, N-, and P/Q-type calcium channels (Borgland, Connor, & Christie, 2001; Connor & Christie, 1998; Connor, Vaughan, Chieng, & Christie, 1996; Knoflach, Reinscheid, Civelli, & Kemp, 1996).

2.2 Complex modulation of nociceptive processing by N/OFQ

Although both N/OFQ and its receptor share many structural and cellular action similarities with the opioid peptides and their receptors, when injected intracerebroventricularly (icv) into mice, N/OFQ resulted in a decrease in hotplate escape jumping latency and in tail flick latency, suggesting that it is pronociceptive (Meunier, 1997; Meunier et al., 1995; Reinscheid et al., 1995). Subsequent studies with additional controls showed that icv injection of saline *per se* caused an opioid-mediated SIA that could be reversed by either naloxone or N/OFQ (Mogil, Grisel, Zhangs, Belknap, & Grandy, 1996). Further studies found that SIA is mediated by at least two distinct neuronal mechanisms: opioid- (naloxone-reversible) and nonopioid-mediated pain inhibitory pathways (naloxone-irreversible), depending on the intensity of the stressor. If the severity of the stress is great enough, nonopioid-mediated pain inhibitory pathways will be activated.

SIA plays a role in adaptation to threats and is a component of the defensive behavioral response to prepare for "fight or flight." However, exaggerated or prolonged SIA phenomenon has been regarded as a performance deficit (Blair, Galina, Holmes, & Amit, 1982). Recently it has been suggested SIA may serve as a measure of stress severity and as a model of "depersonalization disorder" in humans. Thus, SIA must be critically regulated. (Kenunen & Prakh'e, 2005). SIA is partially depressed by naloxone, but is completely blocked by centrally administered N/OFQ. Using targeted disruption of the N/OFQ or with an NOP antagonist, researchers have shown that endogenous N/OFQ plays a role in tonic inhibition of SIA and other stress responses (Calo et al., 2000; Koster et al., 1999; Rizzi et al., 2001). Since N/OFQ exerts inhibitory effects on both opioid- and nonopioid-mediated SIA, it is not strictly an antiopiate. The neural pathways modulated by N/OFQ that could account for antagonism of nonopioid analgesia were unknown. Although N/OFQ produces antiopioid and pronociceptive effects when injected icv, N/OFQ was found to be analgesic when administered in the spinal cord and was also reported to be an effective analgesic in several models of chronic pain when injected intrathecally. Both N/OFQ itself and the potent and selective antagonist SB-612111 have been reported to possess antihyperalgesic activity in models of chronic pain

(Hao, Yu, & Xu, 1998; Zaratin et al., 2004). Nonpeptide SR14150 is a high-affinity NOP partial agonist, while SR16835 is a full agonist with slightly lower affinity. The selective NOP antagonists SR14148 and SR16430 were also recently identified. All these small-molecule ligands produce complex modulation on pain in rodents (Khroyan et al., 2011; Zaveri et al., 2004; Zaveri, Jiang, Olsen, Polgar, & Toll, 2005). These small-molecule ligands can cross the blood–brain barrier and thus facilitate the research into NOP activation or inhibition in modulation of pain.

2.3 N/OFQ and the stress response

In addition to reversing SIA and modulating other nociceptive effects, anxiolytic-like activity after central administration of N/OFQ has been demonstrated using a variety of behavioral test paradigms. The anxiolytic-like effects of N/OFQ were consistent across several behavioral paradigms, generating different types of anxiety states in animals, and these effects were observed at low, nonsedative doses (Jenck et al., 1997). Studies using transgenic mice lacking expression of the N/OFQ precursor protein confirmed the physiological functions of N/OFQ in the neuronal modulation of stress responses (Koster et al., 1999; Reinscheid & Civelli, 2002). The absence of N/OFQ increased stress-related variables in behavior and sensory processing in null mutant mice. Basal and poststress plasma corticosterone levels were found to be elevated in N/OFQ-deficient animals, indicating a chronic activation of the hypothalamic-pituitary-adrenal (HPA) system that might contribute to the observed phenotypic changes. These results suggest that the N/OFQ system may have important functions in the neural circuitry involved in the neurobiological regulation of behavior to cope with stress and fear (Koster et al., 1999; Reinscheid & Civelli, 2002).

Mice lacking the NOP receptor appeared to have similar anxiety-like behavior compared to WT mice in standard group-housing conditions. However, under single housing conditions the anxiolytic effects of the endogenous N/OFQ system became apparent (Ouagazzal, Moreau, Pauly-Evers, & Jenck, 2003). Moreover, another study reported that icv administration of N/OFQ in rats increased anxiety-related behavior during neophobic tests of anxiety and elevated circulating levels of corticosterone (Fernandez et al., 2004). While the reason for the discrepancy is unknown, it raises questions about how N/OFQ neurotransmission participates in processing emotionally salient and stressful stimuli and indicates that N/OFQ likely interacts with other neuropeptides. Similar to nonopioid-mediated SIA, neural substrates modulated by N/OFQ that could account for anxiolytic effects *in vivo* are elusive.

3. THE HYPOCRETINS/OREXINS SYSTEM

3.1 The discovery of Hcrts

Using a differential cloning approach hypocretins 1 and 2 (Hcrt1 and Hcrt2) were discovered by de Lecea and Kilduff, and their colleagues in 1998 (de Lecea et al., 1998). Soon thereafter, Sakurai and his colleagues independently identified these peptides, termed orexin A and orexin B as ligands binding to two orphan G protein-coupled receptors (Sakurai et al., 1998). Hcrt1 and Hcrt2 are pairs of neuropeptides derived from a single precursor, *prepro-orexin,* which undergoes subsequent proteolytic cleavage. In contrast to the widespread expression of *prepro-N/OFQ*, the *prepro-orexin* is almost exclusively expressed in the posterior lateral and perifornical hypothalamic areas. In this chapter, we use "Hcrt1" to denote orexin-A—a 33-amino acid residue polypeptide with an N-terminal pyroglutamyl residue and C-terminal amidation; and we use "Hcrt2" to denote orexin-B—a 28-amino acid linear polypeptide with C-terminal amidation. To refer to the receptors for these peptides, we use "HcrtR1" and "HcrtR2" to denote orexin A and orexin B receptors, respectively. However, for consistency with the literature, we maintain the original names for the three transgenic mouse lines orexin/EGFP (Yamanaka et al., 2003), orexin/ataxin-3 (Hara et al., 2001), and orexin/YC2.1 (Tsujino et al., 2005).

In contrast with N/OFQ, the cellular action of the Hcrt peptides has been uniformly reported as excitatory, eliciting depolarization and increasing firing of action potentials of many cell types in a variety of brain regions. Several ionic mechanisms have been reported to underlie the excitatory effects of Hcrt. A calcium imaging study (van den Pol, Gao, Obrietan, Kilduff, & Belousov, 1998) showed that the Hcrt peptides raised cytoplasmic Ca^{2+} in hypothalamic cells via a G_q-protein enhancement of Ca^{2+} influx through plasma membrane channels. Patch clamp studies revealed that Hcrt could induce a Na^{2+}-dependent nonspecific cation current (Davis, Williams, Xu, Glatzer, & Smith, 2003; Hwang, Chen, & Dun, 2001; Yang & Ferguson, 2002), decrease G protein-coupled inward rectifier conductance (GIRK) (Hoang, Bajic, Yanagisawa, Nakajima, & Nakajima, 2003), increase Na^{2+}/Ca^{2+} exchanger conductance (Eriksson, Sergeeva, Stevens, & Haas, 2002), and reduce the slow component of after hyperpolarization (Horvath, Diano, & van den Pol, 1999). These varied physiological mechanisms, combined with Hcrt's modulation of synaptic transmission, support the notion that the peptide has a potent neuromodulatory effect on a variety of physiological systems in the CNS.

3.2 Hcrt-induced analgesia

The Hcrts play an important role in regulation of wakefulness and sleep, particularly the integration of adaptive behavior: arousal and vigilance, stress adaptive and threat defense responses, drinking and food seeking, and nociceptive processing (Kilduff, 2005; Li et al., 2014; Peyron et al., 1998; Sakurai & Mieda, 2011). Defects in this system, either loss of the Hcrt neurons, deletion of Hcrts or mutation of HcrtR, result in the sleep disorder narcolepsy in both humans and animal models (Hungs & Mignot, 2001; Kroeger & de Lecea, 2009; Peyron et al., 2000; Ripley et al., 2001). Furthermore, several lines of evidence have suggested that the Hcrts play a role in nociceptive processing. Localization of Hcrt fibers to the hypothalamus, thalamus, and periaqueductal gray is consistent with a role in sensory processing. Moreover, there is evidence for the presence of robust projections from the hypothalamus to lamina I of the spinal cord, an area associated with nociceptive transmission (van den Pol, 1999).

Behavioral pain studies show that Hcrt1 is analgesic when given icv or intravenous injection, but not subcutaneous injection, in mouse and rat models of nociception and hyperalgesia (Bingham et al., 2001). The efficacy of Hcrt1 was similar to that of morphine in the hotplate test and the carrageenan-induced thermal hyperalgesia test. Furthermore, Hcrt1-induced analgesic effects were mediated via HcrtR1, and involvement of the opiate system was ruled out because Hcrt1-induced analgesia was blocked by the orexin-A antagonist SB-334867, but not by naloxone in rats and in mice (Bingham et al., 2001). Orexin-A antagonists had no effect in acute nociceptive tests, but under inflammatory conditions, they were prohyperalgesic, suggesting a tonic inhibitory Hcrt drive under these pathological conditions. A study employing *prepro-orexin* KO mice (Watanabe et al., 2005) found that although the baseline pain thresholds of KO and wild-type (WT) mice were not different, KO mice displayed a greater degree of hyperalgesia caused by carrageenan-induced inflammation and less SIA than WT mice did (Watanabe et al., 2005). Double staining of Hcrt and Fos in WT mice revealed activation of Hcrt neurons under both experimental conditions. These results suggest that persistent pain and stress activate Hcrt neurons, which act to inhibit pain transmission. Furthermore, recent studies show that icv administration of Hcrt-1 before middle cerebral artery occlusion (MCAO) in rats (Kitamura et al., 2010; Yuan et al., 2011) and mice (Harada, Fujita-Hamabe, & Tokuyama, 2011) decrease the infarct size. Conversely, using orexin/ataxin-3 mice, in which Hcrt-producing neurons degenerate in early adulthood, we found that infarct volumes after MCAO were significantly larger, with worse neurological scores, and decreased spontaneous activity in transgenic AT compared to WT mice. Further studies reveal that endogenous Hcrt plays a protective role against ischemia by modulating inflammatory response following focal cerebral ischemia (Xiong et al., 2013).

3.3 Hcrts and stress responses

Behavioral arousal and alertness are prerequisites of the stress response. The Hcrt system sets the baseline of arousal and vigilance states under normal conditions, and also plays a role in defense responses to a threat (Kayaba et al., 2003). During stress, activation of the HPA axis is initiated by corticotropin-releasing hormone (CRH), which is synthesized in the paraventricular hypothalamic nucleus (PVN). Lateral hypothalamic *prepro-orexin* mRNA expression is decreased by 50% after adrenalectomy, whereas peripheral glucocorticoid treatment with dexamethasone restores its expression to normal levels (Stricker-Krongrad & Beck, 2002). CRH-immunoreactive terminals make direct synaptic contacts with Hcrt-expressing neurons, and numerous Hcrt neurons express the CRH-R1/2 receptors. Furthermore, CRH excites a subpopulation of Hcrt cells through CRH-R1 receptors. Activation of Hcrt neurons in response to acute stress is severely impaired in CRH-R1 KO mice (Winsky-Sommerer, Boutrel, & De Lecea, 2003; Winsky-Sommerer et al., 2004). Moreover, dual immunostaining indicated that immobilized stress or cold exposure induced Fos expression on Hcrt-containing neurons in the rat hypothalamus (Sakamoto, Yamada, & Ueta, 2004). These results provide evidence that Hcrt expression and neuronal activity in the LHA are modulated by stress-induced elevations of CRH and glucocorticoids (Stricker-Krongrad & Beck, 2002).

Conversely, central administration of Hcrt1 caused a rapid increase in plasma adrenocorticotropic hormone and corticosterone concentrations, as well as increased CRH and arginine vasopressin (AVP) mRNA levels in the parvocellular cells of the PVN, indicating that Hcrt1 activates the HPA axis involving both CRH and AVP expression (Ida et al., 2000; Sakamoto et al., 2004). The stress response is impaired in Hcrt-deficient mice (Zhang, Shimoyama, Fukuda, & Kuwaki, 2006). These data suggest that, in response to stressor stimuli, CRH stimulates the release of Hcrts, which reciprocally stimulate CRH release. This positive feedback circuit contributes to activation and maintenance of arousal associated with the stress response.

4. INTERACTION BETWEEN THE N/OFQ AND HCRT SYSTEMS

Physiological responses to stress or noxious stimuli include changes in sensory processing, endocrine and metabolic homeostasis, and behavior, which ultimately leads to the advantageous behavioral strategy during stress and the development of analgesia induced by the stress (Kenunen & Prakh'e, 2005). Such complex and integrated adaptive behaviors must be critically regulated by a multitude of neuronal circuits. Characterization of these pathways may identify novel stress-related CNS mechanisms for the regulation of SIA and stress adaptive response that may have implications for the management of stress in humans. Hypersensitivity to a nonnoxious stimulus is manifested as persistent pain following a transient noxious stimulus. Similarly, increased vulnerability to stress is thought to be a major contributing factor in human psychiatric disorders, such as general anxiety disorder and possibly related to development of posttraumatic stress disorder.

It is now recognized that Hcrts not only play a crucial role in the stability of arousal and alertness but also contribute to the normal stress adaptive response under physiological conditions and may be involved in abnormal anxiogenesis and addiction under "pathological" conditions (Johnson, Molosh, Fitz, Truitt, & Shekhar, 2012; Paneda, Winsky-Sommerer, Boutrel, & de Lecea, 2005; Winsky-Sommerer et al., 2004). In contrast, N/OFQ consistently produces inhibitory effects on neuronal activity and attenuates stress responses, manifesting as attenuation of SIA and anxiolytic effects as described above. It is emerging that these two systems regulate physiological and adaptive behavioral responses, particularly in SIA and anxiety, in opposing directions. These specific actions are consistent with their differential modulations in more general global behaviors such as activity/inactivity, wake/sleep, locomotion, feeding, and rewarding. Therefore, it is plausible that a significant interaction between these two neuropeptide systems may occur. A few studies have explored this potential direct interaction. An electrophysiological study demonstrated that Hcrt suppresses GIRK currents that are activated by N/OFQ in nucleus tuberomammillaris cells, indicating differential modulation on K conductance by these two neuropeptides (Hoang et al., 2003). Another study showed that N/OFQ does not inhibit exogenously applied Hcrt-evoked norepinephrine release in rat cerebrocortical slices, indicating a lack of effects of N/OFQ at Hcrt projection sites in the cerebral cortex (Hirota et al., 2003).

Based on the growing evidence that suggests the Hcrt and N/OFQ systems play positive and negative modulations, respectively, in a variety of behavioral responses to external stimuli, particularly stressors, a series of studies have been conducted to test *the hypothesis that N/OFQ exerts its in vivo effects on neurobehavior, primarily SIA and anxiety-related behavior, through modulation of the Hcrt system via direct inhibition of neuronal activity in the in the LHA*. Several criteria have been met to establish the previously unrecognized neuronal circuit of interaction between N/OFQ and the Hcrt systems, which is now reviewed in this chapter. One such neuronal circuit that integrates this dual modulation has recently been illustrated by the author and his colleagues (Gerashchenko et al., 2011; Xie et al., 2008; Zou et al., 2010, APSS Abstract #0216).

4.1 A local and direct neuronal circuit between N/OFQ- and Hcrt-producing neurons

Physical contact of N/OFQ-containing afferents with Hcrt neurons or N/OFQ co-localization in the Hcrt neurons provides a structural basis for direct interaction. A local neuronal circuit between N/OFQ- and Hcrt-producing neurons has been initially established using immunohistochemical studies (Gerashchenko et al., 2011; Xie et al., 2008). Using fluorescent light microscopy and a confocal microscope, the presence of N/OFQ-immunoreactive fibers in the vicinity of, and in putative contact with, Hcrt-immunoreactive neurons in the hypothalamus of mice and rats is shown in Fig. 1A and B.

Electron microscopic (EM) techniques would be the definitive way to establish synaptic contacts between N/OFQ fibers and Hcrt neurons. As shown in Fig. 1C, the color light micrograph illustrates a light brown immunolabeled hypocretin neuron (Hcrt) contacted by a dark black bouton (black arrow) representing immunolabeling for N/OFQ. Under the EM it appears that the N/OFQ-immunolabeled axon terminal is in synaptic contact with the Hcrt-immunolabeled dendrite in the posterior lateral and perifornical hypothalamic areas of the mouse (Fig. 1C).

To confirm the direct neuronal connection between the N/OFQ and Hcrt systems in the brain, another species, the rat, was used to identify

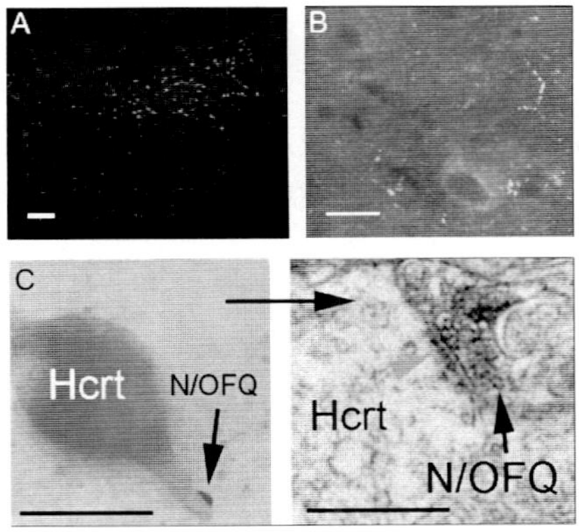

Figure 1 (A) A low-power micrograph shows Hcrt-immunoreactive neurons at the level of the perifornical area in the mouse hypothalamus. Scale bar 100 μm. (B) A confocal image shows that N/OFQ-immunoreactive fibers are seen in the vicinity of, and are in putative contact with Hcrt-immunoreactive neurons in the same section as (A). N/OFQ (green) fibers are in close proximity to Hcrt-immunoreactive (red) cells. Scale bar 10 μm. (C) A color light micrograph of a light brown immunolabeled Hcrt neuron contacted by a dark black bouton (black arrow) representing immunolabeling for N/OFQ. Bar scale 10 μm. Right panel, an electron micrograph taken from ultrathin sections of the bouton indicated on the light micrograph on the left panel. Black arrow indicated the N/OFQ immunolabeled axon terminal that is in synaptic contact (red arrowheads) with the Hcrt-immunolabeled dendrite. Bar scale 1 μm. *Figure was modified from Xie et al. (2008).* (See the color plate.)

the sources of N/OFQ input to the Hcrt neurons. Although Hcrt neurons do not constitute a nucleus, they are exclusively located in the LHA. In contrast, N/OFQ-containing neurons and fibers are widely distributed throughout the brain of rodents (Anton et al., 1996; Neal et al., 1999). N/OFQ-immunoreactive cells (labelled in green) are observed in the lateral hypothalamic area and in the area dorsal to the core distribution of Hcrt-immunoreactive cells (labelled in red). Both neuropeptides apparently are not co-localized in the same neurons, which appear to be connected and form a local circuit (Fig. 2A and B). These two peptidergic systems are synaptically connected and this connection was confirmed using EM analysis. Fig. 2C presents three examples of N/OFQ-immunolabeled axon terminals evidently in synaptic contact with Hcrt-immunolabeled dendrites of Hcrt neurons. Interestingly, another study conducted in a similar period showed that most N/OFQ-immunoreactive cells in the LHA also contain Hcrts and activity of this group of neurons is associated with the control of feeding behavior and body weight in rats (Maolood & Meister, 2010). Whether N/OFQ and Hcrts can be co-released is currently unknown. Nevertheless,

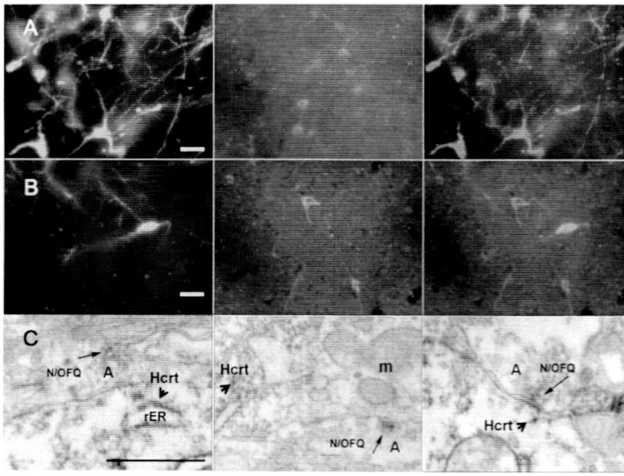

Figure 2 (A) Hcrt-immunoreactive cells (green color) and N/OFQ-immunoreactive cells (red color) are located in the perifornical hypothalamic area of the rat. (B) Hcrt-immunoreactive cells (green color) and N/OFQ-immunoreactive cells (red color) are located in the area dorsal to perifornical hypothalamic area of the rat. The right column shows merged images of left and middle columns. Scale bar 20 μm. (C) Electron micrograph showing an asymmetrical synaptic membrane specialization (black arrow) between the N/OFQ bouton-like structure and the Hcrt-immunolabeled dendrite in the rat lateral hypothalamus. Annotations to the photographs indicate the different cellular elements: A to indicate axon terminals, arrowheads to indicate Hcrt IHC postsynaptically, rER to point to rough endoplasmic reticulum, and m to indicate mitochondria. Scale bar 1 μm. *Figure was modified from Gerashchenko et al. (2011).* (See the color plate.)

the finding that N/OFQ-synthesizing neurons in the LHA are located proximately to Hcrt neurons or even co-localized in the same cells and the N/OFQ-containing fibers make synaptic contact with Hcrt cells provide the neuroanatomical circuit for a potential direct functional interaction between the N/OFQ and Hcrt systems.

4.2 Cellular physiological and pharmacological actions of N/OFQ on Hcrt neurons

At the cellular function level, N/OFQ should exert robust modulation of Hcrt neuronal activity and synaptic inputs onto the Hcrt neurons with defined mechanisms to support the hypothesis of direct functional interaction. The modulation of Hcrt neurons by N/OFQ was studied using whole-cell patch clamp recordings of Hcrt neurons from transgenic mice orexin/EGFP in which EGFP was linked to the Hcrt promoter (Xie et al., 2006b; Yamanaka et al., 2003). In current-clamp mode, most orexin/EGFP neurons recorded displayed spontaneous firing of action potentials superimposed on spontaneous synaptic activity at resting potentials (approximately −50 mV). The spontaneous spikes were reversibly blocked by tetrodotoxin (TTX; Fig. 3A and B), indicating that TTX-sensitive Na^+ channels mediate these action potentials. Bath application of N/OFQ (0.1–3 μM) caused a long-lasting hyperpolarization, decreased input resistance, and blocked spontaneous firing of action potentials of the Hcrt neuron. The postsynaptic inhibition on Hcrt neurons persevered in the presence of TTX, suggesting direct postsynaptic effects. The novel selective NOP receptor antagonist SR14148 (10 μM) blocked the effect of N/OFQ, confirming the mediation by NOP activation (Fig. 3C).

Furthermore, N/OFQ has been shown to inhibit voltage-gated Ca^{2+} channels and depress intracellular Ca^{2+} levels in Hcrt neurons. First, under conditions that block voltage-gated Na^+ and K^+ channels, depolarizing the

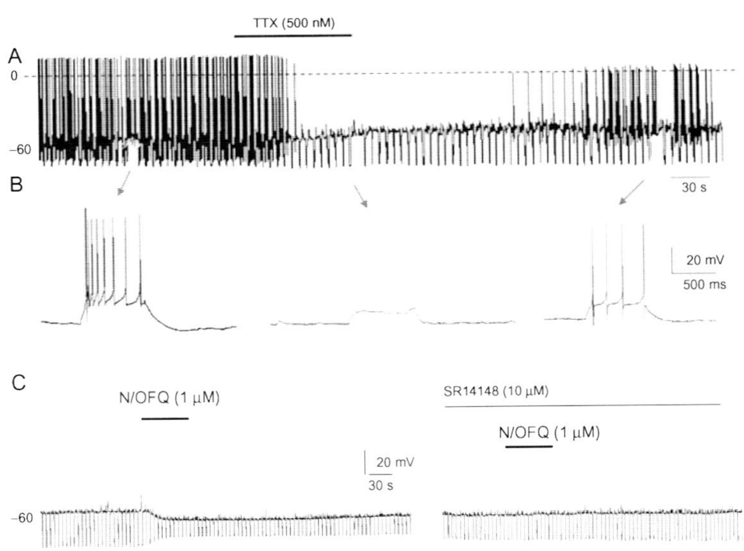

Figure 3 Hcrt neurons displayed spontaneous firing of action potentials, typically at 0.5–5 Hz frequencies, and were blocked by TTX. (A) Bath application of TTX (0.1–1 μM) reversibly blocked spike firing. (B) A train of action potentials elicited by a depolarizing current injection (50 pA, 500 ms) was also reversibly blocked by TTX. The mean resting potential of orexin/EGFP neurons was −51 ± 3 mV (mean ± SEM, $n = 45$). Membrane potential was held at −60 mV with dc injection, and hyperpolarizing current pulses (−0.3 nA, 800 ms) were delivered every 5 s throughout the experiments. (C) N/OFQ-induced hyperpolarization and input resistance decrease on a Hcrt-containing cell in the presence of TTX. SR14148 (10 μM) completely blocked the effects of N/OFQ. (unpublished data).

membrane evoked a high-voltage-activated Ca^{2+} current. N/OFQ (1 μM) inhibited the Ca^{2+} current to 88 ± 4.5% of control. Second, Ca^{2+} imaging of Hcrt neurons from the transgenic *orexin/YC2.1* mice in which Hcrt neurons specifically express the yellow cameleon calcium-sensing protein (Tsujino et al., 2005), revealed that N/OFQ inhibited cytoplasmic Ca^{2+} levels of the Hcrt neurons in a concentration-dependent manner with an EC_{50} of approximately 50 nM N/OFQ (Xie et al., 2008). Furthermore, selective nonpeptide NOP agonist SR14150 (10 μM) also decreases cytoplasmic Ca^{2+} levels, which was reversed by the NOP antagonist SB612111 (Tsujino, Sakurai & Xie, unpublished observation).

The presynaptic effects of N/OFQ on spontaneous excitatory and inhibitory postsynaptic currents (EPSCs and IPSCs) were investigated using whole-cell patch clamp recordings under voltage-clamp mode. N/OFQ (1 μM) decreased the frequency of Spontaneous EPSCs, with little effect on the amplitude of these events. Similarly, N/OFQ also significantly inhibited the frequency of sIPSCs, without an effect on the amplitude of these events. NOP-mediated activation, like the $GABA_B$ receptor, reduces both excitatory and inhibitory inputs into Hcrt neurons. A decrease in inhibitory synaptic inputs by N/OFQ is not consistent with the inhibitory effect of N/OFQ on Hcrt neurons. However, the net effect of N/OFQ on Hcrt neuronal activity will first depend on the tonic modulation of Hcrt neurons from these two synaptic inputs. Based on EM and functional studies, it seems that Hcrt neurons receive more glutamatergic excitatory inputs than GABAergic inhibitory inputs (Horvath & Gao, 2005; Xie et al., 2006b). Therefore, it is expected that the overall effect of N/OFQ modulation of synaptic inputs into Hcrt neurons would be predominantly inhibitory. Furthermore, the overall effects of N/OFQ on Hcrt cell excitability will also depend on both pre- and postsynaptic effects. The postsynaptic hyperpolarization is larger and longer lasting. Therefore, the overall N/OFQ modulation on the Hcrt system *in vivo* should be inhibitory.

4.3 Coordinated modulation of SIA

A link between N/OFQ and Hcrt systems should also be established at the whole organism level to ultimately support the hypothesis. N/OFQ and Hcrt exert opposite modulations on a variety of global behaviors, such as in regulation of sleep/wake, locomotion, and rewarding behaviors. In this chapter, we focus on the interaction between the Hcrt and N/OFQ systems in the regulation of stress responses, such as SIA and anxiety because these aspects were previously overlooked. Uncovering such a neuronal circuit that provides dual modulation to balance positive and negative responses to stress could have implications for the management of stress and pain in humans. Initial studies were undertaken to assess whether any differences exist at baseline in thermal pain thresholds and anxiety-like behavior between orexin/ataxin-3 and age-matched littermate WT mice (Xie et al., 2008). Using the hotplate and tail flick tests the thermal pain threshold did not differ between orexin/ataxin-3 mice and WT mice in both tests (Fig. 4). The observation of a lack of thermal pain threshold changes at baseline between the two genotypes is consistent with the result obtained with *prepro-orexin* KO mice (Watanabe et al., 2005).

The SIA phenomenon can be induced by a variety of stressors. The restraint immobilization-induced SIA is indicated by a significant increase in the hindpaw-withdrawal latency compared to the control unrestrained mice. The SIA typically disappeared 30 min after the termination of restraint (Fig. 4A). To investigate whether endogenous Hcrt is necessary for the generation of SIA, SIA in adult orexin/ataxin-3 and WT controls was assessed. As shown in Fig. 4B, restraint caused hindpaw-withdrawal latency to increase significantly in restraint compared to non-restraint WT mice. In contrast, SIA was absent in orexin/ataxin-3 mice following restraint, indicating the critical role of Hcrt neurons in generation of SIA. Central administration of exogenous Hcrt produced a trend toward increased hotplate latency in unrestrained animals, indicating an analgesic effect (Fig. 4C), which is consistent with the literature (Bingham et al., 2001). Hcrt appears not to further augment restraint SIA in WT mice, suggesting that restraint stress may induce maximal activation of the endogenous Hcrt system. In contrast, N/OFQ blocked SIA, consistent with previously reported results (Calo et al., 2000; Rizzi et al., 2001). Studies using the *N/OFQ* gene KO mouse model or *NOP* receptor gene KO model (Koster et al., 1999; Rizzi et al., 2011), and a peptide NOP antagonist (Rizzi et al., 2001) have demonstrated that endogenous N/OFQ exerts a tonic inhibitory modulation of SIA. If acute suppression of Hcrt signaling is involved in the inhibition of SIA by N/OFQ, exogenous Hcrt should restore SIA. Indeed, co-administration of Hcrt in animals treated with N/OFQ overcomes N/OFQ-produced inhibition of SIA (Fig. 4C). In addition to that, Hcrt may antagonize N/OFQ by blocking the NOP-activation-induced GIRK current (Hoang et al., 2003). Hcrt might also trigger downstream neurotransmitter systems (such as serotonin and histamine, etc.) that participate in the generation of SIA. Exogenous Hcrt induced similar analgesic effects

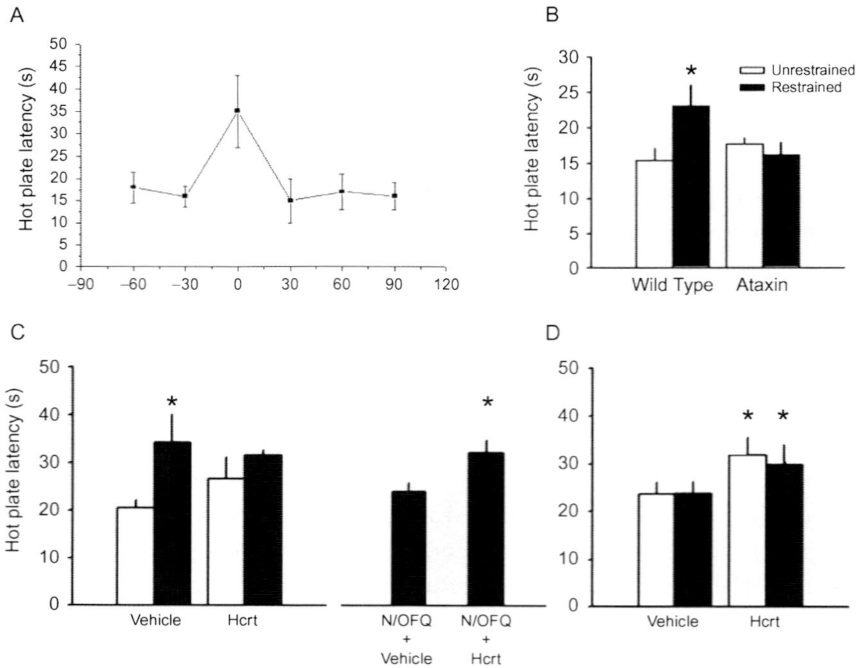

Figure 4 (A) Time course of the development of SIA in WT mice. Mice were subjected to 30 min restraint (time point −30 to time point 0) followed immediately by a hotplate test at time point 0. ANOVA indicated significant effects of time on hotplate latency ($p \leq 0.002$). (B) Hotplate latency was significantly increased in WT mice immediately after 30 min of restraint stress (*$P < 0.05$ vs. WT controls). A significant effect of restraint was not observed in orexin/ataxin-3 mice. (C) N/OFQ blocks SIA, and Hcrt restores the SIA in WT mice. Mice were subjected to the 30 min restraint immediately after the icv injection of either vehicle, N/OFQ, Hcrt, or a combination of both neuropeptides. ANOVA indicated significant effects of restraint in WT mice ($n = 8$ per group, $p \leq 0.002$). Hcrt (1.5 nmol/mouse) produced insignificant analgesia in the unrestrained group. Hcrt (1.5 nmol/mouse) did not further increase SIA. N/OFQ (1 nmol/mouse) inhibits the SIA (compared with vehicle restraint, $p \geq 0.3$). However, co-application of Hcrt (1.5 nmol/mouse) with N/OFQ (1 nmol/mouse) restores SIA. *Significantly different from vehicle-unrestrained control, Fisher's PLSD. (D) Lack of SIA in orexin/ataxin-3 mice. Neither icv nor restraint had a significant effect on orexin/ataxin-3 mice (ANOVA $p > 0.10$). Hcrt (1.5 nmol/mouse) caused acute analgesia in both unrestrained and restrained animals compared to vehicle groups ($p \leq 0.05$). *Significantly different from vehicle-unrestrained control, Fisher's PLSD. *Figure was modified from Xie et al. (2008).*

in both unrestrained and restrained orexin/ataxin-3 animals, mimicking SIA (Fig. 4D; Xie et al., 2008).

Although icv injection of N/OFQ blocking SIA appears to exert its effect through modulation of Hcrt neurons, the site of N/OFQ action on modulation of SIA was elusive. To further define N/OFQ modulation of the SIA directly via the inhibition of the Hcrt neurons, both fluorescent microscopy and EM were employed. N/OFQ-containing neurons are located in the lateral hypothalamus and the N/OFQ-containing fibers make direct contacts with the Hcrt neurons of the rat. Paw thermal nociceptive tests revealed that the immobilization restraint of the rat increased the thermal pain threshold by $20.5 \pm 7.6\%$. Bilateral microinjection of N/OFQ into the rat perifornical area of the lateral hypothalamus, the brain area in which the Hcrt neurons are exclusively located, abolished the SIA (Fig. 5A). Activity of Hcrt neurons in the same animals was assessed using Fos immunohistochemistry, an *in vivo* neuronal activity marker. Percentage of Fos^{+}/Hcrt neurons was lower in rats injected with N/OFQ than rats injected with saline (Fig. 5B). These results further confirm that N/OFQ modulation of SIA is mediated by direct inhibition of Hcrt neuronal activity in the perifornical area of the rat.

4.4 Coordinated modulation of anxiety-related behavior

As reviewed above, Hcrt neurons play a crucial role in stress defense responses, for example SIA, under physiological conditions. Hcrt neurons can be activated by CRH and in turn stimulate CRH neurons involving anxiogenesis under pathological conditions (Johnson et al., 2012; Paneda et al., 2005; Winsky-Sommerer et al., 2004). In contrast, activation of the NOP receptor inhibits neuronal activity and produces anxiolytic effects. Since the N/OFQ-induced inhibition of the Hcrt neurons underlies its attenuation of SIA (Gerashchenko et al., 2011; Xie et al., 2008), whether N/OFQ exerts its anxiolytic effects on anxiety-related behavior also via direct inhibition of Hcrt neuronal activity was investigated using unconditioned (light/dark preference and elevated plus-maze) and conditioned fear (Vogel conflict drinking) tests in adult orexin/ataxin-3 transgenic and WT mice (Zou et al., 2010, APSS Abstract #0216). Since animal activity levels influence anxiety-like behavior and confounds the evaluation of drug-induced anxiolytic effects, it is important to first evaluate basal spontaneous activity. Using the homecage behavioral monitoring system SmartCage™ (AfaSci, Inc. Redwood City, CA), it was found that transgenic mice were significantly less active than WT in the early dark phase as indicated by the measure of distance traveled and rearing. However, there were no significant differences in locomotion between the genotypes in the light-phase, especially during the anxiety-like behavior tests. These general behavioral

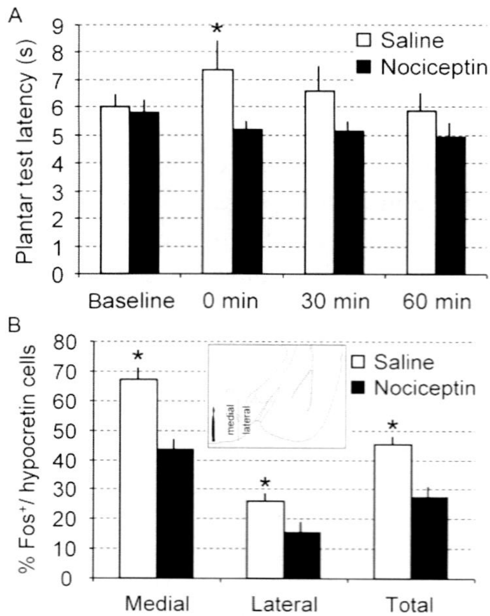

Figure 5 (A) Microinjection of N/OFQ into the rat lateral hypothalamus abolishes stress-induced analgesia. Plantar test in rats before and after restraint stress. The plantar test was repeated in four measurements. The first baseline measurement was performed 30 min before the restraint, the second measure was done immediately after the release from the restraint (0 min), the third and fourth measures were made at 30 min and 60 min after the release from the restraint, respectively. $*P < 0.05$ between N/OFQ and saline groups at the same time point. (B) Local microinjection of N/OFQ depresses Fos expression in Hcrt neurons in the rats. Percentage of Fos^{+}Hcrt neurons is shown in rats injected with N/OFQ or saline. The counts were analyzed separately for the Hcrt neurons located medially and laterally to the fornix (f), as schematically shown in the insert. Total counts include Hcrt neurons located on both sides of the fornix. Paired comparisons versus corresponding saline group, Mann Whitney U test: $*p < 0.05$. *Figure was modified from Gerashchenko et al. (2011).*

features are consistent with previous reports (Hara et al., 2001; Xiong et al., 2013; Yang et al., 2013).

N/OFQ peptide cannot cross the blood–brain barrier and icv injection of N/OFQ was confounded by effects of anesthesia in transgenic mice due to delayed emergence from anesthesia (Hara et al., 2009; Kelz et al., 2008). Therefore, the nonpeptide NOP agonists SR14150 and SR16835 and NOP antagonists SR14148 and SB612111 were used because they permit systemic administration (Khroyan et al., 2011; Zaveri et al., 2004, 2005). The doses of the NOP agonists used were carefully chosen to avoid causing significant sedative effects on spontaneous activity and locomotion.

The light/dark preference, elevated plus-maze exploring, and Vogel conflict drinking represent a test panel assessing a spectrum of graded anxiety-related behaviors in rodents. The unconditioned fear tests are based on the natural aversion of rodents to light or elevated open spaces, whereas the Vogel test creates a strong conflict between thirst-driven drinking and fear of electric shocks. Different degrees of fear may activate different components or levels of endogenous stress response regulation such as the Hcrt and N/OFQ systems. There were no significant differences in anxiety-like behavior between the two genotypes under the unconditioned fear tests (Fig. 6). The NOP agonist SR14150 did not influence light/dark preference of either genotype (Fig. 6A). In the elevated plus-maze, SR14150 caused a significant anxiolytic effect in WT mice, as indicated by an increase in time spent in open arms. The anxiolytic effect was blocked by the NOP antagonist SR14150. Notably, the NOP agonist did not produce anxiolytic effects in transgenic mice (Fig. 6B).

In the Vogel test, there were no significant differences in fear-induced anxiety-like behavior between transgenic and WT mice either, as indicated by the shock number received every 20th licks during a conflict-drinking test. The NOP agonist SR14150 did not increase free-drinking at baseline (Fig. 6C), while SR14150 produced a dramatic and significant increase in shock number received every 20th licks during the conflict-drinking test. However, the anxiolytic actions of the NOP agonist observed in WT mice were virtually absent in transgenic mice (Fig. 6D). Another NOP agonist SR16835 produces similar anxiolytic effects as SR14150 in WT mice, and its anxiolytic action is also substantially attenuated in the transgenic mice. Furthermore, using patch-clamp recordings and Ca^{2+} imaging of Hcrt neurons, SR14150 inhibited Hcrt neuronal activity and depressed intracellular Ca^{2+} levels. Both of these effects were blocked by the NOP antagonist SB612111, further confirming they are mediated by the NOP receptors. These findings suggest that Hcrt neurons are key neuronal substrates for the anxiolytic action of NOP agonists (Fig. 6).

5. CONCLUSIONS

Recent studies have provided evidence that N/OFQ-containing fibers contact Hcrt neurons and some cells located in the LHA co-produce N/OFQ and Hcrts. N/OFQ peptide and nonpeptide NOP agonists inhibit Hcrt neuronal activity, and depress cytoplasmic Ca^{2+} in Hcrt neurons. SIA caused by restraint is blocked by icv injection of N/OFQ,

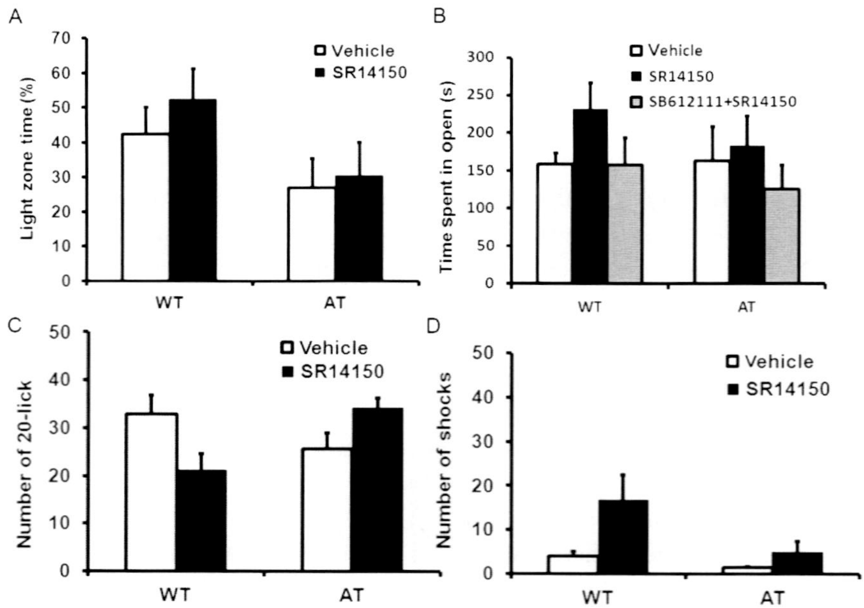

Figure 6 (A) Effects of the NOP agonists on anxiety-like behavior. (A) There were no significant differences in time spent in the light compartment between the two genotypes and between vehicle control and treatment with a NOP agonist SR14150 (1 mg/kg, i.p.). (B) There were no significant differences in time spent in the open arms between the orexin/ataxin-3 and WT mice. SR14150 (1 mg/kg, i.p.) increased time spent in the open arms in WT mice ($p=0.052$), which was blocked by the NOP antagonist SB612111. NOP against or antagonist did not have any significant effects in orexin/ataxin-3 mice in the elevated plus-maze test. (C) Both WT and orexin/ataxin-3 were given free-drink (without shocks) for 10 min following 48-h water deprivation in the presence and absence of SR14150 (1 mg/kg, i.p.). There were no significant differences between the two genotypes and between drug treatment and control ($n=14$ per group, data pooled from two separate experiments). Number of licks is expressed in multiples of 20-licks for visual comparison with conflict-drinking data. (D) In conflict-drinking (with a shock delivered every 20th lick), SR14150 (1 mg/kg, i.p.) significantly increased the number of shocks received in WT mice ($n=7$, ** $p=0.008$) but not in orexin/ataxin-3 mice ($n=7$ per group). (unpublished data).

while co-application of Hcrt can restore the phenomenon of SIA in WT mice. In contrast, SIA phenomenon was absent in orexin/ataxin-3 mice; whereas exogenously applied Hcrt induces acute analgesia mimicking SIA. Restraint also causes SIA in rats. Direct microinjection of N/OFQ into the LHA attenuates SIA and depresses Fos expression in Hcrt neurons.

The Hcrt system sets the arousal and vigilance states, and plays a dual role in stress adaptive responses and anxiogenesis, as illustrated in Fig. 7. The

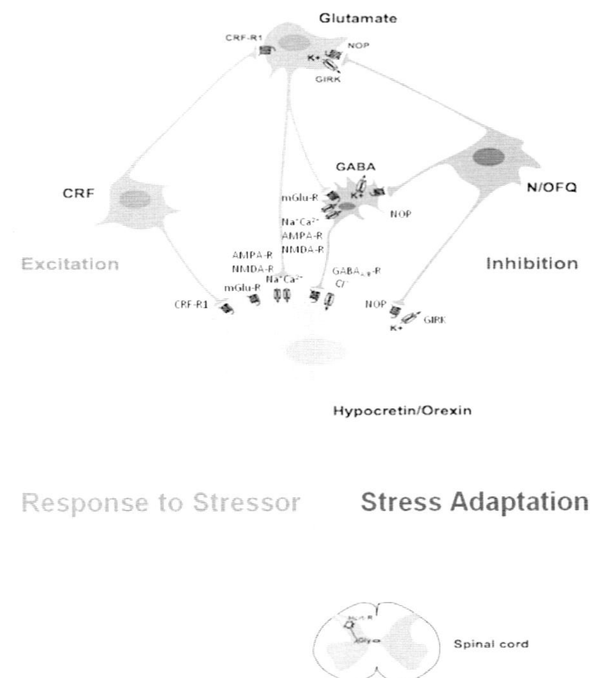

Figure 7 Schematic diagram illustrating Hcrt and N/OFQ coordinately regulating stress responses and adaptation. In addition to glutamate and GABA playing opposite roles in modulation of Hcrt neuronal activity, multiple neuropeptides have been shown to directly modulate Hcrt neurons. For example, CRH stimulates Hcrt neuronal activity enhancing stress responses; while N/OFQ (nociceptin/orphanin FQ) inhibits Hcrt neuronal activity, facilitating stress adaptation. (See the color plate.)

extent of anxiolytic action induced by N/OFQ or the NOP agonists depends on stress levels in WT mice, suggesting that endogenous Hcrt neurons and the N/OFQ system are activated under higher levels of stress. Anxiolytic effects of the NOP agonists observed in WT mice were significantly attenuated or even absent in the transgenic mice. These observations suggest that Hcrt neurons are a key neuronal target for the anxiolytic action of NOP activation. An alternative and complementary interaction between the N/OFQ and Hcrt systems could occur at the projected sites that receive both N/OFQ and Hcrt efferent inputs presynaptically and co-express their cognate receptors postsynaptically. NOP and HcrtR selective antagonists would provide a powerful tool to address these questions in the future.

If the N/OFQ system directly interacts with the Hcrt neurons in the LHA and/or differentially modulates the Hcrt system projected neurons throughout the CNS, then the two neuropeptides would provide an effective coordinated-modulation to achieve a fine balance of stress adaptive behaviors (Fig. 7). Furthermore, several new lines of evidence from transgenic phenotypic analyses and pharmacological studies also suggest differential modulation of reward, learning and memory, and drug dependence between the N/OFQ and Hcrt systems (Harris & Aston-Jones, 2006; Harris, Wimmer, & Aston-Jones, 2005; Olszewski & Levine, 2004; Yang et al., 2013). The Hcrt system acts as an integrator of multiple physiological functions (Li et al., 2014) as a result from a myriad of modulations by "classical" neurotransmitters (e.g., glutamate, GABA, 5-HT and norepinephrine) (Carter, de Lecea, & Adamantidis, 2013; Sergeeva, Andreeva, Garret, Scherer, & Haas, 2005; Xie et al., 2006b; Yamanaka et al., 2003) and a variety of neuropeptides. In addition to N/OFQ and CRH, several other neuropeptides have been found to influence the activity of Hcrt neurons; for example, ghrelin, leptin (Yamanaka et al., 2003), glucagon-like peptide-1 (Acuna-Goycolea & van den Pol, 2004), neuropeptide Y (Fu, Acuna-Goycolea, & van den Pol, 2004), and dynorphin (Li & van den Pol, 2006). The neuropeptide S (Xu et al., 2004) can excite a small subpopulation of Hcrt neurons (Xie et al., 2006a, b). Lastly, the Hcrt system can also sense energy balance and be stimulated by glucose and nutritionally relevant mixtures of amino acids (Karnani et al., 2011; Yamanaka et al., 2003). Knowledge obtained from these interaction studies will further our understanding of how the Hcrt system acts as a central mediator for N/OFQ and other neuropeptides in the integration and regulation of adaptive behaviors such as nociceptive processing, stress responses, rewarding and beyond.

ACKNOWLEDGMENTS

This work was supported by NIH grants R01MH078194, R43NS065555, R43 NS073311R44, DA026363, R43 AG043203 and R44 DA031578. I would like to thank Bende Zou, Liya Yang, Conrado Pascual and Christopher Lieu in AfaSci, Inc. for their contributions to unpublished data, James Xie and Julian Xie for editing, and Junko Hara and Zhaohui Li for assistance in preparation of Figures.

REFERENCES

Acuna-Goycolea, C., & van den Pol, A. (2004). Glucagon-like peptide 1 excites hypocretin/orexin neurons by direct and indirect mechanisms: Implications for viscera-mediated arousal. *The Journal of Neuroscience*, *24*(37), 8141–8152.

Anton, B., Fein, J., To, T., Li, X., Silberstein, L., & Evans, C. J. (1996). Immunohistochemical localization of ORL-1 in the central nervous system of the rat. *The Journal of Comparative Neurology*, *368*(2), 229–251.

Bingham, S., Davey, P. T., Babbs, A. J., Irving, E. A., Sammons, M. J., Wyles, M., et al. (2001). Orexin-A, an hypothalamic peptide with analgesic properties. *Pain*, *92*(1–2), 81–90.

Blair, R., Galina, Z. H., Holmes, L. J., & Amit, Z. (1982). Stress-induced analgesia: A performance deficit or a change in pain responsiveness? *Behavioral and Neural Biology*, *34*(2), 152–158.

Borgland, S. L., Connor, M., & Christie, M. J. (2001). Nociceptin inhibits calcium channel currents in a subpopulation of small nociceptive trigeminal ganglion neurons in mouse. *The Journal of Physiology*, *536*(Pt. 1), 35–47.

Bunzow, J. R., Saez, C., Mortrud, M., Bouvier, C., Williams, J. T., Low, M., et al. (1994). Molecular cloning and tissue distribution of a putative member of the rat opioid receptor gene family that is not a mu, delta or kappa opioid receptor type. *FEBS Letters*, *347*(2–3), 284–288.

Burdakov, D., Karnani, M. M., & Gonzalez, A. (2013). Lateral hypothalamus as a sensor-regulator in respiratory and metabolic control. *Physiology & Behavior*, *121*, 117–124.

Calo, G., Bigoni, R., Rizzi, A., Guerrini, R., Salvadori, S., & Regoli, D. (2000). Nociceptin/orphanin FQ receptor ligands. *Peptides*, *21*(7), 935–947.

Carter, M. E., de Lecea, L., & Adamantidis, A. (2013). Functional wiring of hypocretin and LC-NE neurons: Implications for arousal. *Frontiers in Behavioral Neuroscience*, *7*, 43.

Chiou, L. C., Liao, Y. Y., Fan, P. C., Kuo, P. H., Wang, C. H., Riemer, C., et al. (2007). Nociceptin/orphanin FQ peptide receptors: Pharmacology and clinical implications. *Current Drug Targets*, *8*(1), 117–135.

Connor, M., & Christie, M. J. (1998). Modulation of Ca2+ channel currents of acutely dissociated rat periaqueductal grey neurons. *The Journal of Physiology*, *509*(Pt. 1), 47–58.

Connor, M., Vaughan, C. W., Chieng, B., & Christie, M. J. (1996). Nociceptin receptor coupling to a potassium conductance in rat locus coeruleus neurones in vitro. *British Journal of Pharmacology*, *119*(8), 1614–1618.

Davis, S. F., Williams, K. W., Xu, W., Glatzer, N. R., & Smith, B. N. (2003). Selective enhancement of synaptic inhibition by hypocretin (orexin) in rat vagal motor neurons: Implications for autonomic regulation. *The Journal of Neuroscience*, *23*(9), 3844–3854.

de Lecea, L., Kilduff, T. S., Peyron, C., Gao, X., Foye, P. E., Danielson, P. E., et al. (1998). The hypocretins: Hypothalamus-specific peptides with neuroexcitatory activity. *Proceedings of the National Academy of Sciences of the United States of America*, *95*(1), 322–327.

Eriksson, K. S., Sergeeva, O. A., Stevens, D. R., & Haas, H. L. (2002). Neurotransmitter-induced activation of sodium-calcium exchange causes neuronal excitation. *Annals of the New York Academy of Sciences*, *976*, 405–407.

Fernandez, F., Misilmeri, M. A., Felger, J. C., & Devine, D. P. (2004). Nociceptin/orphanin FQ increases anxiety-related behavior and circulating levels of corticosterone during neophobic tests of anxiety. *Neuropsychopharmacology*, *29*(1), 59–71.

Fu, L. Y., Acuna-Goycolea, C., & van den Pol, A. N. (2004). Neuropeptide Y inhibits hypocretin/orexin neurons by multiple presynaptic and postsynaptic mechanisms: Tonic depression of the hypothalamic arousal system. *The Journal of Neuroscience*, *24*(40), 8741–8751.

Gerashchenko, D., Horvath, T. L., & Xie, X. S. (2011). Direct inhibition of hypocretin/orexin neurons in the lateral hypothalamus by nociceptin/orphanin FQ blocks stress-induced analgesia in rats. *Neuropharmacology*, *60*(4), 543–549.

Hao, J. X., Yu, W., & Xu, X. J. (1998). Evidence that spinal endogenous opioidergic systems control the expression of chronic pain-related behaviors in spinally injured rats. *Experimental Brain Research*, *118*(2), 259–268.

Hara, J., Beuckmann, C. T., Nambu, T., Willie, J. T., Chemelli, R. M., Sinton, C. M., et al. (2001). Genetic ablation of orexin neurons in mice results in narcolepsy, hypophagia, and obesity. *Neuron*, *30*(2), 345–354.

Hara, J., Gerashchenko, D., Wisor, J. P., Sakurai, T., Xie, X., & Kilduff, T. S. (2009). Thyrotropin-releasing hormone increases behavioral arousal through modulation of hypocretin/orexin neurons. *The Journal of Neuroscience*, *29*(12), 3705–3714.

Harada, S., Fujita-Hamabe, W., & Tokuyama, S. (2011). Effect of orexin-A on post-ischemic glucose intolerance and neuronal damage. *Journal of Pharmacological Sciences*, *115*(2), 155–163.

Harris, G. C., & Aston-Jones, G. (2006). Arousal and reward: A dichotomy in orexin function. *Trends in Neurosciences*, *29*(10), 571–577.

Harris, G. C., Wimmer, M., & Aston-Jones, G. (2005). A role for lateral hypothalamic orexin neurons in reward seeking. *Nature*, *437*(7058), 556–559.

Hirota, K., Kushikata, T., Kudo, M., Kudo, T., Smart, D., & Matsuki, A. (2003). Effects of central hypocretin-1 administration on hemodynamic responses in young-adult and middle-aged rats. *Brain Research*, *981*(1–2), 143–150.

Hoang, Q. V., Bajic, D., Yanagisawa, M., Nakajima, S., & Nakajima, Y. (2003). Effects of orexin (hypocretin) on GIRK channels. *Journal of Neurophysiology*, *90*(2), 693–702.

Horvath, T. L., Diano, S., & van den Pol, A. N. (1999). Synaptic interaction between hypocretin (orexin) and neuropeptide Y cells in the rodent and primate hypothalamus: A novel circuit implicated in metabolic and endocrine regulations. *The Journal of Neuroscience*, *19*(3), 1072–1087.

Horvath, T. L., & Gao, X. B. (2005). Input organization and plasticity of hypocretin neurons: Possible clues to obesity's association with insomnia. *Cell Metabolism*, *1*(4), 279–286.

Hungs, M., & Mignot, E. (2001). Hypocretin/orexin, sleep and narcolepsy. *Bioessays*, *23*(5), 397–408.

Hwang, L. L., Chen, C. T., & Dun, N. J. (2001). Mechanisms of orexin-induced depolarizations in rat dorsal motor nucleus of vagus neurones in vitro. *The Journal of Physiology*, *537*(Pt 2), 511–520.

Ida, T., Nakahara, K., Murakami, T., Hanada, R., Nakazato, M., & Murakami, N. (2000). Possible involvement of orexin in the stress reaction in rats. *Biochemical and Biophysical Research Communications*, *270*(1), 318–323.

Jenck, F., Moreau, J. L., Martin, J. R., Kilpatrick, G. J., Reinscheid, R. K., Monsma, F. J., Jr., et al. (1997). Orphanin FQ acts as an anxiolytic to attenuate behavioral responses to stress. *Proceedings of the National Academy of Sciences of the United States of America*, *94*(26), 14854–14858.

Johnson, P. L., Molosh, A., Fitz, S. D., Truitt, W. A., & Shekhar, A. (2012). Orexin, stress, and anxiety/panic states. *Progress in Brain Research*, *198*, 133–161.

Karnani, M. M., Apergis-Schoute, J., Adamantidis, A., Jensen, L. T., de Lecea, L., Fugger, L., et al. (2011). Activation of central orexin/hypocretin neurons by dietary amino acids. *Neuron*, *72*(4), 616–629.

Kayaba, Y., Nakamura, A., Kasuya, Y., Ohuchi, T., Yanagisawa, M., Komuro, I., et al. (2003). Attenuated defense response and low basal blood pressure in orexin knockout mice. *American Journal of Physiology. Regulatory, Integrative and Comparative Physiology*, *285*(3), R581–R593.

Kelz, M. B., Sun, Y., Chen, J., Cheng Meng, Q., Moore, J. T., Veasey, S. C., et al. (2008). An essential role for orexins in emergence from general anesthesia. *Proceedings of the National Academy of Sciences of the United States of America*, *105*(4), 1309–1314.

Kenunen, O. G., & Prakh'e, I. V. (2005). Is stress-induced analgesia a measure of stress severity? *Rossiĭskii Fiziologicheskiĭ Zhurnal Imeni I.M. Sechenova*, *91*(3), 277–285.

Khroyan, T. V., Polgar, W. E., Orduna, J., Montenegro, J., Jiang, F., Zaveri, N. T., et al. (2011). Differential effects of nociceptin/orphanin FQ (NOP) receptor agonists in acute versus chronic pain: Studies with bifunctional NOP/mu receptor agonists in the sciatic nerve ligation chronic pain model in mice. *The Journal of Pharmacology and Experimental Therapeutics*, *339*(2), 687–693.

Kilduff, T. S. (2005). Hypocretin/orexin: Maintenance of wakefulness and a multiplicity of other roles. *Sleep Medicine Reviews*, *9*(4), 227–230.

Kitamura, E., Hamada, J., Kanazawa, N., Yonekura, J., Masuda, R., Sakai, F., et al. (2010). The effect of orexin-A on the pathological mechanism in the rat focal cerebral ischemia. *Neuroscience Research*, *68*(2), 154–157.

Knoflach, F., Reinscheid, R. K., Civelli, O., & Kemp, J. A. (1996). Modulation of voltage-gated calcium channels by orphanin FQ in freshly dissociated hippocampal neurons. *The Journal of Neuroscience*, *16*(21), 6657–6664.

Koster, A., Montkowski, A., Schulz, S., Stube, E. M., Knaudt, K., Jenck, F., et al. (1999). Targeted disruption of the orphanin FQ/nociceptin gene increases stress susceptibility and impairs stress adaptation in mice. *Proceedings of the National Academy of Sciences of the United States of America*, *96*(18), 10444–10449.

Kroeger, D., & de Lecea, L. (2009). The hypocretins and their role in narcolepsy. *CNS & Neurological Disorders: Drug Targets*, *8*(4), 271–280.

Leon-Olea, M., Miller-Perez, C., Sanchez-Islas, E., Mendoza-Sotelo, J., Garduno-Gutierrez, R., de Gortari, P., et al. (2013). The nociceptin/orphanin FQ-like opioid peptide in nervous periesophageal ganglia of land snail Helix aspersa. *Brain Research*, *1505*, 22–46.

Li, J., Hu, Z., & de Lecea, L. (2014). The hypocretins/orexins: Integrators of multiple physiological functions. *British Journal of Pharmacology*, *171*(2), 332–350.

Li, Y., & van den Pol, A. N. (2006). Differential target-dependent actions of coexpressed inhibitory dynorphin and excitatory hypocretin/orexin neuropeptides. *The Journal of Neuroscience*, *26*(50), 13037–13047.

Maolood, N., & Meister, B. (2010). Nociceptin/orphanin FQ peptide in hypothalamic neurones associated with the control of feeding behaviour. *Journal of Neuroendocrinology*, *22*(2), 75–82.

Meunier, J. C. (1997). Nociceptin/orphanin FQ and the opioid receptor-like ORL1 receptor. *European Journal of Pharmacology*, *340*(1), 1–15.

Meunier, J. C., Mollereau, C., Toll, L., Suaudeau, C., Moisand, C., Alvinerie, P., et al. (1995). Isolation and structure of the endogenous agonist of opioid receptor-like ORL1 receptor. *Nature*, *377*(6549), 532–535.

Mogil, J. S., Grisel, J. E., Zhangs, G., Belknap, J. K., & Grandy, D. K. (1996). Functional antagonism of mu-, delta- and kappa-opioid antinociception by orphanin FQ. *Neuroscience Letters*, *214*(2–3), 131–134.

Mollereau, C., & Mouledous, L. (2000). Tissue distribution of the opioid receptor-like (ORL1) receptor. *Peptides*, *21*(7), 907–917.

Neal, C. R., Jr., Mansour, A., Reinscheid, R., Nothacker, H. P., Civelli, O., & Watson, S. J., Jr. (1999). Localization of orphanin FQ (nociceptin) peptide and messenger RNA in the central nervous system of the rat. *The Journal of Comparative Neurology*, *406*(4), 503–547.

Olszewski, P. K., & Levine, A. S. (2004). Minireview: Characterization of influence of central nociceptin/orphanin FQ on consummatory behavior. *Endocrinology*, *145*(6), 2627–2632.

Ouagazzal, A. M., Moreau, J. L., Pauly-Evers, M., & Jenck, F. (2003). Impact of environmental housing conditions on the emotional responses of mice deficient for nociceptin/orphanin FQ peptide precursor gene. *Behavioural Brain Research*, *144*(1–2), 111–117.

Paneda, C., Winsky-Sommerer, R., Boutrel, B., & de Lecea, L. (2005). The corticotropin-releasing factor-hypocretin connection: Implications in stress response and addiction. *Drug News & Perspectives*, *18*(4), 250–255.

Peyron, C., Faraco, J., Rogers, W., Ripley, B., Overeem, S., Charnay, Y., et al. (2000). A mutation in a case of early onset narcolepsy and a generalized absence of hypocretin peptides in human narcoleptic brains. *Nature Medicine*, *6*(9), 991–997.

Peyron, C., Tighe, D. K., van den Pol, A. N., de Lecea, L., Heller, H. C., Sutcliffe, J. G., et al. (1998). Neurons containing hypocretin (orexin) project to multiple neuronal systems. *The Journal of Neuroscience*, *18*(23), 9996–10015.

Reinscheid, R. K., & Civelli, O. (2002). The orphanin FQ/nociceptin knockout mouse: A behavioral model for stress responses. *Neuropeptides*, *36*(2–3), 72–76.

Reinscheid, R. K., Nothacker, H. P., Bourson, A., Ardati, A., Henningsen, R. A., Bunzow, J. R., et al. (1995). Orphanin FQ: A neuropeptide that activates an opioidlike G protein-coupled receptor. *Science*, *270*(5237), 792–794.

Ripley, B., Overeem, S., Fujiki, N., Nevsimalova, S., Uchino, M., Yesavage, J., et al. (2001). CSF hypocretin/orexin levels in narcolepsy and other neurological conditions. *Neurology*, *57*(12), 2253–2258.

Rizzi, A., Marzola, G., Bigoni, R., Guerrini, R., Salvadori, S., Mogil, J. S., et al. (2001). Endogenous nociceptin signaling and stress-induced analgesia. *Neuroreport*, *12*(14), 3009–3013.

Rizzi, A., Molinari, S., Marti, M., Marzola, G., & Calo, G. (2011). Nociceptin/orphanin FQ receptor knockout rats: In vitro and in vivo studies. *Neuropharmacology*, *60*(4), 572–579.

Sakamoto, F., Yamada, S., & Ueta, Y. (2004). Centrally administered orexin-A activates corticotropin-releasing factor-containing neurons in the hypothalamic paraventricular nucleus and central amygdaloid nucleus of rats: Possible involvement of central orexins on stress-activated central CRF neurons. *Regulatory Peptides*, *118*(3), 183–191.

Sakurai, T., Amemiya, A., Ishii, M., Matsuzaki, I., Chemelli, R. M., Tanaka, H., et al. (1998). Orexins and orexin receptors: A family of hypothalamic neuropeptides and G protein-coupled receptors that regulate feeding behavior. *Cell*, *92*(4), 573–585.

Sakurai, T., & Mieda, M. (2011). Connectomics of orexin-producing neurons: Interface of systems of emotion, energy homeostasis and arousal. *Trends in Pharmacological Sciences*, *32*(8), 451–462.

Sergeeva, O. A., Andreeva, N., Garret, M., Scherer, A., & Haas, H. L. (2005). Pharmacological properties of GABAA receptors in rat hypothalamic neurons expressing the epsilon-subunit. *The Journal of Neuroscience*, *25*(1), 88–95.

Shoblock, J. R. (2007). The pharmacology of Ro 64–6198, a systemically active, nonpeptide NOP receptor (opiate receptor-like 1, ORL-1) agonist with diverse preclinical therapeutic activity. *CNS Drug Reviews*, *13*(1), 107–136.

Stricker-Krongrad, A., & Beck, B. (2002). Modulation of hypothalamic hypocretin/orexin mRNA expression by glucocorticoids. *Biochemical and Biophysical Research Communications*, *296*(1), 129–133.

Tsujino, N., Yamanaka, A., Ichiki, K., Muraki, Y., Kilduff, T. S., Yagami, K., et al. (2005). Cholecystokinin activates orexin/hypocretin neurons through the cholecystokinin A receptor. *The Journal of Neuroscience*, *25*(32), 7459–7469.

van den Pol, A. N. (1999). Hypothalamic hypocretin (orexin): Robust innervation of the spinal cord. *The Journal of Neuroscience*, *19*(8), 3171–3182.

van den Pol, A. N., Gao, X. B., Obrietan, K., Kilduff, T. S., & Belousov, A. B. (1998). Presynaptic and postsynaptic actions and modulation of neuroendocrine neurons by a new hypothalamic peptide, hypocretin/orexin. *The Journal of Neuroscience*, *18*(19), 7962–7971.

Watanabe, S., Kuwaki, T., Yanagisawa, M., Fukuda, Y., & Shimoyama, M. (2005). Persistent pain and stress activate pain-inhibitory orexin pathways. *Neuroreport*, *16*(1), 5–8.

Winsky-Sommerer, R., Boutrel, B., & De Lecea, L. (2003). The role of the hypocretinergic system in the integration of networks that dictate the states of arousal. *Drug News & Perspectives*, *16*(8), 504–512.

Winsky-Sommerer, R., Yamanaka, A., Diano, S., Borok, E., Roberts, A. J., Sakurai, T., et al. (2004). Interaction between the corticotropin-releasing factor system and hypocretins (orexins): A novel circuit mediating stress response. *The Journal of Neuroscience*, *24*(50), 11439–11448.

Witkin, J. M., Statnick, M. A., Rorick-Kehn, L. M., Pintar, J. E., Ansonoff, M., Chen, Y., et al. (2013). The biology of Nociceptin/Orphanin FQ (N/OFQ) related to obesity, stress, anxiety, mood, and drug dependence. *Pharmacology & Therapeutics*, *141*(3), 283–299.

Xie, X., Crowder, T. L., Sakurai, T., de Lecea, L., & Kilduff, T. S. (2006a). Neuropeptide S modulates hypocretin/orexin neurons via pre- and post-synaptic mechanisms in the mouse hypothalamus. Associated Professional Sleep Societies (APSS) Meeting Abstract #0013.

Xie, X., Crowder, T. L., Yamanaka, A., Morairty, S. R., Lewinter, R. D., Sakurai, T., et al. (2006b). GABA(B) receptor-mediated modulation of hypocretin/orexin neurones in mouse hypothalamus. *The Journal of Physiology*, *574*(Pt. 2), 399–414.

Xie, X., Wisor, J. P., Hara, J., Crowder, T. L., LeWinter, R., Khroyan, T. V., et al. (2008). Hypocretin/orexin and nociceptin/orphanin FQ coordinately regulate analgesia in a mouse model of stress-induced analgesia. *The Journal of Clinical Investigation*, *118*(7), 2471–2481.

Xiong, X., White, R. E., Xu, L., Yang, L., Sun, X., Zou, B., et al. (2013). Mitigation of murine focal cerebral ischemia by the hypocretin/orexin system is associated with reduced inflammation. *Stroke*, *44*(3), 764–770.

Xu, Y. L., Reinscheid, R. K., Huitron-Resendiz, S., Clark, S. D., Wang, Z., Lin, S. H., et al. (2004). Neuropeptide S: A neuropeptide promoting arousal and anxiolytic-like effects. *Neuron*, *43*(4), 487–497.

Yamanaka, A., Beuckmann, C. T., Willie, J. T., Hara, J., Tsujino, N., Mieda, M., et al. (2003). Hypothalamic orexin neurons regulate arousal according to energy balance in mice. *Neuron*, *38*(5), 701–713.

Yang, B., & Ferguson, A. V. (2002). Orexin-A depolarizes dissociated rat area postrema neurons through activation of a nonselective cationic conductance. *The Journal of Neuroscience*, *22*(15), 6303–6308.

Yang, L., Zou, B., Xiong, X., Pascual, C., Xie, J., Malik, A., et al. (2013). Hypocretin/orexin neurons contribute to hippocampus-dependent social memory and synaptic plasticity in mice. *The Journal of Neuroscience*, *33*(12), 5275–5284.

Yuan, L. B., Dong, H. L., Zhang, H. P., Zhao, R. N., Gong, G., Chen, X. M., et al. (2011). Neuroprotective effect of orexin-A is mediated by an increase of hypoxia-inducible factor-1 activity in rat. *Anesthesiology*, *114*(2), 340–354.

Zaratin, P. F., Petrone, G., Sbacchi, M., Garnier, M., Fossati, C., Petrillo, P., et al. (2004). Modification of nociception and morphine tolerance by the selective opiate receptor-like orphan receptor antagonist (−)-cis-1-methyl-7-[[4-(2,6-dichlorophenyl) piperidin-1-yl]methyl]-6,7,8,9-tetrahy dro-5H-benzocyclohepten-5-ol (SB-612111). *The Journal of Pharmacology and Experimental Therapeutics*, *308*(2), 454–461.

Zaveri, N. T., Jiang, F., Olsen, C. M., Deschamps, J. R., Parrish, D., Polgar, W., et al. (2004). A novel series of piperidin-4-yl-1,3-dihydroindol-2-ones as agonist and antagonist ligands at the nociceptin receptor. *Journal of Medicinal Chemistry*, *47*(12), 2973–2976.

Zaveri, N., Jiang, F., Olsen, C., Polgar, W., & Toll, L. (2005). Small-molecule agonists and antagonists of the opioid receptor-like receptor (ORL1, NOP): Ligand-based analysis of structural factors influencing intrinsic activity at NOP. *The AAPS Journal*, 7(2), E345–E352.

Zhang, Y., Gandhi, P. R., & Standifer, K. M. (2012). Increased nociceptive sensitivity and nociceptin/orphanin FQ levels in a rat model of PTSD. *Molecular Pain*, *8*, 76.

Zhang, W., Shimoyama, M., Fukuda, Y., & Kuwaki, T. (2006). Multiple components of the defense response depend on orexin: Evidence from orexin knockout mice and orexin neuron-ablated mice. *Autonomic Neuroscience*, *126–127*, 139–145.

Zou, B., Pasumarthi, R., Xie, J., Gerashchenko, D., Black, S. W. T., Khroyan, T., Toll, L., et al. (2010). Nociceptin/orphanin FQ receptor-mediated anxiolytic effects through interaction with hypocretin/orexin neurons in mice. Associated Professional Sleep Societies (APSS) Meeting Abstract #0216.

CHAPTER TWELVE

Nociceptin/Orphanin-FQ Modulation of Learning and Memory

Ouagazzal Abdel-Mouttalib[1]
IGBMC, (UMR7104), CNRS, Illkirch, France
[1]Corresponding author: e-mail address: abdel-mouttalib.ouagazzal@univ-amu.fr

Contents

Abstract

In the past decades, a large number of neuropeptides with unknown functions have been identified in the brain. Among the newly discovered peptides, nociceptin or orphanin-FQ (N/OFQ) peptide has attracted considerable attention because of its sequence homology with the opioid peptide family. N/OFQ and its cognate receptor (NOP receptor) are distributed widely in the mammalian central nervous system, though particularly intense expression is found in corticolimbic structures. Such distinctive pattern of expression suggests a key role of N/OFQ system in higher brain functions, such as cognition and emotion. In this chapter, we will outline the findings supporting the role played by N/OFQ and NOP receptors in learning and memory and discuss the underlying mechanisms.

1. INTRODUCTION

The nociceptin/orphanin-FQ (N/OFQ) peptide represents a recent addition to the opioid peptides family. N/OFQ is a 17-amino acid neuropeptide generated from a large 181-amino acid precursor peptide

Vitamins and Hormones, Volume 97
ISSN 0083-6729
http://dx.doi.org/10.1016/bs.vh.2014.10.006

(prepro-N/OFQ) (Meunier et al., 1995; Reinscheid et al., 1995). The *prepro-N/OFQ* gene is located on the p21 short arm of human chromosome 8 and encodes a sequence that is highly conserved across mouse, rats, bovine, and human species and has hallmarks of opioid peptides precursor proteins (Lambert, 2008; Reinscheid, Nothacker, & Civelli, 2000). Although N/OFQ shares some structural similarities with endogenous opioid peptides, it does not bind to traditional opioid receptors (mu, delta, and kappa receptors) and its pharmacological effects are not sensitive to naloxone. N/OFQ has been identified as a natural ligand of the orphan NOP receptor (also known as opioid receptor-like 1, KOR3, OP4, or LC132) (Henderson & McKnight, 1997; Meunier et al., 1995; Reinscheid et al., 1995). The *NOP* gene is in the q13.2–13.3 region of human chromosome 20. Introns and exons are organized like in *mu, delta, and kappa opioid receptor* genes, suggesting that all four genes have a common ancestor and belong to the same family (Stevens, 2009). Like classical opioid receptors, NOP receptor is a prototypical G protein-coupled receptor that couples to pertussis toxin-sensitive and -insensitive G proteins. At cellular level, NOP receptor negatively regulates the activity of adenylate cyclases, inhibits voltage-gated Ca^{2+} channels, and activates inward rectifying K^{+} channels (Hawes, Graziano, & Lambert, 2000; New & Wong, 2002). Despite its structural and functional homology with opioid receptors, NOP receptor has a little or no affinity for opioid peptides or synthetic opiates. As such, N/OFQ and NOP receptor constitute a novel neuromodulatory system that is pharmacologically distinct from the opioid family.

N/OFQ and NOP receptor are more densely located in the central nervous system than in the periphery and their distribution pattern is distinct from that of opioid peptides and receptors. High expression of N/OFQ and NOP receptor transcripts is found in the cortex, limbic regions (e.g., septum, hippocampus, and amygdala), central gray, hypothalamic nuclei, brainstem aminergic nuclei (substantia nigra, ventral tegmental area, raphe complex, and locus coeruleus) and the dorsal horn of the spinal cord (Darland, Heinricher, & Grandy, 1998; Neal et al., 1999a, 1999b). The distinctive pattern of distribution of N/OFQ in the brain and its homology with opioids peptides has generated considerable excitement regarding possible physiological and behavioral functions mediated by this neuropeptide. The wealth of data accumulated during the past decade has implicated N/OFQ and NOP receptor in a host of behavioral processes. In this chapter, we will provide an overview of the evidence that substantiates the important role played by N/OFQ in learning and memory and discuss the underlying mechanisms of action of this neuropeptide.

2. N/OFQ MODULATION OF MNEMONIC FUNCTIONS

As illustrated by its anatomical distribution, N/OFQ system is an integral part of the corticolimbic circuits subserving cognition. Very high densities of N/OFQ and NOP receptor are found in the neocortex, cingulate, and piriform cortex. In the amygdaloid complex, the expression is ranging from intense in the central and medial amygdala to very low in the lateral amygdala. In the hippocampal formation, a high expression of N/OFQ and NOP receptor is found in the dentate gyrus, CA1 and CA3 subregions (Cavallini, Marino, Beani, Bianchi, & Siniscalchi, 2003; Darland et al., 1998; Neal, Mansour, Reinscheid, Nothacker, Civelli, et al., 1999a, 1999b). Such anatomical expression suggests that major cortical and limbic structures may be sensitive to the action of N/OFQ. Indeed, electrophysiological and neurochemical studies have shown that N/OFQ and NOP receptor exert a potent inhibitory actions on neuronal excitability and neurotransmitter release in the neocortex, the septum, the hippocampus, and the amygdala (Meis, 2003; Roberto & Siggins, 2006; Schlicker & Morari, 2000; Uezu et al., 2005; Wang et al., 2010; Yoshitake, Ijiri, Kehr, & Yoshitake, 2013). N/OFQ was also reported to inhibit various forms of synaptic plasticity, such as long-term potentiation (LTP) and long-term depression, which are considered as key cellular mechanisms underlying information storage and memory (Bongsebandhu-phubhakdi & Manabe, 2007; Manabe et al., 1998; Wei & Xie, 1999; Yu & Xie, 1998). Consistent with these findings, behavioral studies have shown that N/OFQ modulates different forms of mammalian learning and cognitive processes, including spatial learning, aversive associative conditioning, recognition memory, working memory, and pre-attentive sensorimotor gating mechanisms (Tables 1 and 2).

2.1 N/OFQ modulation of spatial learning

Initial insight into the role of N/OFQ in spatial learning comes from series of studies using the water-maze navigation task, a widely used procedure for assessing hippocampal-dependent functions (Table 1). In this task, animals learn to use spatial cues (extra-maze visual cues) to locate an escape platform hidden below the surface of the opaque water (the hidden platform or place learning version). The visible platform variant can be used as a control to assess the animal capacity to navigate using nonspatial cues. Sandin et al. (1997) have been the first to show that bilateral infusion of N/OFQ into dorsal hippocampus impairs acquisition of spatial learning in rats without

Table 1 Effects of NOP receptor agonists on learning and memory in rodents

Drug	Species	Dose (route of injection)	Task and effect		Reference
Spatial learning					
			WTZ		
N/OFQ	Rats	10 nmol (DHP)	↓		Sandin, Georgieva, Schott, Ogren, and Terenius (1997)
	Rats	5 nmol (DHP)	↓		Redrobe, Calo, Guerrini, Regoli, and Quirion (2000)
	Rats	3.3 nmol (DHP)	↓		Sandin, Ogren, and Terenius (2004)
	Mice	10 nmol (i.c.v.)	↓		Kuzmin, Madjid, Johansson, Terenius, and Ogren (2009)
Ro64-6198	Mice	1 mg/kg (i.p.)	↓		Higgins et al. (2002)
	Mice	0.3 mg/kg (i.p.)	↓		Kuzmin et al. (2009)
	Mice	10 nmol (i.c.v.)	↓		
Conditioned fear learning					
			CXT	Cue	
N/OFQ	Mice	0.01–1 nmol (i.c.v.)	↓	∅	Mamiya et al. (2003)
	Rats	1–2.5 nmol (i.c.v.)	↓	∅	Fornari, Soares, Ferreira, Moreira, and Oliveira (2008)
		5 nmol (i.c.v.)	↓	↓	
Ro64-6198	Mice	0.3–1 mg/kg (i.p.)	↓	∅	Goeldner, Reiss, Wichmann, Kieffer, and Ouagazzal (2009)
SR-8993	Mice	3 mg/kg (i.p.)	N/A	↓	Andero et al. (2013)
	Mice	100 ng (CEA)	N/A	↓	
			PAAV		
N/OFQ	Mice	5 nmol (i.c.v.)	↓		Hiramatsu and Inoue (1999a, 1999b)
	Mice	10 nmol (i.c.v.)	↓		Mamiya, Noda, Nishi, Takeshima, and Nabeshima (1999)
	Mice	1–4 nmol (i.c.v.)	↓		Liu, Lee, Nishiuchi, Kimura, and Tachibana (2007)
	Rats	1–10 pmol (BLA)	↓		Roozendaal, Lengvilas, McGaugh, Civelli, and Reinscheid (2007)

Table 1 Effects of NOP receptor agonists on learning and memory in rodents—cont'd

Drug	Species	Dose (route of injection)	Task and effect	Reference
Recognition memory				
			NOR	
N/OFQ	Mice	1 nmol (i.c.v.)	↓	Goeldner et al. (2008)
	Mice	3 nmol (DHP)	↓	
Ro64-6198	Mice	1 mg/kg (i.p.)	↓	
Working memory				
			SPA	
N/OFQ	Mice	1–10 pmol (i.c.v.)	∅	Mamiya et al. (1999)
	Mice	1.5–5 nmol (i.c.v.)	↓	Hiramatsu and Inoue (1999a, 1999b)
			DMP and DNMP	
Ro64-6198	Rats	6 mg/kg (i.p.)	↓	Higgins et al. (2002)
Sensorimotor gating				
			PPI	
Ro64-6198	Mice	1.5 mg/kg (i.p.)	↓	Ces et al. (2012)

WTZ: water maze navigation, CXT: contextual fear conditioning, Cued: cued fear conditioning, PAAV: passive avoidance learning, NOR: novel object recognition, SPA: Y-maze spontaneous alternations, DMP and DNMP: delayed matching and nonmatching to position, PPI: prepulse inhibition of acoustic startle reflex, DHP: dorsal hippocampus, i.c.v.: intracerebroventricular, i.p.: intraperitoneal, CEA: central amygdala, BLA: basolateral amygdala, ∅: no effect; ↓: impairment, N/A: effect not addressed in cited study.

affecting basic swimming abilities. This observation was later confirmed and extended by several other groups showing that intra-cerebroventricular or intra-hippocampal infusions of exogenous N/OFQ induce spatial learning deficits in rats and mice (Kuzmin et al., 2009; Redrobe et al., 2000; Sandin et al., 2004). The amnesic effects of N/OFQ could be blocked by co-infusion of the NOP receptors antagonist, [$Nphe^1$]-nociceptin (1–13)-NH_2 (Nph^1), confirming that they are mediated by NOP receptors (Kuzmin et al., 2009; Redrobe et al., 2000). Other groups used the synthetic and brain-penetrant selective agonist, Ro64-6198, which represents an

Table 2 Effects of antagonism of N/OFQ function on learning and memory in rodents

Manipulation	Species	Dose (route of injection)	Task and effect		Reference
Spatial learning					
			WTZ		
NOPR knockout	Mice		↑		Manabe et al. (1998)
N/OFQ knockout	Mice		∅		Koster et al. (1999)
			↑		Higgins et al. (2002)
			↑		Kuzmin et al. (2009)
Nphe[1]	Rats	50 nmol (DHP)	∅		Redrobe et al. (2000)
	Mice	10 nmol (i.c.v.)	∅		Kuzmin et al. (2009)
$NC-NH_2$	Rats	10 nmol (DHP)	∅		Sandin et al. (2004)
Conditioned fear learning					
			CXT	Cue	
NOPR knockout	Mice		↑	∅	Mamiya et al. (2003)
N/OFQ knockout	Mice		↑		Higgins et al. (2002)
			PAAV		
NOPR knockout	Mice		↑		Mamiya et al. (1999)
N/OFQ knockout	Mice		↑		Higgins et al. (2002)
Nphe[1]	Rats	10 mol (BLA)	↑		Roozendaal et al. (2007)
Working memory					
			SPA		
NOPR knockout	Mice		∅		Mamiya et al. (1999)

NOPR: NOP receptor, WTZ: water maze navigation, CXT: contextual fear conditioning, Cued: cued fear conditioning, PAAV: passive avoidance learning, SPA: Y-maze spontaneous alternations, DHP: dorsal hippocampus; i.c.v.: intracerebroventricular, BLA: basolateral amygdala, Nphe[1]: [Nphe[1]]-nociceptin (1-13)-NH_2 (Nph[1]); $NC-NH_2$: [$Phe^1\Psi(CH_2-NH)GLY_2$]-nociceptin(1-13)-NH_2, ∅: no effect, ↑: improvement.

alternative to N/OFQ peptide for probing the function of NOP receptors. Like N/OFQ, systemic administration of Ro64-6198 before training impaired spatial learning in the water maze task. At the amnesic doses, Ro64-6198 had no effect on cued learning in the visible platform task, thus ruling out a nonspecific effect of this agonist on the ability of the animals to execute the task (Higgins et al., 2002; Kuzmin et al., 2009). The above studies consistently show that pharmacological stimulations of NOP receptor impair spatial memory formation.

Studies on knockout mice lacking NOP receptor and N/OFQ peptide provided further support for the inhibitory role of endogenous N/OFQ system in spatial learning (Table 2). When tested in the water maze task, mice deficient in NOP receptor displayed improved place learning and enhanced memory retention during the probe trial (Manabe et al., 1998). Studies on mice lacking the *prepro-N/OFQ* gene, and thus the endogenous N/OFQ peptide, have given disparate results. The initial study on the behavioral phenotype of N/OFQ knockout mice described these animals as normal (Koster et al., 1999), but later studies revealed a cognitive improvement. Higgins et al. (2002) reported learning facilitation during the place training but not the probe trial, while in the study of Kuzmin et al. (2009) the cognitive improvement was evident at the probe trial and during reversal learning task. Such variations in learning performances of N/OFQ knockout mice may be attributed to differences in the training protocols and/or the genetic background of the mice. To date, much less is known about the effects of acute NOP receptor blockade on spatial learning. Few studies have used the peptidic NOP receptor antagonists, [Phe$^1\Psi$(CH_2-NH)GLY$_2$]-nociceptin (1–13)-NH_2 and [Nphe^1]-nociceptin (1–13)-NH_2, at a single dose to demonstrate the pharmacological selectivity of N/OFQ effects (Table 2). In all cases, these compounds were reported to be inactive on their own when administered either i.c.v or directly into the dorsal hippocampus (Kuzmin et al., 2009; Redrobe et al., 2000; Sandin et al., 2004). Thus, more studies with broader dose-ranges of NOP receptor antagonists are required to determine whether acute blockade of N/OFQ transmission impacts spatial learning.

2.2 N/OFQ modulation of fear learning and memory

Aversive conditioning is perhaps the form of learning in which the contribution of N/OFQ and NOP receptors has been extensively studied. The available findings demonstrate an important role of N/OFQ system in various forms of aversive learning (Tables 1 and 2).

2.2.1 Fear conditioning learning

Classical fear conditioning is the most widely used paradigms for studying the brain mechanisms subserving associative emotional learning and memory across species, including rodents. The standard procedure consists of one phase learning session, in which an auditory cue (conditioned stimuli, CS) is paired with a noxious unconditioned stimulus (US), such as foot shock (US), immediately after familiarization of the animals to the training context. As a result of this CS–US pairing, the CS acquires aversive properties. Animals are also capable of forming configural representations of the collection of cues that form the spatial context, which can also be associated with the aversive US. During the testing, re-exposure to the conditioning chamber or CS alone elicits a fear reaction (immobility or freezing behavior) in the absence of the US. Despite sharing some similarities, contextual and cued fear conditioning represent different forms of aversive associative learning. Cued fear conditioning involves discrete unisensory information processing that is amygdala-dependent but hippocampus-independent. On the other hand, contextual fear conditioning involves configural or spatial learning that heavily depends on hippocampal functions (Fanselow, 2009; LeDoux, 2003; Rudy, Huff, & Matus-Amat, 2004).

Mamiya et al. (2003) have shown that i.c.v. administration of N/OFQ immediately after conditioning impairs long-term consolidation of contextual but not cued fear memory in mice. This memory deficit was reversed by co-infusion of the NOP receptor antagonist, [Nphe^{1}]-nociceptin (1–13)-NH_2, thus confirming the pharmacological specificity of N/OFQ effect. Similar differential effects of N/OFQ on contextual and cued fear conditioning were reported by Fornari et al. (2008) in rats. However, in their experiments N/OFQ was effective when administered before but not after conditioning, thus leading to the conclusion that the peptide interferes with the acquisition processes. In line with the later findings, we found that systemic administration of the NOP receptor agonist, Ro64-6198, was more potent in reducing acquisition of contextual than cued fear conditioning in BL6N mice (Goeldner et al., 2009). These amnesic effects of Ro64-6198 were not linked to changes in pain threshold or to the state-dependent learning. However, one should keep in mind that NOP receptor agonists also possess anxiolytic-like activity. Thus, when administered prior to the conditioning, they may produce their amnesic effects simply by reducing fear levels. To demonstrate that Ro64-6198 specifically disrupts mnemonic processing we used the immediate shock-deficit procedure, which offers the means to evaluate drug effects on contextual learning in the absence of the foot-shock. The procedure consists of a two-phases training session in which context learning and context-shock association are temporally separated. In this task, naïve mice shocked briefly after being placed in the context show no sign of enhanced fear like non-shocked animals. By contrast, those pre-exposed 24 h earlier to the spatial context display a robust associative conditioning indicating that development of contextual fear involves sequential learning processes: the formation of a context representation and its association with the aversive unconditional stimulus (Fanselow, 2009; Rudy, 2009). Interestingly, mice pretreated with Ro64-6198 before pre-exposure to the training context show no sign of enhanced fear upon conditioning. These findings clearly show that NOP receptor stimulation produces specific disruption of contextual memory. More recently, Andero et al. (2013) have reported the effect of a novel-selective NOP receptor agonist, SR-8993, on cued fear learning. They showed that SR-8993 administered systemically either before or after conditioning reduces formation of cued fear memory in mice. SR-8993 was also effective when injected directly into the central amygdala, indicating that NOP receptor inhibits consolidation of cued fear memory in the former structure. These findings are however at odd with earlier studies showing a lack of effect of N/OFQ and Ro64-6198 on cued fear conditioning (Fornari et al., 2008; Goeldner et al., 2009; Mamiya et al., 2003). This discrepancy may be explained by the differences in the mouse strains (or species) and/or the training protocols used.

Both NOP receptor and N/OFQ knockout mice were assessed in the fear conditioning paradigm (Table 2). Mamiya et al. (2003) reported an improvement of contextual but not cued fear learning in NOP receptor knockout mice. The testing delay of the animals proved to be a critical factor for revealing this cognitive phenotype. When animals were tested 2 days after conditioning, wild-type and NOP receptor knockout mice displayed a comparable level of contextual fear. When the retention delay was prolonged to 7 days, wild types exhibited a marked decline in fear level, while no changes were detected in knockout mice. Comparable improvement in fear conditioning was reported by Higgins et al. (2002) in mice lacking N/OFQ peptide. However, in their study the testing was performed using compounds cues (CS presented in the original conditioning context), and therefore they did not attempt to distinguish between contextual and cued fear conditioning. To the best of our knowledge, the effect of pharmacological blockade of NOP receptors on fear conditioning learning have not been evaluated or reported to date. In a recent preliminary study we evaluated the

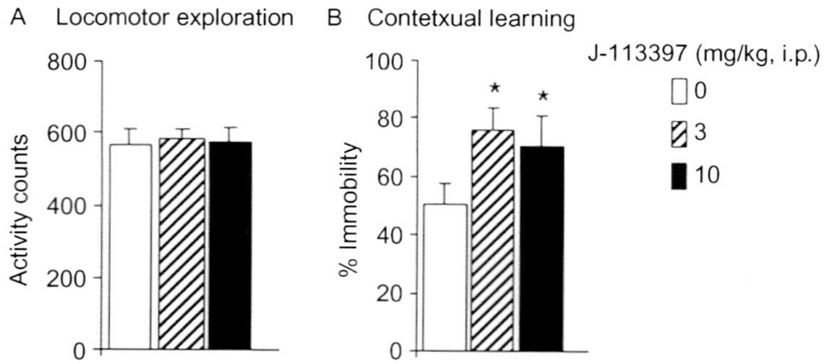

Figure 1 The NOP receptor antagonist, J-113397, improves contextual learning in mice. BL6N mice received an injection of J-113397 prior to pre-exposure session (4 min period, A) and submitted the following day drug free to the conditioning session (B). See Goeldner et al. (2009) for further details about the immediate shock-deficit procedure. *$p < 0.05$, significantly different from vehicle-treated group.

effect of the brain-penetrant NOP receptor antagonists, J-113397, in the immediate shock deficit paradigm. We found that J-113397 improves contextual learning without affecting basal locomotor activity (Fig. 1). These findings corroborate those obtained with knockout mice suggesting that endogenous N/OFQ contributes to associative fear conditioning.

2.2.2 Passive avoidance learning

Passive avoidance task is another popular paradigm that has been used by many laboratories to study the contribution of N/OFQ system to associative aversive learning (Table 1 and 2). In this task, animals learn to avoid an environment in which they previously received mild a foot-shock. In the step-through passive avoidance task, the animals are placed into an aversive brightly lit compartment and after stepping completely into the preferred dimly lit compartment a mild foot shock is delivered. In the step-down version the animals are placed on a small elevated platform and given a mild foot-shock when they step off the platform on the grid bellow. During training or testing (carried after a retention delay) the latency to enter the dark compartment or to step off the platform is used as an index of the ability of the animal to learn and remember the association between the aversive stimulus and the specific environmental context.

Earlier studies showed that i.c.v. administration of exogenous N/OFQ before conditioning severely impairs acquisition of passive avoidance learning in mice (Hiramatsu & Inoue, 1999b; Mamiya et al., 1999). This impairment could not be blocked by naloxone or U-50,488H, thus ruling out an involvement of classical opioid receptors. In a later study, Liu et al. (2007) used a multi-trial conditioning protocol to evaluate the effect of N/OFQ on learning performance of mice. They showed that N/OFQ pretreatment produces learning deficit that could be blocked by the NOP receptor antagonist, [Nphe[1]]-nociceptin (1–13)-NH_2. However, because N/OFQ was administered prior to the training the later results may be compromised by an effect of the peptide on fear expression. More convincing evidence for the contribution of N/OFQ to passive avoidance learning was provided by Roozendaal et al. (2007) using post-training injections of the exogenous peptide. They showed that local injection of N/OFQ into the basolateral amygdala either immediately or 3 h after conditioning impairs long-term retention performance (48 h delay). By contrast, more delayed injection of the peptide (6 h after conditioning) was ineffective. These findings unambiguously show that N/OFQ interferes with mnemonic processes associated to long-term consolidation of passive avoidance learning.

Genetic or pharmacological antagonism of endogenous N/OFQ transmission has been found to improve passive avoidance conditioning (Table 2). Using multi-trial training protocol, Mamiya et al. (1999) have shown that NOP receptor knockout mice display a better learning performance than wild types. This improvement was not accountable by differences in nociceptive threshold because no differences in the step-through latencies or the number of training trials were detected between genotypes during the conditioning phase. Comparable cognitive improvement was reported by Higgins et al. (2002) in N/OFQ knockout mice using a single learning trial protocol. They showed that knockout mice display a normal step-though latency during the conditioning phase but better memory retention than wild types when tested 24 h and 72 h post-training. Further evidence that endogenous N/OFQ contribute to passive avoidance learning comes from the study of Roozendaal et al. (2007) using central administration of the NOP receptor antagonist, [Nphe[1]]-nociceptin (1–13)-NH_2. They showed that immediate post-training blockade of NOP in the basolateral amygdala facilitates long-term (48 h delay) retention of passive avoidance conditioning. On the other hand, delayed blockade of NOP receptors (3 h after training) was ineffective indicating that endogenous N/OFQ is released during early phases of consolidation. These series of findings nicely show that endogenous N/OFQ system inhibits consolidation of passive avoidance learning.

While central administration of N/OFQ at high doses (nmol dose-ranges) was consistently reported to produce learning deficits in the passive avoidance task, promnesic effects were also reported in the literature with very low doses of the peptide (Table 1). In series of studies Hiramatsu and coworkers have shown that N/OFQ administered i.c.v. or into dorsal hippocampus at fmol dose-ranges had no effect by itself, but it was able to reduce the learning deficits induced by cholinergic receptor antagonists (Hiramatsu & Inoue, 1999b, 2000; Hiramatsu, Miwa, Hashimoto, Kawai, & Nomura, 2008). However, neither pharmacological blockade nor genetic ablation of NOP receptors was able to reverse these promnesic effects of N/OFQ (Hiramatsu et al., 2008; Miwa et al., 2010). Blockade of opioid receptors with naloxone benzoylhydrazone or nor-bi-naltorphimine was also ineffective indicating that N/OFQ effects were not mediated by opioid mechanisms (Hiramatsu & Inoue, 2000). Additional work with N/OFQ metabolites led to the identification of N/OFQ(14–17) peptide as the potential fragment mediating the paradoxical promnesic effects of exogenous N/OFQ (Miwa et al., 2010).

Taken together the findings from the fear conditioning and passive avoidance paradigms provide converging evidence that N/OFQ and NOP receptors play an important inhibitory role in fear learning. They show that this neuropeptide–receptor system modulates in general way fear memory formation by interfering with acquisition as well as consolidation processes.

2.3 N/OFQ modulation of recognition memory

Recognition memory is the ability to judge that a currently present object, person, place, or event, has previously been encountered or experienced. Recognition memory is an important aspect of human declarative memory and is one of the routine memory abilities altered in patients with amnesic syndrome and Alzheimer's disease. To examine whether N/OFQ system modulates recognition memory we used the novel object exploration procedure (Table 1), which resembles the visual paired comparisons task given to human subjects (Ennaceur, 2010). This test capitalizes on the rodents' innate preference for novelty, and it does not require food or water restriction or the application of reinforcing stimuli (electric shock or food delivery). The novel object recognition task is particularly suitable for studying the effects of NOP receptor ligands on mnemonic functions because it requires a little training and is, by far, less stressful than other learning procedures, such as the water maze, the fear conditioning and the passive avoidance tasks. The experimental procedure consists of prehabituation to spatial context alone followed by an acquisition session in which mice were familiarized with one object (sample object). In the choice trial, a novel object is presented together with the previously experienced sample object and recognition memory is reflected by greater exploration of the novel object than the previously seen familiar object.

We showed that i.c.v. administration of N/OFQ or systemic administration of the NOP receptor agonist, Ro64-6198, before the acquisition session severely impairs recognition memory in mice (Goeldner et al., 2008). The amnesic effect of Ro64-6198 could be detected at a long (24 h) but not at a short (3 h) retention delay, suggesting that this compound alters consolidation rather than acquisition processes. At amnesic doses, Ro64-6198 had no effect on recall performance irrespective of the retention delay (3 or 24 h), which rules out nonspecific actions of this compound on perception and attention. Given the key role of the hippocampus in recognition memory we examined whether the amnesic effects of N/OFQ and Ro64-6198 were mediated by this structure. Local infusion of N/OFQ into dorsal hippocampus produced the expected recognition memory deficit, while blockade of hippocampal NOP receptor with selective NOP receptor antagonist, UFP-101, reversed the amnesic effect of Ro64-6198. These data complement previous studies showing that dorsal hippocampus is a brain region in which N/OFQ-NOP receptor system plays an important inhibitory role on mnemonic functions.

2.4 N/OFQ modulation of working memory

Working memory refers to a short-term memory system that maintains and processes information relevant to an ongoing task, often under conditions of competing demands. To date the role of N/OFQ in working memory remains largely unexplored. Few studies have evaluated the effects of exogenous N/OFQ on spontaneous alternation behavior in a y-maze, which provides a gross evaluation of spatial working memory performance (Table 1). The y-maze procedure is based on the natural tendency of rodents to explore a novel environment and does not involve preliminary training or reinforcers. When placed in a y-maze, mice generally explore the least recently visited arm and thus tend to alternate their visit between the three arms. To explore successfully the three arms successively the mouse must maintain an ongoing record of most recently visited arms,

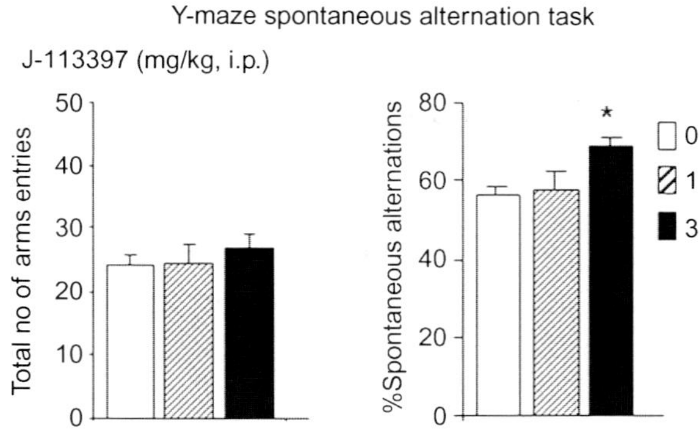

Figure 2 The NOP receptor antagonist, J-113397, improves spatial working memory in mice. *$p < 0.05$, significantly different from vehicle-treated group.

2.5 N/OFQ modulation of sensorimotor gating

Gating or filtering mechanisms are central aspect of the information processing function involved in the initial screening of the chaotic flow of extero- and interoceptive information to prevent sensory overload. Prepulse inhibition of startle reflex (PPI) is an example of operational measure of sensorimotor gating, a form of pre-attentive sensory filtering mechanism that regulates sensory input to the brain and allows the early stage of information processing to occur without disruption. PPI is a multimodal phenomenon and refers to the inhibition of a startle reflex response to a loud acoustic startling stimulus (pulse) when it is preceded by a barely detectable stimulus (prepulse) of the same (intra-modal) or different (cross-modal) sensory modality. Deficits in PPI have been reported in schizophrenia, schizotypal personality disorder, and affective bipolar disorder and are thought to contribute to sensory flooding and cognitive fragmentation observed in these neuropsychiatric diseases (Braff, Geyer, & Swerdlow, 2001; Geyer, 2006). In a recent study, we showed that Ro64-6198 administered systemically was more potent in reducing visual than acoustic PPI in mice (Table 1). At effective doses, Ro64-6198 had no effect on the startle response to the pulse or the detection of the visual prepulse, which indicates that this agonist produces a specific disruption of the gating process. The PPI deficit induced by Ro64-6198 was blocked by co-administration of the NOP receptor antagonist, compound 24, confirming that it is mediated by NOP receptors (Ces et al., 2012). Overall, these findings show that excessive stimulation of NOP receptor impairs sensorimotor gating, which may have widespread consequences on many aspect of cognitive behavior.

3. MECHANISMS OF N/OFQ-MEDIATED MODULATION OF COGNITIVE FUNCTIONS

The precise neural substrates and cellular mechanisms underlying the modulatory actions of N/OFQ system on mnemonic functions are still not completely understood. Electrophysiological and neurochemical studies have shown that the dominant effect of N/OFQ in the limbic circuits is to suppress neuronal excitability and neurotransmitter release (Meis, 2003; Moran, Abdulla, & Smith, 2000). One neurotransmitter that has been extensively studied in relation to N/OFQ modulation of cognition is the glutamate, which plays a central role in synaptic plasticity and learning and memory. N/OFQ has been shown to inhibit glutamate release and multiple forms of NMDA dependent-synaptic plasticity in the hippocampus (Bongsebandhu-phubhakdi & Manabe, 2007; Manabe et al., 1998; Wei & Xie, 1999; Yu & Xie, 1998). In hippocampal slices, exogenously applied N/OFQ reduces LTP induced by postsynaptic NMDA receptor stimulation (Yu & Xie, 1998), while pharmacological blockade of NOP receptors has the opposite effects (Bongsebandhu-phubhakdi & Manabe, 2007; Manabe et al., 1998). Accordingly, the improvement of contextual fear conditioning and hippocampal LTP facilitation revealed in NOP receptor knockout mice could be blocked by the NMDA receptor antagonist, MK-801 (Mamiya et al., 2003). These findings strongly suggest that NOP receptors modulate hippocampus-dependent learning through functional cross-talk with NMDA receptors. In support of this conclusion, we showed that Ro64-6198 and MK-801 administered together at subthreshold doses synergistically impair contextual fear learning and recognition memory (Goeldner et al., 2008, 2009). More importantly, we demonstrated that NOP and NMDA receptors converge on common MAPK/ERK pathway in the hippocampus to modulate long-term recognition memory formation (Goeldner et al., 2008). ERK processes several events relevant for establishing lasting memories. In particular, it translocates to the nucleus to initiate the transcriptional activity that is necessary for synaptic plasticity and long-term stabilization of the mnemonic trace (Peng, Zhang, Ren, Zhang, & Wang, 2010; Sweatt, 2004; Xia & Storm, 2012). Interestingly, exogenous applied N/OFQ was also found to reduce the activation of Ca2+/calmodulin-dependent protein kinase II (CaMKII) evoked by stimulation of NMDA receptors in hippocampal slice (Mamiya et al., 2003). CaMKII has been proposed to act as a trigger of Ras/Raf/MEK cascade

(Sweatt, 2004; Xia & Storm, 2012), and may thus be one potential upstream target that could relay the opposite influences of NOP and NMDA receptors on ERK activity.

Beside glutamate, another neurotransmitter that has been historically associated with learning and memory is acetylcholine. However, it remains unclear whether N/OFQ modulates mnemonic functions through cholinergic mechanisms. Uezu et al. (2005) have reported an elevated basal level of hippocampal acetylcholine and theta rhythm in NOP receptor knockout mice, thus suggesting that cholinergic component may contribute to the learning improvement documented in these mice. We have also conducted series of interaction studies with Ro64-6198 and cholinergic receptor antagonists (mecamylamine and scopolamine) to verify whether NOP and cholinergic receptors modulate recognition memory through common neural mechanisms (Reiss, Prinssen, Wichmann, Kieffer, & Ouagazzal, 2012). However, we failed to reveal a synergistic interaction, as we previously found with Ro64-6198 and MK-801. These negative findings were somewhat unexpected given that neurochemical studies have shown that N/OFQ inhibits acetylcholine release in many brain structures, including the hippocampus (Cavallini et al., 2003; Itoh, Konya, Takai, Masuda, & Nagai, 1999). One possible explanation could be that NOP and cholinergic receptors may modulate distinct mnemonic processes that support recognition memory formation. Numerous studies in human and animal have shown that cholinergic systems play crucial role during the initial stages of memory formation (e.g., attention and encoding processes) (Furey, 2011; Robinson, Platt, & Riedel, 2011; Winters, Saksida, & Bussey, 2008), while in our condition NOP receptor activation preferentially interfered with long-term consolidation of recognition memory (Goeldner et al., 2008). It should be also stressed that activation of NOP receptors reduces intracellular Ca^{2+} influx by inhibiting voltage-gated Ca^{2+} channels (Hawes et al., 2000; New & Wong, 2002), while activation of muscarinic M_1 receptors results in the production of IP_3 (inositol 1,4,5,-trisphosphate) and subsequent increase in cytoplasmic Ca^{2+} concentration via Ca^{2+} release from IP_3-sensitive endoplasmic reticulum stores (Digby, Shirey, & Conn, 2010; Jones, Byun, & Bubser, 2012). Such dissociation between the action of NOP and muscarinic M_1 receptors on intracellular Ca^{2+} signaling may thus be one potential mechanism through which the former receptors could control distinct signaling cascades involved in synaptic plasticity and recognition memory formation. Future studies are therefore warranted to determine whether the functional dissociation revealed between NOP and cholinergic receptors extends to other forms of learning (e.g., spatial and conditioned fear learning).

Another neurotransmitter that has been demonstrated to play an important role in the N/OFQ modulation of fear memories is noradrenaline. NOP receptors have been reported to exert a tonic inhibitory control on basal noradrenaline release in the basolateral amygdala (Kawahara, Hesselink, van Scharrenburg, & Westerink, 2004). In important study, Roozendaal et al. (2007) have shown that co-infusion of the β1-adrenoceptor antagonist, atenolol, potentiate the impairment of passive avoidance retention induced by N/OFQ injection into the basolateral amygdala. Conversely, the memory improvement induced by intra-amygdala infusion of the NOP receptor antagonist, [Nphe^1]-nociceptin (1–13)-NH_2, could be blocked by atenolol. The findings nicely show that N/OFQ inhibits formation of fear memory though modulation of noradrenergic function in the basolateral amygdala. The modulatory influence of N/OFQ on conditioned fear learning may likely be mediated by multiple other mechanisms. Modulation of glutamatergic function is a more likely possibility since N/OFQ has been shown to inhibit glutamate transmission in the different amygdala nuclei (Kallupi et al., 2014; Meis & Pape, 2001).

4. CONCLUSION AND REMARKS

The behavioral evidence outlined here emphasizes the important role played by N/OFQ and NOP receptors in many aspects of cognition. They show that N/OFQ negatively modulates mnemonic functions by opposing the functions of neurotransmitters, such as glutamate and noradrenaline, which promote synaptic plasticity and memory formation. Although remarkable progress has been made in understanding the contribution of N/OFQ to cognition several pieces of the puzzle are still missing. So far, all work on cognition has focused on the function of N/OFQ in the hippocampus and the amygdala. N/OFQ and NOP receptors are present in several other corticolimbic regions (e.g., prefrontal cortex, entorhinal cortex, septum, etc.) in which the exact functional role of N/OFQ system remains unexplored. Perhaps the most important question that remains to be addressed in the future concerns the role played by N/OFQ in the pathophysiological processes of cognitive disorders. The functional antagonism revealed between NOP and NMDA receptors at behavioral and cellular levels suggests that under pathological conditions normal endogenous N/OFQ release may exacerbate cognitive impairments linked to deficiency of glutamatergic signaling. Dysfunction of glutamatergic signaling is a central event in the pathogenic cascades of Alzheimer's disease and possibly other forms of dementias. Failure of glutamatergic synaptic function and signaling cascades (e.g., CaMKII and MAPK/ERK pathways) caused by neuronal injury (e.g., toxic beta-amyloid species) is thought to lead to the initial cognitive impairments and progressively to the build-up of neurotoxicity that ultimately causes neurodegeneration (Knobloch & Mansuy, 2008; Lacor, 2007). It is worth noting that recruitment of endogenous N/OFQ is one component of the inflammatory responses triggered by neuronal insults. Many proinflammatory mediators (cytokines, interleukin-β and tumor necrosis factor-α, oxygen species) and brain insults (brain trauma and exposure to LPS and neurotoxic agents) promote biosynthesis of N/OFQ in astrocytes and glial cells (Buzas, Rosenberger, Kim, & Cox, 2002; Buzas, Symes, & Cox, 1999; Di Benedetto et al., 2009; Leggett, Dawe, Jessop, & Fulford, 2009). It is therefore plausible that enhanced secretion of N/OFQ may be one component of the pathogenic processes underlying the cognitive deterioration seen in Alzheimer's disease and other dementias. Until recently, the new possibilities that targeting of NOP receptor may offer in the treatment of cognitive decline has stayed largely unexplored because of the lack of selective and brain-penetrant NOP receptor antagonists. The recent development of synthetic selective NOP receptor antagonists (J-113397, Trap-101, SB-612111, Ly2940094), opens up the way not only for enhancing our understanding of NOP receptor function but also for exploring its utility as a target for cognitive enhancement.

ACKNOWLEDGMENTS

This work was supported by the Centre National de la Recherche Scientifique (CNRS), the Institut National de la Santé et de la Recherche Médicale (INSERM) and the Université de Strasbourg.

REFERENCES

Andero, R., Brothers, S. P., Jovanovic, T., Chen, Y. T., Salah-Uddin, H., Cameron, M., et al. (2013). Amygdala-dependent fear is regulated by Oprl1 in mice and humans with PTSD. *Science Translational Medicine*, *5*(188), 173–188.

Bongsebandhu-phubhakdi, S., & Manabe, T. (2007). The neuropeptide nociceptin is a synaptically released endogenous inhibitor of hippocampal long-term potentiation. *The Journal of Neuroscience*, *27*(18), 4850–4858.

Braff, D. L., Geyer, M. A., & Swerdlow, N. R. (2001). Human studies of prepulse inhibition of startle: Normal subjects, patient groups, and pharmacological studies. *Psychopharmacology*, *156*(2–3), 234–258.

Buzas, B., Rosenberger, J., Kim, K. W., & Cox, B. M. (2002). Inflammatory mediators increase the expression of nociceptin/orphanin FQ in rat astrocytes in culture. *Glia*, *39*(3), 237–246.

Buzas, B., Symes, A. J., & Cox, B. M. (1999). Regulation of nociceptin/orphanin FQ gene expression by neuropoietic cytokines and neurotrophic factors in neurons and astrocytes. *Journal of Neurochemistry*, *72*(5), 1882–1889.

Cavallini, S., Marino, S., Beani, L., Bianchi, C., & Siniscalchi, A. (2003). Nociceptin inhibition of acetylcholine efflux from different brain areas. *Neuroreport*, *14*(17), 2167–2170.

Ces, A., Reiss, D., Walter, O., Wichmann, J., Prinssen, E. P., Kieffer, B. L., et al. (2012). Activation of nociceptin/orphanin FQ peptide receptors disrupts visual but not auditory sensorimotor gating in BALB/cByJ mice: Comparison to dopamine receptor agonists. *Neuropsychopharmacology*, *37*(2), 378–389.

Darland, T., Heinricher, M. M., & Grandy, D. K. (1998). Orphanin FQ/nociceptin: A role in pain and analgesia, but so much more. *Trends in Neurosciences*, *21*(5), 215–221.

Di Benedetto, M., Cavina, C., D'Addario, C., Leoni, G., Candeletti, S., Cox, B. M., et al. (2009). Alterations of N/OFQ and NOP receptor gene expression in the substantia nigra and caudate putamen of MPP+and 6-OHDA lesioned rats. *Neuropharmacology*, *56*(4), 761–767.

Digby, G. J., Shirey, J. K., & Conn, P. J. (2010). Allosteric activators of muscarinic receptors as novel approaches for treatment of CNS disorders. *Molecular BioSystems*, *6*(8), 1345–1354.

Ennaceur, A. (2010). One-trial object recognition in rats and mice: Methodological and theoretical issues. *Behavioural Brain Research*, *215*(2), 244–254.

Fanselow, M. S. (2009). From contextual fear to a dynamic view of memory systems. *Trends in Cognitive Sciences*, *14*(1), 7–15.

Fornari, R. V., Soares, J. C., Ferreira, T. L., Moreira, K. M., & Oliveira, M. G. (2008). Effects of nociceptin/orphanin FQ in the acquisition of contextual and tone fear conditioning in rats. *Behavioral Neuroscience*, *122*(1), 98–106.

Furey, M. L. (2011). The prominent role of stimulus processing: Cholinergic function and dysfunction in cognition. *Current Opinion in Neurology*, *24*(4), 364–370.

Geyer, M. A. (2006). The family of sensorimotor gating disorders: Comorbidities or diagnostic overlaps? *Neurotoxicity Research*, *10*(3–4), 211–220.

Goeldner, C., Reiss, D., Wichmann, J., Kieffer, B. L., & Ouagazzal, A. M. (2009). Activation of nociceptin opioid peptide (NOP) receptor impairs contextual fear learning in mice through glutamatergic mechanisms. *Neurobiology of Learning and Memory*, *91*(4), 393–401.

Goeldner, C., Reiss, D., Wichmann, J., Meziane, H., Kieffer, B. L., & Ouagazzal, A. M. (2008). Nociceptin receptor impairs recognition memory via interaction with NMDA receptor-dependent mitogen-activated protein kinase/extracellular signal-regulated kinase signaling in the hippocampus. *The Journal of Neuroscience*, *28*(9), 2190–2198.

Hawes, B. E., Graziano, M. P., & Lambert, D. G. (2000). Cellular actions of nociceptin: Transduction mechanisms. *Peptides*, *21*(7), 961–967.

Henderson, G., & McKnight, A. T. (1997). The orphan opioid receptor and its endogenous ligand–nociceptin/orphanin FQ. *Trends in Pharmacological Sciences*, *18*(8), 293–300.

Higgins, G. A., Kew, J. N., Richards, J. G., Takeshima, H., Jenck, F., Adam, G., et al. (2002). A combined pharmacological and genetic approach to investigate the role of orphanin FQ in learning and memory. *The European Journal of Neuroscience*, *15*(5), 911–922.

Hiramatsu, M., & Inoue, K. (1999a). Effects of nocistatin on nociceptin-induced impairment of learning and memory in mice. *European Journal of Pharmacology*, *367*(2–3), 151–155.

Hiramatsu, M., & Inoue, K. (1999b). Nociceptin/orphanin FQ and nocistatin on learning and memory impairment induced by scopolamine in mice. *British Journal of Pharmacology*, *127*(3), 655–660.

Hiramatsu, M., & Inoue, K. (2000). Improvement by low doses of nociceptin on scopolamine-induced impairment of learning and/or memory. *European Journal of Pharmacology, 395*(2), 149–156.
Hiramatsu, M., Miwa, M., Hashimoto, K., Kawai, S., & Nomura, N. (2008). Nociceptin/orphanin FQ reverses mecamylamine-induced learning and memory impairment as well as decrease in hippocampal acetylcholine release in the rat. *Brain Research, 1195*, 96–103.
Itoh, K., Konya, H., Takai, E., Masuda, H., & Nagai, K. (1999). Modification of acetylcholine release by nociceptin in conscious rat striatum. *Brain Research, 845*(2), 242–245.
Jones, C. K., Byun, N., & Bubser, M. (2012). Muscarinic and nicotinic acetylcholine receptor agonists and allosteric modulators for the treatment of schizophrenia. *Neuropsychopharmacology, 37*(1), 16–42.
Kallupi, M., Varodayan, F. P., Oleata, C. S., Correia, D., Luu, G., & Roberto, M. (2014). Nociceptin/orphanin FQ decreases glutamate transmission and blocks ethanol-induced effects in the central amygdala of naive and ethanol-dependent rats. *Neuropsychopharmacology, 39*, 1081–1092.
Kawahara, Y., Hesselink, M. B., van Scharrenburg, G., & Westerink, B. H. (2004). Tonic inhibition by orphanin FQ/nociceptin of noradrenaline neurotransmission in the amygdala. *European Journal of Pharmacology, 485*(1–3), 197–200.
Knobloch, M., & Mansuy, I. M. (2008). Dendritic spine loss and synaptic alterations in Alzheimer's disease. *Molecular Neurobiology, 37*(1), 73–82.
Koster, A., Montkowski, A., Schulz, S., Stube, E. M., Knaudt, K., Jenck, F., et al. (1999). Targeted disruption of the orphanin FQ/nociceptin gene increases stress susceptibility and impairs stress adaptation in mice. *Proceedings of the National Academy of Sciences of the United States of America, 96*(18), 10444–10449.
Kuzmin, A., Madjid, N., Johansson, B., Terenius, L., & Ogren, S. O. (2009). The nociceptin system and hippocampal cognition in mice: A pharmacological and genetic analysis. *Brain Research, 1305*(Suppl.), S7–S19.
Lacor, P. N. (2007). Advances on the understanding of the origins of synaptic pathology in AD. *Current Genomics, 8*(8), 486–508.
Lambert, D. G. (2008). The nociceptin/orphanin FQ receptor: A target with broad therapeutic potential. *Nature Reviews Drug Discovery*, 7(8), 694–710.
LeDoux, J. (2003). The emotional brain, fear, and the amygdala. *Cellular and Molecular Neurobiology, 23*(4–5), 727–738.
Leggett, J. D., Dawe, K. L., Jessop, D. S., & Fulford, A. J. (2009). Endogenous nociceptin/orphanin FQ system involvement in hypothalamic-pituitary-adrenal axis responses: Relevance to models of inflammation. *Journal of Neuroendocrinology, 21*(11), 888–897.
Liu, E. H., Lee, T. L., Nishiuchi, Y., Kimura, T., & Tachibana, S. (2007). Nocistatin and its derivatives antagonize the impairment of short-term acquisition induced by nociceptin. *Neuroscience Letters, 416*(2), 155–159.
Mamiya, T., Noda, Y., Nishi, M., Takeshima, H., & Nabeshima, T. (1999). Nociceptin system plays a role in the memory retention: Involvement of naloxone benzoylhydrazone binding sites. *Neuroreport, 10*(6), 1171–1175.
Mamiya, T., Yamada, K., Miyamoto, Y., Konig, N., Watanabe, Y., Noda, Y., et al. (2003). Neuronal mechanism of nociceptin-induced modulation of learning and memory: Involvement of N-methyl-D-aspartate receptors. *Molecular Psychiatry, 8*(8), 752–765.
Manabe, T., Noda, Y., Mamiya, T., Katagiri, H., Houtani, T., Nishi, M., et al. (1998). Facilitation of long-term potentiation and memory in mice lacking nociceptin receptors. *Nature, 394*(6693), 577–581.
Meis, S. (2003). Nociceptin/orphanin FQ: Actions within the brain. *The Neuroscientist, 9*(2), 158–168.
Meis, S., & Pape, H. C. (2001). Control of glutamate and GABA release by nociceptin/orphanin FQ in the rat lateral amygdala. *The Journal of Physiology, 532*(Pt. 3), 701–712.

Meunier, J. C., Mollereau, C., Toll, L., Suaudeau, C., Moisand, C., Alvinerie, P., et al. (1995). Isolation and structure of the endogenous agonist of opioid receptor-like ORL1 receptor. *Nature*, *377*(6549), 532–535.

Miwa, M., Uchida, S., Horiba, F., Takeshima, H., Nabeshima, T., & Hiramatsu, M. (2010). Nociceptin and its metabolite attenuate U0126-induced memory impairment through a nociceptin opioid peptide (NOP) receptor-independent mechanism. *Neurobiology of Learning and Memory*, *93*(3), 396–405.

Moran, T. D., Abdulla, F. A., & Smith, P. A. (2000). Cellular neurophysiological actions of nociceptin/orphanin FQ. *Peptides*, *21*(7), 969–976.

Neal, C. R., Jr., Mansour, A., Reinscheid, R., Nothacker, H. P., Civelli, O., Akil, H., et al. (1999a). Opioid receptor-like (ORL1) receptor distribution in the rat central nervous system: Comparison of ORL1 receptor mRNA expression with (125)I-[(14)Tyr]-orphanin FQ binding. *The Journal of Comparative Neurology*, *412*(4), 563–605.

Neal, C. R., Jr., Mansour, A., Reinscheid, R., Nothacker, H. P., Civelli, O., & Watson, S. J., Jr. (1999b). Localization of orphanin FQ (nociceptin) peptide and messenger RNA in the central nervous system of the rat. *The Journal of Comparative Neurology*, *406*(4), 503–547.

New, D. C., & Wong, Y. H. (2002). The ORL1 receptor: Molecular pharmacology and signalling mechanisms. *Neurosignals*, *11*(4), 197–212.

Peng, S., Zhang, Y., Ren, B., Zhang, J., & Wang, H. (2010). Effect of ketamine administration on memory consolidation, p-CREB and c-fos expression in the hippocampal slices of minor rats. *Molecular Biology Reports*, *38*(4), 2401–2407.

Redrobe, J. P., Calo, G., Guerrini, R., Regoli, D., & Quirion, R. (2000). [Nphe(1)]-Nociceptin (1–13)-NH(2), a nociceptin receptor antagonist, reverses nociceptin-induced spatial memory impairments in the Morris water maze task in rats. *British Journal of Pharmacology*, *131*(7), 1379–1384.

Reinscheid, R. K., Nothacker, H. P., Bourson, A., Ardati, A., Henningsen, R. A., Bunzow, J. R., et al. (1995). Orphanin FQ: A neuropeptide that activates an opioidlike G protein-coupled receptor. *Science*, *270*(5237), 792–794.

Reinscheid, R. K., Nothacker, H., & Civelli, O. (2000). The orphanin FQ/nociceptin gene: Structure, tissue distribution of expression and functional implications obtained from knockout mice. *Peptides*, *21*(7), 901–906.

Reiss, D., Prinssen, E. P., Wichmann, J., Kieffer, B. L., & Ouagazzal, A. M. (2012). The nociceptin orphanin FQ peptide receptor agonist, Ro64-6198, impairs recognition memory formation through interaction with glutamatergic but not cholinergic receptor antagonists. *Neurobiology of Learning and Memory*, *98*(3), 254–260.

Roberto, M., & Siggins, G. R. (2006). Nociceptin/orphanin FQ presynaptically decreases GABAergic transmission and blocks the ethanol-induced increase of GABA release in central amygdala. *Proceedings of the National Academy of Sciences of the United States of America*, *103*(25), 9715–9720.

Robinson, L., Platt, B., & Riedel, G. (2011). Involvement of the cholinergic system in conditioning and perceptual memory. *Behavioural Brain Research*, *221*(2), 443–465.

Roozendaal, B., Lengvilas, R., McGaugh, J. L., Civelli, O., & Reinscheid, R. K. (2007). Orphanin FQ/nociceptin interacts with the basolateral amygdala noradrenergic system in memory consolidation. *Learning & Memory*, *14*(1–2), 29–35.

Rudy, J. W. (2009). Context representations, context functions, and the parahippocampal-hippocampal system. *Learning & Memory*, *16*(10), 573–585.

Rudy, J. W., Huff, N. C., & Matus-Amat, P. (2004). Understanding contextual fear conditioning: Insights from a two-process model. *Neuroscience and Biobehavioral Reviews*, *28*(7), 675–685.

Sandin, J., Georgieva, J., Schott, P. A., Ogren, S. O., & Terenius, L. (1997). Nociceptin/orphanin FQ microinjected into hippocampus impairs spatial learning in rats. *The European Journal of Neuroscience*, *9*(1), 194–197.

Sandin, J., Ogren, S. O., & Terenius, L. (2004). Nociceptin/orphanin FQ modulates spatial learning via ORL-1 receptors in the dorsal hippocampus of the rat. *Brain Research*, *997*(2), 222–233.

Schlicker, E., & Morari, M. (2000). Nociceptin/orphanin FQ and neurotransmitter release in the central nervous system. *Peptides*, *21*(7), 1023–1029.

Stevens, C. W. (2009). The evolution of vertebrate opioid receptors. *Frontiers in Bioscience*, *14*, 1247–1269.

Sweatt, J. D. (2004). Mitogen-activated protein kinases in synaptic plasticity and memory. *Current Opinion in Neurobiology*, *14*(3), 311–317.

Uezu, K., Sano, A., Sei, H., Toida, K., Houtani, T., Sugimoto, T., et al. (2005). Enhanced hippocampal acetylcholine release in nociceptin-receptor knockout mice. *Brain Research*, *1050*(1–2), 118–123.

Wang, W., Cui, Q., Li, Y., Li, B., Yang, X., Cui, L., et al. (2010). The role of ERK-1/2 in the N/OFQ-induced inhibition of delayed rectifier potassium currents. *Biochemical and Biophysical Research Communications*, *394*(4), 1058–1062.

Wei, W. Z., & Xie, C. W. (1999). Orphanin FQ suppresses NMDA receptor-dependent long-term depression and depotentiation in hippocampal dentate gyrus. *Learning & Memory*, *6*(5), 467–477.

Winters, B. D., Saksida, L. M., & Bussey, T. J. (2008). Object recognition memory: Neurobiological mechanisms of encoding, consolidation and retrieval. *Neuroscience and Biobehavioral Reviews*, *32*(5), 1055–1070.

Xia, Z., & Storm, D. R. (2012). Role of signal transduction crosstalk between adenylyl cyclase and MAP kinase in hippocampus-dependent memory. *Learning & Memory*, *19*(9), 369–374.

Yoshitake, S., Ijiri, S., Kehr, J., & Yoshitake, T. (2013). Concurrent modulation of extracellular levels of noradrenaline and cAMP during stress and by anxiogenic- or anxiolytic-like neuropeptides in the prefrontal cortex of awake rats. *Neurochemistry International*, *62*(3), 314–323.

Yu, T. P., & Xie, C. W. (1998). Orphanin FQ/nociceptin inhibits synaptic transmission and long-term potentiation in rat dentate gyrus through postsynaptic mechanisms. *Journal of Neurophysiology*, *80*(3), 1277–1284.

INDEX

Note: Page numbers followed by "*f*" indicate figures and "*t*" indicate tables.

and continuously update such a record. Spontaneous alternation behavior is therefore used as an index of spatial working memory, while the total number of entries into the different arms is used an index of locomotor exploration in the maze. Mamiya et al. (1999) found no effect of N/OFQ on spontaneous alternations of mice following i.c.v. administration of low doses (1–10 pmol/5 μl). In another study, Hiramatsu and Inoue (1999a) used higher doses (0.5–5.0 nmol/5 μl) and found an impairment of working memory. At effective doses, N/OFQ had no effect on the total number of arm entries indicating that the working memory deficit was not attributed to an alteration of locomotor exploration. Finally, Higgins et al. (2002) used a delayed response task in rats to evaluate the effect of Ro64-6198 on working memory (Table 1). The task is conducted in Skinner chamber that contains two retractable levers situated either side of a central food tray and the rats are trained to lever press for a food reward. In the sample phase, a single lever (sample lever) is inserted into the chamber and it retracts as soon it is pressed by the rat. After a retention delay (varying from 2 to 24 s), the two levers are presented and the rat has to either select the sample lever if it is a matching task or select the other lever if it is a nonmatching task to receive the food reward. They showed that systemic administration of Ro64-6198 produces a delay dependent impairment of the percent correct responses in both the delayed match and nonmatch to position tasks. Analysis of other performance measures (e.g., number of omission, correct response latencies, etc.) confirmed that declines in working memory induced by Ro64-6198 were not attributed to an effect on response bias or neurological dysfunction. Collectively, the above findings suggest that NOP receptor may contribute to modulation of working memory.

So far, only NOP receptor knockout mice have been assessed in the y-maze spontaneous alternation task (Table 1), and any abnormality was reported in these mice compared to their wild-type counterparts (Mamiya et al., 1999). In an effort to clarify further the role of endogenous N/OFQ in working memory we recently evaluated the effects of the NOP receptor, J-113397, in the y-maze spontaneous alternations task. We found that this compound improves working memory performance in mice, thus revealing the inhibitory role of endogenous N/OFQ on working memory (Fig. 2). This promising positive data need however to be confirmed with other NOP receptor antagonists (e.g., SB-612111, compound 24, UFP-101) and refined working memory tasks (e.g., the radial arm maze, delayed match, and nonmatch to position).

G

H

I

K

L

M

N

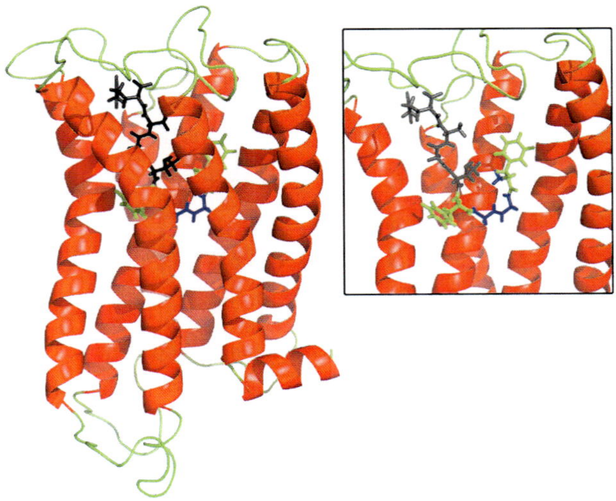

Rink-Jan Lohman *et al.*, Figure 1 Modeled structure of ORL-1 and putative binding of nociceptin. The membrane-spanning domain of ORL-1 is typical of other rhodopsin-like GPCRs. The nociceptin-binding site for ORL-1 consists of two adjacent hydrophobic pockets in a crevice formed by transmembrane helices 3, 5, 6, and 7, corresponding to the conserved opioid-binding site in opioid receptors. Further profiling has identified the absence of lipid-facing charged residues in TM helices 2, 3, and 4 in ORL-1, which is atypical for GPCRs (Topham, Mouledous, Poda, Maigret, & Meunier, 1998). The transmembrane helical domains are represented by red ribbon while extracellular and intracellular loops are depicted as green tube. Nociceptin is depicted in lines (black: address domain, blue: message domain) where Phe1 and Phe4 from the message domain is depicted in green. Inset; Phe1 (green) docks deep into one of the pockets and interacts with Asp130 (highlighted in black), a highly conserved residue in TM2. By contrast, Thr5 and Gly6 bind to a nonconserved region in EL2, where the basic side chain of Thr5 makes favorable contact with Gln286 in the acidic EL2 loop. Residues 5 and 6 thus might serve as one of the determinants of selective binding of ORL-1 as other more classical opioids will encounter unfavorable binding due to the presence of cationic residues at the same positions (Mollereau et al., 1999; Topham et al., 1998). The crystal structure of ORL-1 in complex with a peptidomimetic was recently reported (Thompson et al., 2012).

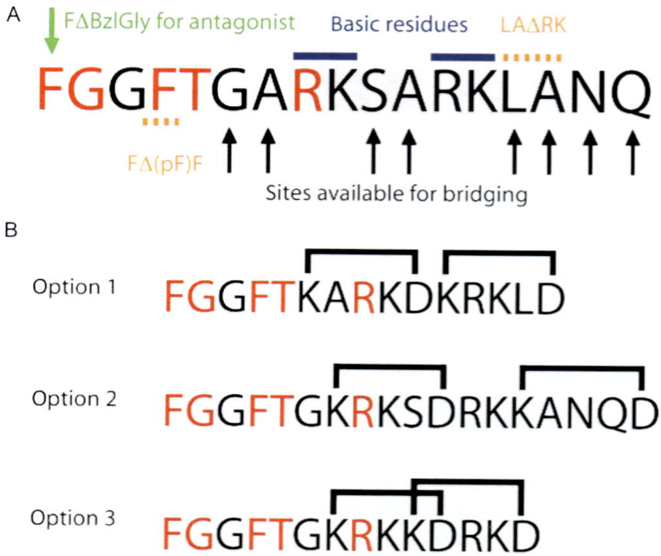

Rink-Jan Lohman *et al.*, Figure 7 Design strategies for nociceptin(1–17)-NH_2 analogues. (A) Amino acids important for activity are highlighted in red and the modification sites for Phe1BzlGly (green), Phe4(pF)Phe, Leu12Arg, and Ala13Lys (orange). (B) Three prospective cyclization strategies show the relative bridge positions and conservation of Phe1, Gly2, Phe4, Thr5, and Arg8.

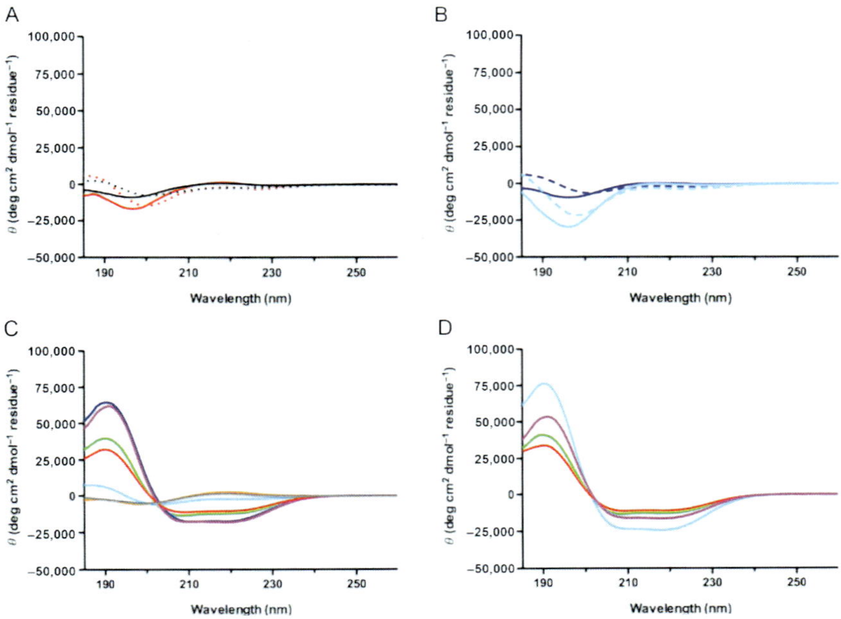

Rink-Jan Lohman *et al.*, Figure 8 CD spectra of nociceptin peptides. (A) Unconstrained agonists of **2** in buffer (—) and 30% TFE (· · ·), and compound **3** in buffer (—) and 30% TFE (· · ·). (B) Unconstrained antagonists of compound **4** in buffer (—) and 30% TFE (– –). Compound **5** in buffer (—) and 30% TFE (– –). (C) Constrained compounds **11** (—), **12** (—), **13** (—), **14** (—), **15** (—), **17** (—), and **16** (—). (D) Constrained compounds **18** (—), **19** (—), **20** (—), and **21** (—).

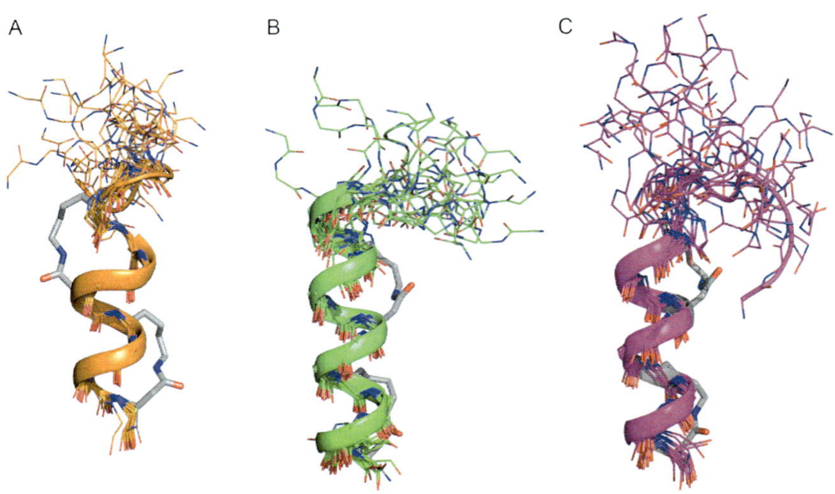

Rink-Jan Lohman *et al.*, Figure 13 NMR-derived solution structures for compounds **11**, **12**, and **17**. Shown are the 20 lowest energy conformations of (A) **11** (RMSD [K6-D15] 0.31 Å^2), (B) **12** (RMSD [K7-D18] 0.75 Å^2), and (C) **17** (RMSD [K7-D18] 0.68 Å^2). Lactam bridges are shown as gray sticks.

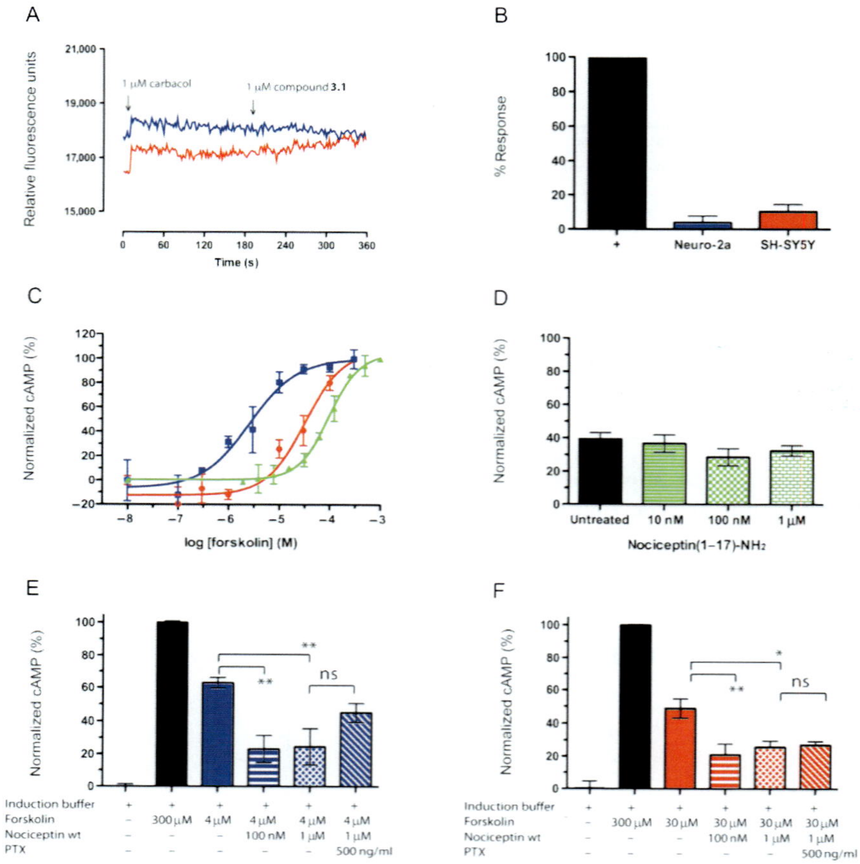

Rink-Jan Lohman *et al.*, Figure 14 Signaling of nociceptin(1–17)-OH activation of ORL-1. (A, B) No intracellular calcium release was induced from Neuro-2a (blue) or SH-SY5Y cells (red) by carbacol or nociceptin(1–17)-OH (compound **1**). (C) Dose–response curves to forskolin in U937 cells (green), Neuro-2a (blue), and SH-SY5Y (red) with EC_{50} 94, 32, and 2.6 μM, respectively, $n = 2$. (D) U937 preincubated with increasing concentrations of nociceptin(1–17)-OH showed no inhibit forskolin-induced cAMP (20 μM). (E) Neuro-2a cells preincubated with increasing concentration of nociceptin(1–17)-OH inhibited 4 μM forskolin-induced cAMP. (F) SH-SY5Y preincubated with nociceptin(1–17)-OH inhibited 30 μM forskolin-induced cAMP. Treatment with PTX was not able to modulate any cAMP effects. $*p \leq 0.05$; $**p \leq 0.01$; one-way ANOVA with Tukey's multiple comparison. ns, not significant.

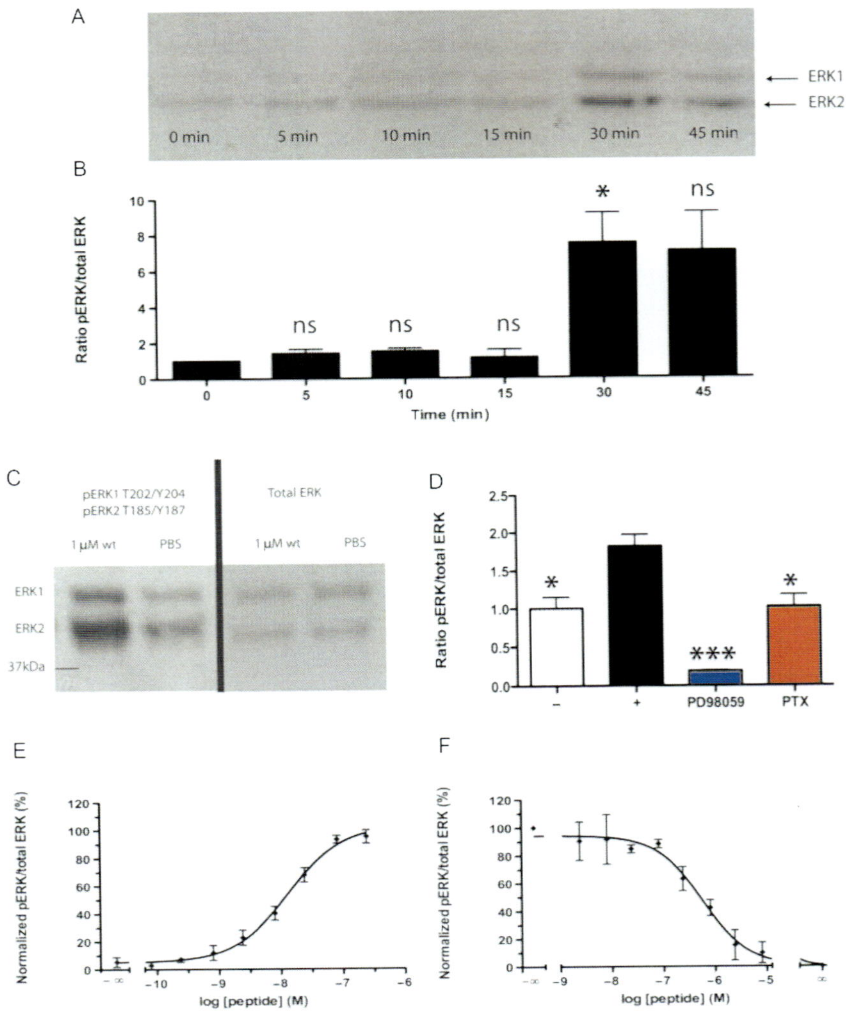

Rink-Jan Lohman *et al.*, Figure 15 Serum-starved Neuro-2a pERK response to nociceptin(1–17)-NH_2 (compound **2**). (A) Western blot of time-dependent pERK response to peptide **2** (rabbit polyclonal ERK1+2 [pTpY185/187]) and (B) quantification of the pERK response showing maximal response at 30–45 min ($n=2$). (C) Western blot of pERK response to 1 µM peptide **2** (rabbit IgG phospho-ERK1[T202/Y204]/ERK2[T185/Y187]). (D) Peptide **2** pERK response (300 nM) mediated through MEK and pertussis toxin-sensitive G-proteins. (E) Peptide **2** dose response with an EC_{50} of 11 nM ($n \geq 3$). (F) Antagonist response of [BzlGly1]nociceptin(1–13)-NH_2 with an IC_{50} of 570 nM ($n \geq 3$), challenged with 100 nM nociceptin(1–17)-NH_2. pERK assessed through AlphaScreen Surefire Kit.

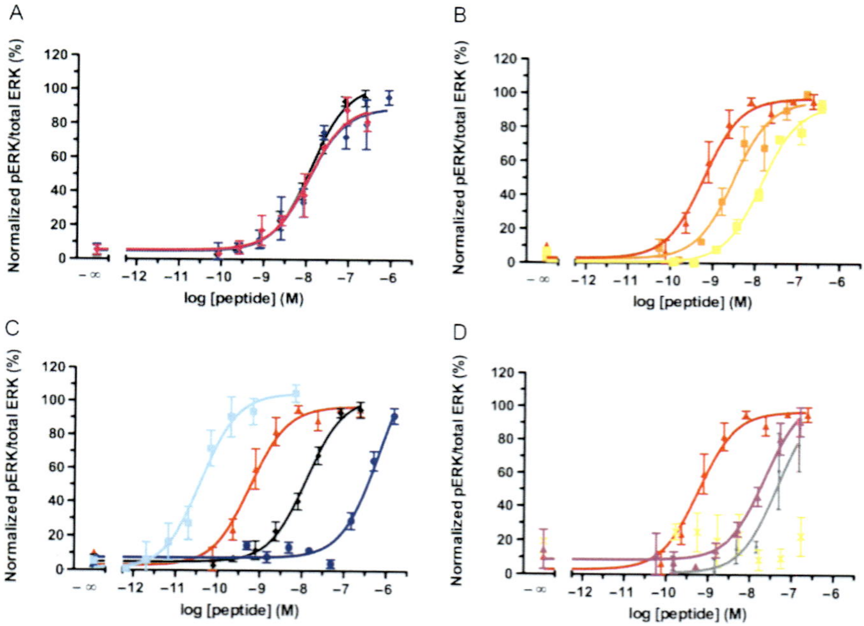

Rink-Jan Lohman *et al.*, Figure 16 pERK concentration–response profiles for agonist compounds. (A) Compounds **1** (—), **2** (—), and **3** (—). (B) Compounds **11** (—), **12** (—), and **13** (—). (C) Compounds **17** (—), **12** (—), **2** (—), and **8** (—). (D) Compounds **12** (—), **14** (—), **15** (—), and **16** (—). EC_{50} for all compounds reported in Table 8. All data normalized to PBS control (0%) and 1 μM compound **1** (100%), all data points represent $n \geq 3$. For illustration purposes only, hillslope was set to 1.0.

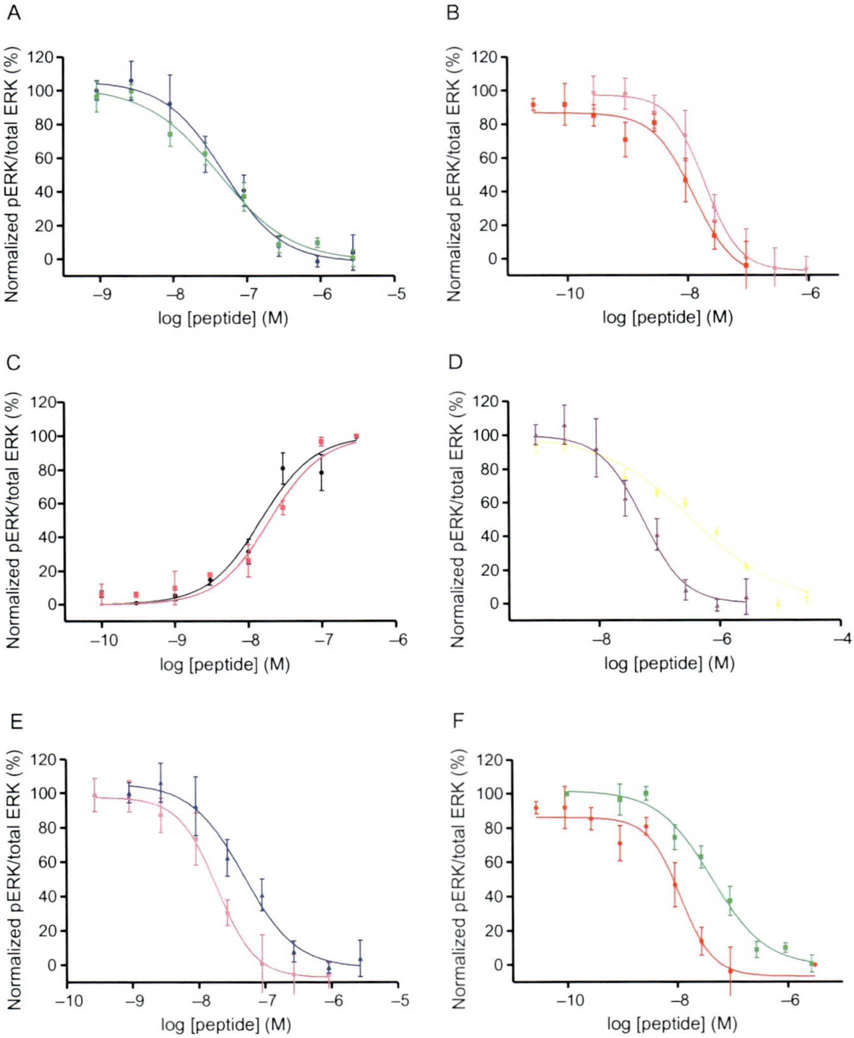

Rink-Jan Lohman *et al.*, Figure 17 Comparison of nociceptin(1–17)-NH_2 and its truncated analogues for agonism, or antagonism. (A) Compound **10** (---), compound **7** (---); (B) compound **21** (---), compound **22** (---); (C) nociceptin(1–17)-NH_2 (compound **2** (---) versus nociceptin(1–13)-NH_2 compound **3** (---)); (D) [BzlG[1]]nociceptin(1–17)-NH_2 (---) versus [BzlG[1]]nociceptin(1–13)-NH_2 (---); (E) compound **22** (---), compound **10** (---); (F) compound **21** (---), compound **7** (---). All data normalized to PBS control (0%) and 100 nM compound **2** (100%), and all data points represent $n \geq 3$. Statistical analysis ($p \leq 0.001$) using Student's *t*-test (Prism 5, Graphpad Software).

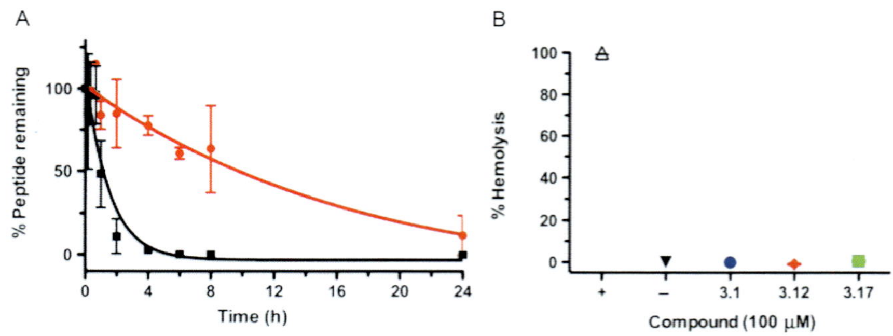

Rink-Jan Lohman *et al.*, Figure 18 Stability and hemolytic studies of active nociceptin compounds. (A) Time-dependent degradation of **1** (—) and **12** (—) in normal human serum at 37 °C. Data represent $n=2$, $p \leq 0.01$ at the 6-h time point. (B) Negligible hemolysis of human erythrocytes by compounds **1**, **12**, and **17** at 37 °C over 1 h, data normalized using 1% Triton X-100 (100% hemolysis) and PBS (0% hemolysis). Data represent $n \geq 3$.

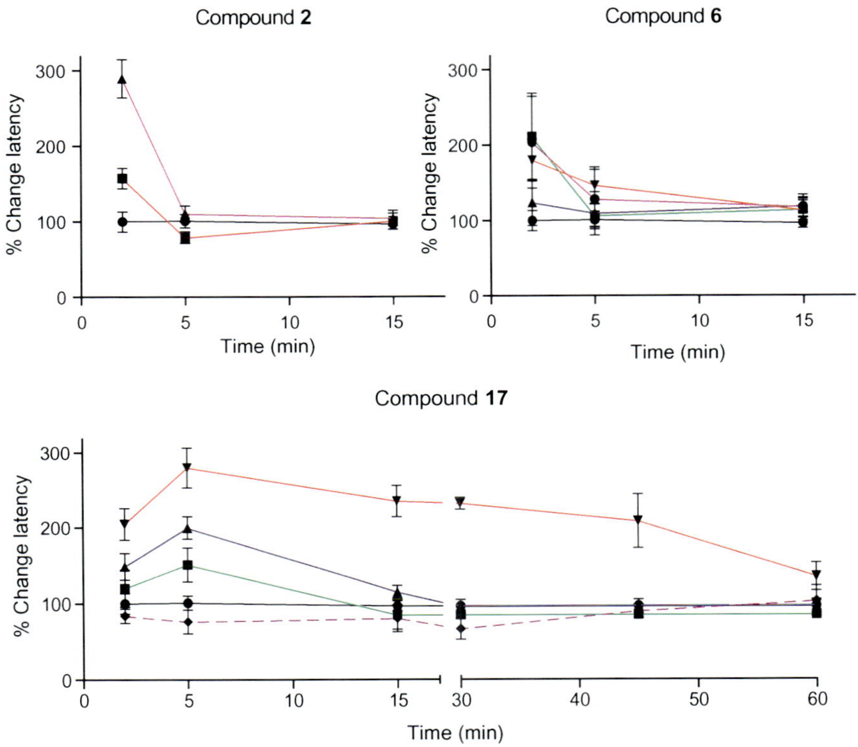

Rink-Jan Lohman *et al.*, Figure 19 Antinociception activity of constrained peptides *in vivo*. Dose-dependent paw withdrawal latency effects in rats exposed to 53 °C hot plate at various time points postcompound administration. Compounds **2** and **6** have similar potencies and duration of activity at the doses given (EC_{50} 185 and 160 pmol, respectively), whereas compound **17** showed significant improvements in both potency (EC_{50} 70 pmol, $p < 0.05$) and duration of activity, with activity still recordable after 60 min at the highest dose. The antagonist compound **21** (200 nmol) when coadministered with agonist compound **17** (200 pmol) blocked all antinociceptive activity induced by the agonist **17**. These data clearly show an increased potency and stability of the agonist **17** at the ORL-1 receptor *in vivo* compared to the native nociceptin (compound **2**) and the linear constrained peptide **6**. It also provides evidence that the antagonist, compound **21**, has efficacy *in vivo*. All compounds were administered intraplantar in sterile saline (20 μL). Agonist compounds **2**, **6**, and **17**; saline control (0 pmol peptide, —), 50 pmol (—), 100 pmol (—), 200 pmol (—), and 400 pmol (—). Antagonist compound **21** (200 nmol, - -) administered together with agonist compound **21** (200 pmol).

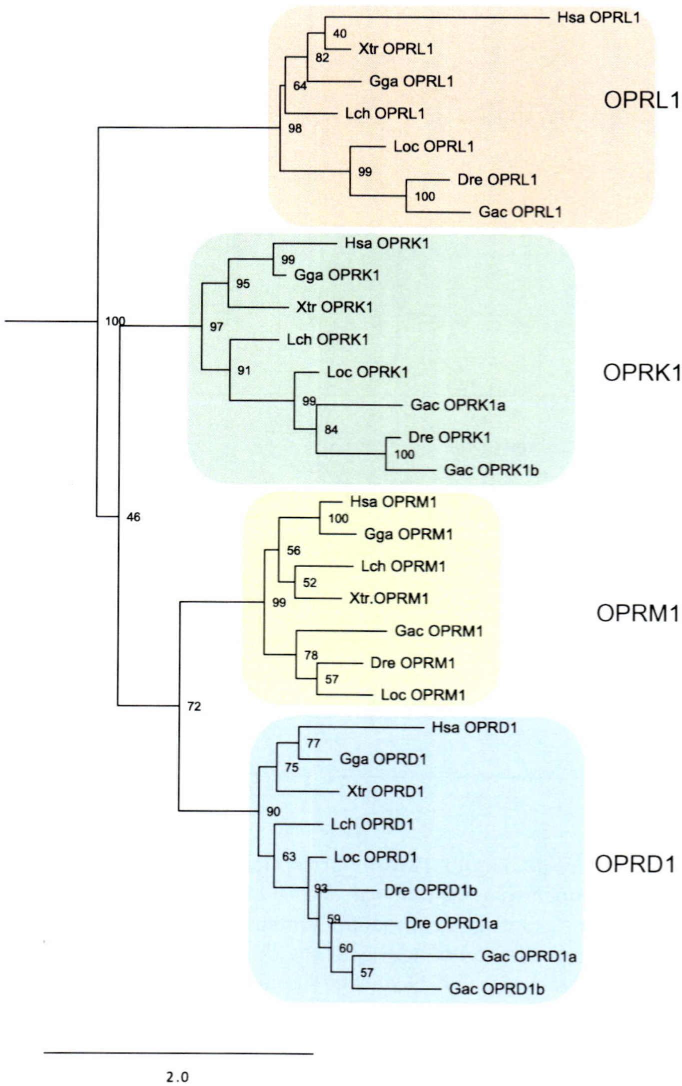

Dan Larhammar *et al.*, Figure 2 Phylogenetic tree for the opioid receptor family calculated with the maximum likelihood method using the PhyML3.0 algorithm (Guindon et al., 2010). The background colors mark the four receptor clades. The tree was rooted with the five human somatostatin receptors. OPRD1 is the delta receptor (DOP), OPRK1 is kappa (KOP), OPRM1 is mu (MOP), and OPRL1 is the nociceptin (NOP) receptor. Species abbreviations: Hsa, *Homo sapiens*; Gga, *Gallus gallus*, chicken; Xtr, *Xenopus* (Silurana) *tropicalis*, western clawed frog; Lch, *Latimeria chalumnae*, African coelacanth; Dre, *Danio rerio*, zebrafish; Gac, *Gasterosteus aculeatus*, three-spined stickleback; Loc, *Lepisosteus oculatus*, spotted gar. Note that OPRK1 is present in two copies in Gac and that OPRD1 is present in two copies in both Dre and Gac. The alignment of the opioid receptor sequences used to calculate the phylogenetic tree in Figure 2 can be obtained from the corresponding author upon request.

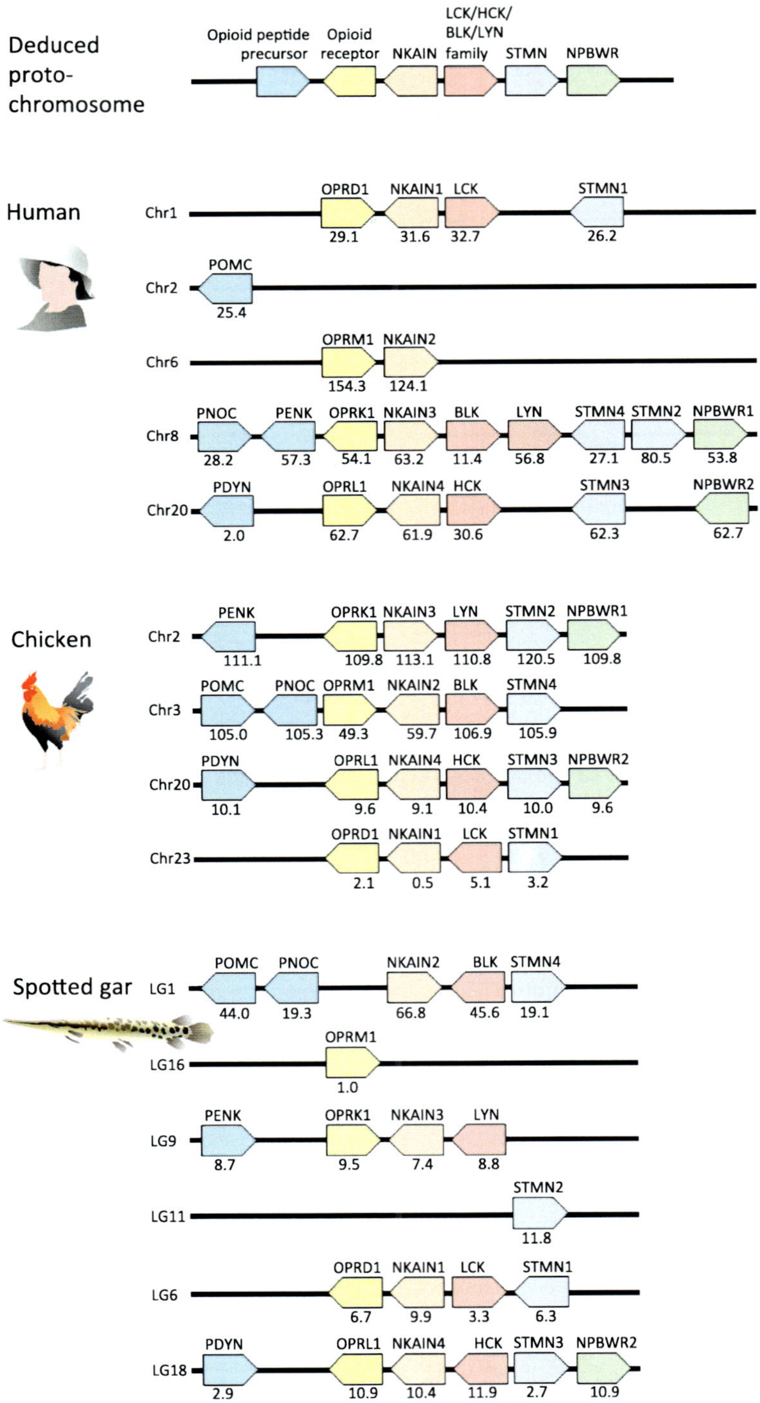

Dan Larhammar *et al.*, Figure 3 See legend on next page.

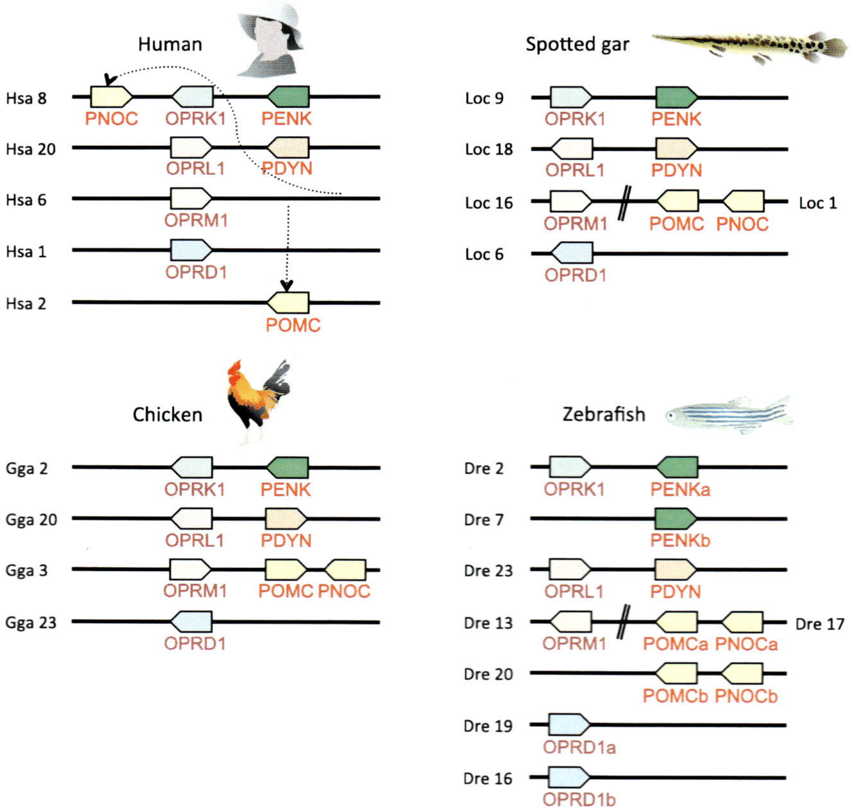

Dan Larhammar *et al.*, Figure 4 Chromosomal regions for the opioid peptide precursor genes and the opioid family receptor genes in human, chicken, spotted gar, and zebrafish. Horizontal lines symbolize chromosomes. The figure is a modified version of the figure shown by Sundstrom et al. (2010) with addition of PDYN in chicken and the complete repertoire of opioid genes in the spotted gar.

Dan Larhammar *et al.*, Figure 3 Chromosomal regions for the four opioid peptide precursor genes and the four opioid family receptor genes in human, chicken, and spotted gar. Horizontal lines symbolize chromosomes. The top line shows the deduced protochromosome prior to the two basal vertebrate tetraploidizations. Four neighboring gene families are shown: NKAIN, the Na^{+}/K^{+} transporting ATPase interacting family; the LCK/HCK/BLK/LYN family of Src-related tyrosine kinases; STMN, stathmin-like proteins; NPBWR, neuropeptide B/W receptors. Each gene family is marked with a color or shade of gray. Chr stands for chromosome, LG stands for linkage group. The numbers below the gene boxes show positions in megabases. The orientation of each gene box is indicated by its pointed end, pointing in the 5′-to-3′ direction, and the relative orientations of the genes within each chromosome have been maintained although the gene order has been modified to highlight the similarities. The order and orientations of the genes on the deduced protochromosome are speculative as these may have shifted multiple times during evolution.

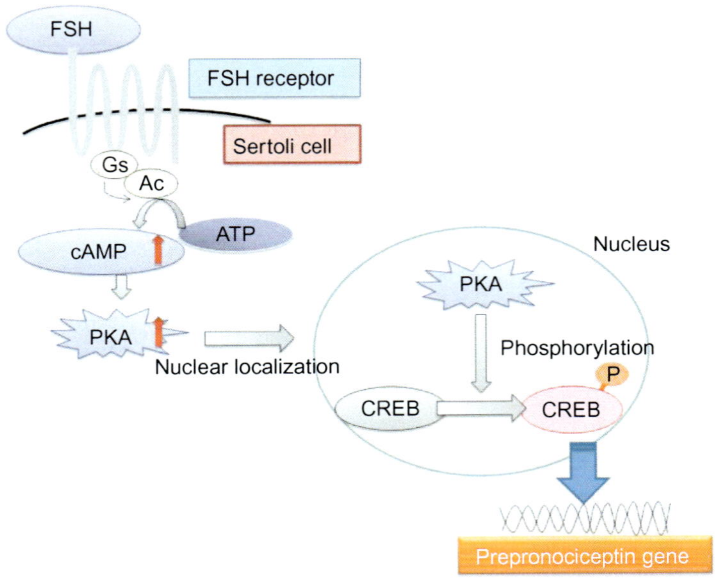

Ko Eto, Figure 1 The FSH intracellular signal transduction inducing prepronociceptin gene transcription in Sertoli cells. Gs, Gs protein coupled to the seven-transmembrane receptor for FSH; Ac, adenylyl cyclase; PKA, cAMP-dependent protein kinase, or protein kinase A; CREB, cAMP-response element-binding protein; P, phosphorylated. Upward arrows indicate increases in the concentration and the activity.

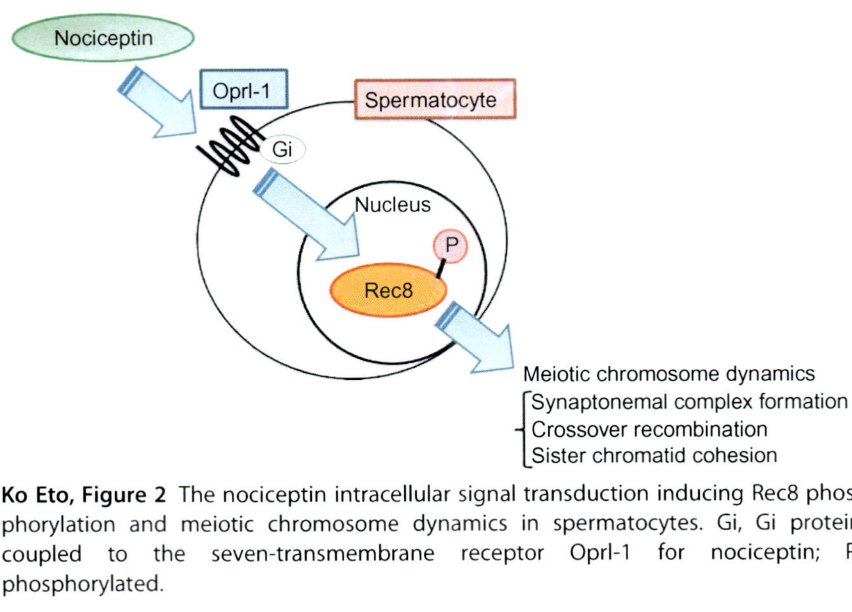

Ko Eto, Figure 2 The nociceptin intracellular signal transduction inducing Rec8 phosphorylation and meiotic chromosome dynamics in spermatocytes. Gi, Gi protein coupled to the seven-transmembrane receptor Oprl-1 for nociceptin; P, phosphorylated.

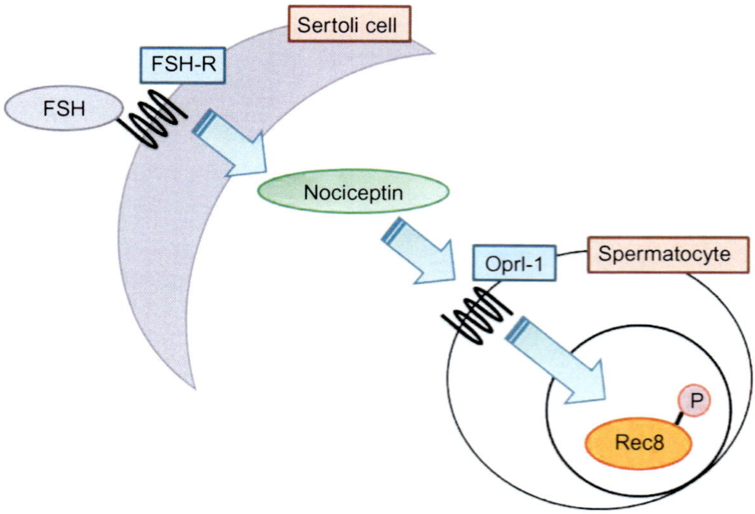

Ko Eto, Figure 3 The nociceptin-mediated intercellular signal transduction communicating between Sertoli cells and spermatocytes within testicular tubules.

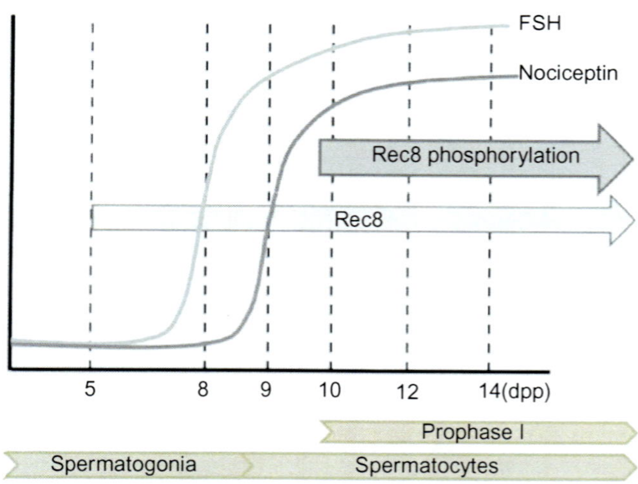

Ko Eto, Figure 4 The temporal and spatial relationships among serum FSH levels, testicular nociceptin levels, germ Rec8 levels and phosphorylation, germ cell development, and meiosis initiation during the early phase of spermatogenesis in postnatal murine testes. After birth, spermatogonia exist as the only germ cells in testes until 9 dpp; serum FSH levels increase from 8 dpp onward; after further 24 h testicular nociceptin levels increase and spermatocytes appear in addition to spermatogonia; and Rec8 phosphorylation begins to happen around 10 dpp when meiosis I begins with a prophase (prophase I) although Rec8 expression is already observed from 5 dpp.

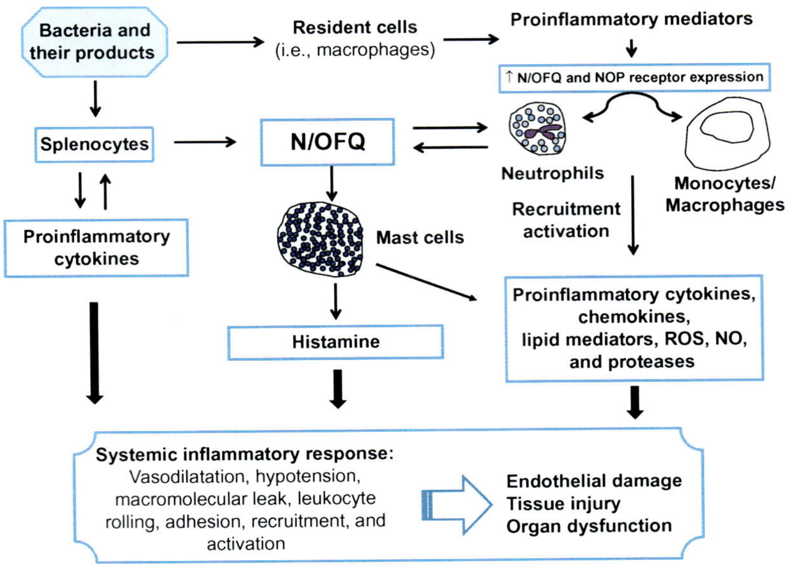

Elaine C. Gavioli *et al.*, Figure 1 Relationship between N/OFQ-NOP receptor system, systemic inflammation, and sepsis. In response to microbial invasion, macrophages release inflammatory mediators, which induce NOP receptor expression and N/OFQ release from mononuclear and polymorphonuclear cells. Splenocytes are also stimulated by bacterial products and then releasing N/OFQ, besides proinflammatory cytokines. N/OFQ may activate neutrophils, monocytes, macrophages, and mast cells, then contributing to the release of histamine, proinflammatory cytokines, chemokines, lipid mediators, reactive oxygen species (ROS), nitric oxide (NO), and proteases. All these mediators acting together evoke a systemic inflammation response, thus contributing to endothelial damage and organ dysfunction in sepsis.

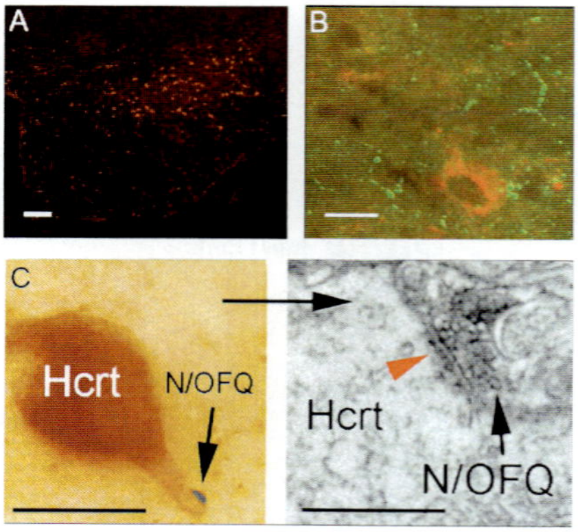

Xinmin (Simon) Xie, Figure 1 (A) A low-power micrograph shows Hcrt-immunoreactive neurons at the level of the perifornical area in the mouse hypothalamus. Scale bar 100 μm. (B) A confocal image shows that N/OFQ-immunoreactive fibers are seen in the vicinity of, and are in putative contact with Hcrt-immunoreactive neurons in the same section as (A). N/OFQ (green) fibers are in close proximity to Hcrt-immunoreactive (red) cells. Scale bar 10 μm. (C) A color light micrograph of a light brown immunolabeled Hcrt neuron contacted by a dark black bouton (black arrow) representing immunolabeling for N/OFQ. Bar scale 10 μm. Right panel, an electron micrograph taken from ultrathin sections of the bouton indicated on the light micrograph on the left panel. Black arrow indicated the N/OFQ immunolabeled axon terminal that is in synaptic contact (red arrowheads) with the Hcrt-immunolabeled dendrite. Bar scale 1 μm. *Figure was modified from Xie et al. (2008).*

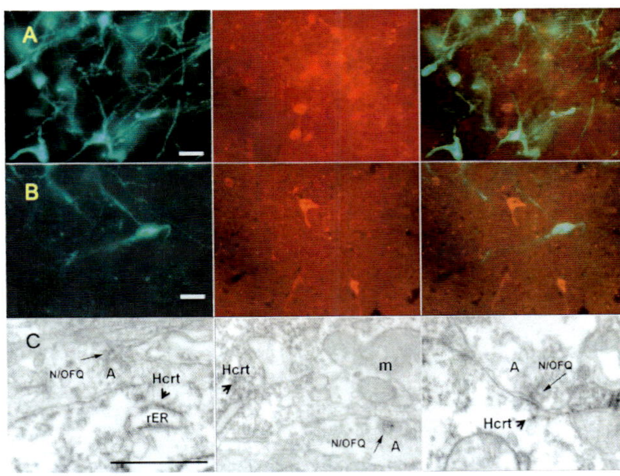

Xinmin (Simon) Xie, Figure 2 (A) Hcrt-immunoreactive cells (green color) and N/OFQ-immunoreactive cells (red color) are located in the perifornical hypothalamic area of the rat. (B) Hcrt-immunoreactive cells (green color) and N/OFQ-immunoreactive cells (red color) are located in the area dorsal to perifornical hypothalamic area of the rat. The right column shows merged images of left and middle columns. Scale bar 20 μm. (C) Electron micrograph showing an asymmetrical synaptic membrane specialization (black arrow) between the N/OFQ bouton-like structure and the Hcrt-immunolabeled dendrite in the rat lateral hypothalamus. Annotations to the photographs indicate the different cellular elements: A to indicate axon terminals, arrowheads to indicate Hcrt IHC postsynaptically, rER to point to rough endoplasmic reticulum, and m to indicate mitochondria. Scale bar 1 μm. *Figure was modified from Gerashchenko et al. (2011).*

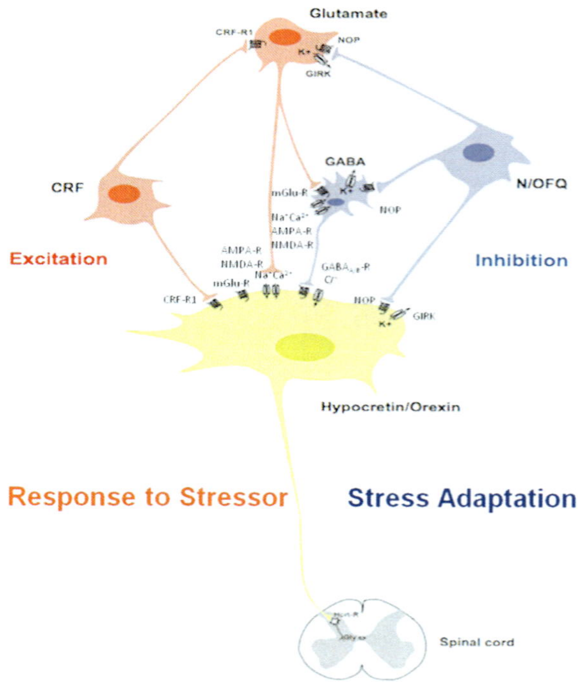

Xinmin (Simon) Xie, Figure 7 Schematic diagram illustrating Hcrt and N/OFQ coordinately regulating stress responses and adaptation. In addition to glutamate and GABA playing opposite roles in modulation of Hcrt neuronal activity, multiple neuropeptides have been shown to directly modulate Hcrt neurons. For example, CRH stimulates Hcrt neuronal activity enhancing stress responses; while N/OFQ (nociceptin/orphanin FQ) inhibits Hcrt neuronal activity, facilitating stress adaptation.

CPI Antony Rowe
Eastbourne, UK
April 28, 2015